Non-Equilibrium Ferrohydrodynamics

Ferrofluids, synthetic intelligent liquid materials, exhibit fascinating flow phenomena when subjected to magnetic fields. This book systematically examines the fundamentals, applications, mathematical derivations, and properties of non-equilibrium ferrohydrodynamics and explores the magnetization and viscosity properties of ferrofluids.

The book begins with an introduction to the basic concepts and applications of non-equilibrium ferrohydrodynamics and transport theory. It then goes on to discuss the equations of non-equilibrium and quasi-equilibrium ferrohydrodynamics and the magnetic levitation force in ferrofluids. The author discusses two crucial properties of ferrofluids: Magnetization properties and viscosity properties, presenting the ferrofluid magnetization equations and the magneto viscous effect, an important physical manifestation of non-equilibrium theory. The different types of flows of ferrofluids under the application of different magnetic field configurations are also elucidated, illustrating specific manifestations of non-equilibrium ferrohydrodynamics in different flow patterns.

This book will be a valuable reference for researchers and students of mechanical engineering with an interest in fluid dynamics, computational mechanics, and electromagnetism. Professionals working in areas of ferrofluid applications such as mechanics, petrochemical engineering, environmental sciences, and biomedicine will also benefit.

Wenming Yang is currently an associate professor at the School of Mechanical Engineering, University of Science and Technology Beijing, China. His research interests include ferrohydrodynamics and computational multi-physics, especially the basic mechanical problems of ferrofluids in applications such as seals, vibration damping, and sensors.

Non-Equilibrium Ferrohydrodynamics

Wenming Yang

CRC Press
Taylor & Francis Group
Boca Raton London New York

CRC Press is an imprint of the
Taylor & Francis Group, an **informa** business

First edition published 2025
by CRC Press
2385 NW Executive Center Drive, Suite 320, Boca Raton FL 33431

and by CRC Press
4 Park Square, Milton Park, Abingdon, Oxon, OX14 4RN

CRC Press is an imprint of Taylor & Francis Group, LLC

ISBN: 978-1-032-88906-1 (hbk)
ISBN: 978-1-032-88918-4 (pbk)
ISBN: 978-1-003-54034-2 (ebk)

DOI: 10.1201/9781003540342

Typeset in Minion
by SPi Technologies India Pvt Ltd (Straive)

Contents

List of Illustrations, ix

List of Symbols, xvi

Preface, xxii

CHAPTER 1 ▪ Introduction 1

1.1 FERROFLUIDS 1

1.2 APPLICATIONS OF FERROFLUIDS 3

1.3 BASIC CONTENT OF NON-EQUILIBRIUM FERROHYDRODYNAMICS 6

REFERENCES 8

CHAPTER 2 ▪ Basic Equations of Non-Equilibrium Ferrohydrodynamics 11

2.1 REYNOLDS' TRANSPORT THEORY 11

2.2 CONSERVATIONS EQUATION OF MASS 14

2.3 CAUCHY'S PRINCIPLE OF STRESS AND CAUCHY'S EQUATION OF MOTION 14

2.4 CONSERVATION EQUATIONS OF ANGULAR MOMENTUM 16

2.5 CONSERVATION EQUATION OF ENERGY 21

2.6 EQUATION OF ENTROPY GROWTH RATE 29

2.6.1 Equilibrium Thermodynamic Relationship 29

2.6.2 Equation of Entropy Growth Rate 30

2.7 CONSTITUTIVE RELATIONSHIPS 34

2.7.1 Clausius–Duhem Inequality 34

2.7.2 Constitutive Equations of $\boldsymbol{q}$ and $\boldsymbol{j}'$ 35

2.7.3 Constitutive Equations for Stress Tensor and Couple Stress Tensor 35

2.7.4 Expansion of the Equations of Motion 39

2.7.5 Expansion of the Angular Momentum Equation 41

2.8 SIMPLIFIED FORMS OF THE EQUATION OF MOTION AND ANGULAR MOMENTUM EQUATION IN THE ABSENCE OF ELECTRIC FIELD 43
2.9 OTHER SIMPLIFIED FORMS OF THE EQUATION OF MOTION AND ANGULAR MOMENTUM EQUATION 44
2.10 STRESS TENSOR OF FERROFLUIDS 46
REFERENCES 48

CHAPTER 3 ■ Introduction to Quasi-Equilibrium Ferrohydrodynamics 49
3.1 EQUATIONS AND STRESS TENSOR FOR QUASI-EQUILIBRIUM FERROHYDRODYNAMICS 49
3.2 BERNOULLI EQUATION FOR FERROFLUIDS 51
3.3 THE KELVIN FORCE IN QUASI-EQUILIBRIUM FERROFLUIDS 52
3.4 MAGNETIC LEVITATION FORCES IN FERROFLUIDS 53
3.4.1 Radial Magnetic Levitation Force on a Cylindrical Permanent Magnet in a Ferrofluid 54
3.4.2 Radial Magnetic Levitation Force Exerted on Cylindrical Composite Magnets Immersed in Ferrofluids 74
3.4.3 Calculation of Axial Magnetic Levitation Force on Permanent Magnets in Ferrofluids Using the Magnetic Charge Image Method 82
3.4.4 Calculation of Axial Magnetic Levitation Force on Permanent Magnets in Ferrofluids Using the Current Image Method 89
REFERENCES 96

CHAPTER 4 ■ Magnetic Properties of Ferrofluids 98
4.1 EQUILIBRIUM MAGNETIZATION EQUATION 98
4.2 MAGNETIC RELAXATION 101
4.3 PHENOMENOLOGICAL MAGNETIZATION EQUATION I 103
4.4 MAGNETIZATION EQUATION DERIVED MICROSCOPICALLY 107
4.5 PHENOMENOLOGICAL MAGNETIZATION EQUATION II 111
REFERENCES 112

CHAPTER 5 ■ Viscosity and Magnetoviscous Effects of Ferrofluids 114
5.1 INTRINSIC VISCOSITY OF FERROFLUIDS 114
5.2 ROTATIONAL VISCOSITY IN STATIC MAGNETIC FIELD 116
5.3 "NEGATIVE" VISCOSITY IN ALTERNATING MAGNETIC FIELDS 125

5.3.1 Rotational Viscosity under the Action of a Low-amplitude Alternating Magnetic Field at Low Shear Rates 125

5.3.2 Rotational Viscosity under the Action of a Limited-Amplitude Alternating Magnetic Field at Low Shear Rates 134

REFERENCES 140

CHAPTER 6 ▪ Ferrofluid Planar Couette–Poiseuille Flows 142

6.1 COUETTE–POISEUILLE FLOW OF REGULAR FLUIDS 143

6.2 FLOW CHARACTERISTICS IN CONSTANT UNIFORM MAGNETIC FIELDS 144

6.2.1 Normalized Ferrohydrodynamic Equations 144

6.2.2 Governing Equations of Ferrofluid Couette–Poiseuille Flows in Constant Uniform Magnetic Fields 146

6.2.3 Asymptotic Solutions for Flow Velocity Distributions in Weak Fields 148

6.2.4 Methods and Parameters Applied in Numerically Solving the System of Ferrohydrodynamic Equations 152

6.2.5 Magnetization Component in the Direction Perpendicular to the Magnetic Field 153

6.2.6 Magnetization Relaxation Effects in Couette–Poiseuille Flows in Constant Magnetic Fields 154

6.2.7 Effects of Physical Quantities on Flow Characteristics 157

6.3 FLOW CHARACTERISTICS IN MAGNETIC FIELDS WITH CONSTANT GRADIENT 163

6.3.1 Governing Equations 163

6.3.2 Magnetic Field and Flow Velocity Distributions Within Ferrofluids 166

6.3.3 Flow Characteristics of Ferrofluids 166

6.4 FLOW CHARACTERISTICS IN TIME-VARYING MAGNETIC FIELDS 179

6.4.1 Ferrofluid Couette Flows in a Weak Rotating Magnetic Fields 179

6.4.2 Flow Characteristics under the Influence of Alternating and Rotating Magnetic Fields When Disregarding the Effects of Spin Viscosity 184

6.4.3 Flow Characteristics under the Influence of Alternating and Rotating Magnetic Fields When Considering the Effects of Spin Viscosity 199

REFERENCES 220

Chapter 7 ▪ Ferrofluid Flow in Pipes 221

7.1 STEADY FLOW IN CONSTANT UNIFORM MAGNETIC FIELDS 221

7.1.1 Flow Characteristics under the Influence of Axial Magnetic Fields 221

7.1.2 Flow Characteristics under the Influence of Radial Magnetic Fields 231

7.1.3 Flow Rate and Drag Coefficient 233

7.2 STEADY FLOW IN CONSTANT NON-UNIFORM MAGNETIC FIELDS 235

7.2.1 Flow Characteristics at Weak Non-Equilibrium States 235

7.2.2 Flow Characteristics at Non-Equilibrium States 241

7.3 FLOW IN AN OSCILLATING MAGNETIC FIELD 244

7.4 SPONTANEOUS ROTATION IN FERROFLUID PIPE FLOWS 249

7.4.1 Governing Equations and Their Solutions 249

7.4.2 Conditions for Generating Rotational Motion 254

REFERENCES 257

Chapter 8 ▪ Laminar Circular Flow of Ferrofluids 259

8.1 VELOCITY AND VORTICITY OF FERROFLUIDS 259

8.2 MAGNETIC FIELD DISTRIBUTION AND EQUILIBRIUM MAGNETIZATION IN FERROFLUIDS WHEN THE CYLINDERS ARE STATIONARY 262

8.3 MAGNETIZATION IN FERROFLUIDS WHILE THE CYLINDERS ARE ROTATING 266

8.4 MAGNETIC FIELD INTENSITY PERPENDICULAR TO THE EXTERNAL MAGNETIC FIELD INSIDE THE CYLINDER WHEN THE CYLINDERS ROTATE 268

REFERENCES 270

Chapter 9 ▪ Spin-Up Flow of Ferrofluids 271

9.1 ROTATIONAL VELOCITY AT THE FERROFLUID SURFACE 271

9.1.1 Magnetic Torque Acting on the Ferrofluid 272

9.1.2 Rotational Velocity at the Ferrofluid Surface 274

9.2 TORQUE ACTING ON THE FERROFLUID 276

9.2.1 Rotational Velocity of Ferrofluid and Particle Spin Velocity Based on Spin-Diffusion Theory 276

9.2.2 Torque Acting on the Container Wall 287

REFERENCES 290

INDEX, 291

Illustrations

FIGURES

Figure 1.1	A typical ferrofluid seal structure	3
Figure 1.2	A ferrofluid rotary inertial damper	4
Figure 1.3	A ferrofluid active damper	5
Figure 1.4	Schematic illustration of a ferrofluid tactile sensor	6
Figure 3.1	Model for calculating radial magnetic levitation forces in the bipolar coordinate	54
Figure 3.2	Radial magnetic levitation force varies with eccentricity and radius ratio ($\mu_r = 1.1$, $R_1 = 5$ mm)	72
Figure 3.3	Relationship between the initial radial magnetic levitation force and the ratio of the outer shell to the permanent magnet radius ($b/R_1 = 0.0$, $R_1 = 5$ mm)	73
Figure 3.4	A composite structure composed of multiple cylindrical permanent magnets	74
Figure 3.5	A model for calculating magnetic field distribution	74
Figure 3.6	A model for calculating the magnetic levitation force of a composite magnet in a ferrofluid	77
Figure 3.7	Magnetic levitation force exerted on composite magnets in a ferrofluid	80
Figure 3.8	Variation of the maximum magnetic levitation force experienced by the composite structure with the relative permeability of the ferrofluid	81
Figure 3.9	Variation of the initial magnetic levitation force with the spacing between the magnetic pole faces of permanent magnets at unit eccentricity	82
Figure 3.10	A cylindrical permanent magnet and equivalent magnetic charges on its two end faces	83
Figure 3.11	Model of calculating the image magnetic charge intensity	84
Figure 3.12	Model of calculating the magnetic levitation force using the magnetic charge image method	86
Figure 3.13	The axial magnetic levitation force calculated using the magnetic charge image method	89

Figure 3.14 A cylindrical permanent magnet and magnetization current on its cylindrical surface 90
Figure 3.15 Model of calculating the magnetization current 91
Figure 3.16 Model of calculating the magnetic levitation force using the current image method 93
Figure 3.17 Axial magnetic levitation force calculated by the current image method 96
Figure 4.1 Magnetization curve of a typical ferrofluid (APG513A, Ferrotec) 99
Figure 4.2 Brownian mechanism and Néel mechanism in the magnetization process of ferrofluids 101
Figure 4.3 Comparison of the magnetization relaxation time of ferrofluids (surfactant thickness 2 nm, $\eta_0 = 0.108$ Pa·s, $K = 7.8 \times 10^4$ J/m^3, $T = 298$ K, $\tau_0 = 10^{-9}$ s) 102
Figure 5.1 Couette flow of a ferrofluid 117
Figure 5.2 Rotational viscosity of a ferrofluid under the condition of $\Omega\tau_B \ll 1$ 120
Figure 5.3 Rotational viscosity varies with the external magnetic field ($\Omega\tau_B = 2$) 123
Figure 5.4 Rotational viscosity varies with the external magnetic field ($\Omega\tau_B = 4$) 124
Figure 5.5 Rotational viscosity varies with the external magnetic field ($\Omega\tau_B = 6$) 124
Figure 5.6 Variation of relative rotational viscosity with magnetic field frequency under weak field conditions 132
Figure 5.7 Decomposition of the magnetization vector 135
Figure 5.8 Variation of γ with α_0 for different values of $w\tau_B$ 139
Figure 5.9 Iso-γ lines in the ($w\tau_B$, α_0) plane 140
Figure 6.1 Ferrofluid planar Couette–Poiseuille flow 142
Figure 6.2 Velocity distributions of Couette–Poiseuille flow for a Newtonian fluid 143
Figure 6.3 Comparison of the asymptotic solutions of Couette–Poiseuille flow with numerical results 151
Figure 6.4 Schematic diagram of mesh generation in the fluid domain 152
Figure 6.5 Comparison of the magnetization perpendicular to the magnetic field in the Couette–Poiseuille flows when different magnetization equations are applied. (a) $P^* = 0$, $Re = 1$ and (b) $P^* = 2$, $Re = 20$ 154
Figure 6.6 Magnetic field intensity and magnetization in a Couette–Poiseuille flow of a ferrofluid ($Re = 10$, $P^* = 2$, $Mn_f = 5$) 155
Figure 6.7 Deviation of flow velocity distribution caused by magnetization relaxation from that of ordinary fluids ($Re = 1$, $Mn_f = 100$) 155
Figure 6.8 Deviation of flow vorticity caused by magnetization relaxation from that of ordinary fluids. ($Re = 1$, $Mn_f = 100$) 155
Figure 6.9 Distribution of magnetization perpendicular to the direction of the external magnetic field within the flow cross-section ($Re = 1$, $Mn_f = 100$) 156

Figure 6.10 Spin velocity distribution in the flow cross-section ($Re = 1, Mn_f = 100$) 156
Figure 6.11 Variation of tangential stress on the upper plate and the overall flow rate with the dimensionless pressure gradient ($Re = 1, Mn_f = 100$) 158
Figure 6.12 Variation of M_x with dimensionless pressure in a ferrofluid near the upper plate 158
Figure 6.13 Distribution of flow properties within a cross-section at different M_{nf} ($Re = 1, P^* = 2$). (a) Flow velocity, (b) flow vorticity, (c) magnetization perpendicular to the direction of the magnetic field, and (d) particle spin velocity 159
Figure 6.14 Variation of (a) flow rate, (b) mean vorticity, (c) dimensionless stress, and (d) M_x near the upper plate ($P^* = 2$) 160
Figure 6.15 Flow characteristics at different Re ($P^* = 2, M_{nf} = 1$). (a) Particle spin velocity within the flow cross-section and (b) magnetization component perpendicular to the direction of the external magnetic field 161
Figure 6.16 Variation in (a) M_x and (b) dimensionless total stress with Re 161
Figure 6.17 Distributions of velocity, particle spin, and M_x under the influence of different magnetization relaxation time ($P^* = 2, M_{nf} = 20, Re = 10$) 162
Figure 6.18 Variation of (a) dimensionless stress, (b) M_x, (c) mean vorticity, and (d) flow rate with magnetization relaxation time ($P^* = 2, M_{nf} = 20, Re = 10$) 162
Figure 6.19 Ferrofluid Couette–Poiseuille flows under the application of field gradients. The direction of the magnetic field gradient is (a) the same as or (b) opposite to the velocity of the wall movement 163
Figure 6.20 Streamlines of magnetic field intensity and magnetization under the conditions of $P^* = 2$ and (a) $G_H = 9$, or (b) $G_H = -9$ 166
Figure 6.21 Comparison of flow velocity distributions at different longitudinal positions within the ferrofluid under a gradient magnetic field ($P^* = 2$). (a) $G_H = 9$ and (b) $G_H = -9$ 167
Figure 6.22 Velocity profiles of a ferrofluid Couette flow ($P^* = 0$) subject to (a) favourable and (b) unfavourable field gradients 167
Figure 6.23 Vorticity profiles of ferrofluid Couette flow ($P^* = 0$) subject to (a) favourable and (b) unfavourable field gradients 168
Figure 6.24 Velocity profiles of a ferrofluid Couette–Poiseuille flow ($P^* = 2$) subject to (a) positive and (b) negative gradient fields 169
Figure 6.25 Flow velocity distribution of a ferrofluid in Couette–Poiseuille flows under the action of a gradient magnetic field at $P^* = 2$ compared to that under the action of a uniform magnetic field. (a) $|G_H| = 7$, (b) $|G_H| = 9$ 169
Figure 6.26 Effect of magnetic field gradient on the location of maximum flow velocity ($P^* = 2$) 170
Figure 6.27 Velocity profiles of ferrofluid Couette–Poiseuille flow ($P^* = -2$) subject to (a) favourable and (b) unfavourable field gradients 170
Figure 6.28 Comparison of flow velocity profiles between gradient and uniform magnetic fields ($P^* = -2$). (a) $|G_H| = 7$, (b) $|G_H| = 9$ 171

Figure 6.29 Distributions of vorticity in ferrofluid. (a) $P^* = 2$, Averaged fields; (b) $P^* = -2$, Averaged fields; (c) $P^* = 2$, Positive gradient fields; (d) $P^* = -2$, Positive gradient fields; (e) $P^* = 2$, Negative gradient fields; (f) $P^* = -2$, Negative gradient fields 171

Figure 6.30 Variation of the root-mean-squared vorticity over the entire domain against the field gradient for different pressure gradients: (a) $P^* = 0$, (b) $P^* = 2$, and (c) $P^* = -2$. —■—, - ● -, and · ▲ · represent averaged, favourable gradient, and unfavourable gradient fields, respectively 172

Figure 6.31 Distribution of M_x in ferrofluid Couette flow ($P^* = 0$) subject to (a) favourable and (b) unfavourable field gradients 173

Figure 6.32 Distribution of M_x in ferrofluid Couette–Poiseuille flow ($P^* = 2$) subject to (a) favourable and (b) unfavourable field gradients 173

Figure 6.33 Variation of M_x near the upper plate against field gradient when (a) $P^* = 0$, (b) $P^* = 2$, and (c) $P^* = -2$. —■—, - ● -, and · ▲ · represent averaged, favourable gradient, and unfavourable gradient fields, respectively 174

Figure 6.34 M_x profiles under the conditions of $P^* = 2$ and $G_H = 7$. - - - -, ……, and —— represent favourable gradient, unfavourable gradient, and averaged fields, respectively 174

Figure 6.35 Distribution of spin velocity in ferrofluid Couette flow ($P^* = 0$) subject to (a) positive and (b) negative gradient fields. ——, - - - -, ……, —·—·—, and ……… represent $|G_H| = 1$, $|G_H| = 3$, $|G_H| = 5$, $|G_H| = 7$, and $|G_H| = 9$, respectively 175

Figure 6.36 Distribution of spin velocity in ferrofluid Couette–Poiseuille flows subject to gradient magnetic fields for the cases of (a) $P^* = 2$ and (b) $P^* = -2$, here solid lines are for averaged fields, dashed lines are for favourable field gradients, and dotted lines are for unfavourable field gradients 176

Figure 6.37 Effect of field gradient on the volumetric flow rate for the cases of (a) $P^* = 0$, (b) $P^* = 2$, and (c) $P^* = -2$. —■—, - ● -, and · ▲ · represent averaged, favourable gradient, and unfavourable gradient fields, respectively 176

Figure 6.38 Effect of field gradient on the tangential stress acting on the moving plate (from top to bottom: $P^* = 0$, $P^* = 2$, and $P^* = -2$; from left to right: magnetic stress, shear stress, and total stress. —■—, - ● -, and · ▲ · represent the averaged field, favourable gradient, and unfavourable gradient fields, respectively) 178

Figure 6.39 Time-averaged magnetic moment density versus dimensionless spin velocity (where $\chi_0 = 1$, solid lines: $w_\tau = 0$, dashed lines: $w_\tau = 1$, dotted line: $w_\tau = 5$, and dotted dash line: $w_\tau = 10$). (a) Uniform magnetic field in x-direction. (b) Uniform magnetic field in y-direction. (c) Rotating magnetic field 189

Figure 6.40 Comparison of the flow velocity distribution within the cross-section of a ferrofluid Couette–Poiseuille flow in a time-varying magnetic field with the flow velocity in a uniform constant magnetic field (the magnitude of the magnetic field is the same as the amplitude of the time-averaged magnetic field intensity and the direction is along the y-direction) ($\chi_0 = 1.01$, $w\tau = 10$, $\eta^*/\zeta^* = 5.65$). (a) $P^* = 2$, (b) $P^* = -2$ 198

Figure 6.41 Flow velocity and spin velocity distributions within the cross-section of ferrofluid Couette–Poiseuille flows under time-varying magnetic fields of different frequencies ($\chi_0 = 3.5$, $\eta^*/\zeta^* = 5.65$, $P^* = 1$, $Re = 0.1$, $Mn_f = 800$, $Pe = 0.00625$). (a) and (b) Uniform time-varying magnetic fields in the x-direction. (c) and (d) Uniform time-varying magnetic fields in the y-direction. (e) and (f) Rotating magnetic fields 199

Figure 6.42 Variation of the zero-order magnetic moment term with w^*. ($\chi_0 = 3.5$, $\hat{H}_x^* = (1+\mathrm{i})/2$) 205

Figure 6.43 Variation of the first-order magnetic moment term with w^*. ($\chi_0 = 3.5$, $\hat{H}_x^* = (1+\mathrm{i})/2, \hat{B}_y^* = (1\pm\mathrm{i})/2$) 207

Figure 6.44 Flow velocity distributions within the flow cross-section of a ferrofluid Couette–Poiseuille flow under the action of time-varying magnetic fields of two frequencies and different magnitudes ($\chi_0 = 3.5$, $\eta^*/\zeta^* = 5.65$, $P^* = 1$, $Re = 0.1$, $Pe = 0.00625$, $\eta' = 10^{-20}$ N·s) 210

Figure 6.45 Spin velocity distributions within the flow cross-section of a ferrofluid Couette–Poiseuille flow under the action of time-varying magnetic fields of two frequencies and different magnitudes ($\chi_0 = 3.5$, $\eta^*/\zeta^* = 5.65$, $P^* = 1$, $Re = 0.1$, $Pe = 0.00625$, $\eta' = 10^{-20}$ N·s) 211

Figure 6.46 Velocity and spin velocity distributions within the flow cross-section of a ferrofluid Couette–Poiseuille flow under time-varying magnetic fields of different frequencies ($\chi_0 = 3.5$, $\eta^*/\zeta^* = 5.65$, $P^* = 1$, $Re = 0.1$, $Mn_f = 800$, $Pe = 0.00625$, $\eta' = 10^{-20}$ N·s) 212

Figure 6.47 Spin velocity distributions within the flow cross-section of a ferrofluid Couette–Poiseuille flow in the uniform time-varying magnetic fields in the x and y directions under the velocity/vorticity matching boundary conditions ($\chi_0 = 3.5$, $\eta/\zeta = 5.65$, $\partial p^*/\partial x^* = -2$, $Re = 0.1$, $Mn_f = 800$, $Pe = 0.00625$, $\eta' = 10^{-20}$ N·s) 214

Figure 6.48 Variation of rotational viscosity with magnitude of magnetic field intensity in a ferrofluid Couette–Poiseuille flow under the action of time-varying magnetic fields at two frequencies ($\chi_0 = 3.5$, $\eta^*/\zeta^* = 5.65$, $P^* = 1$, $Re = 0.1$, $Pe = 0.00625$, $\eta' = 10^{-20}$ N·s) (a) $w^* = 0.1$; (b) $w^* = 5$ 214

Figure 6.49 Variation of rotational viscosity with magnetic field frequency in a ferrofluid Couette–Poiseuille flow for $\tau = 4 \times 10^{-4}$ s ($\chi_0 = 3.5$, $\eta^*/\zeta^* = 5.65$, $P^* = 1$, $Re = 0.1$, $Pe = 0.00625$, $Mn_f = 800$, $\eta' = 10^{-20}$ N·s) 215

Figure 6.50 Variation of rotational viscosity with magnetic field frequency in a ferrofluid Couette–Poiseuille flow for $\tau = 4 \times 10^{-3}$ s ($\chi_0 = 3.5$, $\eta^*/\zeta^* = 5.65$, $P^* = 1$, $Re = 0.1$, $Pe = 0.0625$, $Mn_f = 800$, $\eta' = 10^{-20}$ N·s) 216

Figure 6.51 Effect of the magnitude of the time-varying magnetic field on the tangential stress on the upper plate in ferrofluid Couette–Poiseuille flows ($\chi_0 = 3.5$, $\eta/\zeta = 5.65$, $\partial p^*/\partial x^* = -2$, $Re = 0.1$, $Pe = 0.00625$, $Mn_f = 800$, $\eta' = 10^{-20}$ N·s) 217
Figure 6.52 Effect of the field frequency on the tangential stress on the upper plate in ferrofluid Couette–Poiseuille flows ($\chi_0 = 3.5$, $\eta/\zeta = 5.65$, $\partial p^*/\partial x^* = -2$, $Re = 0.1$, $Pe = 0.00625$, $Mn_f = 800$, $\eta' = 10^{-20}$ N·s) 219
Figure 7.1 A model for the flow of ferrofluids in a pipe 221
Figure 7.2 Velocity distribution of ferrofluid pipe flows under the influence of a uniform constant axial magnetic field varies with the magnetization relaxation time ($H_z = 10^4$ A/m, $Pe = 1$) 229
Figure 7.3 Velocity distribution of ferrofluid pipe flows varies with the axial magnetic field intensity ($\tau = 2 \times 10^{-6}$ s, $Pe = 1$) 230
Figure 7.4 Variation of rotational viscosity in the ferrofluid pipe flows as a function of axial magnetic field intensity ($\tau = 2 \times 10^{-6}$ s, $Pe = 1$) 230
Figure 7.5 Variations in rotational viscosity with magnetization relaxation time in the ferrofluid pipe flows under an application of a uniform constant axial magnetic field ($H_z = 10^4$ A/m, $Pe = 1$) 231
Figure 7.6 Comparison of velocity distributions in ferrofluid pipe flows under the influence of radial and axial magnetic fields ($\tau = 2 \times 10^{-6}$ s, $Pe = 1$) 233
Figure 7.7 Flow rate of ferrofluid pipe flows under the action of radial and axial uniform magnetic fields ($\tau = 2 \times 10^{-6}$ s, $Pe = 1$) 234
Figure 7.8 Variation of relative rotational viscosity with Pe calculated by different methods ($Re_m = 100$, $Re = 0.1$, $\partial p^*/\partial z^* = -1$, $\mathrm{d}H_z^* / \mathrm{d}z^* = 0.01$) 239
Figure 7.9 Flow velocity distributions at different Re_m ($Pe = 0.1$, $Re = 0.1$, $\partial p^*/\partial z^* = -1$, $\mathrm{d}H_z^* / \mathrm{d}z^* = 0.01$ 240
Figure 7.10 Flow velocity distributions at different magnetic field gradients ($Pe = 0.1$, $Re = 0.1$, $\partial p^*/\partial z^* = -1$, $Re_m = 10$ 240
Figure 7.11 Variation of rotational viscosity with Re_m ($\mathrm{d}H_z^* / \mathrm{d}z^* = 0.01$) and magnetic field gradient ($Re_m = 10$) ($Pe = 0.1$, $Re = 0.1$, $\partial p^*/\partial z^* = -1$) 241
Figure 7.12 Ferrofluid pipe flow in an axial gradient magnetic field 242
Figure 7.13 Pressure distribution along the axial direction in ferrofluid pipe flows under the action of axial gradient magnetic fields (solid line: pressure distribution and dashed line: magnetic field intensity distribution) 243
Figure 7.14 Contour lines of relative flow rates (with the numbers below the curves indicating the values of relative flow rate $\Delta Q/Q_0$, $\Omega_0\tau_B = 5$) 247
Figure 7.15 Relationships between the relative variation in flow rate and $\Omega_0\tau_B$ for different values of $w\tau_B$ ($\alpha_0 = 3$) 247
Figure 7.16 Variation of relative rotational viscosity along the radial direction of the circular pipe ($w\tau_B = 4$) 248
Figure 7.17 Distributions of additional velocity v_1 at different time instances ($\alpha_0 = 3$, $w\tau_B = 4$). (a) $\mathbf{\Omega x}_0\tau_B = 0.5$; (b) $\Omega_0\tau_B = 1.8$; and (c) $\Omega_0\tau_B = 3$ 248
Figure 7.18 Rotational flow in a ferrofluid pipe flow 256

Figure 7.19 Variation of volumetric flow rate with pressure gradient P' in ferrofluid pipe flows ($\mu_r = 9$, the solid line represents pure axial flow and the dashed line represents rotational motion) 256

Figure 7.20 Axial and circumferential velocity components in a ferrofluid pipe flow ($\mu_r = 9$, $P' = 11$, the dashed line represents the calculation results when the circumferential speed is set to zero) 257

Figure 7.21 Components of magnetization of ferrofluids ($\mu_r = 9$, $P' = 11$, and $\hat{h} = 8.8$; the dashed line represents the calculation results when the circumferential speed is set to zero) 257

Figure 8.1 Schematic diagram of ferrofluid concentric flows 259

Figure 9.1 Ferrofluid spin-up flows (when the surface is concave) 272

Figure 9.2 Top-down view of the ferrofluid spin-up flows 272

Figure 9.3 Geometric relationships among variables 273

Figure 9.4 Rotational velocity of the ferrofluid surface in a spin-up flow. ($R_0 = 23.3$ mm and $w_0/2\pi = 60$ Hz. The solid lines represent the calculated results from Eq. (9.22), while the symbols O and Δ represent experimental results corresponding to $H_0 = -7.96 \times 10^3$ A/m and $H_0 = -1.59 \times 10^4$ A/m, respectively 276

Figure 9.5 Torque experienced by the cylindrical container in a spin-up flow of ferrofluids. (a) Ferrofluid EMG705 and (b) ferrofluid EMG900 289

TABLE

Table 6.1 Parameter Values Used in Numerical Calculations 153

Symbols

Symbol	Meaning	Unit	First Appearance
a	Coordinate of the focus located on the right side of the y-axis in a bipolar coordinate system	m	3.4.1
A	$A = \Omega - \omega_p$	1/s	2.7.4
$\mathbf{A}, \mathbf{A}_1, \mathbf{A}_2$	An arbitrary second-order tensor	—	2.7.3
$\mathbf{A}^{(a)}$	Anti-symmetric tensor of tensor **A**	—-	2.7.3
$\mathbf{A}^{(s)}$	Skew tensor of tensor **A**	—	2.7.3
$\mathbf{A}^{(sm)}$	Symmetric tensor of tensor **A**	—	2.7.3
b	Axial distance between the cylindrical shell and the cylindrical permanent magnet	m	3.4.1
$\boldsymbol{B}$	Magnetic induction	T	2.5
B_r	Remanent magnetic induction	T	3.4.3
c	Speed of light	m/s	2.5
$\boldsymbol{c}_n$	Force couple acting on a system surface, or surface couple density	N/m	2.4
$\mathbf{C}$	Couple stress tensor	N/m	2.4
CS	External surface of a control volume	—	2.1
CV	Control volume	—	2.1
d_m	Diameter of the particle cores in a ferrofluid	m	2.4
d_p	Diameter of particles in a ferrofluid	m	4.1
$\overline{d}_p$	Average diameter of particles in a ferrofluid	m	2.4
$\boldsymbol{D}$	Electric displacement	C/m^2	2.5
$\mathbf{D}^{(v)}$	Strain rate tensor	1/s	2.6.2
$\boldsymbol{e}$	Unit vector	—	2.4
$\boldsymbol{e}_H$	Unit vector along the magnetic field direction	—	5.3.2
$\boldsymbol{e}_m$	Unit vector along the direction of particle's magnetic moment	—	4.4
$\boldsymbol{e}_x$ 、 $\boldsymbol{e}_y$ 、 $\boldsymbol{e}_z$	Unit vectors in the x, y, and z directions in a Cartesian coordinate system	—	3.4.1
$\boldsymbol{e}_r$ 、 $\boldsymbol{e}_\theta$ 、 $\boldsymbol{e}_z$	Unit vectors in the r, θ, and z directions in a cylindrical coordinate system	—	7.4.1
$\boldsymbol{e}_\gamma$ 、 $\boldsymbol{e}_\theta$	Unit vectors in the directions of coordinates γ and θ in a bipolar coordinate system	—	3.4.1
$\boldsymbol{E}$	Electric field intensity	N/C	2.5
$\boldsymbol{E}'$	Apparent electric field intensity	N/C	2.5
$\boldsymbol{f}$	Body force density of a ferrofluid	N/m^3	3.4.2

(*Continued*)

Symbol	Meaning	Unit	First Appearance
$\boldsymbol{f}_m$	Magnetic body force density	N/m^3	3.1
f_τ	Tangential force exerted by ferrofluid on a wall	N/m^2	2.10
$\boldsymbol{F}$	Body force per unit mass	m/s^2 or N/kg	2.3
$\boldsymbol{F}_{\text{Kelvin}}$	Kelvin force	N/m^3	3.1
$\boldsymbol{F}_m$	Magnetic levitation force acting on a suspended body within a ferrofluid	N	3.4.1
$\mathcal{F}$	Intensity quantity	—	2.1
$\boldsymbol{g}$	Momentum of electromagnetic field	$kg/(m^2{\cdot}s)$	2.3
$\boldsymbol{g}_0$	Gravitational acceleration	m/s^2	3.1
$\mathbf{G}$	Body couple	m^2/s^2	2.4
$\boldsymbol{G}_H$	Gradient of magnetic field intensity	A/m^2	6.3.1
G_H	Magnitude of dimensionless magnetic field intensity gradient	1	6.3.1
h_0	Height dimension	m	6
h_c	Spacing between the composite structure of magnets and wall	m	3.4.2
$\boldsymbol{h}_e$	Unit vector in the same direction as $\boldsymbol{\varsigma}$	—	5.3.2
h_g	Reference height in the direction of gravity	m	3.2
h_s	Climbing distance of ferrofluids along the wall in a spin-up flow	m	9.1
h_θ	Lamé coefficient for the θ coordinate in a bipolar coordinate system	—	3.4.1
h_γ	Lamé coefficient for the γ coordinate in a bipolar coordinate system	—	3.4.1
h'	Distance between the end face of a magnet and the interface of ferrofluid	m	3.4.3
$\boldsymbol{H}$	Magnetic field intensity	A/m	2.5
$\boldsymbol{H}'$	Apparent magnetic field intensity	A/m	2.5
H_0	Amplitude of alternating magnetic field, magnitude of constant uniform magnetic field, or the minimum intensity of gradient magnetic field	A/m	5.3.1/6.2.2/6.3.1
$\boldsymbol{H}_e$	Effective field	A/m	4.3
i	Imaginary unit	—	6.4.2
$\boldsymbol{i}_m$	Surface magnetization current density	A/m	3.4.4
$\mathbf{I}$	Unit tensor	—	2.5
I	Moment of inertia of unit mass ferrofluid	m^2	2.4
I'	Density of particle's moment of inertia in a ferrofluid	kg/m	2.4
$\boldsymbol{I}_m$	Distributed current	A·m	3.4.4
$\boldsymbol{j}_0$	Conductive current density	A/m^2	2.5
$\boldsymbol{j}'$	$\boldsymbol{j}' = \boldsymbol{j}_0 - \rho_e \boldsymbol{v}$	A/m^2	2.5
$\boldsymbol{J}$	Magnetic polarization	$N{\cdot}A/m^3$	3.4.3
$\mathcal{J}$	Extensive property of a fluid	—	2.1
k_B	Boltzmann constant, 1.38×10^{-23}	N·m/K	3.1
k_m	Wave number of periodically arranged magnets in a composite structure	1/m	3.4.2
K	Anisotropy constant of particlulate materials	J/m^3	4.2

(Continued)

Symbol	Meaning	Unit	First Appearance
l	Characteristic length scale of fluid flow	m	5.3.1
l_0	Length of circular tube in a ferrofluid pipe flow	m	7.3
l_m	Curve along the cross-sectional contour of a magnet	m	3.4.1
$L(\,)$	Langevin function	—	3.1
$\boldsymbol{L}$	Total local angular momentum density of a system	kg/(m·s)	2.4
L_0	Length dimension of the flat plate in a Couette–Poiseuille flow	m	6.1
$\boldsymbol{L}_m$	Magnetic moment (density)	Pa	6.4.2
L_p	Inherent characteristic length of a ferrofluid	m	6.4.3
m	Mass	kg	2.1
$\boldsymbol{m}_p$	Magnetic moment of a particle	A·m^2	3.1
$\boldsymbol{M}$	Magnetization	A/m	2.5
M_0	Equilibrium magnetization of a ferrofluid	A/m	3.1
M_d	Saturation magnetization of particulate materials	A/m	4.1
M_S	Saturation magnetization of a ferrofluid	A/m	3.1
$\bar{M}$	Field-averaged magnetization	A/m	3.2
$\tilde{M}_0$	Two times the magnetization on a composite magnet surface	A/m	3.4.2
Mn_f	A dimensionless characteristic number, the ratio of magnetic moment to viscous moment	1	6.2.1
MV	Material volume	m^3	2.1
n	A positive integer	—	3.4.1
$\boldsymbol{n}$	Unit vector pointing outwards perpendicular to a surface	—	2.1
$\boldsymbol{n}'$	Normal unit vector along the γ-contour lines in a bipolar coordinate system	—	3.4.1
N	Number concentration of magnetic particles in a ferrofluid	1/m^3	2.4
p	Pressure in the presence of an external field	Pa	2.6.1
p_0	Thermodynamic pressure in general	Pa	2.6.1
p_m	Magnetic pressure	Pa	3.3
p_s	Magnetostrictive pressure	Pa	3.4.2
$\boldsymbol{P}$	Polarization	C/m^2	2.5
P^*	Dimensionless pressure gradient	—	6.1
Pe	Péclet number, the ratio of magnetization relaxation time to flow characteristic time	1	6.2.1
$\boldsymbol{q}$	Heat flux density vector	W/m^2	2.5
q_m	Magnetic charge intensity	N·m/A	3.4.3
Q	Flow rate	m^2/s	6.1
$\boldsymbol{r}$	Position vector of a fluid element relative to a fixed point	m	2.1
r_0	Radius of particles without surfactant coating	m	5.1
R	Volumetric heat release rate	m^2/s^3	2.5
R_0	Radius of circular pipe in a ferrofluid pipe flow	m	7.1.1
R_1	Radius of the cylindrical magnet in calculating magnetic levitation force	m	3.4.1
R_2	Inner radius of shell in calculating magnetic levitation force	m	3.4.1
Re	Reynolds number	1	6.2.1

(Continued)

Symbol	Meaning	Unit	First Appearance
Re_m	Magnetic Reynolds number	1	7.2.1
$\boldsymbol{s}$	Intrinsic or spin angular momentum density	m^2/s	2.4
S	Area	m^2	2.1
$\mathbf{S}$	Deformation velocity tensor of a ferrofluid	1/s	8.1
S_e	Entropy per unit mass	J/(kg·K) or $m^2/(K{\cdot}s^2)$	2.6.1
$S_e^{(s)}$	Entropy provided by external environment (per unit mass)	J/(kg·K) or $m^2/(K{\cdot}s^2)$	2.7.1
$S_e^{(i)}$	Entropy generation within a system (per unit mass)	J/(kg·K) or $m^2/(K{\cdot}s^2)$	2.7.1
S_m	Surface of the suspended body in a ferrofluid	m^2	3.4
t	Time	s	2.1
t_m	Period of a time-varying magnetic field	s	6.4.2
$\boldsymbol{t}_n$	Stress vector defined on a surface	N/m^2	2.3
T	Temperature	K	2.6.1
$\mathbf{T}$	Stress tensor, $\mathbf{T} = \tilde{\mathbf{T}} - \boldsymbol{vg}$	N/m^2	2.3
$\tilde{\mathbf{T}}$	Total stress tensor	N/m^2	2.3
$\mathbf{T}_m$	Magnetic stress tensor	N/m^2	2.6.2
$\mathbf{T}_m'$	Equivalent magnetic stress tensor	N/m^2	3.1
$\mathbf{T}_v$	Viscous stress tensor	N/m^2	2.6.2
u	Internal energy per unit mass	m^2/s^2	2.5
u_m	Internal energy other than electromagnetic energy in the space occupied by a medium	m^2/s^2	2.5
U	Energy potential	N·m	3.3
$\boldsymbol{v}$	Velocity	m/s	2.1
v_0	Velocity of the plate in a Couette–Poiseuille flow	m/s	5.2
$\boldsymbol{v}_1$	Slowly varying component of velocity	m/s	5.3.1
$\boldsymbol{v}_2$	Rapidly varying component of velocity	m/s	5.3.1
v_m	Average flow rate in a pipe flow	m/s	7.1.1
V	Volume	m^3	2.1
V_h	Volume of a single particle containing surfactant layer	m^3	4.2
V_p	Volume of a single magnetic particle in a ferrofluid	m^3	4.1
w	2π times the frequency of a magnetic field	1/s	5.3.1
w_0	Rotational velocity of a magnetic field in ferrofluid spin-up flow	1/s	9.1.1
W	Orientation distribution function of particle magnetic moment	—	4.4

GREEK SYMBOLS

Symbols	Meaning	Unit	First Appearance
α	Langevin parameter	—	3.1
α_0	Langevin parameter corresponding to alternating magnetic field amplitude	—	5.3.1
β	Angle between vectors $\boldsymbol{H}$ and $\boldsymbol{\Omega}$	rad	5.1
δ_0	Thickness of surfactant layer on a particle	m	5.1

(Continued)

Symbols	Meaning	Unit	First Appearance
δ_m	Surface-to-wall distance of a composite magnet when centred in a ferrofluid	m	3.4.2
δ_{ij}	Kronecker symbol	—	2.4
Δ	Eccentricity of a composite magnet in a ferrofluid	m	3.4.2
ε_0	Vacuum dielectric constant	$C^2/(N\cdot m^2)$	2.5
$\boldsymbol{\epsilon}$	Polyadic tensor	—	2.4
ϵ_{ijk}	Component of polyadic tensor	—	2.4
ϕ	Volume fraction of particles in a ferrofluid	%	2.4
ϕ_c	Volume fraction corresponding to the close-packed state of particles in a ferrofluid	%	5.1
Φ	Dissipation function	$kg/(m\cdot s^3)$	2.7.1
$\tilde{\gamma}$	Coefficient of the Lorentz transformation	—	2.5
γ'	Scaling factor between $\text{tr}\mathbf{T}_v$ and $\text{tr}\mathbf{D}^{(v)}$	Pa·s	2.7.3
Γ	Viscous drag on flat plate in a Couette–Poiseuille flow	Pa	6.1
η	Intrinsic viscosity, or shear viscosity coefficient, of a ferrofluid	Pa·s	2.7.3
η_0	Dynamic viscosity of the carrier of a ferrofluid	Pa·s	2.9
η_{eff}	Effective or apparent viscosity of a ferrofluid in the presence of a magnetic field	Pa·s	5.2
η_r	Rotational viscosity of a ferrofluid	Pa·s	5.1
η'	Shear spin viscosity coefficient	N·s	2.7.3
φ_m	Scalar magnetic potential	A	3.4.1
ϑ	Angle at which magnetization lags behind magnetic field	—	5.3.1
κ_T	Thermal conductivity	W/(m·K)	2.7.2
ℓ	Distance between particles in a ferrofluid	m	2.9
λ	Bulk viscosity coefficient	Pa·s	2.7.3
λ_m	Two times the pole face spacing of magnets in a composite magnet	m	3.4.2
λ_r	Resistance coefficient in a pipe flow	1	7.1.3
λ'	Bulk spin viscosity coefficient	N·s	2.7.3
μ	Permeability of a ferrofluid	N/A^2	3.4.1
μ_0	Vacuum permeability, $4\pi \times 10^{-7}$	N/A^2	2.5
μ_r	Relative permeability of a ferrofluid	1	3.4.1
ν	Kinematic viscosity of a ferrofluid	m^2/s	5.3.1
$\boldsymbol{\nu}$	A small correction field	—	4.4
ψ	Thermodynamic potential	J	4.3
θ_m	Angle between effective field $\boldsymbol{\varsigma}$ and actual field $\boldsymbol{\alpha}$	—	5.2
$\mathcal{R}$	A general field variable	—	2.1
ρ	Density	kg/m^3	2.1
ρ_{coat}	Density of the surfactant on a particle	kg/m^3	2.4
ρ_e	Volume density of free charges	C/m^3	2.5
ρ_{mag}	Density of particle nuclei	kg/m^3	2.4
ρ_s	Density of the particulate material in a ferrofluid	kg/m^3	2.4
σ_E	Conductivity	S/m	2.7.2
σ_m	Surface density of magnetic charges	N/(A·m)	3.4.3

(Continued)

Symbols	Meaning	Unit	First Appearance
$\boldsymbol{\tau}$	Tangential unit vector on walls	—	2.10
τ	Total relaxation time of a ferrofluid	s	4.2
τ_0	Inverse of the Larmor frequency of the magnetic moment of a particle in an anisotropic field	s	4.2
τ_B	Brownian relaxation time	s	2.9
τ_h	Characteristic timescale	s	5.3.1
τ_N	Néel relaxation time	s	4.2
τ_s	Time for the angular velocity of particles in a ferrofluid to shift to flow vorticity	s	5.1
υ	Specific volume, inverse of density	m^3/kg	2.6.1
ω_0	Characteristic angular velocity or characteristic spin velocity	1/s	6.2.1
ω'	Average rotational velocity of magnetic particles in the reference system Σ'	1/s	4.3
ω_p	Average rotational velocity of magnetic particles in a ferrofluid	1/s	2.4
$\boldsymbol{\Omega}$	Fluid rotation rate or vorticity	1/s	2.6.2
$\boldsymbol{\Omega}_0$	Characteristic vorticity	1/s	6.4.3
$\boldsymbol{\Omega}_{rms}$	Root mean square vorticity	1/s	6.2.4
χ	Susceptibility of a ferrofluid	—	8.2
χ_0	Initial susceptibility of a ferrofluid	—	4.1
ς	Dimensionless effective field	—	4.4
ζ	Vortex viscosity	Pa·s	2.7.3

OTHER SYMBOLS

$\dfrac{D}{Dt}$	Material derivative
$\dfrac{\partial}{\partial t}$	Partial derivative
(e)	Superscript, indicating the equilibrium value
T	Superscript, indicating the transpose of a matrix or tensor
Σ'	A local reference system at rest relative to a magnetic particle in a ferrofluid
i, j, k, l, m, n, p	Subscript, indicating one of x, y, or z in a Cartesian coordinate system
(x, y, z)	Coordinates in a Cartesian coordinate system
(r, θ, z)	Coordinates in a cylindrical coordinate system
(θ, γ)	Coordinates in a plane bipolar coordinate system
*	Superscript, indicating a dimensionless physical quantity
$\langle\rangle$	Time-averaged value of a variable
^	Magnitude of a complex variable
//	Component parallel to the relative velocity of reference system motion
$\perp$	Component perpendicular to the relative velocity of reference system motion
tr	Trace of a tensor
‾	Conjugate complex of a complex variable

Preface

SINCE THE PUBLICATION OF the first paper on ferrohydrodynamics, "Ferrohydrodynamics", by Neuringer and Rosensweig in 1964, the field of ferrohydrodynamic research has spanned a remarkable journey of 60 years. Progressing from initial debates over the expression of magnetic volume forces to the universally valid expression of Maxwell's stress tensor, and then to the validity conditions of Helmholtz and Kelvin forces, ferrohydrodynamics has gradually matured through continuous debate among scholars worldwide. Some of the vague concepts that once existed have become clearer, leading to the establishment of a systematic theory of ferrohydrodynamics. However, theoretical frameworks, exemplified by Rosensweig's monograph *Ferrohydrodynamics*, are founded on the assumption of equilibrium magnetization in ferrofluids, limiting their ability to describe non-equilibrium flows and behaviours.

As ferrofluids become increasingly utilized in diverse fields such as mechanics, petrochemical engineering, environmental engineering, and biomedicine, the complexity and variability of applied magnetic fields and operational environments also increase. Therefore, it is imperative to consider the magnetization relaxation processes of ferrofluids, necessitating the application of non-equilibrium ferrohydrodynamic principles to explain the underlying physics. For instance, in high-speed rotating ferrofluid seals, the radial gradient of fluid velocity within the sealing gap is comparable to the inverse of the magnetization relaxation time, requiring the utilization of non-equilibrium ferrofluid Couette–Poiseuille flow theory to predict flow characteristics. Similarly, in active ferrofluid viscous dampers, the alteration of the apparent viscosity of ferrofluid within throttling orifices through the application of alternating magnetic fields necessitates the prediction of throttling resistance using non-equilibrium pipe flow theory. The original purpose of this book is to establish a theoretical foundation for addressing these engineering challenges posed by ferrofluids.

This book is divided into nine chapters. Chapter 1 introduces the basic concepts of ferrofluids, their main applications, and the fundamental ideas of non-equilibrium ferrohydrodynamics. Chapter 2 begins with the theory of transport, deriving in detail the equations of non-equilibrium ferrohydrodynamics, their simplified forms, and presenting the stress tensor of ferrofluids. Chapter 3 covers the fundamental equations of quasi-equilibrium ferrofluid dynamics and the magnetic levitation force in ferrofluids, which can be viewed as a special case of non-equilibrium ferrohydrodynamics when magnetization relaxation is neglected. Chapters 4 and 5 discuss two crucial characteristics of ferrofluids: Magnetization properties and viscosity properties. They also present the magnetization equations of ferrofluids and an important physical manifestation of non-equilibrium theory – the magneto viscous effect.

In Chapters 6–9, various types of flows of ferrofluids under different magnetic field configurations are discussed, including planar Couette–Poiseuille flow, pipe flow, laminar circular flow, and spin-up flow. They represent the specific manifestations of non-equilibrium ferrohydrodynamics in different flow patterns.

A distinguishing feature of this book in addressing the issues at hand is its provision of detailed mathematical derivations for crucial outcomes, enabling readers to draw parallels and apply relevant mathematical theories to explore other fluid dynamic problems. Given the nascent nature of the non-equilibrium ferrohydrodynamic theory and its applications across different flow configurations, the content of this book focuses solely on well-established or experimentally validated flow problems.

I am deeply grateful for the support received from the National Natural Science Foundation of China (Grant Nos. 52375164, 52005033), the Natural Science Foundation of Guangdong Province (Grant No. 2024A1515010860), and the Open Research Subject of Guangxi Key Laboratory of Automobile Components and Vehicle Technology (Grant No. 2023GKLACVTKF02) for funding some of the research presented in this book.

A deeper gratitude in this respect goes to my parents and wife, supporting me over all these years and understanding my work.

Wenming Yang
Beijing

CHAPTER 1

Introduction

FERROFLUIDS ARE ALSO KNOWN as magnetic fluids or magnetic liquids. In order to distinguish from other media that can respond to magnetic fields, such as magnetorheological fluids, this book uses ferrofluids uniformly. Ferrofluids are widely used in engineering and technology because of their combination of magnetic properties and liquid-like fluidity. In their applications, if the velocity gradient of ferrofluid flow is substantial and subjected to strong magnetic fields, the theory of non-equilibrium ferrohydrodynamics is essential for the analysis of engineering challenges.

1.1 FERROFLUIDS

A ferrofluid is a colloidal suspension composed of nano-sized ferro- or ferrimagnetic solid particles coated with a surfactant layer and dispersed in a carrier liquid. Common carrier liquids encompass a diverse range of substances, including water, synthetic hydrocarbons, synthetic esters, fatty acids, hydrocarbons, and various others. Oleic acid, linoleic acid, and numerous other fatty acids belong to the class of surfactants that can be effectively utilized. Typical particulate materials encompass iron tetraoxide and cobalt compounds. Ferrofluids can be categorized depending on the type of carrier liquid or particle utilized. Typically, the diameter of magnetic particles within ferrofluids measures 10 nanometres. However, when coated with surfactants, these particle diameters can vary significantly, ranging from the teens to tens of nanometres (Rosensweig, 1985).

Ferrofluids were first documented in 1938 when Elmore from Massachusetts Institute of Technology successfully synthesized stable magnetic colloids. These colloids were subsequently employed as a replacement for Bitter's magnetic powder method in visualizing the magnetic fields of permanent magnets (Elmore, 1938). Subsequently, in 1964, Neuringer and Rosensweig presented their research on ferrofluids for magneto-thermal conversion, highlighting their synthesis techniques and potential applications (Neuringer and Rosensweig, 1964). Papell from NASA secured the initial patent for the preparation of ferrofluids in 1965 (Papell, 1965). He developed ferrofluids for the purpose of controlling fuel flow in rockets under weightless conditions in space utilizing an external magnetic field. The preparation of magnetic particles plays a pivotal role in the synthesis of

DOI: 10.1201/9781003540342-1

ferrofluids, which are currently produced via chemical co-precipitation, thermal decomposition, and mechanical grinding methods (Hamzah et al., 2021; Oscar et al., 2022; Odenbach, 2009).

Due to the minute size of particles within ferrofluids, they undergo Brownian motion resulting from random collisions with carrier fluid molecules, maintaining the colloidal stability of ferrofluids. Additionally, the particles are coated with surfactants that possess long-chain molecules. These molecules adhere to the particle surface at one end while freely oscillating at the other end within the carrier fluid. This mechanism prevents particles from approaching each other, thus avoiding the agglomeration of magnetic particles caused by long-range magnetic attraction. Consequently, ferrofluids exhibit neither particle precipitation nor significant concentration gradients, regardless of the presence of gravity or magnetic fields (Silfhout et al., 2020).

When exposed to a magnetic field, static ferrofluids undergo magnetization, exhibiting superparamagnetic characteristics without hysteresis. Magnetization in these fluids occurs through two mechanisms: Intrinsic magnetization resulting from the rotation of magnetic domains within the particles and extrinsic magnetization induced by the rotation of the particles themselves. For the extrinsic magnetization mechanism, the angle of particle rotation under the influence of an external magnetic field is determined by the equilibrium between magnetic field energy and thermal motion energy, while the speed of this rotation is influenced by the equilibrium between the magnetic torque acting on the particles and the viscous torque exerted by the carrier. Regardless of whether the magnetization is intrinsic or extrinsic, the magnetization in a stationary ferrofluid is always aligned with the applied magnetic field intensity. In this case, the limiting state of magnetization is characterized by the saturation magnetization. Commercially available ferrofluids currently exhibit saturation magnetization values up to 600 Gs.

Ferrofluids exhibit significantly higher viscosity compared to their carrier fluids. Without the presence of an external magnetic field, the viscosity of ferrofluids primarily relies on the volume fraction of solid particles within them. However, upon exposure to a magnetic field, the apparent viscosity undergoes alterations, leading to magneto viscous effects. Additionally, different kinds of magnetic fields affect the apparent viscosity in distinct manners (Odenbach, 2002).

Ferrofluids are opaque. In the absence of an external magnetic field, particles within ferrofluids are uniformly distributed, resulting in isotropic optical properties. However, upon the introduction of a magnetic field, the particles align along its direction, inducing anisotropic phenomena such as polarization and birefringence within ferrofluids. Notably, these optical properties can be modulated by manipulating the applied magnetic field (Raikher et al., 2002).

The thermal properties of ferrofluids, including density, specific heat, thermal conductivity, diffusion coefficient, etc., are usually expressed as macroscopic equivalent mean values. These mean values are calculated from the relative proportions of solid and liquid phases within ferrofluids. Nevertheless, it is imperative to recognize that in the presence of a magnetic field, ferrofluids exhibit an uneven concentration distribution, also referred to as anisotropy. Therefore, when analyzing the thermal properties in such scenarios, it is necessary to treat ferrofluids as multiphase mediums, where the solid and liquid phases are treated independently.

Attenuation of sound waves occurs in ferrofluids due to the absorption of acoustic energy by the particles that form chains within them. This attenuation can be influenced by an applied magnetic field, which modifies the sound wave attenuation characteristics in an anisotropic manner (Shliomis et al., 2008).

1.2 APPLICATIONS OF FERROFLUIDS

Since the advent of stable ferrofluids, technological advancements have facilitated the development of numerous ferrofluidic devices, thanks to the establishment of ferrohydrodynamic theory and the commercialization following the incorporation of the first ferrofluid company in 1968 (Glossop, 1995). These fluids have opened up new horizons in terms of applications, with one of the earliest marketable products being a dynamic seal proposed in 1969, capable of operating under both vacuum and pressure conditions (Buschow, 2006). By the early 1990s, the commercialization of ferrofluidics had reached a global scale, resulting in the production of tens of millions of related devices annually (Raj and Moskowitz, 1990). Nowadays, ferrofluids have become reliable materials that can tackle intricate engineering challenges, finding applications across diverse fields such as aerospace, mechanics, and medicine (Berkovsky et al., 1993; Zhang et al., 2019; Kole and Khandekar, 2021; Philip, 2023).

The most successful application of ferrofluids in the mechanical field is the dynamic seals (Li, 2010; Yang et al., 2019; Li et al., 2023), as exemplified in Figure 1.1, which depicts a schematic of a typical ferrofluid rotary seal structure (Yang et al., 2024). Due to their responsive properties towards magnetic fields, ferrofluids are effectively attracted and fill the sealing gap between the stationary pole teeth and the rotating shaft, ensuring a robust sealing effect capable of withstanding substantial pressure differences. While ferrofluids are primarily employed in rotary seals operating at medium to low rotational speeds,

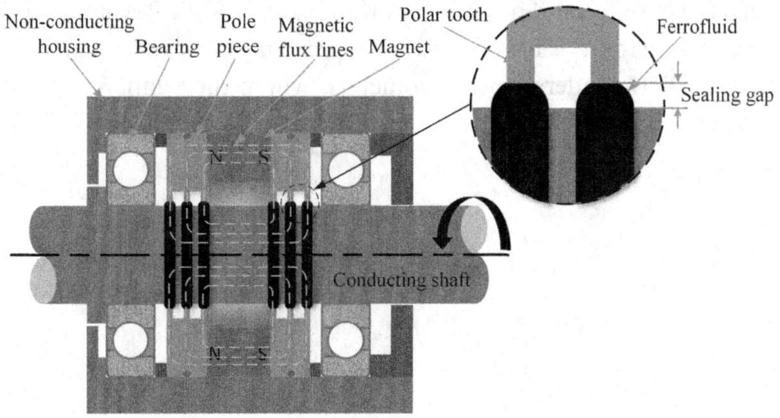

FIGURE 1.1 A typical ferrofluid seal structure. (Reproduced with permission from Elsevier from the literature Yang et al. (2024).)

recent attempts have focused on adapting them to high-speed rotational applications and sealing liquids (Mitamura et al., 2022). Ferrofluid seals offer numerous advantages, including a zero leakage rate, reduced friction, and a straightforward yet compact design, making them a popular choice for seals in vacuum pumps, vacuum coating machines, and numerous other applications.

Ferrofluids can be utilized as lubricants in mechanical sliding applications, or provide gas–liquid mixing support to minimize friction (Hu et al., 2018). Sliding bearings with the applications of ferrofluids exhibit superior load-bearing capacity, vibration resistance, and extended lifespan compared to traditional plain bearings. Additionally, end leakage is eliminated thanks to the confinement by magnetic field (Nagaraj, 1988; Patel et al., 2017).

By leveraging the principle of suspension in ferrofluid, grinding and polishing can be efficiently achieved. In this process, an abrasive is suspended on the ferrofluid's surface, where a magnetic levitation force exerts pressure on the target surface. Ferrofluid polishing utilizes a film encapsulating the ferrofluid, upon which an abrasive is applied. When subjected to a magnetic field, the film moulds the fluid's shape, facilitating adaptable and flexible polishing that contours to the surface being polished. (Zhong, 2008)

Ferrofluids can be used for both passive damping and active control of vibration. By harnessing the viscous dissipation of energy when the fluid is sheared, ferrofluids effectively dissipate excess mechanical energy stemming from vibrations (Yang et al., 2013). A ferrofluid rotary inertial damper, as shown in Figure 1.2, leverages the self-suspension principle of ferrofluids, enabling levitation of the mass block. When a device equipped with this damper experiences accelerating or decelerating rotational vibration, the mass block generates a relative velocity with the housing due to inertia. This relative motion viscously shears the ferrofluid within the gap, creating a damping effect. This method has been effectively used for vibration damping in stepper motors, significantly reducing the motor's stabilization time (Raj et al., 1995). Moreover, ferrofluids are also employed in active damping systems. As depicted in Figure 1.3, a cylinder filled with ferrofluid houses a piston mounted with a mass block. The piston and mass block are connected by a spring. By applying an external magnetic field, the resonance characteristics of the vibration system can be effectively reduced (Shimada and Kamiyama, 1991). Beyond their application in vibration control, ferrofluids have found widespread use in various instruments such as loudspeakers, CD/DVD readers, and viscometers, where they enhance performance by providing damping.

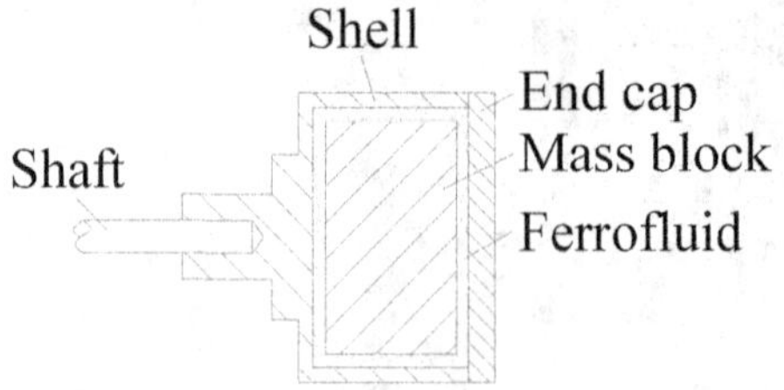

FIGURE 1.2 A ferrofluid rotary inertial damper.

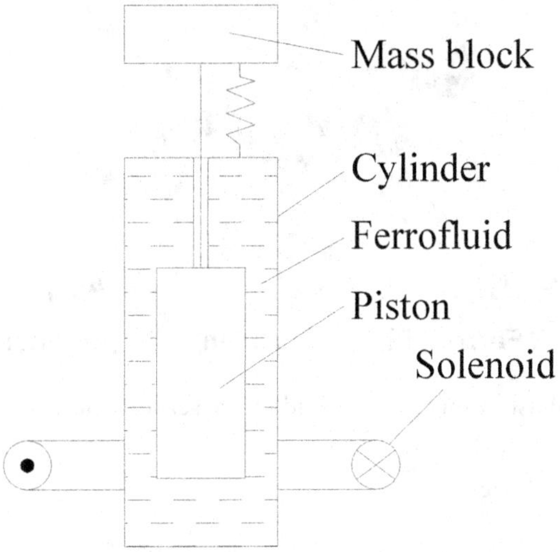

FIGURE 1.3 A ferrofluid active damper.

Ferrofluids can be effectively utilized for separation purposes, leveraging the suspension properties of non-magnetic solids within them. By adjusting the applied magnetic field, one can manipulate the apparent density of a ferrofluid, enabling the flotation of solid materials with different densities to its surface. Subsequently, these diverse materials can be efficiently separated (Zhu et al., 2010). This flotation technique offers numerous advantages, including high operational efficiency and the potential for reuse of the ferrofluid.

In the realm of robotics, the utilization of ferrofluids in the fabrication of magnetic microrobots holds promising potential for applications in minimally invasive surgery and precise drug delivery. These robots can leverage the magneto-thermal properties of ferrofluids to effectively eliminate tumour cells in targeted regions during drug delivery (Gonella et al., 2021; Ji et al., 2022). Additionally, soft capsule robots equipped with oscillating modules can be crafted to enhance drug release and other functionalities (Hua et al., 2022; Zhang et al., 2022).

In the field of sensors, ferrofluids have found widespread applications in acceleration, level, pressure difference, and fibre-optic sensors, serving as highly sensitive components. Additionally, ferrofluids are employed in tactile sensors, leveraging their interfacial deformation characteristics induced by a magnetic field, as depicted in Figure 1.4 (Liu et al., 2022). Furthermore, ferrofluids have been incorporated into surface plasma resonance magnetic field sensors, leveraging their thermal-optical effect (Cennamo et al., 2020; Kaur et al., 2021). Moreover, ferrofluids have found utility in electrochemical biosensors, exploiting their controllability via magnetic fields (Jiao et al., 2022).

In the petrochemical industry, magnetically controlled porous media flow employing ferrofluids has been introduced. The utilization of ferrofluids as substitute fluids for oil extraction holds the potential to significantly improve oil replacement efficiency, as demonstrated by Dou et al. (2022).

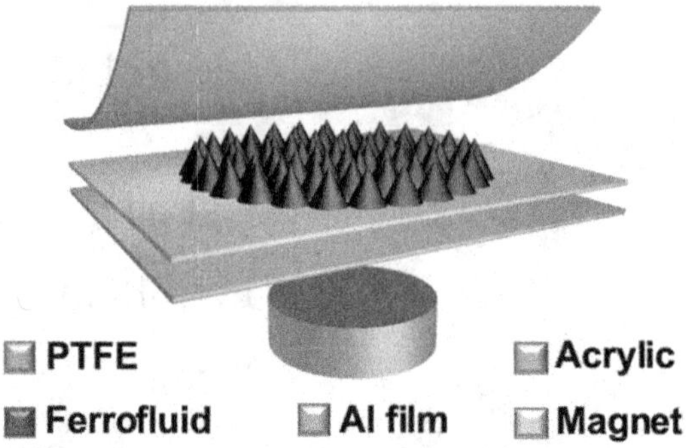

FIGURE 1.4 Schematic illustration of a ferrofluid tactile sensor. (Licence: https://creativecommons.org/licenses/by/4.0/)

In the realm of environmental protection, ferrofluids demonstrate exceptional ion selectivity during the absorption of radioactive elements, leveraging their capabilities for nuclear wastewater treatment as detailed by Chen et al. (2022). Furthermore, the suspension characteristics of non-magnetic particles within ferrofluids are harnessed for the elimination of micron and nano-sized plastic particles from ocean waters (Pramanik et al., 2021). Additionally, ferrofluids act as effective extractants for the recovery of analytes during the extraction process (Nayebi and Shemirani, 2021). It has also been proposed to use ferrofluids for carbon dioxide capture, leveraging their ability to enhance mass transfer efficiency when subjected to a magnetic field (Samadi et al., 2014).

In the biomedical realm, ferrofluids play a pivotal role as contrast agents for lesion detection, enabling photography and precise analysis of affected areas (Socoliuc et al., 2022). Moreover, these fluids can be injected into the body, serving as magnetically guided drug carriers that target lesions, delivering the agents directly to the affected sites for enhanced therapeutic outcomes (Scherer and Neto, 2005; Gonella et al., 2021). When integrated with microfluidics, ferrofluids facilitate the separation of diverse cells, thereby improving cell detection sensitivity (Kosea et al., 2009; Kosea and Koser, 2012; Yang et al., 2023). In addition, ferrofluids are employed in medical thermotherapy, leveraging their heat-generating capabilities under high-frequency alternating magnetic fields to achieve elevated specific absorption rates and facilitate cancer treatment (Jordan et al., 1999).

1.3 BASIC CONTENT OF NON-EQUILIBRIUM FERROHYDRODYNAMICS

The advent of ferrohydrodynamics and its underlying physics was heralded by the publication of the paper "Ferrohydrodynamics" by Neuringer and Rosensweig in *The Physics of Fluids* in 1964. Since then, Rosensweig and numerous other researchers have conducted extensive, long-term investigations into this field. Notably, Rosensweig's monograph *Ferrohydrodynamics*, completed in 1985, offers a comprehensive exposition of the equation systems governing the dynamics of ferrofluids, Bernoulli's theory, and the fundamental principles that govern the impact of magnetic fields on the equilibrium state and motion of ferrofluids. This comprehensive work serves as a solid foundation for the continued advancement of ferrofluid dynamics, greatly facilitating its applications in various fields such as industry, technology, and medicine.

The kinetic theory of ferrofluids, summarized by Rosensweig, relies on the presumption of equilibrium magnetization in ferrofluids. This magnetization occurs instantaneously, resulting in a magnetization relaxation time of zero. Consequently, it is occasionally referred to as quasi-equilibrium ferrohydrodynamics. In this quasi-equilibrium framework, the motion equation of ferrofluids incorporates a magnetization term, indicating that the fluid motion is influenced by magnetization. However, this magnetization is explicitly represented by the equilibrium magnetization equation, expressed as the Langevin function, which solely depends on the magnetic field intensity and is unaffected by the fluid's motion.

Non-Equilibrium ferrohydrodynamics investigates the laws governing the motion of ferrofluids under various forces in flowing ferrofluids, and the interaction between the ferrofluids and solid boundaries during relative motion, when the magnetization vector does not align with the magnetic field intensity. In a flowing ferrofluid, the magnetization relaxation effect becomes significant. Specifically, when the product of flow vorticity and magnetization relaxation time is not significantly less than one, the magnetization lags behind the magnetic field's changes. This lag induces a magnetic moment within the fluid, which is proportional to the vector product of magnetization and magnetic field intensity. This magnetic moment modifies the rotational state of magnetic particles, leading to alterations in their rotational behaviour. Consequently, a disparity between the particles' angular velocity of rotation and the flow vorticity can be maintained. This discrepancy introduces a term representing the magnetic moment into the ferrofluid's equation of motion. Thus, even in a uniform magnetic field where the magnetic bulk force vanishes, the ferrofluid's flow characteristics differ significantly from those observed in the absence of a magnetic field.

Non-Equilibrium ferrohydrodynamics continues to treat ferrofluids as continuous mediums, utilizing the corresponding average values of the thermal parameters of the constituent phases. The differential equations that govern the non-equilibrium flow of ferrofluids encompass conservation laws for mass, momentum, and energy, along with additional equations specific to magnetization, reflecting alterations in magnetization (Fang, 2022; de Carvalho and Gontijo, 2020). Throughout this book, the phenomenological magnetization equations or the microscopically derived magnetization equations are employed as one of the governing equations for ferrofluid non-equilibrium flows.

This book classifies the non-equilibrium flow of ferrofluids based on the concentration of content in current related research, grouping them into flow patterns including planar Couette–Poiseuille flow, pipe flow, concentric flow, and spin-up flow. While the research on non-equilibrium ferrohydrodynamics remains in its nascent stages, certain phenomena observed in these flow patterns lack a definitive understanding of their underlying physical mechanisms. Therefore, this book focuses solely on the established aspects of the aforementioned flow patterns.

REFERENCES

Berkovsky B. M., Medvedev V. F., Krakov M. S., *Magnetic fluids engineering applications*. Oxford: Oxford University Press. 1993.

Buschow K. H. J., *Handbook of magnetic materials*. North-Holland: Elsevier. 2006.

Cennamo N., Arcadio F., Marletta V., Baglio S., Zeni L., Andò B., A magnetic field sensor based on SPR-POF platforms and ferrofluids, *IEEE Transactions on Instrumentation and Measurement*, 70(1): 9504010, 2020.

Chen J., Xia L., Cao Q., Water-based ferrofluid with tunable stability and its significance in nuclear wastewater treatment, *Journal of Hazardous Materials*, 434: 128893, 2022.

de Carvalho D. D., Gontijo R. G., Reconstructing a continuous magnetization field based on local vorticity cells, CFD and Langevin dynamics: A new numerical scheme, *Journal of Magnetism and Magnetic Materials*, 514: 167135, 2020.

Dou X., Chen Z., Cao X., Ma C., Liu J., Oil displacement by the magnetic fluid inside a cylindrical sand filled Sample: Experiments and numerical simulations, *ACS Omega*, 7: 26473–26482, 2022.

Elmore W. C., Ferromagnetic colloid for studying magnetic structures, *Physical Review*, 54: 309, 1938.

Fang A., Consistent hydrodynamics of ferrofluids, *Physics of Fluids*, 34: 013319, 2022.

Glossop M., Ferrofluids enter new domains, *Magnetic Materials*, 1995, 3(11): 537–538.

Gonella V. C., Hanser F., Vorwerk J., Odenbach S., Baumgarten D., Influence of local particle concentration gradient forces on the flow-mediated mass transport in a numerical model of magnetic drug targeting, *Journal of Magnetism and Magnetic Materials*, 525: 167490, 2021.

Hamzah S., Ying L. Y., Abd A. A., Azmi R., Razali N. A., Hairom N. H. H., Mohamad N. A., Harun M. H. C., Synthesis, characterisation and evaluation on the performance of ferrofluid for microplastic removal from synthetic and actual wastewater, *Journal of Environmental Chemical Engineering*, 9: 105894, 2021.

Hu Z., Wang Z., Huang W., Wang X., Supporting and friction properties of magnetic fluids bearings, *Tribology International*, 130: 334–338, 2018.

Hua D., Liu X., Lu H., Sun S., Sotelo M. A., Li Z., Li W., Design, fabrication, and testing of a novel ferrofluid soft capsule robot, *IEEE/ASME Transactions on Mechatronics*, 27(3): 1403–1413, 2022.

Ji Y., Gan C., Dai Y., Bai X., Zhu Z., Song L., Wang L., Chen H., Zhong J., Feng L., Deformable ferrofluid microrobot with omnidirectional self-adaptive mobility, *Journal of Applied Physics*, 131: 064701, 2022.

Jiao J., Zhang H., Zheng J., Ferrofluids transport in bioinspired nanochannels: Application to electrochemical biosensing with magnetic-controlled detection, *Biosensors and Bioelectronics*, 201: 113963, 2022.

Jordan A., Scholz R., Wust P., Fähling H., Felix R., Magnetic fluid hyperthermia (MFH): Cancer treatment with AC magnetic field induced excitation of biocompatible superparamagnetic nanoparticles, *Journal of Magnetism and Magnetic Materials* 201: 413–419, 1999.

Kaur B., Sharma A. K., Prajapati Y. K., Plasmonic sensor for magnetic field detection with chalcogenide glass and ferrofluid materials under thermal variation in near infrared, *Optical Materials*, 117: 111175, 2021.

Kole M., Khandekar S., Engineering applications of ferrofluids: A review, *Journal of Magnetism and Magnetic Materials*, 537: 168222, 2021.

Kosea A. R., Fischer B., Mao L., Koser H., Label-free cellular manipulation and sorting via biocompatible ferrofluids, *Proceedings of the National Academy of Sciences of the United States of America*, 106(51): 21478–21483, 2009.

Kosea A. R., Koser H., Ferrofluid mediated nanocytometry, *Lab on a Chip*, 12: 190–196, 2012.

Li D., *Magnetic fluid sealing theory and application*. Beijing: Science Press. 2010.

Li D., Li Y., Li Z., Wang Y., Theory analyses and applications of magnetic fluids in sealing, *Friction*, 10: 1771–1793, 2023.

Liu J., Wen Z., Lei H., Gao Z., Sun X., A liquid-solid interface-based triboelectric tactile sensor with ultrahigh sensitivity of 21.48 kPa^{-1}, *Nano-Micro Letters*, 14: 88, 2022.

Mitamura Y., Nishimura I., Yano T., Thermal analysis of a miniature magnetic fluid seal installed in an implantable rotary pump, *Journal of Magnetism and Magnetic Materials*, 548: 168977, 2022.

Nagaraj H. S., Investigation of magnetic fields and forces arising in open-circuit-type magnetic bearings. *ASLE Tribology Transactions*, 31(2): 192–201, 1988.

Nayebi R., Shemirani F., Ferrofluids-based microextraction systems to process organic and inorganic targets: The state-of-the-art advances and applications, *Trends in Analytical Chemistry*, 138: 116232, 2021.

Neuringer J. L., Rosensweig R. E., Ferrohydrodynamics, *The Physics of Fluids*, 7: 1927, 1964.

Odenbach S., *Magnetoviscous effects in ferrofluids*. Berlin: Springer. 2002.

Odenbach S., *Colloidal magnetic fluids: Basics, development, and application of ferrofluids*. Berlin: Springer Verlag. 2009.

Oscar O., Sussy I. C. R., Pabel C. A., Illya A. M. V., Approaches on ferrofluid synthesis and applications: Current status and future perspectives, *ACS Omega*, 7: 3134–3150, 2022.

Papell S. S., Low viscosity magnetic fluid obtained by the colloidal suspension of magnetic particles. US Patent: 3215572, 1965.

Patel N. S., Vakharia D. P., Deheri G. M., Patel H. C., Experimental performance analysis of ferrofluid based hydrodynamic journal bearing with different combination of materials, *Wear*, 376–377: 1877–1884, 2017.

Philip J., Magnetic nanofluids (ferrofluids): Recent advances, applications, challenges, and future directions, *Advances in Colloid and Interface Science*, 311: 102810, 2023.

Pramanik B. K., Pramanik S. K., Monira S., Understanding the fragmentation of microplastics into nano-plastics and removal of nano/microplastics from wastewater using membrane, air flotation and nano-ferrofluid processes, *Chemosphere*, 282: 131053, 2021.

Raikher Y. L., Stepanov V. I., Bacri J. C., Perzynski R., Orientational dynamics of ferrofluids with finite magnetic anisotropy of the particles: Relaxation of magneto-birefringence in crossed fields, *Physical Review E*, 66(2): 021203, 2002.

Raj K., Moskowitz R., Commercial applications of ferrofluids, *Journal of Magnetism and Magnetic Materials*, 85(1–3): 233–245, 1990.

Raj K., Moskowitz R., Casciari R., Advances in ferrofluid technology, *Journal of Magnetism and Magnetic Materials*, 149(1–2): 174–180, 1995.

Rosensweig R. E., *Ferrohydrodynamics*. Cambridge: Cambridge University Press. 1985.

Samadi Z., Haghshenasfard M., Moheb A., CO_2 absorption using nanofluids in a wetted-wall column with external magnetic field, *Chemical Engineering & Technology*, 37(3): 462–470, 2014.

Scherer C., Neto A. M. F., Ferrofluids: Properties and applications, *Brazilian Journal of Physics*, 35(3A): 718–727, 2005.

Shimada K., Kamiyama S., A basic study on oscillatory characteristics of magnetic fluid viscous damper, *Transactions of the Japan Society of Mechanical Engineers, Part B*, 57(544): 4111–4115, 1991.

Shliomis M. I., Mond M., Morozov K., Ultrasound attenuation in ferrofluids, *Physical Review Letters*, 101: 04505, 2008.

Silfhout A. M. V., Engelkamp H., Erné B. H., Colloidal stability of aqueous ferrofluids at 10 T, *Journal of Physical Chemistry Letters*, 11(15): 5908–5912, 2020.

Socoliuc V., Avdeev M. V., Kuncser V., Turcu R., Tombácz E., Vékás L., Ferrofluids and bio-ferrofluids: Looking back and stepping forward, *Nanoscale*, 14: 4786, 2022.

Liu Yang, Vieira R. M. S., Mao L., Simultaneous and multimodal antigen binding profiling and isolation of rare cells via 1uantitative ferrohydrodynamic cell separation, *ACS Nano*, 17: 94–110, 2023.

Yang W., Li D., Feng Z., Hydrodynamics and energy dissipation in a ferrofluid damper, *Journal of Vibration and Control*, 2013, 19(2): 183–190.

Yang W., Ren J., Li Y., Liu B., Modeling shape deformation of ferrofluid ring section and secondary flow in ferrofluid seals with rectangular polar teeth using BEM and FVM, *Communications in Nonlinear Science and Numerical Simulation*, 128: 107640, 2024.

Yang X., Sun P., Chen F., Hao F., Li D., Thomas P. J., Numerical and experimental studies of a novel converging stepped ferrofluid seal, *IEEE Transactions on Magnetics*, 55(3): 1–6, 2019.

Zhang X., Sun L., Yu Y., Zhao Y., Flexible ferrofluids: Design and applications, *Advanced Materials*, 31(51): 1903497, 2019.

Zhang Y., Qin L., Wang J., Xu W., A liquid-solid mixed robot based on ferrofluid with high flexibility and high controllability, *Applied Physics Letters*, 121: 122402, 2022.

Zhong Z. W., Recent advances in polishing of advanced materials, *Materials and Manufacturing Processes*, 23: 449–456, 2008.

Zhu T., Marrero F., Mao L., Continuous separation of non-magnetic particles inside ferrofluids, *Microfluidics and Nanofluidics*, 9(4–5): 1003–1009, 2010.

CHAPTER 2

Basic Equations of Non-Equilibrium Ferrohydrodynamics

FERROHYDRODYNAMICS IS THE DISCIPLINE that delves into the flow patterns exhibited by ferrofluids when subjected to magnetic fields. Central to it are the fundamental equations: The conservation of mass, momentum, and energy (Yang, 2023; Xie, 2019). Notably, these equations incorporate magnetic forces and magnetic moments derived from the magnetic stress tensor, differing significantly from those governing regular fluids (Landau and Lifshitz, 1960; Jansons, 1983; Liu, 1993, 1995). The objective of this chapter is to develop a comprehensive set of equations within the framework of irreversible thermodynamics, specifically tailored to describe ferrofluid flows (Felderhof and Kroh, 1999). This derivation starts with the most general setting: A conductive, compressible polar fluid and utilizes the integral format of the conservation equations for mass, momentum, angular momentum, and energy.

2.1 REYNOLDS' TRANSPORT THEORY

The axiom of conservation states that measurable physical quantities within any isolated system remain conserved within a localized region. Fluid flow and its associated transfer phenomena are manifestations of this axiom. Describing these phenomena mathematically involves two approaches: The Lagrangian, which focuses on a fixed unit of mass and treats the system as a whole, and the Eulerian, which focuses on a fixed region, considering the control volume (CV) as the focal point. For instance, when dealing with a field variable $\mathcal{R}\left(t,\boldsymbol{r}\left(t\right)\right)$, where t denotes time and $\boldsymbol{r}$ represents a spatial position vector, the derivative or rate of change of this variable is referred to as the Eulerian derivative, denoted as $\partial\mathcal{R}/\partial t$, when calculated for a fixed spatial position. Conversely, if the derivative is determined by tracking a specific moving fluid unit, it is termed the Lagrangian derivative, or material derivative, denoted as $\mathrm{D}\mathcal{R}/\mathrm{D}t$. In theoretical mechanics, the mathematical representation

DOI: 10.1201/9781003540342-2

used to describe Newton's second law, applicable to masses or rigid bodies, falls under the Lagrangian methodology. The linkage between these two representations can be derived using the chain rule for derivatives

$$\frac{D\mathcal{R}}{Dt} = \frac{\partial \mathcal{R}}{\partial t} + \boldsymbol{v} \cdot \nabla \mathcal{R} \tag{2.1}$$

where the first term at the right-hand side of the equation represents the local rate of change of the field variable $\mathcal{R}$, while the second term represents its convective rate of change. The equation indicates that, as a fluid cell traverses the flow field characterized by velocity $\boldsymbol{v}$, the total rate of change of the field variable $\mathcal{R}$ within that cell is equivalent to the summation of its local and convective rates of change.

To enable the Eulerian method to express conservation laws effectively and allow for the seamless application of established system-based physical laws within the Eulerian framework, it is necessary to derive Eulerian-format integrals over the material volume (MV) of the fluid in motion according to the Reynolds' transport theorem. MV denotes a specific volume that consistently encompasses identical fluid particles at any given moment. Now, let $\mathcal{J}$ be an extensive property of the fluid, one that is directly related to the amount of material, such as mass, momentum, volume, energy, and the like. The corresponding intensity quantity corresponding to this extensive quantity is

$$\mathcal{F}(\boldsymbol{r},t) = \frac{d\mathcal{J}}{dm} \tag{2.2}$$

The intensity quantity $\mathcal{F}$, which is independent of mass, represents quantities, such as density, velocity, temperature, and others, on a per-unit-mass basis. $\mathcal{F}$ can be a scalar, vector, or tensor function, while m denotes mass. The relationship between the extensive property and intensity quantity can be expressed via an integral relation,

$$\mathcal{J} = \int_{V(t)} \mathcal{F}(\boldsymbol{r},t)\,dV \tag{2.3}$$

where $V(t)$ represents the MV.

The Reynolds' transport theorem states that for any CV undergoing motion or deformation, if it coincides with the MV at a specific moment, then the overall instantaneous change in extensive property $\mathcal{J}$ within the MV is equal to the sum of the total instantaneous change of $\mathcal{J}$ within the CV and the net convective outflow of $\mathcal{J}$ across the CV's surfaces.

$$\frac{D}{Dt}\int_{MV} \mathcal{F}(\boldsymbol{r},t)\,dV = \frac{d}{dt}\int_{CV} \mathcal{F}(\boldsymbol{r},t)\,dV + \int_{CS} \mathcal{F}(\boldsymbol{r},t)\,\boldsymbol{v}\cdot\boldsymbol{n}\,dS \tag{2.4}$$

where $\boldsymbol{n}$ is the unit vector pointing outwards perpendicular to the surfaces, CS is the outer surfaces of the CV, and S is the area of these outer surfaces.

If the CV remains constant and does not vary with time, the first term at the right-hand side of Eq. (2.4) can be expressed according to Leibniz's formula as

$$\frac{\mathrm{d}}{\mathrm{d}t}\int_{\mathrm{CV}}\mathcal{F}(\boldsymbol{r},t)\mathrm{d}V=\int_{\mathrm{CV}}\frac{\partial}{\partial t}\mathcal{F}(\boldsymbol{r},t)\mathrm{d}V \tag{2.5}$$

Then, the Reynolds' transport theorem becomes

$$\frac{\mathrm{D}}{\mathrm{D}t}\int_{\mathrm{MV}}\mathcal{F}(\boldsymbol{r},t)\mathrm{d}V=\int_{\mathrm{CV}}\frac{\partial}{\partial t}\mathcal{F}(\boldsymbol{r},t)\mathrm{d}V+\int_{\mathrm{CS}}\mathcal{F}(\boldsymbol{r},t)\boldsymbol{v}\cdot\boldsymbol{n}\mathrm{d}S \tag{2.6}$$

Since the CV and the MV overlap at this moment, the integral at the left-hand side of Eq. (2.6) is identical to the integral computed over the CV, denoted as V throughout the subsequent discussion. Then Eq. (2.6) becomes

$$\frac{\mathrm{D}}{\mathrm{D}t}\int_{V}\mathcal{F}(\boldsymbol{r},t)\mathrm{d}V=\int_{V}\frac{\partial}{\partial t}\mathcal{F}(\boldsymbol{r},t)\mathrm{d}V+\int_{S}\mathcal{F}(\boldsymbol{r},t)\boldsymbol{v}\cdot\boldsymbol{n}\mathrm{d}S \tag{2.7}$$

According to Gauss's theorem and the vector relationship

$$\nabla\cdot(\mathcal{F}\boldsymbol{v})=\mathcal{F}(\nabla\cdot\boldsymbol{v})+\boldsymbol{v}\cdot\nabla\mathcal{F}$$

the surface integral at the right-hand side of Eq. (2.7) can be expanded as

$$\int_{S}\mathcal{F}(\boldsymbol{r},t)\boldsymbol{v}\cdot\boldsymbol{n}\mathrm{d}S=\int_{V}\mathcal{F}(\nabla\cdot\boldsymbol{v})\mathrm{d}V+\int_{V}\boldsymbol{v}\cdot\nabla\mathcal{F}\mathrm{d}V \tag{2.8}$$

which leads to

$$\frac{\mathrm{D}}{\mathrm{D}t}\int_{V}\mathcal{F}(\boldsymbol{r},t)\mathrm{d}V=\int_{V}\left[\frac{\mathrm{D}\mathcal{F}}{\mathrm{D}t}+\mathcal{F}(\nabla\cdot\boldsymbol{v})\right]\mathrm{d}V \tag{2.9}$$

Let $\mathcal{F}=\rho\mathcal{R}$, where $\mathcal{R}$ is any field variable and ρ is the density of a fluid. Then, Eq. (2.9) becomes

$$\frac{\mathrm{D}}{\mathrm{D}t}\int_{V}\rho\mathcal{R}\mathrm{d}V=\int_{V}\left[\rho\frac{\mathrm{D}\mathcal{R}}{\mathrm{D}t}+\mathcal{R}\left(\frac{\mathrm{D}\rho}{\mathrm{D}t}+\rho\nabla\cdot\boldsymbol{v}\right)\right]\mathrm{d}V \tag{2.10}$$

Utilizing the continuity equation (Eq. (2.13) in Section 2.2), one can deduce that the terms enclosed in the parentheses at the right-hand side of Eq. (2.10) vanishes, ultimately resulting in an alternate formulation of Reynolds' transport theorem:

$$\frac{\mathrm{D}}{\mathrm{D}t}\int_{V}\rho\mathcal{R}\mathrm{d}V=\int_{V}\rho\frac{\mathrm{D}\mathcal{R}}{\mathrm{D}t}\mathrm{d}V \tag{2.11}$$

2.2 CONSERVATION EQUATIONS OF MASS

For a fluid with a constant mass and volume V in the flow field, the mass remains constant throughout its deformation over time, expressed through the integral equation as

$$\frac{\mathrm{D}}{\mathrm{D}t}\int_{V}\rho \mathrm{d}V = 0 \tag{2.12}$$

This is equivalent to substituting ρ for $\mathcal{F}$ in Eq. (2.9), resulting in the continuity equation

$$\frac{\mathrm{D}\rho}{\mathrm{D}t} + \rho(\nabla\cdot\boldsymbol{v}) = 0 \tag{2.13}$$

Alternatively, a material derivative can be expanded to obtain another representation of the continuity equation

$$\frac{\partial\rho}{\partial t} + \nabla\cdot(\rho\boldsymbol{v}) = 0 \tag{2.14}$$

Furthermore, a third representation of the continuity equation can be derived directly from Eq. (2.13)

$$\frac{1}{\rho}\frac{\mathrm{D}\rho}{\mathrm{D}t} = -\nabla\cdot\boldsymbol{v} \tag{2.15}$$

2.3 CAUCHY'S PRINCIPLE OF STRESS AND CAUCHY'S EQUATION OF MOTION

By assuming a continuous medium, Newton's second law for a rigid body can be generalized to a continuum, stating that the rate of change in momentum of a substance with a MV is equal to the sum of the volume forces acting upon that volume and the surface forces exerted on the enclosing surfaces MS of that volume,

$$\frac{\mathrm{D}}{\mathrm{D}t}\int_{\mathrm{MV}}(\rho\boldsymbol{v} + \boldsymbol{g})\mathrm{d}V = \int_{\mathrm{MV}}\rho\boldsymbol{F}\mathrm{d}V + \oint_{\mathrm{MS}}\boldsymbol{t}_n\mathrm{d}S \tag{2.16}$$

where $\boldsymbol{g}$ is the electromagnetic momentum vector, $\boldsymbol{F}$ is the volume force per unit mass, except the electromagnetic volume force, $\boldsymbol{t}_n$ is the stress vector defined on the surfaces, including compressive stress, viscous stress, and electromagnetic stress. Given the overlap between the MV and the CV, Eq. (2.16) can be reformulated by substituting MV with CV, leading to

$$\frac{\mathrm{D}}{\mathrm{D}t}\int_{V}(\rho\boldsymbol{v} + \boldsymbol{g})\mathrm{d}V = \int_{V}\rho\boldsymbol{F}\mathrm{d}V + \oint_{S}\boldsymbol{t}_n\mathrm{d}S \tag{2.17}$$

The law of conservation of momentum, as expressed by this equation, clarifies the principles governing the exchange of momentum between a system and its surroundings. In Eq. (2.17), the first term on the right-hand side signifies the inflow of momentum from areas beyond the CV, while the second term represents the rate at which momentum accumulates or dissipates within the system due to the surface forces exerted by the surrounding medium. For the sake of simplicity, we shall refrain from distinguishing between MV and CV and henceforth, referring to them uniformly as V.

The surface stress vector $\boldsymbol{t}_n$ can be expressed as $\boldsymbol{t}_n = \boldsymbol{n}\cdot\tilde{\mathbf{T}}$ in terms of the total stress tensor $\tilde{\mathbf{T}}$. This tensor is equal to the sum of the compressive stress tensor, the viscous stress tensor, and the electromagnetic stress tensor. Then, the surface integral at the right-hand side of Eq. (2.17) can be represented as

$$\oint_S \boldsymbol{t}_n \mathrm{d}S = \oint_S \boldsymbol{n}\cdot\tilde{\mathbf{T}}\mathrm{d}S = \int_V \nabla\cdot\tilde{\mathbf{T}}\mathrm{d}V \tag{2.18}$$

The term on the left-hand side of Eq. (2.17) can be expressed in terms of Eq. (2.11) as

$$\frac{\mathrm{D}}{\mathrm{D}t}\int_V (\rho\boldsymbol{v}+\boldsymbol{g})\mathrm{d}V = \int_V \left[\rho\frac{\mathrm{D}\boldsymbol{v}}{\mathrm{D}t} + \rho\frac{\mathrm{D}}{\mathrm{D}t}\left(\frac{\boldsymbol{g}}{\rho}\right)\right]\mathrm{d}V \tag{2.19}$$

where the second term of the integrand at the right-hand side can be further expanded as

$$\rho\frac{\mathrm{D}}{\mathrm{D}t}\left(\frac{\boldsymbol{g}}{\rho}\right) = \rho\frac{1}{\rho}\frac{\mathrm{D}\boldsymbol{g}}{\mathrm{D}t} - \frac{1}{\rho}\boldsymbol{g}\frac{\mathrm{D}\rho}{\mathrm{D}t} = \frac{\partial \boldsymbol{g}}{\partial t} + \boldsymbol{v}\cdot\nabla\boldsymbol{g} + \boldsymbol{g}(\nabla\cdot\boldsymbol{v}) = \frac{\partial \boldsymbol{g}}{\partial t} + \nabla\cdot(\boldsymbol{v}\boldsymbol{g}) \tag{2.20}$$

By substituting the derived results from Eqs. (2.18), (2.19), and (2.20) into Eq. (2.17) and leveraging the arbitrariness of the integral volume, one obtains

$$\rho\frac{\mathrm{D}\boldsymbol{v}}{\mathrm{D}t} + \frac{\partial \boldsymbol{g}}{\partial t} + \nabla\cdot(\boldsymbol{v}\boldsymbol{g}) = \rho\boldsymbol{F} + \nabla\cdot\tilde{\mathbf{T}} \tag{2.21}$$

Letting

$$\mathbf{T} = \tilde{\mathbf{T}} - \boldsymbol{v}\boldsymbol{g} \tag{2.22}$$

and the momentum conservation equation satisfied by fluid cells is obtained.

$$\rho\frac{\mathrm{D}\boldsymbol{v}}{\mathrm{D}t} + \frac{\partial \boldsymbol{g}}{\partial t} = \rho\boldsymbol{F} + \nabla\cdot\mathbf{T} \tag{2.23}$$

Using continuity Eq. (2.14) and the divergence expansion of the product between two vectors, the first term on the left-hand side of Eq. (2.23) can be expressed as

$$\rho\frac{\mathrm{D}\boldsymbol{v}}{\mathrm{D}t} = \rho\frac{\partial \boldsymbol{v}}{\partial t} + (\rho\boldsymbol{v}\cdot\nabla)\boldsymbol{v} + \boldsymbol{v}\frac{\partial \rho}{\partial t} + \boldsymbol{v}\nabla\cdot(\rho\boldsymbol{v}) = \frac{\partial(\rho\boldsymbol{v})}{\partial t} + \nabla\cdot(\rho\boldsymbol{v}\boldsymbol{v}) \tag{2.24}$$

This leads to another form of the momentum conservation equation

$$\frac{\partial(\rho \boldsymbol{v})}{\partial t}+\nabla\cdot(\rho \boldsymbol{v}\boldsymbol{v})+\frac{\partial \boldsymbol{g}}{\partial t}=\rho \boldsymbol{F}+\nabla\cdot\mathrm{T} \tag{2.25}$$

where $\rho\boldsymbol{v}\boldsymbol{v}$ is the flux of the volume density of momentum $\rho\boldsymbol{v}$.

2.4 CONSERVATION EQUATIONS OF ANGULAR MOMENTUM

Beyond the momentum exchange between a system and its surroundings, there is also an exchange of angular momentum. To be precise, the system receives angular momentum from its surroundings through the body couple $\boldsymbol{G}$ (volume couple density), while the neighbouring media modifies the angular momentum within the system via the surface couple $\boldsymbol{c}_n$ (force couple moment per unit area) acting on the system's surfaces. Let the total local angular momentum density of the system (per unit volume) be expressed as

$$\boldsymbol{L}=\boldsymbol{r}\times(\rho\boldsymbol{v}+\boldsymbol{g})+\rho\boldsymbol{s} \tag{2.26}$$

where $\boldsymbol{r}$ is the position vector of the fluid cell relative to a fixed point and $\boldsymbol{r}\times\boldsymbol{v}$ represents the external or orbital angular momentum. For ferrofluids, the orbital angular momentum is related to the translational motion of magnetic particles and solvent molecules. Additionally, $\boldsymbol{s}$ signifies the density of the intrinsic or spin angular momentum (per unit mass), stemming from the rotation of the magnetic particles themselves. For non-polar fluids, the intrinsic angular momentum vanishes. Conversely, for polar fluids like ferrofluids, the spin field arises due to the rotation or spin of the magnetic particles and the rotation of the surrounding carrier fluid induced by particle rotation (Odenbach, 2002). Furthermore, there exists a coupling effect between the intrinsic and orbital angular momenta. The angular momentum theorem states that the rate of change in time of a system's total angular momentum is equal to the vector sum of the external moments resulting from both the surface and volume forces acting on it, which is expressed as

$$\frac{\mathrm{D}}{\mathrm{D}t}\int_V\left[\boldsymbol{r}\times(\rho\boldsymbol{v}+\boldsymbol{g})+\rho\boldsymbol{s}\right]\mathrm{d}V=\int_V(\rho\boldsymbol{G}+\boldsymbol{r}\times\rho\boldsymbol{F})\mathrm{d}V+\oint_S(\boldsymbol{c}_n+\boldsymbol{r}\times\boldsymbol{t}_n)\mathrm{d}S \tag{2.27}$$

For non-polar fluids, the terms $\rho\boldsymbol{G}$ and $\boldsymbol{c}_n$ at the right-hand side of Eq. (2.27) vanish, indicating that the overall angular momentum balance equation for non-polar fluids is

$$\frac{\mathrm{D}}{\mathrm{D}t}\int_V\rho(\boldsymbol{r}\times\boldsymbol{v})\mathrm{d}V=\int_V(\boldsymbol{r}\times\rho\boldsymbol{F})\mathrm{d}V+\oint_S(\boldsymbol{r}\times\boldsymbol{t}_n)\mathrm{d}S \tag{2.28}$$

The terms in Eq. (2.27) are simplified as follows. For the first term at its left-hand side, an expression is obtained with the use of Eq. (2.11)

$$\begin{aligned}\frac{\mathrm{D}}{\mathrm{D}t}\int_V \boldsymbol{r}\times\rho\boldsymbol{v}\mathrm{d}V &= \int_V \rho\frac{\mathrm{D}}{\mathrm{D}t}(\boldsymbol{r}\times\boldsymbol{v})\mathrm{d}V \\ &= \int_V \rho\left(\frac{\mathrm{D}\boldsymbol{r}}{\mathrm{D}t}\times\boldsymbol{v}+\boldsymbol{r}\times\frac{\mathrm{D}\boldsymbol{v}}{\mathrm{D}t}\right)\mathrm{d}V \\ &= \int_V \left(\boldsymbol{r}\times\rho\frac{\mathrm{D}\boldsymbol{v}}{\mathrm{D}t}\right)\mathrm{d}V\end{aligned} \tag{2.29}$$

where in obtaining the third equality use was made of

$$\frac{\mathrm{D}\boldsymbol{r}}{\mathrm{D}t}=\boldsymbol{v} \tag{2.30}$$

For the second term at the left-hand side of Eq. (2.27), one has

$$\begin{aligned}\frac{\mathrm{D}}{\mathrm{D}t}\int_V \boldsymbol{r}\times\boldsymbol{g}\mathrm{d}V &= \frac{\mathrm{D}}{\mathrm{D}t}\int_V\left(\rho\boldsymbol{r}\times\frac{\boldsymbol{g}}{\rho}\right)\mathrm{d}V \\ &= \int_V \rho\frac{\mathrm{D}}{\mathrm{D}t}\left(\boldsymbol{r}\times\frac{\boldsymbol{g}}{\rho}\right)\mathrm{d}V \\ &= \int_V\left[\boldsymbol{v}\times\boldsymbol{g}+\boldsymbol{r}\times\left(\frac{\mathrm{D}\boldsymbol{g}}{\mathrm{D}t}-\frac{\boldsymbol{g}}{\rho}\frac{\mathrm{D}\rho}{\mathrm{D}t}\right)\right]\mathrm{d}V \\ &= \int_V\left[\boldsymbol{v}\times\boldsymbol{g}+\boldsymbol{r}\times\left(\frac{\partial\boldsymbol{g}}{\partial t}+\boldsymbol{v}\cdot\nabla\boldsymbol{g}+\boldsymbol{g}(\nabla\cdot\boldsymbol{v})\right)\right]\mathrm{d}V \\ &= \int_V\left\{\boldsymbol{v}\times\boldsymbol{g}+\boldsymbol{r}\times\left[\frac{\partial\boldsymbol{g}}{\partial t}+\nabla\cdot(\boldsymbol{v}\boldsymbol{g})\right]\right\}\mathrm{d}V\end{aligned} \tag{2.31}$$

where continuity Eq. (2.15) is applied in obtaining the fourth equality. For the integrand $\boldsymbol{v}\times\boldsymbol{g}$ in Eq. (2.31), a simplification with the use of tensor representations and the polyadic tensor $\boldsymbol{\epsilon}$ gives

$$\begin{aligned}\boldsymbol{v}\times\boldsymbol{g} &= v_i\boldsymbol{e}_i\times g_j\boldsymbol{e}_j \\ &= v_i g_j \epsilon_{ijk}\boldsymbol{e}_k \\ &= \boldsymbol{\epsilon}:\boldsymbol{v}\boldsymbol{g}\end{aligned} \tag{2.32}$$

where $\boldsymbol{e}$ is the unit vector. In the Cartesian coordinate system, the subscripts i, j, and k represent one of x, y, and z, respectively. ϵ_{ijk} is the component of the polyadic tensor. Substitution of (2.32) in (2.31) gives

$$\frac{\mathrm{D}}{\mathrm{D}t}\int_V \boldsymbol{r}\times\boldsymbol{g}\mathrm{d}V = \int_V\left\{\boldsymbol{\epsilon}:\boldsymbol{v}\boldsymbol{g}+\boldsymbol{r}\times\left[\frac{\partial\boldsymbol{g}}{\partial t}+\nabla\cdot(\boldsymbol{v}\boldsymbol{g})\right]\right\}\mathrm{d}V \tag{2.33}$$

For the third term on the left-hand side of Eq. (2.27), use of Eq. (2.11) gives

$$\frac{\mathrm{D}}{\mathrm{D}t}\int_V \rho s \mathrm{d}V = \int_V \rho \frac{\mathrm{D}s}{\mathrm{D}t}\mathrm{d}V \tag{2.34}$$

The surface couple density $\boldsymbol{c}_n$ be expressed as the dot product of the surface normal vector and a couple stress tensor

$$\boldsymbol{c}_n = \boldsymbol{n}\cdot\mathbf{C} \tag{2.35}$$

The surface density of the traction force $\boldsymbol{t}_n$ and its moment can be represented by

$$\boldsymbol{t}_n = \boldsymbol{n}\cdot\tilde{\mathbf{T}} \tag{2.36}$$

$$\boldsymbol{r}\times\boldsymbol{t}_n = \boldsymbol{r}\times\left(\boldsymbol{n}\cdot\tilde{\mathbf{T}}\right) = -\boldsymbol{n}\cdot\left(\tilde{\mathbf{T}}\times\boldsymbol{r}\right) \tag{2.37}$$

Thus, with the aid of the Gauss theorem, the second integral at the right-hand side of Eq. (2.27) can be written as

$$\oint_S \left(\boldsymbol{c}_n + \boldsymbol{r}\times\boldsymbol{t}_n\right) dS = \int_V \left[\nabla\cdot\mathbf{C} - \nabla\cdot\left(\tilde{\mathbf{T}}\times\boldsymbol{r}\right)\right] dV \tag{2.38}$$

The term $\nabla\cdot\left(\tilde{\mathbf{T}}\times\boldsymbol{r}\right)$ in Eq. (2.38) can be calculated according to the law of tensor component derivation in the Cartesian coordinate system and then one has

$$\begin{aligned}
\nabla\cdot\left(\tilde{\mathbf{T}}\times\boldsymbol{r}\right) &= \boldsymbol{e}_i \frac{\partial}{\partial x_i}\cdot\left(\boldsymbol{e}_j\boldsymbol{e}_k\tilde{T}_{jk}\times\boldsymbol{e}_l x_l\right) \\
&= \left(\boldsymbol{e}_i\cdot\boldsymbol{e}_j\right)\left(\boldsymbol{e}_k\times\boldsymbol{e}_l\right)\left(x_l\frac{\partial\tilde{T}_{jk}}{\partial x_i} + T_{jk}\frac{\partial x_l}{\partial x_i}\right) \\
&= \delta_{ij}\epsilon_{klm}\boldsymbol{e}_m\left(x_l\frac{\partial\tilde{T}_{jk}}{\partial x_i} + \delta_{li}\tilde{T}_{jk}\right) \\
&= \epsilon_{klm}\boldsymbol{e}_m x_l\frac{\partial\tilde{T}_{jk}}{\partial x_j} + \epsilon_{kim}\boldsymbol{e}_m\tilde{T}_{ik} \\
&= -\epsilon_{lkm}\boldsymbol{e}_m x_l\frac{\partial\tilde{T}_{jk}}{\partial x_j} - \epsilon_{mik}\boldsymbol{e}_m\tilde{T}_{ik}
\end{aligned} \tag{2.39}$$

where use has been made of $\boldsymbol{r} = \boldsymbol{e}_l x_l$. In the Cartesian coordinate system, the subscripts l, m, and n represent one of x, y, and z, respectively. Owing to the following expressions,

$$\nabla\cdot\tilde{\mathbf{T}} = \boldsymbol{e}_i\frac{\partial}{\partial x_i}\cdot\boldsymbol{e}_j\boldsymbol{e}_k\tilde{T}_{jk} = \delta_{ij}\boldsymbol{e}_k\frac{\partial\tilde{T}_{jk}}{\partial x_i} = \boldsymbol{e}_k\frac{\partial\tilde{T}_{jk}}{\partial x_j} \tag{2.40}$$

$$\boldsymbol{r}\times\left(\nabla\cdot\tilde{\mathrm{T}}\right)=\boldsymbol{e}_l x_l\times\boldsymbol{e}_k\frac{\partial\tilde{T}_{jk}}{\partial x_j}=\epsilon_{lkm}\boldsymbol{e}_m x_l\frac{\partial\tilde{T}_{jk}}{\partial x_j} \tag{2.41}$$

$$\boldsymbol{\epsilon}\cdot\cdot\tilde{\mathbf{T}}=\epsilon_{ijk}\boldsymbol{e}_i\boldsymbol{e}_j\boldsymbol{e}_k\cdot\cdot\tilde{T}_{lm}\boldsymbol{e}_l\boldsymbol{e}_m=\epsilon_{ijk}\tilde{T}_{lm}\boldsymbol{e}_i\delta_{kl}\delta_{jm}=\epsilon_{iml}\tilde{T}_{lm}\boldsymbol{e}_i=-\epsilon_{ilm}\tilde{T}_{lm}\boldsymbol{e}_i \tag{2.42}$$

$$\boldsymbol{\epsilon}:\tilde{\mathbf{T}}=\epsilon_{ijk}\boldsymbol{e}_i\boldsymbol{e}_j\boldsymbol{e}_k:\tilde{T}_{lm}\boldsymbol{e}_l\boldsymbol{e}_m=\epsilon_{ijk}\tilde{T}_{lm}\boldsymbol{e}_i\delta_{km}\delta_{jl}=\epsilon_{ilm}\tilde{T}_{lm}\boldsymbol{e}_i \tag{2.43}$$

where δ_{ij} is the Kronecker symbol, a comparison of the results of Eqs. (2.41)–(2.43) with the results of Eq. (2.39) yields

$$\begin{aligned}\nabla\cdot\left(\tilde{\mathbf{T}}\times\boldsymbol{r}\right)&=-\boldsymbol{r}\times\left(\nabla\cdot\tilde{\mathbf{T}}\right)+\boldsymbol{\epsilon}\cdot\cdot\tilde{\mathbf{T}}\\&=-\boldsymbol{r}\times\left(\nabla\cdot\tilde{\mathbf{T}}\right)-\boldsymbol{\epsilon}:\tilde{\mathbf{T}}\end{aligned} \tag{2.44}$$

Substitution of Eq. (2.44) in Eq. (2.38) gives

$$\oint_S\left(\boldsymbol{c}_n+\boldsymbol{r}\times\boldsymbol{t}_n\right)\mathrm{d}S=\int_V\left[\nabla\cdot\mathbf{C}+\boldsymbol{r}\times\left(\nabla\cdot\tilde{\mathbf{T}}\right)+\boldsymbol{\epsilon}:\tilde{\mathbf{T}}\right]\mathrm{d}V \tag{2.45}$$

Now introducing Eqs. (2.29), (2.33), (2.34), and (2.45) back into Eq. (2.27) gives

$$\int_V\left[\rho\frac{\mathrm{D}\boldsymbol{s}}{\mathrm{D}t}+\boldsymbol{r}\times\left(\rho\frac{\mathrm{D}\boldsymbol{v}}{\mathrm{D}t}+\frac{\partial\boldsymbol{g}}{\partial t}-\nabla\cdot\mathrm{T}-\rho\boldsymbol{F}\right)\right]\mathrm{d}V=\int_V\left(\rho\boldsymbol{G}+\nabla\cdot\mathbf{C}+\boldsymbol{\epsilon}:\mathbf{T}\right)\mathrm{d}V \tag{2.46}$$

where use has been made of Eq. (2.22). The second term of the integrand at the left-hand side of Eq. (2.46) vanishes according to Eq. (2.23). Due to the arbitrariness of the integral volume, the conservation equation for intrinsic angular momentum is obtained

$$\rho\frac{\mathrm{D}\boldsymbol{s}}{\mathrm{D}t}=\rho\boldsymbol{G}+\nabla\cdot\mathbf{C}+\boldsymbol{\epsilon}:\mathbf{T} \tag{2.47}$$

This equation reveals that the sources that influence the rate of change in the intrinsic angular momentum of a ferrofluid cell include the body couple transmitted by the surrounding medium $\boldsymbol{G}$; the surface couple $\nabla\cdot\mathbf{C}$, which signifies the diffusion of intrinsic angular momentum across the cell's surface; and the exchange of angular momentum with the external angular momentum as indicated by $\boldsymbol{\epsilon}:\mathbf{T}$. By applying a similar approach to Eq. (2.23), an alternative formulation of Eq. (2.47) can be derived,

$$\frac{\partial(\rho\boldsymbol{s})}{\partial t}+\nabla\cdot\left(\rho\boldsymbol{v}\boldsymbol{s}\right)=\rho\boldsymbol{G}+\nabla\cdot\mathbf{C}+\boldsymbol{\epsilon}:\mathbf{T} \tag{2.48}$$

where $\rho\boldsymbol{v}\boldsymbol{s}$ can be regarded as the flux of the volume density of intrinsic angular momentum $\rho\boldsymbol{s}$, that is the spin flux.

For monodisperse spherical particles suspended within a ferrofluid, the spin angular momentum density is

$$\boldsymbol{s} = I\boldsymbol{\omega}_{\mathrm{p}} \tag{2.49}$$

where $\boldsymbol{\omega}_{\mathrm{p}}$ is the average angular velocity of the particles and I is the moment of inertia per unit mass of fluid. This moment of inertia is associated with the density of moments of inertia of the particles within a unit volume of ferrofluid,

$$I' = \rho I \tag{2.50}$$

where I' is the moment of inertia per unit volume of fluid. It can be calculated by (Odenbach, 2002)

$$I' = \frac{1}{10}\rho_s \phi \bar{d}_{\mathrm{p}}^2 \tag{2.51}$$

where $\bar{d}_{\mathrm{p}}$ is the average diameter of the particles and ρ_s is the density of the particulate material. Since the particles consist of nuclei and surfactants, ρ_s is generally estimated as (Shliomis, 2021)

$$\rho_s = \rho_{\mathrm{coat}}\left[1+\left(\frac{\rho_{\mathrm{mag}}}{\rho_{\mathrm{coat}}}-1\right)\left(\frac{d_{\mathrm{m}}}{\bar{d}_{\mathrm{p}}}\right)^5\right] \tag{2.52}$$

where ρ_{mag} and ρ_{coat} are the densities of the particle nuclei and surfactant, respectively, and d_{m} is the diameter of the particle nuclei. The parameter ϕ in Eq. (2.51) is the volume fraction of particles. It is feasible to represent I' as

$$I' = \frac{1}{60}\pi\rho_s N \bar{d}_{\mathrm{p}}^5 \tag{2.53}$$

leveraging the relationship

$$\phi = \frac{1}{6}\pi \bar{d}_{\mathrm{p}}^3 N \tag{2.54}$$

that exists between the volume fraction ϕ and the number concentration of particles N.

Using Eq. (2.49) one has

$$\frac{\mathrm{D}\boldsymbol{s}}{\mathrm{D}t} = I\frac{\mathrm{D}\boldsymbol{\omega}_{\mathrm{p}}}{\mathrm{D}t} \tag{2.55}$$

Therefore, the conservation equation for angular momentum (Eq. (2.47)) can also be expressed in an alternative form for a ferrofluid

$$\rho I \frac{\mathrm{D}\boldsymbol{\omega}_{\mathrm{p}}}{\mathrm{D}t} = \rho \mathbf{G} + \nabla \cdot \mathbf{C} + \boldsymbol{\epsilon} : \mathbf{T} \tag{2.56}$$

or, equivalently

$$I' \frac{\mathrm{D}\boldsymbol{\omega}_{\mathrm{p}}}{\mathrm{D}t} = \rho \mathbf{G} + \nabla \cdot \mathbf{C} + \boldsymbol{\epsilon} : \mathbf{T} \tag{2.57}$$

2.5 CONSERVATION EQUATION OF ENERGY

The law of conservation of energy states that the rate of change in total energy over time, within a ferrofluid system with constant mass and variable volume, is equal to the sum of the work performed on the system by external forces and the energy introduced into the system from external sources, which is expressed as

$$\frac{\mathrm{D}}{\mathrm{D}t}\int_V \rho\left(u + \frac{v^2}{2} + \frac{I}{2}\omega_{\mathrm{p}}^2\right)\mathrm{d}V = \int_V \left(\rho \boldsymbol{G}\cdot\boldsymbol{\omega}_{\mathrm{p}} + \rho \boldsymbol{F}\cdot\boldsymbol{v}\right)\mathrm{d}V + \oint_S \left(\boldsymbol{c}_n\cdot\boldsymbol{\omega}_{\mathrm{p}} + \boldsymbol{t}_n\cdot\boldsymbol{v}\right)\mathrm{d}S$$
$$-\oint_S \boldsymbol{n}\cdot\boldsymbol{q}\,\mathrm{d}S - \oint_S \boldsymbol{n}\cdot\left(\boldsymbol{E}'\times\boldsymbol{H}'\right)\mathrm{d}S + \int_V \rho R\,\mathrm{d}V \tag{2.58}$$

where u is the internal energy per unit mass, $\boldsymbol{q}$ is the heat flow density vector (heat conducted per unit area), R stands for the volumetric heat release rate (e.g., radiation, etc.), and $\boldsymbol{\omega}_{\mathrm{p}}$ denotes the angular velocity vector. The Poynting vector $\boldsymbol{E}' \times \boldsymbol{H}'$ represents the electromagnetic field energy flow density in a moving medium, with the apparent electric field intensity $\boldsymbol{E}'$ and apparent magnetic field intensity $\boldsymbol{H}'$ denoted as

$$\boldsymbol{E}' = \boldsymbol{E}_{//} + \tilde{\gamma}\left(\boldsymbol{E}_{\perp} + \boldsymbol{v}\times\boldsymbol{B}\right) \tag{2.59}$$

$$\boldsymbol{H}' = \boldsymbol{H}_{//} + \tilde{\gamma}\left(\boldsymbol{H}_{\perp} - \boldsymbol{v}\times\boldsymbol{D}\right) \tag{2.60}$$

where the subscripts $//$ and $\perp$ represent the components parallel and perpendicular to the relative velocity of the reference system $\boldsymbol{v}$, respectively, $\boldsymbol{E}$ denotes the electric field intensity, $\boldsymbol{H}$ represents the magnetic field intensity, $\boldsymbol{B}$ is the magnetic induction, and $\boldsymbol{D}$ refers to the electric displacement vector. Additionally, the coefficient $\tilde{\gamma}$ represents the coefficient of the Lorentz transformation, which is expressed as

$$\tilde{\gamma} = \left[1 - \left(\frac{v}{c}\right)^2\right]^{-\frac{1}{2}} \tag{2.61}$$

The apparent electric field and apparent magnetic field observed by an observer travelling at a velocity $\boldsymbol{v}$ are represented by the terms $\boldsymbol{v} \times \boldsymbol{B}$ and $\boldsymbol{v} \times \boldsymbol{D}$ in Eqs. (2.59) and (2.60), respectively. In scenarios where the magnitude of this velocity is significantly less than the speed of light c, the coefficients $\tilde{\gamma} \approx 1$, then one has

$$\boldsymbol{E}' = \boldsymbol{E} + \boldsymbol{v} \times \boldsymbol{B}, \boldsymbol{H}' = \boldsymbol{H} - \boldsymbol{v} \times \boldsymbol{D} \tag{2.62}$$

For the term $\oint_S (\boldsymbol{t}_n \cdot \boldsymbol{v}) \mathrm{d}S$ appearing in Eq. (2.58), it can be expanded based on the Cauchy theory and Gauss theorem

$$\oint_S (\boldsymbol{t}_n \cdot \boldsymbol{v}) \mathrm{d}S = \int_S \boldsymbol{n} \cdot \tilde{\mathbf{T}} \cdot \boldsymbol{v} \mathrm{d}S = \int_V \nabla \cdot (\tilde{\mathbf{T}} \cdot \boldsymbol{v}) \mathrm{d}V \tag{2.63}$$

As implied by the tensor notation

$$\nabla \cdot (\tilde{\mathbf{T}} \cdot \boldsymbol{v}) = \frac{\partial}{\partial x_i} \boldsymbol{e}_i \cdot (\tilde{T}_{jk} \boldsymbol{e}_j \boldsymbol{e}_k \cdot v_l \boldsymbol{e}_l) = \frac{\partial}{\partial x_i} \boldsymbol{e}_i \cdot (\tilde{T}_{jk} \boldsymbol{e}_j v_k) = v_k \frac{\partial \tilde{T}_{ik}}{\partial x_i} + \tilde{T}_{ik} \frac{\partial v_k}{\partial x_i} \tag{2.64}$$

$$(\nabla \cdot \tilde{\mathbf{T}}) \cdot \boldsymbol{v} = \frac{\partial}{\partial x_i} \boldsymbol{e}_i \cdot \tilde{T}_{jk} \boldsymbol{e}_j \boldsymbol{e}_k \cdot v_l \boldsymbol{e}_l = \frac{\partial \tilde{T}_{ik}}{\partial x_i} v_k \tag{2.65}$$

$$\tilde{\mathbf{T}}^{\mathrm{T}} \cdot\cdot \nabla \boldsymbol{v} = \tilde{T}_{ji} \boldsymbol{e}_i \boldsymbol{e}_j \cdot\cdot \frac{\partial}{\partial x_k} \boldsymbol{e}_k v_l \boldsymbol{e}_l = \tilde{T}_{ji} \frac{\partial v_i}{\partial x_j} \tag{2.66}$$

$$\tilde{\mathbf{T}} : \nabla \boldsymbol{v} = \tilde{T}_{ij} \boldsymbol{e}_i \boldsymbol{e}_j : \frac{\partial}{\partial x_k} \boldsymbol{e}_k v_l \boldsymbol{e}_l = \tilde{T}_{ij} \frac{\partial v_j}{\partial x_i} \tag{2.67}$$

comparing the results of Eqs. (2.64) to (2.67), it can be obtained that

$$\nabla \cdot (\tilde{\mathbf{T}} \cdot \boldsymbol{v}) = (\nabla \cdot \tilde{\mathbf{T}}) \cdot \boldsymbol{v} + \tilde{\mathbf{T}}^{\mathrm{T}} \cdot\cdot \nabla \boldsymbol{v} = (\nabla \cdot \tilde{\mathbf{T}}) \cdot \boldsymbol{v} + \tilde{\mathbf{T}} : \nabla \boldsymbol{v} \tag{2.68}$$

Thus, Eq. (2.63) can be rewritten as

$$\oint_S (\boldsymbol{t}_n \cdot \boldsymbol{v}) \mathrm{d}S = \int_V \left[(\nabla \cdot \tilde{\mathbf{T}}) \cdot \boldsymbol{v} + \tilde{\mathbf{T}} : \nabla \boldsymbol{v} \right] \mathrm{d}V \tag{2.69}$$

Similarly, one has

$$\oint_S (\boldsymbol{c}_n \cdot \boldsymbol{\omega}_{\mathrm{p}}) \mathrm{d}S = \int_V \left[(\nabla \cdot \mathbf{C}) \cdot \boldsymbol{\omega}_{\mathrm{p}} + \mathbf{C} : \nabla \boldsymbol{\omega}_{\mathrm{p}} \right] \mathrm{d}V \tag{2.70}$$

Applying Gauss theorem to the Poynting vector term in Eq. (2.58), one obtains

$$\oint_S \boldsymbol{n} \cdot (\boldsymbol{E}' \times \boldsymbol{H}') \mathrm{d}S = \int_V \nabla \cdot (\boldsymbol{E}' \times \boldsymbol{H}') \mathrm{d}V \tag{2.71}$$

Substituting Eq. (2.62) into Eq. (2.71) and expanding the integrand gives

$$\begin{aligned}\nabla\cdot(\boldsymbol{E}'\times\boldsymbol{H}') &= \nabla\cdot\left[(\boldsymbol{E}+\boldsymbol{v}\times\boldsymbol{B})\times(\boldsymbol{H}-\boldsymbol{v}\times\boldsymbol{D})\right]\\ &= \nabla\cdot\left[\boldsymbol{E}\times\boldsymbol{H}-\boldsymbol{E}\times(\boldsymbol{v}\times\boldsymbol{D})+\boldsymbol{v}\times\boldsymbol{B}\times\boldsymbol{H}-(\boldsymbol{v}\times\boldsymbol{B})\times(\boldsymbol{v}\times\boldsymbol{D})\right]\\ &= \nabla\cdot(\boldsymbol{E}\times\boldsymbol{H})-\left\{(\boldsymbol{v}\times\boldsymbol{D})\cdot(\nabla\times\boldsymbol{E})-\boldsymbol{E}\cdot\left[\nabla\times(\boldsymbol{v}\times\boldsymbol{D})\right]\right\}\\ &\quad+\left\{\boldsymbol{H}\cdot\left[\nabla\times(\boldsymbol{v}\times\boldsymbol{B})\right]-(\boldsymbol{v}\times\boldsymbol{B})\cdot(\nabla\times\boldsymbol{H})\right\}-\nabla\cdot\left[(\boldsymbol{v}\times\boldsymbol{B})\times(\boldsymbol{v}\times\boldsymbol{D})\right]\\ &= \nabla\cdot(\boldsymbol{E}\times\boldsymbol{H})+\left\{\boldsymbol{E}\cdot\left[\nabla\times(\boldsymbol{v}\times\boldsymbol{D})\right]+\boldsymbol{H}\cdot\left[\nabla\times(\boldsymbol{v}\times\boldsymbol{B})\right]\right\}\\ &\quad-\left[(\boldsymbol{v}\times\boldsymbol{D})\cdot(\nabla\times\boldsymbol{E})+(\boldsymbol{v}\times\boldsymbol{B})\cdot(\nabla\times\boldsymbol{H})\right]-\nabla\cdot\left[(\boldsymbol{v}\times\boldsymbol{B})\times(\boldsymbol{v}\times\boldsymbol{D})\right]\end{aligned} \tag{2.72}$$

Using Maxwell's equations

$$-\frac{\partial \boldsymbol{B}}{\partial t}=\nabla\times\boldsymbol{E},\ \frac{\partial \boldsymbol{D}}{\partial t}+\boldsymbol{j}_0=\nabla\times\boldsymbol{H} \tag{2.73}$$

where $\boldsymbol{j}_0$ is the conductive current density to simplify the term in Eq. (2.72)

$$\begin{aligned}(\boldsymbol{v}\times\boldsymbol{D})\cdot(\nabla\times\boldsymbol{E})+(\boldsymbol{v}\times\boldsymbol{B})\cdot(\nabla\times\boldsymbol{H}) &= -\frac{\partial \boldsymbol{B}}{\partial t}\cdot(\boldsymbol{v}\times\boldsymbol{D})+\left(\frac{\partial \boldsymbol{D}}{\partial t}+\boldsymbol{j}_0\right)\cdot(\boldsymbol{v}\times\boldsymbol{B})\\ &= \boldsymbol{v}\cdot\left(\frac{\partial \boldsymbol{B}}{\partial t}\times\boldsymbol{D}\right)-\boldsymbol{v}\cdot\left(\frac{\partial \boldsymbol{D}}{\partial t}\times\boldsymbol{B}\right)+\boldsymbol{j}_0\cdot(\boldsymbol{v}\times\boldsymbol{B})\\ &= -\boldsymbol{v}\cdot\frac{\partial \boldsymbol{g}}{\partial t}+\boldsymbol{j}_0\cdot(\boldsymbol{v}\times\boldsymbol{B})\end{aligned} \tag{2.74}$$

where use has been made of the mixed product formula and the definition of the electromagnetic field momentum density

$$\boldsymbol{g}=\boldsymbol{D}\times\boldsymbol{B} \tag{2.75}$$

For the following term, it can be simplified as

$$\begin{aligned}\nabla\cdot\left[(\boldsymbol{v}\times\boldsymbol{B})\times(\boldsymbol{v}\times\boldsymbol{D})\right] &= \nabla\cdot\left\{\boldsymbol{v}\left[(\boldsymbol{v}\times\boldsymbol{B})\cdot\boldsymbol{D}\right]-\boldsymbol{D}\left[(\boldsymbol{v}\times\boldsymbol{B})\cdot\boldsymbol{v}\right]\right\}\\ &= \nabla\cdot\left\{\boldsymbol{v}\left[(\boldsymbol{v}\times\boldsymbol{B})\cdot\boldsymbol{D}\right]\right\}\\ &= \boldsymbol{v}\cdot\nabla\left[(\boldsymbol{v}\times\boldsymbol{B})\cdot\boldsymbol{D}\right]+\left[(\boldsymbol{v}\times\boldsymbol{B})\cdot\boldsymbol{D}\right](\nabla\cdot\boldsymbol{v})\\ &= \boldsymbol{v}\cdot\nabla\left[\boldsymbol{v}\cdot(\boldsymbol{B}\times\boldsymbol{D})\right]+\left[\boldsymbol{v}\cdot(\boldsymbol{B}\times\boldsymbol{D})\right](\nabla\cdot\boldsymbol{v})\\ &= -\boldsymbol{v}\cdot\nabla(\boldsymbol{v}\cdot\boldsymbol{g})-(\boldsymbol{v}\cdot\boldsymbol{g})(\nabla\cdot\boldsymbol{v})\\ &= -\boldsymbol{v}\cdot\left[\boldsymbol{v}\cdot\nabla\boldsymbol{g}+\boldsymbol{g}\cdot\nabla\boldsymbol{v}+\boldsymbol{v}\times(\nabla\times\boldsymbol{g})+\boldsymbol{g}\times(\nabla\times\boldsymbol{v})\right]-(\boldsymbol{v}\cdot\boldsymbol{g})(\nabla\cdot\boldsymbol{v})\\ &= -\boldsymbol{v}\cdot(\boldsymbol{v}\cdot\nabla\boldsymbol{g})-\boldsymbol{v}\cdot(\boldsymbol{g}\cdot\nabla\boldsymbol{v})-\boldsymbol{v}\cdot\left[\boldsymbol{g}\times(\nabla\times\boldsymbol{v})\right]-(\boldsymbol{v}\cdot\boldsymbol{g})(\nabla\cdot\boldsymbol{v})\\ &= -\boldsymbol{v}\cdot\left[\nabla\cdot(\boldsymbol{v}\boldsymbol{g})\right]-\boldsymbol{v}\cdot(\boldsymbol{g}\cdot\nabla\boldsymbol{v})-\boldsymbol{v}\cdot\left[\boldsymbol{g}\times(\nabla\times\boldsymbol{v})\right]\end{aligned} \tag{2.76}$$

Applying the tensor expansion to the final two terms on the right-hand side of Eq. (2.76) results in

$$
\begin{aligned}
-\boldsymbol{v}\cdot(\boldsymbol{g}\cdot\nabla\boldsymbol{v})-\boldsymbol{v}\cdot\left[\boldsymbol{g}\times(\nabla\times\boldsymbol{v})\right] &= -v_i\boldsymbol{e}_i\cdot\left(g_j\boldsymbol{e}_j\cdot\frac{\partial}{\partial x_k}\boldsymbol{e}_k v_l\boldsymbol{e}_l\right)-v_i\boldsymbol{e}_i\cdot\left[g_j\boldsymbol{e}_j\times\left(\frac{\partial}{\partial x_k}\boldsymbol{e}_k\times v_l\boldsymbol{e}_l\right)\right]\\
&= -v_i\boldsymbol{e}_i\cdot\left(g_j\frac{\partial v_l}{\partial x_j}\boldsymbol{e}_l\right)-v_i\boldsymbol{e}_i\cdot\left[g_j\boldsymbol{e}_j\times\left(\frac{\partial v_l}{\partial x_k}\epsilon_{klm}\boldsymbol{e}_m\right)\right]\\
&= -v_i g_j\frac{\partial v_i}{\partial x_j}-v_i\boldsymbol{e}_i\cdot\left(g_j\frac{\partial v_l}{\partial x_k}\epsilon_{klm}\epsilon_{jmn}\boldsymbol{e}_n\right)\\
&= -v_i g_j\frac{\partial v_i}{\partial x_j}-v_i g_j\frac{\partial v_l}{\partial x_k}\epsilon_{klm}\epsilon_{jmi}\\
&= -v_i g_j\frac{\partial v_i}{\partial x_j}-v_i g_j\frac{\partial v_l}{\partial x_k}\epsilon_{mlk}\epsilon_{mji}\\
&= -v_i g_j\frac{\partial v_i}{\partial x_j}-v_i g_j\frac{\partial v_l}{\partial x_k}\left(\delta_{lj}\delta_{ki}-\delta_{li}\delta_{kj}\right)\\
&= -v_i g_j\frac{\partial v_i}{\partial x_j}-v_i g_j\frac{\partial v_j}{\partial x_i}+v_i g_j\frac{\partial v_i}{\partial x_j}=-v_i g_j\frac{\partial v_j}{\partial x_i}
\end{aligned}
\tag{2.77}
$$

There is another equality,

$$
(\boldsymbol{v}\boldsymbol{g}):\nabla\boldsymbol{v}=v_i\boldsymbol{e}_i g_j\boldsymbol{e}_j:\frac{\partial}{\partial x_k}\boldsymbol{e}_k v_l\boldsymbol{e}_l=v_i g_j\frac{\partial v_l}{\partial x_k}\delta_{jl}\delta_{ki}=v_i g_j\frac{\partial v_j}{\partial x_i}
\tag{2.78}
$$

Comparing the results of Eq. (2.78) with those of Eq. (2.77), one yields

$$
-\boldsymbol{v}\cdot(\boldsymbol{g}\cdot\nabla\boldsymbol{v})-\boldsymbol{v}\cdot\left[\boldsymbol{g}\times(\nabla\times\boldsymbol{v})\right]=-(\boldsymbol{v}\boldsymbol{g}):\nabla\boldsymbol{v}
\tag{2.79}
$$

Substituting Eq. (2.79) back into Eq. (2.76) gives

$$
\nabla\cdot\left[(\boldsymbol{v}\times\boldsymbol{B})\times(\boldsymbol{v}\times\boldsymbol{D})\right]=-\boldsymbol{v}\cdot\left[\nabla\cdot(\boldsymbol{v}\boldsymbol{g})\right]-(\boldsymbol{v}\boldsymbol{g}):\nabla\boldsymbol{v}
\tag{2.80}
$$

Or just represent the term $-\boldsymbol{v}\cdot(\boldsymbol{g}\cdot\nabla\boldsymbol{v})$ as

$$
-\boldsymbol{v}\cdot(\boldsymbol{g}\cdot\nabla\boldsymbol{v})=-(\boldsymbol{v}\boldsymbol{g})\cdot\cdot\nabla\boldsymbol{v}
\tag{2.81}
$$

To further simplify the term $\nabla\cdot(\boldsymbol{E}\times\boldsymbol{H})$ in Eq. (2.72), vector operators in conjunction with Maxwell's equations

$$
\nabla\cdot\boldsymbol{B}=0,\ \nabla\cdot\boldsymbol{D}=\rho_e
\tag{2.82}
$$

are applied, where ρ_e is the volume density of free charges. Then the following terms can be expanded as

$$\nabla\times(\boldsymbol{v}\times\boldsymbol{D})=\rho_e\boldsymbol{v}-\boldsymbol{D}(\nabla\cdot\boldsymbol{v})+\boldsymbol{D}\cdot\nabla\boldsymbol{v}-\boldsymbol{v}\cdot\nabla\boldsymbol{D} \tag{2.83}$$

$$\nabla\times(\boldsymbol{v}\times\boldsymbol{B})=-\boldsymbol{B}(\nabla\cdot\boldsymbol{v})+\boldsymbol{B}\cdot\nabla\boldsymbol{v}-\boldsymbol{v}\cdot\nabla\boldsymbol{B} \tag{2.84}$$

After rearranging the terms, one obtains

$$\boldsymbol{v}\cdot\nabla\boldsymbol{D}=\rho_e\boldsymbol{v}-\boldsymbol{D}(\nabla\cdot\boldsymbol{v})+\boldsymbol{D}\cdot\nabla\boldsymbol{v}-\nabla\times(\boldsymbol{v}\times\boldsymbol{D}) \tag{2.85}$$

$$\boldsymbol{v}\cdot\nabla\boldsymbol{B}=-\boldsymbol{B}(\nabla\cdot\boldsymbol{v})+\boldsymbol{B}\cdot\nabla\boldsymbol{v}-\nabla\times(\boldsymbol{v}\times\boldsymbol{B}) \tag{2.86}$$

Using Eqs. (2.85), (2.86) and (2.73) yields

$$\frac{\mathrm{D}\boldsymbol{D}}{\mathrm{D}t}=\frac{\partial\boldsymbol{D}}{\partial t}+\boldsymbol{v}\cdot\nabla\boldsymbol{D}=\nabla\times\boldsymbol{H}-\boldsymbol{j}_0+\rho_e\boldsymbol{v}-\boldsymbol{D}(\nabla\cdot\boldsymbol{v})+\boldsymbol{D}\cdot\nabla\boldsymbol{v}-\nabla\times(\boldsymbol{v}\times\boldsymbol{D}) \tag{2.87}$$

$$\frac{\mathrm{D}\boldsymbol{B}}{\mathrm{D}t}=\frac{\partial\boldsymbol{B}}{\partial t}+\boldsymbol{v}\cdot\nabla\boldsymbol{B}=-\nabla\times\boldsymbol{E}-\boldsymbol{B}(\nabla\cdot\boldsymbol{v})+\boldsymbol{B}\cdot\nabla\boldsymbol{v}-\nabla\times(\boldsymbol{v}\times\boldsymbol{B}) \tag{2.88}$$

After performing a rearrangement, one has

$$\nabla\times\boldsymbol{H}=\frac{\mathrm{D}\boldsymbol{D}}{\mathrm{D}t}+\boldsymbol{j}_0-\rho_e\boldsymbol{v}+\boldsymbol{D}(\nabla\cdot\boldsymbol{v})-\boldsymbol{D}\cdot\nabla\boldsymbol{v}+\nabla\times(\boldsymbol{v}\times\boldsymbol{D}) \tag{2.89}$$

$$\nabla\times\boldsymbol{E}=-\frac{\mathrm{D}\boldsymbol{B}}{\mathrm{D}t}-\boldsymbol{B}(\nabla\cdot\boldsymbol{v})+\boldsymbol{B}\cdot\nabla\boldsymbol{v}-\nabla\times(\boldsymbol{v}\times\boldsymbol{B}) \tag{2.90}$$

Thus, one obtains

$$\boldsymbol{H}\cdot(\nabla\times\boldsymbol{E})=-\boldsymbol{H}\cdot\frac{\mathrm{D}\boldsymbol{B}}{\mathrm{D}t}-\boldsymbol{H}\cdot\left[\boldsymbol{B}(\nabla\cdot\boldsymbol{v})\right]+\boldsymbol{H}\cdot(\boldsymbol{B}\cdot\nabla\boldsymbol{v})-\boldsymbol{H}\cdot\left[\nabla\times(\boldsymbol{v}\times\boldsymbol{B})\right] \tag{2.91}$$

$$\boldsymbol{E}\cdot(\nabla\times\boldsymbol{H})=\boldsymbol{E}\cdot\frac{\mathrm{D}\boldsymbol{D}}{\mathrm{D}t}+\boldsymbol{E}\cdot\boldsymbol{j}_0-\rho_e\boldsymbol{E}\cdot\boldsymbol{v}+\boldsymbol{E}\cdot\left[\boldsymbol{D}(\nabla\cdot\boldsymbol{v})\right]-\boldsymbol{E}\cdot(\boldsymbol{D}\cdot\nabla\boldsymbol{v})+\boldsymbol{E}\cdot\left[\nabla\times(\boldsymbol{v}\times\boldsymbol{D})\right] \tag{2.92}$$

From Eqs. (2.91) and (2.92), one has

$$\begin{aligned}
\nabla\cdot(\boldsymbol{E}\times\boldsymbol{H})&=\boldsymbol{H}\cdot(\nabla\times\boldsymbol{E})-\boldsymbol{E}\cdot(\nabla\times\boldsymbol{H})\\
&=-\boldsymbol{H}\cdot\frac{\mathrm{D}\boldsymbol{B}}{\mathrm{D}t}-\boldsymbol{H}\cdot\left[\boldsymbol{B}(\nabla\cdot\boldsymbol{v})\right]+\boldsymbol{H}\cdot(\boldsymbol{B}\cdot\nabla\boldsymbol{v})-\boldsymbol{H}\cdot\left[\nabla\times(\boldsymbol{v}\times\boldsymbol{B})\right]\\
&\quad-\left\{\boldsymbol{E}\cdot\frac{\mathrm{D}\boldsymbol{D}}{\mathrm{D}t}+\boldsymbol{E}\cdot\boldsymbol{j}_0-\rho_e\boldsymbol{E}\cdot\boldsymbol{v}+\boldsymbol{E}\cdot\left[\boldsymbol{D}(\nabla\cdot\boldsymbol{v})\right]-\boldsymbol{E}\cdot(\boldsymbol{D}\cdot\nabla\boldsymbol{v})+\boldsymbol{E}\cdot\left[\nabla\times(\boldsymbol{v}\times\boldsymbol{D})\right]\right\}\\
&=-\boldsymbol{E}\cdot\frac{\mathrm{D}\boldsymbol{D}}{\mathrm{D}t}-\boldsymbol{H}\cdot\frac{\mathrm{D}\boldsymbol{B}}{\mathrm{D}t}-\boldsymbol{E}\cdot(\boldsymbol{j}_0-\rho_e\boldsymbol{v})-\left\{\boldsymbol{E}\cdot\left[\nabla\times(\boldsymbol{v}\times\boldsymbol{D})\right]+\boldsymbol{H}\cdot\left[\nabla\times(\boldsymbol{v}\times\boldsymbol{B})\right]\right\}\\
&\quad+\left[\boldsymbol{E}\cdot(\boldsymbol{D}\cdot\nabla\boldsymbol{v})+\boldsymbol{H}\cdot(\boldsymbol{B}\cdot\nabla\boldsymbol{v})\right]-\left\{\boldsymbol{E}\cdot\left[\boldsymbol{D}(\nabla\cdot\boldsymbol{v})\right]+\boldsymbol{H}\cdot\left[\boldsymbol{B}(\nabla\cdot\boldsymbol{v})\right]\right\}
\end{aligned} \tag{2.93}$$

For the last two terms in Eq. (2.93), the tensor representation shows that

$$\boldsymbol{E}\cdot\left(\boldsymbol{D}\cdot\nabla\boldsymbol{v}\right)=E_i\boldsymbol{e}_i\cdot\left(D_j\boldsymbol{e}_j\cdot\frac{\partial}{\partial x_k}\boldsymbol{e}_k v_l\boldsymbol{e}_l\right)=E_i\boldsymbol{e}_i\cdot\left(D_j\frac{\partial v_l}{\partial x_j}\boldsymbol{e}_l\right)=E_iD_j\frac{\partial v_i}{\partial x_j} \tag{2.94}$$

$$\boldsymbol{E}\cdot\left[\boldsymbol{D}\left(\nabla\cdot\boldsymbol{v}\right)\right]=E_i\boldsymbol{e}_i\cdot\left[D_j\boldsymbol{e}_j\left(\frac{\partial}{\partial x_k}\boldsymbol{e}_k\cdot v_l\boldsymbol{e}_l\right)\right]=E_i\boldsymbol{e}_i\cdot\left(D_j\boldsymbol{e}_j\frac{\partial v_k}{\partial x_k}\right)=E_iD_i\frac{\partial v_k}{\partial x_k} \tag{2.95}$$

In addition, there are

$$\boldsymbol{ED}\cdot\cdot\left(\nabla\boldsymbol{v}\right)=E_i\boldsymbol{e}_iD_j\boldsymbol{e}_j\cdot\cdot\left(\frac{\partial}{\partial x_k}\boldsymbol{e}_k v_l\boldsymbol{e}_l\right)=E_iD_j\frac{\partial v_l}{\partial x_k}\delta_{jk}\delta_{il}=E_iD_j\frac{\partial v_i}{\partial x_j} \tag{2.96}$$

$$\boldsymbol{DE}:\left(\nabla\boldsymbol{v}\right)=D_j\boldsymbol{e}_jE_i\boldsymbol{e}_i:\left(\frac{\partial}{\partial x_k}\boldsymbol{e}_k v_l\boldsymbol{e}_l\right)=D_jE_i\frac{\partial v_l}{\partial x_k}\delta_{jk}\delta_{il}=D_jE_i\frac{\partial v_i}{\partial x_j} \tag{2.97}$$

$$\left(\boldsymbol{E}\cdot\boldsymbol{D}\right)\mathbf{I}\cdot\cdot\left(\nabla\boldsymbol{v}\right)=\left(E_i\boldsymbol{e}_i\cdot D_j\boldsymbol{e}_j\right)\boldsymbol{e}_k\boldsymbol{e}_k\cdot\cdot\left(\frac{\partial}{\partial x_l}\boldsymbol{e}_l v_m\boldsymbol{e}_m\right)=E_iD_i\frac{\partial v_k}{\partial x_k}=\left(\boldsymbol{E}\cdot\boldsymbol{D}\right)\mathbf{I}:\left(\nabla\boldsymbol{v}\right) \tag{2.98}$$

where **I** is the unit dyadic. Comparing the results of Eqs. (2.94), (2.96), and (2.97) individually, as well as comparing the results of Eqs. (2.94) and (2.98) separately, leads to

$$\boldsymbol{E}\cdot\left(\boldsymbol{D}\cdot\nabla\boldsymbol{v}\right)=\boldsymbol{ED}\cdot\cdot\left(\nabla\boldsymbol{v}\right)=\boldsymbol{DE}:\left(\nabla\boldsymbol{v}\right) \tag{2.99}$$

$$\boldsymbol{E}\cdot\left[\boldsymbol{D}\left(\nabla\cdot\boldsymbol{v}\right)\right]=\left(\boldsymbol{E}\cdot\boldsymbol{D}\right)\mathbf{I}\cdot\cdot\left(\nabla\boldsymbol{v}\right)=\left(\boldsymbol{E}\cdot\boldsymbol{D}\right)\mathbf{I}:\left(\nabla\boldsymbol{v}\right) \tag{2.100}$$

After applying the identical treatment for the magnetic field terms in Eq. (2.93), and subsequently substituting the derived result alongside Eqs. (2.99) and (2.100) into Eq. (2.93), one has

$$\begin{aligned}\nabla\cdot\left(\boldsymbol{E}\times\boldsymbol{H}\right)=&-\boldsymbol{E}\cdot\frac{\mathrm{D}\boldsymbol{D}}{\mathrm{D}t}-\boldsymbol{H}\cdot\frac{\mathrm{D}\boldsymbol{B}}{\mathrm{D}t}-\boldsymbol{E}\cdot\left(\boldsymbol{j}_0-\rho_e\boldsymbol{v}\right)-\left\{\boldsymbol{E}\cdot\left[\nabla\times\left(\boldsymbol{v}\times\boldsymbol{D}\right)\right]\right.\\&\left.+\boldsymbol{H}\cdot\left[\nabla\times\left(\boldsymbol{v}\times\boldsymbol{B}\right)\right]\right\}+\left[\left(\boldsymbol{DE}+\boldsymbol{BH}\right)-\left(\boldsymbol{E}\cdot\boldsymbol{D}+\boldsymbol{H}\cdot\boldsymbol{B}\right)\mathbf{I}\right]:\left(\nabla\boldsymbol{v}\right)\end{aligned} \tag{2.101}$$

Substituting the results of Eqs. (2.74), (2.80), and (2.101) back into the expression Eq. (2.72) for the Poynting term, one gets

$$\begin{aligned}\nabla\cdot\left(\boldsymbol{E}'\times\boldsymbol{H}'\right)=&-\boldsymbol{E}\cdot\frac{\mathrm{D}\boldsymbol{D}}{\mathrm{D}t}-\boldsymbol{H}\cdot\frac{\mathrm{D}\boldsymbol{B}}{\mathrm{D}t}-\boldsymbol{E}\cdot\left(\boldsymbol{j}_0-\rho_e\boldsymbol{v}\right)+\left[\left(\boldsymbol{DE}+\boldsymbol{BH}\right)-\left(\boldsymbol{E}\cdot\boldsymbol{D}+\boldsymbol{H}\cdot\boldsymbol{B}\right)\mathbf{I}\right]:\left(\nabla\boldsymbol{v}\right)\\&+\left[\boldsymbol{v}\cdot\frac{\partial\boldsymbol{g}}{\partial t}-\boldsymbol{j}_0\cdot\left(\boldsymbol{v}\times\boldsymbol{B}\right)\right]+\boldsymbol{v}\cdot\left[\nabla\cdot\left(\boldsymbol{vg}\right)\right]+\left(\boldsymbol{vg}\right):\nabla\boldsymbol{v}\end{aligned} \tag{2.102}$$

By inserting the representations of the surface integrals from Eqs. (2.69) and (2.70), and the result of (2.102) into Eq. (2.58), then leveraging the arbitrariness of the integral volume, one can apply Eq. (2.11) to the left-hand side of Eq. (2.58) and obtains the following equation:

$$\begin{aligned}\rho\frac{\mathrm{D}}{\mathrm{D}t}\left(u+\frac{v^2}{2}+\frac{I}{2}\omega_\mathrm{p}^2\right)&=\left(\rho\boldsymbol{G}\cdot\boldsymbol{\omega}_\mathrm{p}+\rho\boldsymbol{F}\cdot\boldsymbol{v}\right)+\left[\left(\nabla\cdot\tilde{\mathrm{T}}\right)\cdot\boldsymbol{v}+\tilde{\mathrm{T}}:\nabla\boldsymbol{v}\right]\\&+\left[\left(\nabla\cdot\mathrm{C}\right)\cdot\boldsymbol{\omega}_\mathrm{p}+\mathrm{C}:\nabla\boldsymbol{\omega}_\mathrm{p}\right]-\nabla\cdot\boldsymbol{q}-\left\{-\boldsymbol{E}\cdot\frac{\mathrm{D}\boldsymbol{D}}{\mathrm{D}t}-\boldsymbol{H}\cdot\frac{\mathrm{D}\boldsymbol{B}}{\mathrm{D}t}-\boldsymbol{E}\cdot\left(\boldsymbol{j}_0-\rho_e\boldsymbol{v}\right)\right.\\&+\left[\left(\boldsymbol{DE}+\boldsymbol{BH}\right)-\left(\boldsymbol{E}\cdot\boldsymbol{D}+\boldsymbol{H}\cdot\boldsymbol{B}\right)\mathrm{I}\right]:\left(\nabla\boldsymbol{v}\right)\\&\left.+\left[\boldsymbol{v}\cdot\frac{\partial\boldsymbol{g}}{\partial t}-\boldsymbol{j}_0\cdot\left(\boldsymbol{v}\times\boldsymbol{B}\right)\right]+\boldsymbol{v}\cdot\left[\nabla\cdot\left(\boldsymbol{vg}\right)\right]+\left(\boldsymbol{vg}\right):\nabla\boldsymbol{v}\right\}+\rho R\end{aligned}\tag{2.103}$$

According to the momentum equation, the second and third terms on the left-hand side of Eq. (2.103) can be written as

$$\rho\frac{\mathrm{D}}{\mathrm{D}t}\left(\frac{v^2}{2}\right)=\rho\boldsymbol{v}\cdot\frac{\mathrm{D}\boldsymbol{v}}{\mathrm{D}t}=-\boldsymbol{v}\cdot\frac{\partial\boldsymbol{g}}{\partial t}+\rho\boldsymbol{v}\cdot\boldsymbol{F}+\boldsymbol{v}\cdot\left(\nabla\cdot\mathrm{T}\right)\tag{2.104}$$

$$\rho\frac{\mathrm{D}}{\mathrm{D}t}\left(\frac{I}{2}\omega_\mathrm{p}^2\right)=\rho I\boldsymbol{\omega}_\mathrm{p}\cdot\frac{\mathrm{D}\boldsymbol{\omega}_\mathrm{p}}{\mathrm{D}t}=\boldsymbol{\omega}_\mathrm{p}\cdot\rho\boldsymbol{G}+\boldsymbol{\omega}_\mathrm{p}\cdot\left(\nabla\cdot\mathrm{C}\right)+\boldsymbol{\omega}_\mathrm{p}\cdot\left(\epsilon:\mathrm{T}\right)\tag{2.105}$$

If the following definition is available,

$$\boldsymbol{j}'=\boldsymbol{j}_0-\rho_e\boldsymbol{v}\tag{2.106}$$

one has

$$\begin{aligned}\boldsymbol{j}'\cdot\boldsymbol{E}'&=\left(\boldsymbol{j}_0-\rho_e\boldsymbol{v}\right)\cdot\left(\boldsymbol{E}+\boldsymbol{v}\times\boldsymbol{B}\right)\\&=\boldsymbol{j}_0\cdot\boldsymbol{E}-\rho_e\boldsymbol{v}\cdot\boldsymbol{E}+\boldsymbol{j}_0\cdot\left(\boldsymbol{v}\times\boldsymbol{B}\right)-\rho_e\boldsymbol{v}\cdot\left(\boldsymbol{v}\times\boldsymbol{B}\right)\\&=\boldsymbol{j}_0\cdot\boldsymbol{E}-\rho_e\boldsymbol{v}\cdot\boldsymbol{E}+\boldsymbol{j}_0\cdot\left(\boldsymbol{v}\times\boldsymbol{B}\right)\end{aligned}\tag{2.107}$$

In addition, using the constitutive relationships for magnetic and electric fields,

$$\boldsymbol{B}=\mu_0\left(\boldsymbol{M}+\boldsymbol{H}\right),\boldsymbol{D}=\varepsilon_0\boldsymbol{E}+\boldsymbol{P}\tag{2.108}$$

where $\boldsymbol{M}$ is the magnetization, μ_0 is the vacuum permeability, and ε_0 is the permittivity in vacuum. The term $\boldsymbol{E}\cdot\frac{\mathrm{D}\boldsymbol{D}}{\mathrm{D}t}$ in Eq. (2.103) can be expressed as

$$\boldsymbol{E}\cdot\frac{\mathrm{D}\boldsymbol{D}}{\mathrm{D}t}=\varepsilon_0\boldsymbol{E}\cdot\frac{\mathrm{D}\boldsymbol{E}}{\mathrm{D}t}+\boldsymbol{E}\cdot\frac{\mathrm{D}\boldsymbol{P}}{\mathrm{D}t}\tag{2.109}$$

Use of expression

$$\rho \frac{\mathrm{D}}{\mathrm{D}t}\left(\frac{\varepsilon_0 E^2}{2\rho}\right) = \varepsilon_0 \boldsymbol{E}\cdot\frac{\mathrm{D}\boldsymbol{E}}{\mathrm{D}t} - \frac{\varepsilon_0 E^2}{2\rho}\frac{\mathrm{D}\rho}{\mathrm{D}t} = \varepsilon_0 \boldsymbol{E}\cdot\frac{\mathrm{D}\boldsymbol{E}}{\mathrm{D}t} + \frac{\varepsilon_0 E^2}{2}(\nabla\cdot\boldsymbol{v}) \tag{2.110}$$

to expand Eq. (2.109) shows

$$\boldsymbol{E}\cdot\frac{\mathrm{D}\boldsymbol{D}}{\mathrm{D}t} = \rho \frac{\mathrm{D}}{\mathrm{D}t}\left(\frac{\varepsilon_0 E^2}{2\rho}\right) - \frac{\varepsilon_0 E^2}{2}(\nabla\cdot\boldsymbol{v}) + \boldsymbol{E}\cdot\frac{\mathrm{D}\boldsymbol{P}}{\mathrm{D}t} \tag{2.111}$$

Similarly, for the term $\boldsymbol{H}\cdot\frac{\mathrm{D}\boldsymbol{B}}{\mathrm{D}t}$ in Eq. (2.103), one has

$$\boldsymbol{H}\cdot\frac{\mathrm{D}\boldsymbol{B}}{\mathrm{D}t} = \mu_0 \boldsymbol{H}\cdot\frac{\mathrm{D}\boldsymbol{M}}{\mathrm{D}t} + \mu_0 \boldsymbol{H}\cdot\frac{\mathrm{D}\boldsymbol{H}}{\mathrm{D}t} \tag{2.112}$$

and

$$\rho \frac{\mathrm{D}}{\mathrm{D}t}\left(\frac{\mu_0 H^2}{2\rho}\right) = \mu_0 \boldsymbol{H}\cdot\frac{\mathrm{D}\boldsymbol{H}}{\mathrm{D}t} - \frac{\mu_0 H^2}{2\rho}\frac{\mathrm{D}\rho}{\mathrm{D}t} = \mu_0 \boldsymbol{H}\cdot\frac{\mathrm{D}\boldsymbol{H}}{\mathrm{D}t} + \frac{\mu_0 H^2}{2}(\nabla\cdot\boldsymbol{v}) \tag{2.113}$$

$$\boldsymbol{H}\cdot\frac{\mathrm{D}\boldsymbol{B}}{\mathrm{D}t} = \mu_0 \boldsymbol{H}\cdot\frac{\mathrm{D}\boldsymbol{M}}{\mathrm{D}t} + \rho \frac{\mathrm{D}}{\mathrm{D}t}\left(\frac{\mu_0 H^2}{2\rho}\right) - \frac{\mu_0 H^2}{2}(\nabla\cdot\boldsymbol{v}) \tag{2.114}$$

Now introducing the results of Eqs. (2.104), (2.105), (2.107), (2.111), and (2.114) into Eq. (2.103) and using Eq. (2.22), gives the final form of the equation for energy conservation in ferrofluids

$$\begin{aligned} \rho \frac{\mathrm{D}u_m}{\mathrm{D}t} &= \mathbf{T} : \nabla \boldsymbol{v} + \mathbf{C} : \nabla \boldsymbol{\omega}_\mathrm{p} - \boldsymbol{\omega}_\mathrm{p} \cdot (\boldsymbol{\epsilon} : \mathbf{T}) - \left(\frac{\varepsilon_0 E^2}{2} + \frac{\mu_0 H^2}{2}\right)(\nabla\cdot\boldsymbol{v}) + \boldsymbol{E}\cdot\frac{\mathrm{D}\boldsymbol{P}}{\mathrm{D}t} \\ &+ \mu_0 \boldsymbol{H}\cdot\frac{\mathrm{D}\boldsymbol{M}}{\mathrm{D}t} - \left[(\boldsymbol{DE} + \boldsymbol{BH}) - (\boldsymbol{E}\cdot\boldsymbol{D} + \boldsymbol{H}\cdot\boldsymbol{B})\mathbf{I}\right] : (\nabla \boldsymbol{v}) + \boldsymbol{j}'\cdot\boldsymbol{E}' - \nabla\cdot\boldsymbol{q} + \rho R \end{aligned} \tag{2.115}$$

where

$$u_m = u - \frac{\varepsilon_0 E^2}{2\rho} - \frac{\mu_0 H^2}{2\rho} \tag{2.116}$$

It represents the internal energy of the medium except the energy of the electromagnetic field in the space occupied by the medium.

2.6 EQUATION OF ENTROPY GROWTH RATE

2.6.1 Equilibrium Thermodynamic Relationship

The state of a polarizable substance is influenced by both polarization $\boldsymbol{P}$ and magnetization $\boldsymbol{M}$, with the volume density of polarization work expressed as $\boldsymbol{E} \cdot d\boldsymbol{D}$ and $\boldsymbol{H} \cdot d\boldsymbol{B}$, respectively. The Gibbs equation that establishes the relationship between internal energy and entropy per unit mass is (Odenbach, 2002)

$$\mathrm{d}u^{(e)} = T\mathrm{d}S_e^{(e)} + \frac{p^{(e)}d\rho}{\rho^2} + \boldsymbol{E}^{(e)} \cdot \mathrm{d}\left(\frac{\boldsymbol{D}^{(e)}}{\rho}\right) + \boldsymbol{H}^{(e)} \cdot \mathrm{d}\left(\frac{\boldsymbol{B}^{(e)}}{\rho}\right) \tag{2.117}$$

The superscript (e) denotes the equilibrium value, S_e is the entropy per unit mass, and T is the temperature. The relationship between the pressure in the presence of a field $p^{(e)}$ and the ordinary thermodynamic pressure in the absence of a field p_0 is (Rosensweig, 1985)

$$\begin{aligned} p^{(e)} = p_0(\rho,T) + \frac{1}{2}\left(\varepsilon_0 E^{(e)2} + \mu_0 H^{(e)2}\right) + \int_0^E \left(\frac{\partial(\upsilon P)}{\partial \upsilon}\right)_{T,H,E}^{(e)} \mathrm{d}E^{(e)} \\ + \mu_0 \int_0^H \left(\frac{\partial(\upsilon M)}{\partial \upsilon}\right)_{T,H,E}^{(e)} \mathrm{d}H^{(e)} = p + \frac{1}{2}\left(\varepsilon_0 E^{(e)2} + \mu_0 H^{(e)2}\right) \end{aligned} \tag{2.118}$$

where $\upsilon = 1/\rho$ is the specific volume. In the absence of fields, $p^{(e)}$ degenerates into $p_0(\rho, T)$. The variable p is expressed as

$$p = p_0(\rho,T) + \int_0^E \left(\frac{\partial(\upsilon P)}{\partial \upsilon}\right)_{T,H,E}^{(e)} \mathrm{d}E^{(e)} + \mu_0 \int_0^H \left(\frac{\partial(\upsilon M)}{\partial \upsilon}\right)_{T,H,E}^{(e)} \mathrm{d}H^{(e)} \tag{2.119}$$

The use of Eq. (2.108) in Eq. (2.117) gives

$$\mathrm{d}\left(\frac{\boldsymbol{D}^{(e)}}{\rho}\right) = \mathrm{d}\left(\frac{\varepsilon_0 \boldsymbol{E}^{(e)} + \boldsymbol{P}^{(e)}}{\rho}\right) = \frac{\varepsilon_0}{\rho}\mathrm{d}\boldsymbol{E}^{(e)} - \frac{\varepsilon_0}{\rho^2}\boldsymbol{E}^{(e)}\mathrm{d}\rho + \frac{\mathrm{d}\boldsymbol{P}^{(e)}}{\rho} - \frac{\boldsymbol{P}^{(e)}}{\rho^2}\mathrm{d}\rho \tag{2.120}$$

$$\mathrm{d}\left(\frac{\boldsymbol{B}^{(e)}}{\rho}\right) = \mathrm{d}\left(\frac{\mu_0(\boldsymbol{M}+\boldsymbol{H})}{\rho}\right) = \frac{\mu_0}{\rho}\mathrm{d}\boldsymbol{M}^{(e)} - \frac{\mu_0}{\rho^2}\boldsymbol{M}^{(e)}\mathrm{d}\rho + \frac{\mu_0}{\rho}\mathrm{d}\boldsymbol{H}^{(e)} - \frac{\mu_0}{\rho^2}\boldsymbol{H}^{(e)}\mathrm{d}\rho \tag{2.121}$$

Substituting Eqs. (2.120) and (2.121) into Eq. (2.117) and using Eq. (2.118), yields

$$\begin{aligned} \rho\mathrm{d}u^{(e)} = \rho T\mathrm{d}S_e^{(e)} + \frac{\mathrm{d}\rho}{\rho}\left[p + \frac{1}{2}\left(\varepsilon_0 E^{(e)2} + \mu_0 H^{(e)2}\right)\right] + \boldsymbol{E}^{(e)} \cdot \left(\varepsilon_0 \mathrm{d}\boldsymbol{E}^{(e)} - \frac{\varepsilon_0}{\rho}\boldsymbol{E}^{(e)}\mathrm{d}\rho + d\boldsymbol{P}^{(e)} - \frac{\boldsymbol{P}^{(e)}}{\rho}\mathrm{d}\rho\right) \\ + \boldsymbol{H}^{(e)} \cdot \left(\mu_0 \mathrm{d}\boldsymbol{M}^{(e)} - \frac{\mu_0}{\rho}\boldsymbol{M}^{(e)}\mathrm{d}\rho + \mu_0 \mathrm{d}\boldsymbol{H}^{(e)} - \frac{\mu_0}{\rho}\boldsymbol{H}^{(e)}\mathrm{d}\rho\right) \end{aligned} \tag{2.122}$$

This leads to another form of Gibbs equation

$$\rho \mathrm{d}u^{(e)} = \rho T \mathrm{d}S_e^{(e)} + \left[p - \frac{\varepsilon_0}{2} E^{(e)2} - \frac{\mu_0}{2} H^{(e)2} - \boldsymbol{E}^{(e)} \cdot \boldsymbol{p}^{(e)} - \mu_0 \boldsymbol{H}^{(e)} \cdot \boldsymbol{M}^{(e)} \right] \frac{\mathrm{d}\rho}{\rho}$$
$$+ \boldsymbol{E}^{(e)} \cdot \mathrm{d}\boldsymbol{D}^{(e)} + \boldsymbol{H}^{(e)} \cdot \mathrm{d}\boldsymbol{B}^{(e)} \quad (2.123)$$

This equation represents the relationship between the changes in internal energy and entropy among different equilibrium states of matter.

2.6.2 Equation of Entropy Growth Rate

Dividing Eq. (2.123) by dt and subsequently substituting the conservation equation of mass (2.15) into the resultant equation, one obtains

$$\rho \frac{\mathrm{d}u^{(e)}}{dt} = \rho T \frac{\mathrm{d}S_e^{(e)}}{dt} - \left[p - \frac{\varepsilon_0}{2} E^{(e)2} - \frac{\mu_0}{2} H^{(e)2} - \boldsymbol{E}^{(e)} \cdot \boldsymbol{p}^{(e)} - \mu_0 \boldsymbol{H}^{(e)} \cdot \boldsymbol{M}^{(e)} \right] (\nabla \cdot \boldsymbol{v})$$
$$+ \boldsymbol{E}^{(e)} \cdot \frac{\mathrm{d}\boldsymbol{D}^{(e)}}{dt} + \boldsymbol{H}^{(e)} \cdot \frac{\mathrm{d}\boldsymbol{B}^{(e)}}{dt} \quad (2.124)$$

Assuming that entropy changes slowly, it is reasonable to postulate that the polarization of a substance attains a state of equilibrium at all instances, eliminating any distinction between $\boldsymbol{P}^{(e)}$ and $\boldsymbol{P}$, as well as between $\boldsymbol{M}^{(e)}$ and $\boldsymbol{M}$. Using the definition of Eq. (2.116) and the expression

$$\frac{\mathrm{d}}{dt}\left(\frac{\varepsilon_0 E^{(e)2}}{2\rho} \right) = \frac{\varepsilon_0}{2} \frac{1}{\rho} \frac{\mathrm{d}\left(\boldsymbol{E}^{(e)} . \boldsymbol{E}^{(e)} \right)}{dt} - \frac{\varepsilon_0}{2} E^{(e)2} \frac{1}{\rho^2} \frac{\mathrm{d}\rho}{dt} = \frac{\varepsilon_0}{\rho} \boldsymbol{E}^{(e)} . \frac{\mathrm{d}\boldsymbol{E}^{(e)}}{dt} + \frac{\varepsilon_0}{2\rho} E^{(e)2} (\nabla \cdot \boldsymbol{v}) \quad (2.125)$$

$$\frac{\mathrm{d}}{dt}\left(\frac{\mu_0 H^{(e)2}}{2\rho} \right) = \frac{\mu_0}{2} \frac{1}{\rho} \frac{\mathrm{d}\left(\boldsymbol{H}^{(e)} . \boldsymbol{H}^{(e)} \right)}{dt} - \frac{\mu_0}{2} H^{(e)2} \frac{1}{\rho^2} \frac{\mathrm{d}\rho}{dt}$$
$$= \frac{\mu_0}{\rho} \boldsymbol{H}^{(e)} . \frac{\mathrm{d}\boldsymbol{H}^{(e)}}{dt} + \frac{\mu_0}{2\rho} H^{(e)2} (\nabla \cdot \boldsymbol{v}) \quad (2.126)$$

write Eq. (2.124) as

$$\rho \frac{\mathrm{d}u_m^{(e)}}{dt} = \rho T \frac{\mathrm{d}S_e^{(e)}}{dt} - \left[p - \boldsymbol{E}^{(e)} . \boldsymbol{p}^{(e)} - \mu_0 \boldsymbol{H}^{(e)} . \boldsymbol{M}^{(e)} \right] (\nabla \cdot \boldsymbol{v})$$
$$+ \boldsymbol{E}^{(e)} . \frac{\mathrm{d}\boldsymbol{P}^{(e)}}{dt} + \mu_0 \boldsymbol{H}^{(e)} . \frac{\mathrm{d}\boldsymbol{M}^{(e)}}{dt} \quad (2.127)$$

The internal energy $u_m = u_m(t)$, as expressed in energy Eq. (2.115), can be regarded as a state of equilibrium within the medium, whereas the fields $\boldsymbol{E}$ and $\boldsymbol{H}$ operate in a non-equilibrium state. Consequently, at any given moment, the condition $u_m(t)=u_m^{(e)}(t)$ holds true, leading to the consequence that

$$\frac{du_m(t)}{dt}=\frac{du_m^{(e)}(t)}{dt} \tag{2.128}$$

To calculate the entropy variable between various states of the medium, one has to substitute Eqs. (2.115) and (2.127) into Eq. (2.128) and obtain

$$\begin{aligned}\rho T\frac{dS_e^{(e)}}{dt} &= \mathbf{T}:\nabla\boldsymbol{v}+\mathbf{C}:\nabla\boldsymbol{\omega}_\mathrm{p}-\boldsymbol{\omega}_\mathrm{p}\cdot(\boldsymbol{\epsilon}:\mathbf{T})-\left(\frac{\varepsilon_0E^2}{2}+\frac{\mu_0H^2}{2}\right)(\nabla\cdot\boldsymbol{v})\\ &+\left(\boldsymbol{E}-\boldsymbol{E}^{(e)}\right)\cdot\frac{\mathrm{D}\boldsymbol{P}}{\mathrm{D}t}+\mu_0\left(\boldsymbol{H}-\boldsymbol{H}^{(e)}\right)\cdot\frac{\mathrm{D}\boldsymbol{M}}{\mathrm{D}t}\\ &-\left[(\boldsymbol{DE}+\boldsymbol{BH})-(\boldsymbol{E}\cdot\boldsymbol{D}+\boldsymbol{H}\cdot\boldsymbol{B})\mathbf{I}\right]:(\nabla\boldsymbol{v})+\boldsymbol{j}'\cdot\boldsymbol{E}'-\nabla\cdot\boldsymbol{q}\\ &+\rho R+\left[p-\boldsymbol{E}^{(e)}\cdot\boldsymbol{P}^{(e)}-\mu_0\boldsymbol{H}^{(e)}\cdot\boldsymbol{M}^{(e)}\right](\nabla\cdot\boldsymbol{v})\end{aligned} \tag{2.129}$$

The stress tensor **T** in Eq. (2.129) can be expressed as

$$\mathbf{T}=\mathbf{T}_v+\mathbf{T}_m-p\mathbf{I}-\boldsymbol{v}\boldsymbol{g} \tag{2.130}$$

where $\mathbf{T}_v$ is the viscous stress tensor and $\mathbf{T}_m$ is the magnetic stress tensor of the polarisable fluid expressed as (Landau and Lifshitz, 1960)

$$\mathbf{T}_m=-\left(\frac{\varepsilon_0E^2}{2}+\frac{\mu_0H^2}{2}\right)\mathbf{I}+(\boldsymbol{DE}+\boldsymbol{BH}) \tag{2.131}$$

Substitution of Eqs. (2.130) and (2.131) into Eq. (2.129) and utilizing the identity

$$\mathbf{I}:\nabla\boldsymbol{v}=\nabla\cdot\boldsymbol{v} \tag{2.132}$$

obtains

$$\begin{aligned}\rho T\frac{dS_e^{(e)}}{dt} &= (\mathbf{T}_v-\boldsymbol{v}\boldsymbol{g}):\nabla\boldsymbol{v}+\mathbf{C}:\nabla\boldsymbol{\omega}_\mathrm{p}-\boldsymbol{\omega}_\mathrm{p}\cdot(\boldsymbol{\epsilon}:\mathbf{T})+\boldsymbol{j}'\cdot\boldsymbol{E}'-\nabla\cdot\boldsymbol{q}+\rho R\\ &+(\boldsymbol{E}\cdot\boldsymbol{D}+\boldsymbol{H}\cdot\boldsymbol{B})\mathbf{I}:(\nabla\boldsymbol{v})-\left[\boldsymbol{E}^{(e)}\cdot\boldsymbol{P}+\mu_0\boldsymbol{H}^{(e)}\cdot\boldsymbol{M}\right]\mathbf{I}:\nabla\boldsymbol{v}\\ &+\left(\boldsymbol{E}-\boldsymbol{E}^{(e)}\right)\cdot\frac{\mathrm{D}\boldsymbol{P}}{\mathrm{D}t}+\mu_0\left(\boldsymbol{H}-\boldsymbol{H}^{(e)}\right)\cdot\frac{\mathrm{D}\boldsymbol{M}}{\mathrm{D}t}-\left(\varepsilon_0E^2+\mu_0H^2\right)(\nabla\cdot\boldsymbol{v})\end{aligned} \tag{2.133}$$

For the term $\boldsymbol{\omega}_\mathrm{p}\cdot(\boldsymbol{\epsilon}:\mathbf{T})$ in Eq. (2.133),

$$\boldsymbol{\omega}_\mathrm{p}\cdot(\boldsymbol{\epsilon}:\mathbf{T})=(\mathbf{T}:\boldsymbol{\epsilon})\cdot\boldsymbol{\omega}_\mathrm{p}=\left[(\mathbf{T}_v+\mathbf{T}_m-p\mathbf{I}-\boldsymbol{v}\boldsymbol{g}):\boldsymbol{\epsilon}\right]\cdot\boldsymbol{\omega}_\mathrm{p} \tag{2.134}$$

with the aid of the identity

$$\mathbf{I} \,\dot{:}\, \boldsymbol{\epsilon} = \mathbf{0} \tag{2.135}$$

The expansion of the term is obtained as

$$\mathrm{T}_m \,\dot{:}\, \boldsymbol{\epsilon} = (\boldsymbol{DE} + \boldsymbol{BH}) \,\dot{:}\, \boldsymbol{\epsilon} = (\varepsilon_0 \boldsymbol{EE} + \boldsymbol{PE} + \mu_0 \boldsymbol{HH} + \mu_0 \boldsymbol{MH}) \,\dot{:}\, \boldsymbol{\epsilon} \tag{2.136}$$

where

$$\boldsymbol{PE} \,\dot{:}\, \boldsymbol{\epsilon} = P_i \boldsymbol{e}_i E_j \boldsymbol{e}_j \,\dot{:}\, \epsilon_{lmn} \boldsymbol{e}_l \boldsymbol{e}_m \boldsymbol{e}_n = P_i E_j \epsilon_{ijn} \boldsymbol{e}_n$$

Since the tensor representation of $\boldsymbol{P} \times \boldsymbol{E}$ is

$$\boldsymbol{P} \times \boldsymbol{E} = P_i \boldsymbol{e}_i \times E_j \boldsymbol{e}_j = P_i E_j \epsilon_{ijn} \boldsymbol{e}_n$$

one has

$$\boldsymbol{PE} \,\dot{:}\, \boldsymbol{\epsilon} = \boldsymbol{P} \times \boldsymbol{E} \tag{2.137}$$

The same derivation leads to

$$\boldsymbol{EE} \,\dot{:}\, \boldsymbol{\epsilon} = \mathbf{0}, \boldsymbol{HH} \,\dot{:}\, \boldsymbol{\epsilon} = \mathbf{0}, \boldsymbol{MH} \,\dot{:}\, \boldsymbol{\epsilon} = \boldsymbol{M} \times \boldsymbol{H} \tag{2.138}$$

Substituting the results of Eqs. (2.137) and (2.138) into Eq. (2.136) yields

$$\mathrm{T}_m \,\dot{:}\, \boldsymbol{\epsilon} = \boldsymbol{P} \times \boldsymbol{E} + \mu_0 \boldsymbol{M} \times \boldsymbol{H} \tag{2.139}$$

Now putting Eq. (2.139) into Eq. (2.134) yields

$$\begin{aligned}
\boldsymbol{\omega}_\mathrm{p} \cdot (\boldsymbol{\epsilon} \,\dot{:}\, \mathrm{T}) &= \left[(\mathrm{T}_v + \mathrm{T}_m - p\mathbf{I} - \boldsymbol{v}\boldsymbol{g}) \,\dot{:}\, \boldsymbol{\epsilon} \right] \cdot \boldsymbol{\omega}_\mathrm{p} \\
&= (\mathrm{T}_v - p\mathbf{I} - \boldsymbol{v}\boldsymbol{g}) : \boldsymbol{\epsilon} \cdot \boldsymbol{\omega}_\mathrm{p} + \boldsymbol{P} \times \boldsymbol{E} \cdot \boldsymbol{\omega}_\mathrm{p} + \mu_0 \boldsymbol{M} \times \boldsymbol{H} \cdot \boldsymbol{\omega}_\mathrm{p} \\
&= (\mathrm{T}_v - p\mathbf{I} - \boldsymbol{v}\boldsymbol{g}) : \boldsymbol{\epsilon} \cdot \boldsymbol{\omega}_\mathrm{p} - \boldsymbol{P} \times \boldsymbol{\omega}_\mathrm{p} \cdot \boldsymbol{E} - \mu_0 \boldsymbol{M} \times \boldsymbol{\omega}_\mathrm{p} \cdot \boldsymbol{H} \\
&= (\mathrm{T}_v - p\mathbf{I} - \boldsymbol{v}\boldsymbol{g}) \,\dot{:}\, \boldsymbol{\epsilon} \cdot \boldsymbol{\omega}_\mathrm{p} - (\boldsymbol{P} \times \boldsymbol{\omega}_\mathrm{p}) \cdot \left(\boldsymbol{E} - \boldsymbol{E}^{(e)} \right) - \mu_0 (\boldsymbol{M} \times \boldsymbol{\omega}_\mathrm{p}) \cdot \left(\boldsymbol{H} - \boldsymbol{H}^{(e)} \right)
\end{aligned} \tag{2.140}$$

where use has been made of equations satisfied at equilibrium conditions (with $\boldsymbol{E}^{(e)}$ parallel to $\boldsymbol{P}$, $\boldsymbol{H}^{(e)}$ parallel to $\boldsymbol{M}$)

$$(\boldsymbol{P} \times \boldsymbol{\omega}_\mathrm{p}) \cdot \boldsymbol{E}^{(e)} = \left(\boldsymbol{E}^{(e)} \times \boldsymbol{P} \right) \cdot \boldsymbol{\omega}_\mathrm{p} = 0, \tag{2.141}$$

$$(\boldsymbol{M} \times \boldsymbol{\omega}_\mathrm{p}) \cdot \boldsymbol{H}^{(e)} = \left(\boldsymbol{H}^{(e)} \times \boldsymbol{M} \right) \cdot \boldsymbol{\omega}_\mathrm{p} = 0 \tag{2.142}$$

Rewriting the tensor $\nabla \boldsymbol{v}$ in Eq. (2.133) as the summation of its symmetric and anti-symmetric components,

$$\nabla \boldsymbol{v} = \frac{1}{2}\left[\nabla \boldsymbol{v} + (\nabla \boldsymbol{v})^{\mathrm{T}}\right] + \frac{1}{2}\left[\nabla \boldsymbol{v} - (\nabla \boldsymbol{v})^{\mathrm{T}}\right] = \mathbf{D}^{(v)} + \boldsymbol{\epsilon}\cdot\boldsymbol{\Omega} \tag{2.143}$$

where $\mathbf{D}^{(v)}$ is the strain rate tensor and $\boldsymbol{\Omega}$ is the fluid rotation rate, or flow vorticity, expressed as

$$\boldsymbol{\Omega} = \frac{1}{2}\nabla \times \boldsymbol{v} \tag{2.144}$$

Using the tensor representation to verify the latter part of Eq. (2.143),

$$\frac{1}{2}\left[\nabla \boldsymbol{v} - (\nabla \boldsymbol{v})^{\mathrm{T}}\right] = \frac{1}{2}\left(\frac{\partial v_j}{\partial x_i}\boldsymbol{e}_i\boldsymbol{e}_j - \frac{\partial v_i}{\partial x_j}\boldsymbol{e}_i\boldsymbol{e}_j\right)$$

$$\boldsymbol{\epsilon}\cdot\boldsymbol{\Omega} = \frac{1}{2}\epsilon_{lmn}\boldsymbol{e}_l\boldsymbol{e}_m\boldsymbol{e}_n \cdot \left(\frac{\partial}{\partial x_i}\boldsymbol{e}_i \times v_j\boldsymbol{e}_j\right)$$

$$= \frac{1}{2}\epsilon_{lmn}\boldsymbol{e}_l\boldsymbol{e}_m\boldsymbol{e}_n \cdot \left(\frac{\partial v_j}{\partial x_i}\epsilon_{ijk}\boldsymbol{e}_k\right) = \frac{1}{2}\epsilon_{lmn}\epsilon_{ijn}\frac{\partial v_j}{\partial x_i}\boldsymbol{e}_l\boldsymbol{e}_m = \frac{1}{2}\epsilon_{nml}\epsilon_{nji}\frac{\partial v_j}{\partial x_i}\boldsymbol{e}_l\boldsymbol{e}_m$$

$$= \frac{1}{2}\left(\delta_{mj}\delta_{li} - \delta_{mi}\delta_{lj}\right)\frac{\partial v_j}{\partial x_i}\boldsymbol{e}_l\boldsymbol{e}_m = \frac{1}{2}\left(\frac{\partial v_j}{\partial x_i}\boldsymbol{e}_i\boldsymbol{e}_j - \frac{\partial v_j}{\partial x_i}\boldsymbol{e}_j\boldsymbol{e}_i\right)$$

Equation (2.143) is verified.

For the terms on the right-hand side of Eq. (2.133),

$$\begin{aligned}
&(\boldsymbol{E}\cdot\boldsymbol{D} + \boldsymbol{H}\cdot\boldsymbol{B})\mathbf{I}:(\nabla\boldsymbol{v}) - \left[\boldsymbol{E}^{(e)}\cdot\boldsymbol{P} + \mu_0\boldsymbol{H}^{(e)}\cdot\boldsymbol{M}\right]\mathbf{I}:\nabla\boldsymbol{v} \\
&= \left(\boldsymbol{E}\cdot\boldsymbol{D} - \boldsymbol{E}^{(e)}\cdot\boldsymbol{P} + \boldsymbol{H}\cdot\boldsymbol{B} - \mu_0\boldsymbol{H}^{(e)}\cdot\boldsymbol{M}\right)\mathbf{I}:(\nabla\boldsymbol{v}) \\
&= \left[\boldsymbol{E}\cdot(\varepsilon_0\boldsymbol{E} + \boldsymbol{P}) - \boldsymbol{E}^{(e)}\cdot\boldsymbol{P} + \mu_0\boldsymbol{H}\cdot(\boldsymbol{M} + \boldsymbol{H}) - \mu_0\boldsymbol{H}^{(e)}\cdot\boldsymbol{M}\right]\mathbf{I}:(\nabla\boldsymbol{v}) \\
&= \left[\varepsilon_0E^2 + \left(\boldsymbol{E} - \boldsymbol{E}^{(e)}\right)\cdot\boldsymbol{P} + \mu_0H^2 + \mu_0\left(\boldsymbol{H} - \boldsymbol{H}^{(e)}\right)\cdot\boldsymbol{M}\right](\nabla\cdot\boldsymbol{v})
\end{aligned} \tag{2.145}$$

Introducing Eqs. (2.140), (2.143) and (2.145) into Eq. (2.133) gives the final form of the entropy growth equation

$$\begin{aligned}
\rho T\frac{\mathrm{d}S_e^{(e)}}{dt} &= (\mathbf{T}_v - \boldsymbol{v}\boldsymbol{g}):\left[\mathbf{D}^{(v)} + \boldsymbol{\epsilon}\cdot(\boldsymbol{\Omega} - \boldsymbol{\omega}_{\mathrm{p}})\right] + \mathbf{C}:\nabla\boldsymbol{\omega}_{\mathrm{p}} + \boldsymbol{j}'\cdot\boldsymbol{E}' - \nabla\cdot\boldsymbol{q} + \rho R \\
&+ \left(\boldsymbol{E} - \boldsymbol{E}^{(e)}\right)\cdot\left[\frac{\mathrm{D}\boldsymbol{P}}{\mathrm{D}t} + \boldsymbol{P}(\nabla\cdot\boldsymbol{v}) - \boldsymbol{\omega}_{\mathrm{p}}\times\boldsymbol{P}\right] \\
&+ \mu_0\left(\boldsymbol{H} - \boldsymbol{H}^{(e)}\right)\cdot\left[\frac{\mathrm{D}\boldsymbol{M}}{\mathrm{D}t} + \boldsymbol{M}(\nabla\cdot\boldsymbol{v}) - \boldsymbol{\omega}_{\mathrm{p}}\times\boldsymbol{M}\right]
\end{aligned} \tag{2.146}$$

where use has been made of $p\mathbf{I} : \boldsymbol{\epsilon} = \mathbf{0}$.

2.7 CONSTITUTIVE RELATIONSHIPS

2.7.1 Clausius–Duhem Inequality

The second law of thermodynamics states that entropy increase dS_e within a system is equal to the sum of the entropy provided by the external environment $\mathrm{d}S_e^{(s)}$ and the entropy generation in the system $\mathrm{d}S_e^{(i)}$, that is

$$\mathrm{d}S_e = \mathrm{d}S_e^{(s)} + \mathrm{d}S_e^{(i)} \tag{2.147}$$

and the relation $\mathrm{d}S_e^{(i)} \geq 0$ is satisfied. This leads to

$$\mathrm{d}S_e - \mathrm{d}S_e^{(s)} \geq 0 \tag{2.148}$$

Applying this inequality to the continuum yields

$$\frac{\mathrm{d}}{\mathrm{d}t}\int_V \rho S_e \mathrm{d}V - \left(-\int_S \boldsymbol{n}\cdot\frac{\boldsymbol{q}}{T}\mathrm{d}S + \int_V \frac{\rho R}{T}\mathrm{d}V \right) \geq 0 \tag{2.149}$$

where $\frac{\boldsymbol{q}}{T}$ is local entropy flux. By applying Reynolds' transport theory and Gauss theorem, and considering the arbitrariness of the integral volume, one obtains

$$\rho T \frac{\mathrm{d}S_e}{\mathrm{d}t} + T\nabla\cdot\left(\frac{\boldsymbol{q}}{T}\right) - \rho R \geq 0 \tag{2.150}$$

where the second term at the left-hand side of Eq. (2.150) can be expanded as

$$\begin{aligned} T\nabla\cdot\left(\frac{\boldsymbol{q}}{T}\right) &= T\left[\frac{1}{T}\nabla\cdot\boldsymbol{q} + \boldsymbol{q}\cdot\nabla\left(\frac{1}{T}\right)\right] \\ &= T\left[\frac{1}{T}\nabla\cdot\boldsymbol{q} + \frac{1}{T}\boldsymbol{q}\cdot\nabla\left(\ln\frac{1}{T}\right)\right] \\ &= \nabla\cdot\boldsymbol{q} + \boldsymbol{q}\cdot\nabla\left(\ln\frac{1}{T}\right) \end{aligned} \tag{2.151}$$

Substituting the result from Eq. (2.151) into inequality Eq. (2.150) and utilizing entropy production rate Eq. (2.146), the Clausius–Duhem inequality for the system can be derived.

$$\boldsymbol{j}'\cdot\boldsymbol{E}' + \boldsymbol{q}\cdot\nabla\left(\ln\frac{1}{T}\right) + \Phi \geq 0 \tag{2.152}$$

where Φ is the dissipation function,

$$\Phi = (\mathbf{T}_v - \mathbf{v}\mathbf{g}) : \left[\mathbf{D}^{(v)} + \epsilon \cdot (\boldsymbol{\Omega} - \boldsymbol{\omega}_\mathrm{p})\right] + \mathbf{C} : \nabla \boldsymbol{\omega}_\mathrm{p}$$
$$+ \left(\boldsymbol{E} - \boldsymbol{E}^{(e)}\right) \cdot \left[\frac{\mathrm{D}\boldsymbol{P}}{\mathrm{D}t} + \boldsymbol{P}(\nabla \cdot \boldsymbol{v}) - \boldsymbol{\omega}_\mathrm{p} \times \boldsymbol{P}\right]$$
$$+ \mu_0 \left(\boldsymbol{H} - \boldsymbol{H}^{(e)}\right) \cdot \left[\frac{\mathrm{D}\boldsymbol{M}}{\mathrm{D}t} + \boldsymbol{M}(\nabla \cdot \boldsymbol{v}) - \boldsymbol{\omega}_\mathrm{p} \times \boldsymbol{M}\right] \geq 0 \tag{2.153}$$

It is assumed here that mechanical dissipation, heat flow dissipation, and current dissipation coexist within the system, but there is no coupling among them. Eq. (2.152) is valid under the condition that

$$\boldsymbol{q} \cdot \nabla \left(\ln \frac{1}{T}\right) \geq 0 \tag{2.154}$$

and

$$\boldsymbol{j}' \cdot \boldsymbol{E}' \geq 0 \tag{2.155}$$

Eqs. (2.154) and (2.155) are the prerequisites for the material's constitutive relationships not to violate thermodynamic principles.

2.7.2 Constitutive Equations of $\boldsymbol{q}$ and $\boldsymbol{j}'$

Assuming that the conducting ferrofluids are isotropic, and given that $\boldsymbol{q}$ is proportional to ∇T, and $\boldsymbol{j}'$ is proportional to $\boldsymbol{E}'$, to satisfy inequality Eqs. (2.154) and (2.155), one gets

$$\boldsymbol{q} = -\kappa_T \nabla T \tag{2.156}$$

$$\boldsymbol{j}' = \sigma_E \boldsymbol{E}' \tag{2.157}$$

where Eq. (2.156) corresponds to Fourier's law of thermal conductivity, Eq. (2.157) corresponds to Ohm's law, κ_T and σ_E are thermal and electrical conductivities, respectively.

Substituting Eqs. (2.62) and (2.106) for $\boldsymbol{E}'$ and $\boldsymbol{j}'$ in Eq. (2.157) yields

$$\boldsymbol{j}_0 = \sigma_E (\boldsymbol{E} + \boldsymbol{v} \times \boldsymbol{B}) + \rho_e \boldsymbol{v} \tag{2.158}$$

where the term $\sigma_E(\boldsymbol{v} \times \boldsymbol{B})$ represents the contribution of the Lorentz force to the conduction current, whereas the term $\rho_e \boldsymbol{v}$ denotes the contribution of the moving free charges.

2.7.3 Constitutive Equations for Stress Tensor and Couple Stress Tensor

Any second-order tensor $\mathbf{A}$ can be decomposed into a spherical tensor $\mathbf{A}^{(s)}$, a skew tensor $\mathbf{A}^{(a)}$, and an anti-symmetric tensor

$$\mathbf{A} = \mathbf{A}^{(s)} + \mathbf{A}^{(a)} + \frac{1}{3}\mathbf{I}\mathrm{tr}\mathbf{A} \tag{2.159}$$

The superscripts (s) and (a), respectively, represent the skew tensor and the anti-symmetric tensor, and tr denotes the trace of the tensor. Additionally, the skew tensor $\mathbf{A}^{(s)}$ can be calculated from the symmetric tensor $\mathbf{A}^{(sm)}$ as

$$\mathbf{A}^{(s)} = \mathbf{A}^{(sm)} - \frac{1}{3}\mathbf{I}\text{tr}\mathbf{A} = \frac{1}{2}\left(\mathbf{A} + \mathbf{A}^{T}\right) - \frac{1}{3}\mathbf{I}\text{tr}\mathbf{A} \tag{2.160}$$

The anti-symmetric tensor is calculated as

$$\mathbf{A}^{(a)} = \frac{1}{2}\left(\mathbf{A} - \mathbf{A}^{T}\right) \tag{2.161}$$

Thus, for any two second-order tensors $\mathbf{A}_1$ and $\mathbf{A}_2$, one has

$$\begin{aligned}
\mathbf{A}_1^{(s)} : \mathbf{A}_2^{(a)} &= \left[\frac{1}{2}\left(\mathbf{A}_1 + (\mathbf{A}_1)^{T}\right) - \frac{1}{3}\mathbf{I}\text{tr}\mathbf{A}_1\right] : \frac{1}{2}\left(\mathbf{A}_2 - (\mathbf{A}_2)^{T}\right) \\
&= \frac{1}{4}\left(\mathbf{A}_1 + (\mathbf{A}_1)^{T}\right) : \left(\mathbf{A}_2 - (\mathbf{A}_2)^{T}\right) - \frac{1}{6}\text{tr}\mathbf{A}_1\mathbf{I} : \left(\mathbf{A}_2 - (\mathbf{A}_2)^{T}\right) \\
&= \frac{1}{4}\left[\mathbf{A}_1 : \mathbf{A}_2 - \mathbf{A}_1 : (\mathbf{A}_2)^{T} + (\mathbf{A}_1)^{T} : \mathbf{A}_2 - (\mathbf{A}_1)^{T} : (\mathbf{A}_2)^{T}\right] \\
&\quad - \frac{1}{6}\text{tr}\mathbf{A}_1\left[\mathbf{I} : \mathbf{A}_2 - \mathbf{I} : (\mathbf{A}_2)^{T}\right] \\
&= \frac{1}{4}\begin{bmatrix} \mathrm{A}_{ij}^{1}\boldsymbol{e}_i\boldsymbol{e}_j : \mathrm{A}_{lm}^{2}\boldsymbol{e}_l\boldsymbol{e}_m - \mathrm{A}_{ij}^{1}\boldsymbol{e}_i\boldsymbol{e}_j : \mathrm{A}_{ml}^{2}\boldsymbol{e}_l\boldsymbol{e}_m \\ + \mathrm{A}_{ji}^{1}\boldsymbol{e}_i\boldsymbol{e}_j : \mathrm{A}_{lm}^{2}\boldsymbol{e}_l\boldsymbol{e}_m - \mathrm{A}_{ji}^{1}\boldsymbol{e}_i\boldsymbol{e}_j : \mathrm{A}_{ml}^{2}\boldsymbol{e}_l\boldsymbol{e}_m \end{bmatrix} \\
&\quad - \frac{1}{6}\text{tr}\mathbf{A}_1\left[\boldsymbol{e}_k\boldsymbol{e}_k : \mathrm{A}_{lm}^{2}\boldsymbol{e}_l\boldsymbol{e}_m - \boldsymbol{e}_k\boldsymbol{e}_k : \mathrm{A}_{ml}^{2}\boldsymbol{e}_l\boldsymbol{e}_m\right] \\
&= \frac{1}{4}\left[\mathrm{A}_{ij}^{1}\mathrm{A}_{ij}^{2} - \mathrm{A}_{ij}^{1}\mathrm{A}_{ji}^{2} + \mathrm{A}_{ji}^{1}\mathrm{A}_{ij}^{2} - \mathrm{A}_{ji}^{1}\mathrm{A}_{ji}^{2}\right] - \frac{1}{6}\text{tr}\mathbf{A}_1\left[\mathrm{A}_{ii}^{2} - \mathrm{A}_{ii}^{2}\right] = 0
\end{aligned} \tag{2.162}$$

Similarly, the following expressions exist:

$$\mathbf{A}_1^{(a)} : \mathbf{A}_2^{(s)} = 0 \tag{2.163}$$

$$\begin{aligned}
\mathbf{A}_1^{(s)} : \frac{1}{3}\mathbf{I}\text{tr}\mathbf{A}_2 &= \left[\frac{1}{2}\left(\mathbf{A}_1 + (\mathbf{A}_1)^{T}\right) - \frac{1}{3}\mathbf{I}\text{tr}\mathbf{A}_1\right] : \frac{1}{3}\mathbf{I}\text{tr}\mathbf{A}_2 \\
&= \frac{1}{6}\mathbf{A}_1 : \mathbf{I}\text{tr}\mathbf{A}_2 + \frac{1}{6}(\mathbf{A}_1)^{T} : \mathbf{I}\text{tr}\mathbf{A}_2 - \frac{1}{9}\text{tr}\mathbf{A}_1\text{tr}\mathbf{A}_2\mathbf{I} : \mathbf{I} \\
&= \frac{1}{3}\text{tr}\mathbf{A}_1\text{tr}\mathbf{A}_2 - 3 \times \frac{1}{9}\text{tr}\mathbf{A}_1\text{tr}\mathbf{A}_2 = 0
\end{aligned} \tag{2.164}$$

$$\mathbf{A}_2^{(s)} : \frac{1}{3}\mathbf{I}\text{tr}\mathbf{A}_1 = 0 \tag{2.165}$$

$$\mathbf{A}_1^{(a)} : \frac{1}{3}\mathbf{I}\mathrm{tr}\mathbf{A}_2 = \left[\frac{1}{2}\left(\mathbf{A}_2 - (\mathbf{A}_2)^{\mathrm{T}}\right)\right] : \frac{1}{3}\mathbf{I}\mathrm{tr}\mathbf{A}_2$$
$$= \frac{1}{6}\mathrm{tr}\mathbf{A}_2\left[\mathbf{A}_2 : \mathbf{I} - (\mathbf{A}_2)^{\mathrm{T}} : \mathbf{I}\right] = 0 \tag{2.166}$$

$$\mathbf{A}_2^{(a)} : \frac{1}{3}\mathbf{I}\mathrm{tr}\mathbf{A}_1 = 0 \tag{2.167}$$

Combining the results from Eqs. (2.159) to (2.167), an equality can be obtained

$$\mathbf{A}_1 : \mathbf{A}_2 = \frac{1}{9}\mathrm{tr}\mathbf{A}_1\mathrm{tr}\mathbf{A}_2 + \mathbf{A}_1^{(a)} : \mathbf{A}_2^{(a)} + \mathbf{A}_1^{(s)} : \mathbf{A}_2^{(s)} \tag{2.168}$$

Expanding the term $\mathbf{T}_v : [\mathbf{D}^{(v)} + \boldsymbol{\epsilon} \cdot (\boldsymbol{\Omega} - \boldsymbol{\omega}_{\mathrm{p}})]$ in Eq. (2.153) in accordance with Eq. (2.168), considering that the tensor $\mathbf{D}^{(v)}$ is symmetric, one obtains the following expressions:

$$\left(\mathbf{D}^{(v)}\right)^{(a)} = 0 \tag{2.169}$$

$$\left(\mathbf{D}^{(v)}\right)^{(s)} = \mathbf{D}^{(v)} - \frac{1}{3}\mathbf{I}\mathrm{tr}\mathbf{D}^{(v)} \tag{2.170}$$

For the tensor $\boldsymbol{\epsilon} \cdot (\boldsymbol{\Omega} - \boldsymbol{\omega}_{\mathrm{p}})$,

$$\boldsymbol{\epsilon}\cdot\left(\boldsymbol{\Omega} - \boldsymbol{\omega}_{\mathrm{p}}\right) = \epsilon_{ijk}\boldsymbol{e}_i\boldsymbol{e}_j\boldsymbol{e}_k\cdot\left(\Omega_l\boldsymbol{e}_l - \omega_m\boldsymbol{e}_m\right) = \epsilon_{ijk}\boldsymbol{e}_i\boldsymbol{e}_j\left(\Omega_k - \omega_k\right) \tag{2.171}$$

It is an anti-symmetric tensor, so

$$\left(\boldsymbol{\epsilon}\cdot\left(\boldsymbol{\Omega} - \boldsymbol{\omega}_{\mathrm{p}}\right)\right)^{(a)} = \boldsymbol{\epsilon}\cdot\left(\boldsymbol{\Omega} - \boldsymbol{\omega}_{\mathrm{p}}\right), \tag{2.172}$$

$$\left(\boldsymbol{\epsilon}\cdot\left(\boldsymbol{\Omega} - \boldsymbol{\omega}_{\mathrm{p}}\right)\right)^{(s)} = \mathbf{0}, \tag{2.173}$$

$$\mathrm{tr}\left(\boldsymbol{\epsilon}\cdot\left(\boldsymbol{\Omega} - \boldsymbol{\omega}_{\mathrm{p}}\right)\right) = 0 \tag{2.174}$$

Applying the results of Eq. (2.168) and Eqs. (2.169) to (2.174), one gets

$$\mathbf{T}_v : \left[\mathbf{D}^{(v)} + \boldsymbol{\epsilon}\cdot\left(\boldsymbol{\Omega} - \boldsymbol{\omega}_{\mathrm{p}}\right)\right] = \frac{1}{9}\mathrm{tr}\mathbf{T}_v\mathrm{tr}\mathbf{D}^{(v)} + \mathbf{T}_v^{(a)} : \left[\boldsymbol{\epsilon}\cdot\left(\boldsymbol{\Omega} - \boldsymbol{\omega}_{\mathrm{p}}\right)\right] + \mathbf{T}_v^{(s)} : \left(\mathbf{D}^{(v)} - \frac{1}{3}\mathbf{I}\mathrm{tr}\mathbf{D}^{(v)}\right)$$
$$= \frac{1}{9}\mathrm{tr}\mathbf{T}_v\mathrm{tr}\mathbf{D}^{(v)} + \mathbf{T}_v^{(a)} : \left[\boldsymbol{\epsilon}\cdot\left(\boldsymbol{\Omega} - \boldsymbol{\omega}_{\mathrm{p}}\right)\right] + \mathbf{T}_v^{(s)} : \mathbf{D}^{(v)} \tag{2.175}$$

To guarantee the validity of Eq. (2.153), it is imperative that each term in Eq. (2.175) remains non-negative. Such that the following equations can be presumed to hold.

$$\mathrm{tr}\mathbf{T}_{v} = \gamma'\mathrm{tr}\mathbf{D}^{(v)} \tag{2.176}$$

$$\mathbf{T}_{v}^{(\mathrm{a})} = 2\zeta\left[\epsilon\cdot\left(\boldsymbol{\Omega}-\boldsymbol{\omega}_{\mathrm{p}}\right)\right] \tag{2.177}$$

$$\mathbf{T}_{v}^{(\mathrm{s})} = 2\eta\mathbf{D}^{(v)} \tag{2.178}$$

where the coefficients γ', ζ, and η are all positive. In addition, the viscous stress tensor can be decomposed as

$$\mathbf{T}_{v} = \mathbf{T}_{v}^{(\mathrm{s})} + \mathbf{T}_{v}^{(\mathrm{a})} + \frac{1}{3}\mathbf{I}\mathrm{tr}\mathbf{T}_{v} \tag{2.179}$$

Introducing equations from Eqs. (2.176) to (2.178) into Eq. (2.179) gives

$$\begin{aligned}\mathbf{T}_{v} &= 2\eta\mathbf{D}^{(v)} + 2\zeta\left[\epsilon\cdot\left(\boldsymbol{\Omega}-\boldsymbol{\omega}_{\mathrm{p}}\right)\right] + \frac{\gamma'}{3}\mathbf{I}\mathrm{tr}\mathbf{D}^{(v)} \\ &= \eta\left[\nabla\boldsymbol{v}+\left(\nabla\boldsymbol{v}\right)^{\mathrm{T}}\right] + 2\zeta\epsilon\cdot\left(\boldsymbol{\Omega}-\boldsymbol{\omega}_{\mathrm{p}}\right) + \lambda\left(\nabla\cdot\boldsymbol{v}\right)\end{aligned} \tag{2.180}$$

where use has been made of

$$\lambda = \frac{\gamma'}{3} \tag{2.181}$$

and

$$\mathrm{tr}\mathbf{D}^{(v)} = \nabla\cdot\boldsymbol{v} \tag{2.182}$$

The coefficient η is the shear viscosity coefficient, ζ is the vortex viscosity, and λ is the bulk viscosity coefficient.

For the second term in Eq. (2.153), the expression for the surface couple stress tensor **C** can be derived by analogy with $\mathbf{T}_{v}$, albeit with the assumption that the anti-symmetric couple stress tensor is zero. As a result, one obtains (Odenbach, 2002)

$$\mathbf{C} = \eta'\left[\nabla\boldsymbol{\omega}_{\mathrm{p}} + \left(\nabla\boldsymbol{\omega}_{\mathrm{p}}\right)^{\mathrm{T}}\right] + \lambda'\left(\nabla\cdot\boldsymbol{\omega}_{\mathrm{p}}\right)\mathbf{I} \tag{2.183}$$

where η' and λ' refer to the shear spin viscosity coefficient and the bulk spin viscosity coefficient, respectively. In ferrohydrodynamics, the surface couple stress tensor represents the relationship between the diffusive transport of intrinsic angular momentum on the flow properties.

2.7.4 Expansion of the Equations of Motion

Substituting the stress tensor expression (2.130) into the equation of motion (2.23), and further incorporating the magnetic stress tensor expression (2.131), viscous stress tensor expressions (2.180), and electromagnetic energy density (2.75), one obtains

$$\nabla\cdot\mathbf{T}=\nabla\cdot\left(\mathbf{T}_v+\mathbf{T}_m-p\mathbf{I}-\boldsymbol{v}\boldsymbol{g}\right) \tag{2.184}$$

where

$$\begin{aligned}\nabla\cdot\mathbf{T}_v&=\nabla\cdot\left\{\eta\left[\nabla\boldsymbol{v}+\left(\nabla\boldsymbol{v}\right)^{\mathrm{T}}\right]+2\zeta\boldsymbol{\epsilon}\cdot\left(\boldsymbol{\Omega}-\boldsymbol{\omega}_{\mathrm{p}}\right)+\lambda\left(\nabla\cdot\boldsymbol{v}\right)\mathbf{I}\right\}\\&=\eta\nabla\cdot\left[\nabla\boldsymbol{v}+\left(\nabla\boldsymbol{v}\right)^{\mathrm{T}}\right]+2\zeta\nabla\cdot\left[\boldsymbol{\epsilon}\cdot\left(\boldsymbol{\Omega}-\boldsymbol{\omega}_{\mathrm{p}}\right)\right]+\lambda\nabla\cdot\left[\left(\nabla\cdot\boldsymbol{v}\right)\mathbf{I}\right]\\&=\eta\nabla^2\boldsymbol{v}+\eta\nabla\left(\nabla\cdot\boldsymbol{v}\right)-2\zeta\nabla\times\left(\boldsymbol{\Omega}-\boldsymbol{\omega}_{\mathrm{p}}\right)+\lambda\nabla\left(\nabla\cdot\boldsymbol{v}\right)\\&=\eta\nabla^2\boldsymbol{v}+\left(\eta+\lambda\right)\nabla\left(\nabla\cdot\boldsymbol{v}\right)-2\zeta\nabla\times\left(\boldsymbol{\Omega}-\boldsymbol{\omega}_{\mathrm{p}}\right)\end{aligned} \tag{2.185}$$

Here used has been made of

$$\nabla\cdot\left[\boldsymbol{\epsilon}\cdot\left(\boldsymbol{\Omega}-\boldsymbol{\omega}_{\mathrm{p}}\right)\right]=-\nabla\times\left(\boldsymbol{\Omega}-\boldsymbol{\omega}_{\mathrm{p}}\right) \tag{2.186}$$

and

$$\nabla\cdot\left[\left(\nabla\boldsymbol{v}\right)^{\mathrm{T}}\right]=\nabla\left(\nabla\cdot\boldsymbol{v}\right) \tag{2.187}$$

Eqs. (2.186) and (2.187) are proved as follows:

$$\nabla\cdot\left(\boldsymbol{\epsilon}\cdot\boldsymbol{A}\right)=\frac{\partial}{\partial x_m}\boldsymbol{e}_m\cdot\left(\epsilon_{ijk}\boldsymbol{e}_i\boldsymbol{e}_j\boldsymbol{e}_k\cdot A_l\boldsymbol{e}_l\right)=\frac{\partial}{\partial x_m}\boldsymbol{e}_m\cdot\left(\epsilon_{ijl}\boldsymbol{e}_i\boldsymbol{e}_jA_l\right)=\frac{\partial A_l}{\partial x_i}\epsilon_{ijl}\boldsymbol{e}_j$$

$$\nabla\times\boldsymbol{A}=\frac{\partial}{\partial x_m}\boldsymbol{e}_m\times A_l\boldsymbol{e}_l=\frac{\partial A_l}{\partial x_m}\epsilon_{mlk}\boldsymbol{e}_k=-\frac{\partial A_l}{\partial x_m}\epsilon_{mkl}\boldsymbol{e}_k$$

$$\nabla\cdot\left[\left(\nabla\boldsymbol{v}\right)^{\mathrm{T}}\right]=\frac{\partial}{\partial x_m}\boldsymbol{e}_m\cdot\left[\left(\frac{\partial}{\partial x_i}\boldsymbol{e}_iv_l\boldsymbol{e}_l\right)^{\mathrm{T}}\right]=\frac{\partial}{\partial x_m}\boldsymbol{e}_m\cdot\left(\frac{\partial v_i}{\partial x_l}\boldsymbol{e}_i\boldsymbol{e}_l\right)=\frac{\partial v_i}{\partial x_i\partial x_l}\boldsymbol{e}_l$$

$$\nabla\left(\nabla\cdot\boldsymbol{v}\right)=\frac{\partial}{\partial x_m}\boldsymbol{e}_m\left(\frac{\partial}{\partial x_i}\boldsymbol{e}_i\cdot v_l\boldsymbol{e}_l\right)=\frac{\partial}{\partial x_m}\boldsymbol{e}_m\left(\frac{\partial v_i}{\partial x_i}\right)=\frac{\partial v_i}{\partial x_m\partial x_i}\boldsymbol{e}_m$$

where

$$\boldsymbol{A}=\boldsymbol{\Omega}-\boldsymbol{\omega}_{\mathrm{p}} \tag{2.188}$$

QED.

The second term at the right-hand side of Eq. (2.185) can also be expanded according to the definition of Eq. (2.144) as

$$\begin{aligned} 2\zeta\nabla\times(\boldsymbol{\Omega}-\boldsymbol{\omega}_{\mathrm{p}}) &= \zeta\nabla\times\nabla\times\boldsymbol{v}-2\zeta\nabla\times\boldsymbol{\omega}_{\mathrm{p}} \\ &= \zeta\nabla(\nabla\cdot\boldsymbol{v})-\zeta\nabla^2\boldsymbol{v}-2\zeta\nabla\times\boldsymbol{\omega}_{\mathrm{p}} \end{aligned} \tag{2.189}$$

This leads to another form of Eq. (2.185):

$$\nabla\cdot\mathrm{T}_v = (\eta+\zeta)\nabla^2\boldsymbol{v}+(\eta+\lambda-\zeta)\nabla(\nabla\cdot\boldsymbol{v})+2\zeta\nabla\times\boldsymbol{\omega}_{\mathrm{p}} \tag{2.190}$$

Expanding the divergence of the magnetic stress tensor in Eq. (2.184), one has

$$\begin{aligned} \nabla\cdot\mathrm{T}_m &= \nabla\cdot\left[-\left(\frac{\varepsilon_0E^2}{2}+\frac{\mu_0H^2}{2}\right)\mathbf{I}+(\boldsymbol{DE}+\boldsymbol{BH})\right] \\ &= -\nabla\left(\frac{\varepsilon_0E^2}{2}+\frac{\mu_0H^2}{2}\right)+\nabla\cdot(\boldsymbol{DE})+\nabla\cdot(\boldsymbol{BH}) \\ &= -\frac{\varepsilon_0}{2}\nabla(\boldsymbol{E}\cdot\boldsymbol{E})-\frac{\mu_0}{2}\nabla(\boldsymbol{H}\cdot\boldsymbol{H})+\boldsymbol{D}\cdot\nabla\boldsymbol{E}+\boldsymbol{E}(\nabla\cdot\boldsymbol{D})+\boldsymbol{B}\cdot\nabla\boldsymbol{H}+\boldsymbol{H}(\nabla\cdot\boldsymbol{B}) \\ &= -\varepsilon_0\boldsymbol{E}\cdot\nabla\boldsymbol{E}-\varepsilon_0\boldsymbol{E}\times(\nabla\times\boldsymbol{E})-\mu_0\boldsymbol{H}\cdot\nabla\boldsymbol{H}-\mu_0\boldsymbol{H}\times(\nabla\times\boldsymbol{H})+\varepsilon_0\boldsymbol{E}\cdot\nabla\boldsymbol{E} \\ &\quad +\boldsymbol{P}\cdot\nabla\boldsymbol{E}+\boldsymbol{E}(\nabla\cdot\boldsymbol{D})+\mu_0\boldsymbol{H}\cdot\nabla\boldsymbol{H}+\mu_0\boldsymbol{M}\cdot\nabla\boldsymbol{H}+\boldsymbol{H}(\nabla\cdot\boldsymbol{B}) \\ &= -\varepsilon_0\boldsymbol{E}\times(\nabla\times\boldsymbol{E})-\mu_0\boldsymbol{H}\times(\nabla\times\boldsymbol{H})+\boldsymbol{P}\cdot\nabla\boldsymbol{E}+\rho_e\boldsymbol{E}+\mu_0\boldsymbol{M}\cdot\nabla\boldsymbol{H} \end{aligned} \tag{2.191}$$

where in obtaining the last and the penultimate equalities use was made of Eq. (2.82) and equality (2.108), respectively. According to Eq. (2.73), Eq. (2.191) can be further simplified as

$$\begin{aligned} \nabla\cdot\mathrm{T}_m &= -\varepsilon_0\boldsymbol{E}\times\left(-\frac{\partial\boldsymbol{B}}{\partial t}\right)-\mu_0\boldsymbol{H}\times\left(\frac{\partial\boldsymbol{D}}{\partial t}+\boldsymbol{j}_0\right)+\boldsymbol{P}\cdot\nabla\boldsymbol{E}+\rho_e\boldsymbol{E}+\mu_0\boldsymbol{M}\cdot\nabla\boldsymbol{H} \\ &= \varepsilon_0\boldsymbol{E}\times\left(\frac{\partial\boldsymbol{B}}{\partial t}\right)+\mu_0\left(\frac{\partial\boldsymbol{D}}{\partial t}\right)\times\boldsymbol{H}-\mu_0\boldsymbol{H}\times\boldsymbol{j}_0+\boldsymbol{P}\cdot\nabla\boldsymbol{E}+\rho_e\boldsymbol{E}+\mu_0\boldsymbol{M}\cdot\nabla\boldsymbol{H} \end{aligned} \tag{2.192}$$

The second term at the left-hand side of Eq. (2.23) is expanded in accordance with the definition of the electromagnetic field momentum density given in Eq. (2.75).

$$\begin{aligned} \frac{\partial\boldsymbol{g}}{\partial t} &= \frac{\partial}{\partial t}(\boldsymbol{D}\times\boldsymbol{B}) = \frac{\partial\boldsymbol{D}}{\partial t}\times\boldsymbol{B}+\boldsymbol{D}\times\frac{\partial\boldsymbol{B}}{\partial t} \\ &= \mu_0\frac{\partial\boldsymbol{D}}{\partial t}\times\boldsymbol{H}+\mu_0\frac{\partial\boldsymbol{D}}{\partial t}\times\boldsymbol{M}+\varepsilon_0\boldsymbol{E}\times\frac{\partial\boldsymbol{B}}{\partial t}+\boldsymbol{P}\times\frac{\partial\boldsymbol{B}}{\partial t} \\ &= \mu_0\frac{\partial\boldsymbol{D}}{\partial t}\times\boldsymbol{H}+\mu_0(\nabla\times\boldsymbol{H}-\boldsymbol{j}_0)\times\boldsymbol{M}+\varepsilon_0\boldsymbol{E}\times\frac{\partial\boldsymbol{B}}{\partial t}-\boldsymbol{P}\times(\nabla\times\boldsymbol{E}) \end{aligned} \tag{2.193}$$

Substitution of Eqs. (2.184), (2.185), (2.192), and (2.193) in Eq. (2.23) gives the motion equation

$$\rho\left(\frac{\partial \boldsymbol{v}}{\partial t}+(\boldsymbol{v}\cdot\nabla)\boldsymbol{v}\right)=-\nabla p+\eta\nabla^2\boldsymbol{v}+(\eta+\lambda)\nabla(\nabla\cdot\boldsymbol{v})-2\zeta\nabla\times(\boldsymbol{\Omega}-\boldsymbol{\omega}_\mathrm{p})$$
$$+\boldsymbol{j}_0\times\boldsymbol{B}+\boldsymbol{P}\cdot\nabla\boldsymbol{E}+\rho_e\boldsymbol{E}+\mu_0\boldsymbol{M}\cdot\nabla\boldsymbol{H}-\nabla\cdot(\boldsymbol{v}\boldsymbol{g})$$
$$+\mu_0\boldsymbol{M}\times(\nabla\times\boldsymbol{H})+\boldsymbol{P}\times(\nabla\times\boldsymbol{E})+\rho\boldsymbol{F} \tag{2.194}$$

Another form of the motion equation is obtained by introducing Eq. (2.190):

$$\rho\left(\frac{\partial \boldsymbol{v}}{\partial t}+(\boldsymbol{v}\cdot\nabla)\boldsymbol{v}\right)=-\nabla p+(\eta+\zeta)\nabla^2\boldsymbol{v}+(\eta+\lambda-\zeta)\nabla(\nabla\cdot\boldsymbol{v})+2\zeta\nabla\times\boldsymbol{\omega}_\mathrm{p}$$
$$+\boldsymbol{j}_0\times\boldsymbol{B}+\boldsymbol{P}\cdot\nabla\boldsymbol{E}+\rho_e\boldsymbol{E}+\mu_0\boldsymbol{M}\cdot\nabla\boldsymbol{H}-\nabla\cdot(\boldsymbol{v}\boldsymbol{g})$$
$$+\mu_0\boldsymbol{M}\times(\nabla\times\boldsymbol{H})+\boldsymbol{P}\times(\nabla\times\boldsymbol{E})+\rho\boldsymbol{F} \tag{2.195}$$

2.7.5 Expansion of the Angular Momentum Equation

An expression is obtained with the use of Eq. (2.183)

$$\nabla\cdot\mathbf{C}=\nabla\cdot\left\{\eta'\left[\nabla\boldsymbol{\omega}_\mathrm{p}+(\nabla\boldsymbol{\omega}_\mathrm{p})^\mathrm{T}\right]+\lambda'(\nabla\cdot\boldsymbol{\omega}_\mathrm{p})\mathbf{I}\right\}$$
$$=\eta'\nabla^2\boldsymbol{\omega}_\mathrm{p}+(\eta'+\lambda')\nabla(\nabla\cdot\boldsymbol{\omega}_\mathrm{p}) \tag{2.196}$$

where use has been made of

$$\nabla\cdot\left[(\nabla\boldsymbol{\omega}_\mathrm{p})^\mathrm{T}\right]=\nabla(\nabla\cdot\boldsymbol{\omega}_\mathrm{p}) \tag{2.197}$$

In terms of Eq. (2.180), one gets

$$\boldsymbol{\epsilon}:\mathrm{T}_v=\boldsymbol{\epsilon}:\left\{\eta\left[\nabla\boldsymbol{v}+(\nabla\boldsymbol{v})^\mathrm{T}\right]+2\zeta\boldsymbol{\epsilon}\cdot(\boldsymbol{\Omega}-\boldsymbol{\omega}_\mathrm{p})+\lambda(\nabla\cdot\boldsymbol{v})\mathbf{I}\right\}$$
$$=4\zeta(\boldsymbol{\Omega}-\boldsymbol{\omega}_\mathrm{p}) \tag{2.198}$$

where the following equalities are used:

$$\boldsymbol{\epsilon}:\left[\nabla\boldsymbol{v}+(\nabla\boldsymbol{v})^\mathrm{T}\right]=\mathbf{0} \tag{2.199}$$

$$\boldsymbol{\epsilon}:(\boldsymbol{\epsilon}\cdot\boldsymbol{A})=2\boldsymbol{A} \tag{2.200}$$

$$\boldsymbol{\epsilon}:\mathbf{I}=\mathbf{0} \tag{2.201}$$

The equalities (2.199) and (2.200) are proved according to tensor representations:

$$\nabla \boldsymbol{v} = \frac{\partial v_l}{\partial x_m} \boldsymbol{e}_m \boldsymbol{e}_l$$

$$(\nabla \boldsymbol{v})^{\mathrm{T}} = \frac{\partial v_m}{\partial x_l} \boldsymbol{e}_m \boldsymbol{e}_l$$

$$\begin{aligned}
\boldsymbol{\epsilon} : \left[\nabla \boldsymbol{v} + (\nabla \boldsymbol{v})^{\mathrm{T}} \right] &= \epsilon_{ijk} \boldsymbol{e}_i \boldsymbol{e}_j \boldsymbol{e}_k : \left(\frac{\partial v_l}{\partial x_m} \boldsymbol{e}_m \boldsymbol{e}_l + \frac{\partial v_m}{\partial x_l} \boldsymbol{e}_m \boldsymbol{e}_l \right) \\
&= \epsilon_{ijk} \boldsymbol{e}_i \left(\frac{\partial v_k}{\partial x_j} + \frac{\partial v_j}{\partial x_k} \right) \\
&= \epsilon_{ijk} \boldsymbol{e}_i \frac{\partial v_k}{\partial x_j} - \epsilon_{ikj} \boldsymbol{e}_i \frac{\partial v_j}{\partial x_k} = 0
\end{aligned}$$

$$\begin{aligned}
\boldsymbol{\epsilon} : (\boldsymbol{\epsilon} \cdot \boldsymbol{A}) &= \epsilon_{ijk} \boldsymbol{e}_i \boldsymbol{e}_j \boldsymbol{e}_k : (\epsilon_{lmn} \boldsymbol{e}_l \boldsymbol{e}_m \boldsymbol{e}_n \cdot A_p \boldsymbol{e}_p) \\
&= \epsilon_{ijk} \boldsymbol{e}_i \boldsymbol{e}_j \boldsymbol{e}_k : (\epsilon_{lmn} \boldsymbol{e}_l \boldsymbol{e}_m A_n) \\
&= \epsilon_{ijk} \epsilon_{jkn} A_n \boldsymbol{e}_i = \epsilon_{jki} \epsilon_{jkn} A_n \boldsymbol{e}_i \\
&= 2 \delta_{in} A_n \boldsymbol{e}_i = 2 A_i \boldsymbol{e}_i = 2 \boldsymbol{A}
\end{aligned}$$

QED.

In accord with Eq. (2.131), one has

$$\begin{aligned}
\boldsymbol{\epsilon} : \mathrm{T}_m &= \boldsymbol{\epsilon} : \left[-\left(\frac{\varepsilon_0 E^2}{2} + \frac{\mu_0 H^2}{2} \right) \mathbf{I} + (\boldsymbol{DE} + \boldsymbol{BH}) \right] \\
&= \boldsymbol{\epsilon} : (\boldsymbol{DE} + \boldsymbol{BH}) \\
&= \boldsymbol{D} \times \boldsymbol{E} + \boldsymbol{B} \times \boldsymbol{H} \\
&= \boldsymbol{P} \times \boldsymbol{E} + \mu_0 \boldsymbol{M} \times \boldsymbol{H}
\end{aligned} \tag{2.202}$$

where the constitutive relationship Eq. (2.108) is used in obtaining the last equality.

Now putting Eqs. (2.196), (2.197), and (2.202) into Eq. (2.56), yields the expansion for the angular momentum equation

$$\begin{aligned}
\rho I \frac{\mathrm{D} \boldsymbol{\omega}_{\mathrm{p}}}{\mathrm{D}t} &= \rho \boldsymbol{G} + \eta' \nabla^2 \boldsymbol{\omega}_p + (\eta' + \lambda') \nabla (\nabla \cdot \boldsymbol{\omega}_{\mathrm{p}}) + 4\zeta (\boldsymbol{\Omega} - \boldsymbol{\omega}_{\mathrm{p}}) \\
&\quad + \boldsymbol{P} \times \boldsymbol{E} + \mu_0 \boldsymbol{M} \times \boldsymbol{H} - \boldsymbol{\epsilon} : (\boldsymbol{v} \boldsymbol{g})
\end{aligned} \tag{2.203}$$

2.8 SIMPLIFIED FORMS OF THE EQUATION OF MOTION AND ANGULAR MOMENTUM EQUATION IN THE ABSENCE OF ELECTRIC FIELD

One simplified form of the equation of motion is obtained by setting the electric field related terms to zero in Eq. (2.195),

$$\rho\left(\frac{\partial \boldsymbol{v}}{\partial t}+(\boldsymbol{v}\cdot\nabla)\boldsymbol{v}\right)=-\nabla p+(\eta+\zeta)\nabla^2\boldsymbol{v}+(\eta+\lambda-\zeta)\nabla(\nabla\cdot\boldsymbol{v})$$
$$+2\zeta\nabla\times\boldsymbol{\omega}_{\mathrm{p}}+\mu_0\boldsymbol{M}\cdot\nabla\boldsymbol{H}+\mu_0\boldsymbol{M}\times(\nabla\times\boldsymbol{H})+\rho\boldsymbol{F} \tag{2.204}$$

If, furthermore, the ferrofluids are presumed to be incompressible, namely,

$$\nabla\cdot\boldsymbol{v}=0 \tag{2.205}$$

then Eq. (2.204) becomes

$$\rho\left(\frac{\partial \boldsymbol{v}}{\partial t}+(\boldsymbol{v}\cdot\nabla)\boldsymbol{v}\right)=-\nabla p+(\eta+\zeta)\nabla^2\boldsymbol{v}+2\zeta\nabla\times\boldsymbol{\omega}_{\mathrm{p}}+\mu_0\boldsymbol{M}\cdot\nabla\boldsymbol{H}+\mu_0\boldsymbol{M}\times(\nabla\times\boldsymbol{H})+\rho\boldsymbol{F} \tag{2.206}$$

If there is no current in the ferrofluids, one has

$$\nabla\times\boldsymbol{H}=\boldsymbol{0} \tag{2.207}$$

In this case, Eq. (2.204) is written as

$$\rho\left(\frac{\partial \boldsymbol{v}}{\partial t}+(\boldsymbol{v}\cdot\nabla)\boldsymbol{v}\right)=-\nabla p+(\eta+\zeta)\nabla^2\boldsymbol{v}+2\zeta\nabla\times\boldsymbol{\omega}_{\mathrm{p}}+\mu_0\boldsymbol{M}\cdot\nabla\boldsymbol{H}+\rho\boldsymbol{F} \tag{2.208}$$

Similarly, by setting the terms pertaining to the electric field in Eq. (2.203) to zero, one simplified form of the angular momentum equation can be expressed as

$$\rho I\frac{\mathrm{D}\boldsymbol{\omega}_{\mathrm{p}}}{\mathrm{D}t}=\rho\boldsymbol{G}+\eta'\nabla^2\boldsymbol{\omega}_{\mathrm{p}}+(\eta'+\lambda')\nabla(\nabla\cdot\boldsymbol{\omega}_{\mathrm{p}})+4\zeta(\boldsymbol{\Omega}-\boldsymbol{\omega}_{\mathrm{p}})+\mu_0\boldsymbol{M}\times\boldsymbol{H} \tag{2.209}$$

If the term $(\eta'+\lambda')\nabla(\nabla\cdot\boldsymbol{\omega}_{\mathrm{p}})$ is neglected, this equation becomes

$$\rho I\frac{\mathrm{D}\boldsymbol{\omega}_{\mathrm{p}}}{\mathrm{D}t}=\rho\boldsymbol{G}+\eta'\nabla^2\boldsymbol{\omega}_{\mathrm{p}}+4\zeta(\boldsymbol{\Omega}-\boldsymbol{\omega}_{\mathrm{p}})+\mu_0\boldsymbol{M}\times\boldsymbol{H} \tag{2.210}$$

In Eq. (2.210), the last term at the right-hand side represents the magnetic moment acting on a ferrofluid when the local magnetization deviates from the direction of the applied magnetic field. The anti-symmetric component of Cauchy stress, which signifies the momentum exchange between the intrinsic angular momentum and the macroscopic angular momentum within a ferrofluid, is represented by the third term. The second term signifies the diffusion of the intrinsic angular momentum among fluid cells, which is frequently disregarded due to the tiny value of η'.

2.9 OTHER SIMPLIFIED FORMS OF THE EQUATION OF MOTION AND ANGULAR MOMENTUM EQUATION

In Eq. (2.203), the terms $\boldsymbol{P} \times \boldsymbol{E}$ and $\boldsymbol{\epsilon} : (\boldsymbol{vg})$ are set to zero when the effect of electric polarization and the electric field is disregarded. Furthermore, if it is assumed that both the body couple term $\rho\boldsymbol{G}$ and the spin-diffusion term $(\eta' + \lambda')\, \nabla\, (\nabla \cdot \boldsymbol{\omega}_{\mathrm{p}})$ can be neglected, in terms of the density of moments of inertia for particles in per unit volume of ferrofluids, as expressed in Eq. (2.53), along with the relaxation time of particle rotation,

$$\tau_s = \frac{\rho_s \bar{d}_{\mathrm{p}}^2}{60\eta_0} \tag{2.211}$$

one obtains

$$\frac{I'}{\tau_s} = 6\eta_0\phi \tag{2.212}$$

where η_0 is the viscosity of ferrofluid carriers. Then, from the definition of vortex viscosity,

$$\zeta = 3\eta_0\phi / 2 \tag{2.213}$$

another simplified form of the angular momentum equation can be obtained

$$\rho\frac{\mathrm{D}\boldsymbol{s}}{\mathrm{D}t} = \eta'\nabla^2\boldsymbol{\omega}_{\mathrm{p}} + \frac{1}{\tau_s}\left(\boldsymbol{\Omega} - \boldsymbol{\omega}_{\mathrm{p}}\right) + \mu_0\boldsymbol{M} \times \boldsymbol{H} \tag{2.214}$$

Here the shear spin viscosity coefficient can generally be expressed as

$$\eta' = \frac{\bar{d}_{\mathrm{p}}^2}{6\tau_B} \tag{2.215}$$

where τ_B is the Brownian relaxation time of ferrofluids. It is evident that $\eta' \sim \eta\ell^2\phi^2$, where ℓ is the distance between particles in ferrofluids. The value of ℓ is comparable to the particle size, typically ranging in the order of 10^{-8}. Furthermore, the value of ϕ is approximately 0.01, resulting in a typically minute value for η'. Shliomis estimated the spin viscosity as (Shliomis, 2021)

$$\eta' = \frac{1}{10}\eta\phi\frac{\rho_s}{\rho}\bar{d}_{\mathrm{p}}^2 \tag{2.216}$$

As an illustrative example, consider a water-based ferrofluid with the following specifications: $\bar{d}_{\mathrm{p}} = 39\mathrm{nm}$, carrier viscosity $\eta_0 = 1.02 \times 10^{-3}$ Pa·s, particle volume fraction of

ϕ = 24%, particle nucleus volume fraction of 2%, saturation magnetization of 6.77 × 10^2 A/m, and spin viscosity of η' = 8.7 × 10^{-20} kg·m/s in the absence of a magnetic field. When a magnetic field with an intensity of 1.19 × 10^4 A/m is applied, the spin viscosity attains η' = 11.4 × 10^{-20} kg·m/s. It shows that the value of η' is extremely small, rendering the term containing it negligible in Eq. (2.214). Additionally, it is assumed that the inertial term $\rho I \frac{\mathrm{D}\boldsymbol{\omega}_\mathrm{p}}{\mathrm{D}t}$ can also be disregarded, ultimately leading to another simplified form of the angular momentum equation

$$4\zeta\left(\boldsymbol{\omega}_\mathrm{p} - \boldsymbol{\Omega}\right) = \mu_0 \boldsymbol{M} \times \boldsymbol{H} \tag{2.217}$$

For the motion equation of ferrofluids, represented as Eq. (2.194), assuming that the fluid is non-conductive, lacks any free charge, and exhibits no internal exothermic processes, the terms $\boldsymbol{j}_0 \times \boldsymbol{B}$, $\rho_e \boldsymbol{E}$, $\mu_0 \boldsymbol{M} \times (\nabla \times \boldsymbol{H})$, and $\boldsymbol{P} \times (\nabla \times \boldsymbol{E})$ will all evaluate to zero. In the absence of an electric field and neglecting polarization effects, the terms $\boldsymbol{P} \cdot \nabla \boldsymbol{E}$ and $\nabla \cdot (\boldsymbol{v}\boldsymbol{g})$ will also vanish. Given that the ferrofluid is incompressible, then Eq. (2.205) remains valid, leading to a simplified form of the equation of motion

$$\rho\left[\frac{\partial \boldsymbol{v}}{\partial t} + (\boldsymbol{v} \cdot \nabla)\boldsymbol{v}\right] = -\nabla p + \eta \nabla^2 \boldsymbol{v} - 2\zeta \nabla \times \left(\boldsymbol{\Omega} - \boldsymbol{\omega}_\mathrm{p}\right) + \mu_0 \boldsymbol{M} \cdot \nabla \boldsymbol{H} + \rho \boldsymbol{F} \tag{2.218}$$

In this equation, the fourth term at the right-hand side signifies the magnetic body force that arises due to the inhomogeneity of the magnetic field. The third term represents the anti-symmetric component of the Cauchy stress, which only emerges when the particles' rotational velocity differs from the local vorticity of ferrofluids.

In addition, another simplified form of the motion equation can be obtained by substituting Eq. (2.217) into (2.218)

$$\rho\left[\frac{\partial \boldsymbol{v}}{\partial t} + (\boldsymbol{v} \cdot \nabla)\boldsymbol{v}\right] = -\nabla p + \eta \nabla^2 \boldsymbol{v} + \rho \boldsymbol{F} + \mu_0 \boldsymbol{M} \cdot \nabla \boldsymbol{H} + \frac{\mu_0}{2} \nabla \times \left(\boldsymbol{M} \times \boldsymbol{H}\right) \tag{2.219}$$

This equation can also be obtained by making the divergence of the stress tensor of a ferrofluid. Motion equation (2.219) is the most commonly utilized formulation. The final term at its right-hand side represents the relaxation rate of the magnetic moment, which plays a pivotal role in numerous flow patterns of ferrofluids.

Drawing from expressions (2.211) to (2.213), a different formulation of the momentum equation can be derived from Eq. (2.218),

$$\rho\left[\frac{\partial \boldsymbol{v}}{\partial t} + (\boldsymbol{v} \cdot \nabla)\boldsymbol{v}\right] = -\nabla p + \eta \nabla^2 \boldsymbol{v} + \frac{1}{2\tau_s} \nabla \times \left(\boldsymbol{s} - I\boldsymbol{\Omega}\right) + \mu_0 \boldsymbol{M} \cdot \nabla \boldsymbol{H} + \rho \boldsymbol{F} \tag{2.220}$$

In Eq. (2.219), if the impact of the magnetic moment is disregarded, meaning that the magnetization of the ferrofluid is presumed to align with the magnetic field (known as the parallel magnetization assumption), then the final term at the right-hand side of this equation vanishes. Under these circumstances, Eq. (2.219) can be simplified to the Neuringer–Rosensweig model for ferrofluids (Rosensweig, 1985). On the other hand, even in a uniform magnetic field where the fourth term at the right-hand side of Eq. (2.219) disappears, if the vorticity of the ferrofluid flow is non-zero, the viscous moment drives the particles to rotate, leading to a deviation in the magnetization direction from the external magnetic field. Consequently, the last term at the right-hand side of Eq. (2.219) is non-zero, giving rise to a distinct flow state from that observed in the absence of a magnetic field.

2.10 STRESS TENSOR OF FERROFLUIDS

By substituting the viscous stress tensor Eq. (2.180), pressure expression Eq. (2.119), and magnetic stress tensor Eq. (2.131) into the ferrofluid stress tensor expression Eq. (2.130), respectively, one can obtain the most general form of the ferrofluid stress tensor

$$\mathbf{T}=\eta\left[\nabla\boldsymbol{v}+(\nabla\boldsymbol{v})^{\mathrm{T}}\right]+2\zeta\boldsymbol{\epsilon}\cdot(\boldsymbol{\Omega}-\boldsymbol{\omega}_{\mathrm{p}})+\lambda(\nabla\cdot\boldsymbol{v})\mathbf{I}-\left(\frac{\varepsilon_0E^2}{2}+\frac{\mu_0H^2}{2}\right)\mathbf{I}+(\boldsymbol{DE}+\boldsymbol{BH})$$
$$-\left[p_0(\rho,T)+\int_0^E\left(\frac{\partial(\upsilon P)}{\partial\upsilon}\right)_{T,H,E}^{(e)}\mathrm{d}E^{(e)}+\mu_0\int_0^H\left(\frac{\partial(\upsilon M)}{\partial\upsilon}\right)_{T,H,E}^{(e)}\mathrm{d}H^{(e)}\right]\mathbf{I}-\upsilon\boldsymbol{g} \tag{2.221}$$

For the non-conducting, incompressible ferrofluids, with no external electric field applied, an application of the simplified form of the angular momentum Eq. (2.217) leads to a simplified expression for the stress tensor

$$\mathbf{T}=-p\mathbf{I}+\eta\left[\nabla\boldsymbol{v}+(\nabla\boldsymbol{v})^{\mathrm{T}}\right]-\frac{\mu_0}{2}\boldsymbol{\epsilon}\cdot(\boldsymbol{M}\times\boldsymbol{H})-\frac{\mu_0H^2}{2}\mathbf{I}+\boldsymbol{BH} \tag{2.222}$$

where the magnetostrictive term is incorporated into p. The component form of Eq. (2.222) is

$$T_{ij}=-p\delta_{ij}+\eta\left(\frac{\partial v_i}{\partial x_j}+\frac{\partial v_j}{\partial x_i}\right)-\frac{\mu_0}{2}(M_iH_j-M_jH_i)-\frac{\mu_0H^2}{2}\delta_{ij}+B_iH_j \tag{2.223}$$

The expansion of the term $\boldsymbol{\epsilon}\cdot(\boldsymbol{M}\times\boldsymbol{H})$ is proved as follows:

$$\begin{aligned}\boldsymbol{\epsilon}\cdot(\boldsymbol{M}\times\boldsymbol{H})&=\epsilon_{ijk}\boldsymbol{e}_i\boldsymbol{e}_j\boldsymbol{e}_k\cdot(M_l\boldsymbol{e}_l\times H_m\boldsymbol{e}_m)=\epsilon_{ijk}\boldsymbol{e}_i\boldsymbol{e}_j\boldsymbol{e}_k\cdot(M_lH_m\epsilon_{lmn}\boldsymbol{e}_n)\\&=M_lH_m\epsilon_{ijk}\epsilon_{lmk}\boldsymbol{e}_i\boldsymbol{e}_j=M_lH_m\epsilon_{kij}\epsilon_{klm}\boldsymbol{e}_i\boldsymbol{e}_j=M_lH_m\boldsymbol{e}_i\boldsymbol{e}_j(\delta_{il}\delta_{jm}-\delta_{im}\delta_{jl})\\&=(M_iH_j-M_jH_i)\boldsymbol{e}_i\boldsymbol{e}_j\end{aligned}$$

QED.

Introducing $B_i=\mu_0(H_i+M_i)$ into Eq. (2.223) gives another component form of the stress tensor

$$T_{ij}=-p\delta_{ij}+\eta\left(\frac{\partial v_i}{\partial x_j}+\frac{\partial v_j}{\partial x_i}\right)+\frac{\mu_0}{2}(M_iH_j+M_jH_i)-\frac{\mu_0H^2}{2}\delta_{ij}+\mu_0H_iH_j \tag{2.224}$$

This representation of the stress tensor for ferrofluids has been validated through numerous experiments (Pshenichnikov and Lebedev, 2000). Within a flowing ferrofluid, the stress tensor is asymmetric due to the coupling between the vortices and particle rotation, which makes the third term at the right-hand side of Eq. (2.224) non-zero (Feng et al., 2006).

By utilizing the stress tensor of a ferrofluid, one can determine the force exerted on a solid wall adjacent to the boundary (with a normal unit vector of $\boldsymbol{n}$). Since $\boldsymbol{n} \cdot \mathbf{T}$ denotes the force per unit area, the stress exerted by the ferrofluid on the wall is denoted as $[\boldsymbol{n} \cdot \mathbf{T}]$, where the middle bracket indicates the difference between the two sides of the fluid–solid interface. Substituting this into Eq. (2.223) obtains the stress difference as

$$\boldsymbol{f} = \left[-p\delta_{nj} + \eta \left(\frac{\partial v_n}{\partial x_j} + \frac{\partial v_j}{\partial x_n} \right) - \frac{\mu_0}{2}\left(M_n H_j - M_j H_n\right) - \frac{\mu_0 H^2}{2}\delta_{nj} + B_n H_j \right] \tag{2.225}$$

Then, the tangential stress on the wall can be obtained

$$\begin{aligned} f_\tau &= \boldsymbol{f} \cdot \boldsymbol{\tau} \\ &= \left[-p\delta_{n\tau} + \eta \left(\frac{\partial v_n}{\partial x_\tau} + \frac{\partial v_\tau}{\partial x_n} \right) - \frac{\mu_0}{2}\left(M_n H_\tau - M_\tau H_n\right) - \frac{\mu_0 H^2}{2}\delta_{n\tau} + B_n H_\tau \right] \end{aligned} \tag{2.226}$$

where $\boldsymbol{\tau}$ is the tangential unit vector of the boundary. Putting the boundary conditions for the magnetic field

$$\left[H_\tau \right] = 0 \tag{2.227}$$

and

$$\left[B_n \right] = 0 \tag{2.228}$$

into Eq. (2.226), and applying the boundary conditions on both sides of the boundary – the fluid's normal velocity remains constant along the tangential direction, the tangential stress exerted on the wall is obtained,

$$\begin{aligned} f_\tau &= \boldsymbol{f} \cdot \boldsymbol{\tau} \\ &= \left[\eta \left(\frac{\partial v_n}{\partial x_\tau} + \frac{\partial v_\tau}{\partial x_n} \right) - \frac{\mu_0}{2}\left(M_n H_\tau - M_\tau H_n\right) \right] \\ &= \eta \frac{\partial v_\tau}{\partial x_n} + \frac{\mu_0}{2}\left(M_\tau H_n - M_n H_\tau\right) \end{aligned} \tag{2.229}$$

where a non-magnetic conductive wall is assumed. It can be observed from Eq. (2.229) that the tangential stress experienced by the wall includes not only the conventional viscous stress, which is represented by the first term on the right-hand side of Eq. (2.229) but also a magnetic stress, denoted by the second term on the right-hand side of Eq. (2.229). This magnetic stress arises due to the abrupt changes in magnetization across the ferrofluid interface.

REFERENCES

Felderhof B. U., Kroh H. J., Hydrodynamics of magnetic and dielectric fluids in interaction with the electromagnetic field, *The Journal of Chemical Physics*, 110: 7403, 1999.

Feng S., Graham A. L., Abbott J. R., Brenner H., Antisymmetric stresses in suspensions: Vortex viscosity and energy dissipation, *Journal of Fluid Mechanics*, 563: 97, 2006.

Jansons K. M., Determination of the constitutive equations for a magnetic fluid, *Journal of Fluid Mechanics*, 137: 187–216, 1983.

Landau L. D., Lifshitz E. M., *Electrodynamics of continuous media*. New York: Pergamon Press, 1960.

Liu M., Hydrodynamic theory of electromagnetic fields in continuous media. *Physical Review Letters*, 70(23): 3580–3583, 1993.

Liu M., Fluid dynamics of colloidal magnetic and electric liquid. *Physical Review Letters*, 70(22): 4535–4538, 1995.

Odenbach S., *Ferrofluids, magnetically controllable fluids and their applications*, Heidelberg: Spring, 2002.

Pshenichnikov A. F., Lebedev A. V., Tangential stresses on the magnetic fluid boundary and rotational effect, *Magnetohydrodynamics*, 36(4): 317–326, 2000.

Rosensweig R. E., *Ferrohydrodynamics*, Cambridge: Cambridge University Press, 1985.

Shliomis M. I., How a rotating magnetic field causes ferrofluid to rotate, *Physical Review fluids*, 6: 043701, 2021.

Xie C., Global strong solutions to the Shliomis system for ferrofluids in a bounded domain, *Mathematical Methods in the Applied Sciences*, 42(18): 6021–6028, 2019.

Yang W., A finite volume solver for ferrohydrodynamics coupled with microscopic magnetization dynamics, *Applied Mathematics and Computation*, 441: 127704, 2023.

CHAPTER 3

Introduction to Quasi-Equilibrium Ferrohydrodynamics

THE EARLY THEORY OF ferrohydrodynamics, alternatively referred to as quasi-equilibrium ferrohydrodynamics, rested upon the premise that the magnetization of a ferrofluid remains consistently aligned with the magnetic field. This theory relies on relatively simple governing equations and stress tensors, with the Kelvin force, Bernoulli equation, and magnetic levitation force serving as its foundations for various applications.

3.1 EQUATIONS AND STRESS TENSOR FOR QUASI-EQUILIBRIUM FERROHYDRODYNAMICS

In quasi-equilibrium ferrohydrodynamics, it is postulated that the ferrofluid is non-conductive and incompressible, the particles within this fluid are constrained from rotating freely, and no external electric field is present within the system. Given these premises, the continuity equation (2.205) and the magnetic field equation (2.207) remain valid. By substituting these equations into Eq. (2.194), the equation of motion for quasi-equilibrium ferrohydrodynamics is obtained

$$\rho\left(\frac{\partial \boldsymbol{v}}{\partial t}+(\boldsymbol{v}\cdot\nabla)\boldsymbol{v}\right)=-\nabla p+\eta\nabla^2\boldsymbol{v}+\mu_0\boldsymbol{M}\cdot\nabla\boldsymbol{H}+\rho\boldsymbol{g}_0 \tag{3.1}$$

where $\boldsymbol{g}_0$ is the acceleration of gravity.

The third term on the right-hand side of Eq. (3.1) is the Kelvin force,

$$\boldsymbol{F}_{\text{Kelvin}}=\mu_0\boldsymbol{M}\cdot\nabla\boldsymbol{H} \tag{3.2}$$

DOI: 10.1201/9781003540342-3

Since magnetization $\boldsymbol{M}$ is assumed to align with the local magnetic field, it can be expressed as

$$\boldsymbol{M} = \left(\frac{M}{H}\right)\boldsymbol{H} \tag{3.3}$$

Therefore, the Kelvin force can also be expressed as

$$\boldsymbol{F}_{\text{Kelvin}} = \mu_0 \frac{M}{H} \boldsymbol{H} \cdot \nabla \boldsymbol{H} \tag{3.4}$$

Applying the vector equality

$$\boldsymbol{H} \cdot \nabla \boldsymbol{H} = \frac{1}{2} \nabla H^2 - \boldsymbol{H} \times \left(\nabla \times \boldsymbol{H}\right)$$

and Eq. (2.207) gives the expression of the Kelvin force (Odenbach and Liu, 2001)

$$\boldsymbol{F}_{\text{Kelvin}} = \mu_0 M \nabla H \tag{3.5}$$

This leads to another form of the equation of motion

$$\rho\left(\frac{\partial \boldsymbol{v}}{\partial t} + \left(\boldsymbol{v} \cdot \nabla\right)\boldsymbol{v}\right) = -\nabla p + \eta \nabla^2 \boldsymbol{v} + \mu_0 M \nabla H + \rho \boldsymbol{g}_0 \tag{3.6}$$

The magnetization in Eqs. (3.1) and (3.6) is identical to equilibrium magnetization M_0, and its relationship with the magnetic field intensity is expressed by the Langevin function

$$M_0 = M_S L\left(\alpha\right) = m_p N L\left(\alpha\right) \tag{3.7}$$

where M_S is the saturation magnetization of ferrofluids, m_p is the magnitude of the magnetic moment of individual particles, N is the number concentration of magnetic particles in the ferrofluid, and $L(\alpha)$ is the Langevin function with the parameter α

$$\alpha = \frac{\mu_0 m_p H}{k_B T} \tag{3.8}$$

Here, k_B is the Boltzmann constant. In quasi-equilibrium ferrohydrodynamics, it is evident that the flow of ferrofluid is influenced by magnetization, whereas the flow velocity has no impact on magnetization.

Substituting Eq. (2.119) for p into Eq. (3.6) and disregarding the polarization of the ferrofluid, one obtains the third form of the quasi-equilibrium ferrohydrodynamic equation

$$\rho\left(\frac{\partial \boldsymbol{v}}{\partial t} + \left(\boldsymbol{v} \cdot \nabla\right)\boldsymbol{v}\right) = -\nabla p_0 + \eta \nabla^2 \boldsymbol{v} + \rho \boldsymbol{g}_0 - \nabla\left[\mu_0 \int_0^H \left(\frac{\partial\left(\upsilon M\right)}{\partial \upsilon}\right)_{T,H} \mathrm{d}H\right] + \mu_0 M \nabla H \tag{3.9}$$

The fourth term at the right-hand side of Eq. (3.9) represents the magnetostrictive force (Rosensweig, 1985), while p_0 represents the ordinary thermodynamic pressure.

The last two terms on the right-hand side of Eq. (3.9) can be regarded as the magnetic body force density $\boldsymbol{f}_m$ under the quasi-equilibrium condition, that is

$$f_m = -\nabla\left[\mu_0\int_0^H\left(\frac{\partial(\upsilon M)}{\partial\upsilon}\right)_{T,H}\mathrm{d}H\right] + \mu_0 M\nabla H \tag{3.10}$$

Equation (3.10) can alternatively be derived from the divergence of the equivalent magnetic stress tensor $\mathbf{T}_m'$ of ferrofluids under quasi-equilibrium conditions (Shliomis et al., 1988), namely,

$$f_m = \nabla\cdot\mathbf{T}_m' \tag{3.11}$$

where

$$\mathbf{T}_m' = -\left[\mu_0\int_0^H\left(\frac{\partial(\upsilon M)}{\partial\upsilon}\right)_{T,H}\mathrm{d}H + \frac{1}{2}\mu_0 H^2\right]\mathbf{I} + \boldsymbol{BH} \tag{3.12}$$

3.2 BERNOULLI EQUATION FOR FERROFLUIDS

Based on Eq. (3.9), it is further assumed that the temperature within ferrofluids remains constant or their magnetization is temperature-independent, indicating that the temperature is significantly below the Curie temperature. Under these conditions, one has

$$\left(\frac{\partial M}{\partial T}\right)_{H,\upsilon} = 0 \tag{3.13}$$

For steady flows, if the flow is irrotational,

$$\nabla\times\boldsymbol{v} = \mathbf{0} \tag{3.14}$$

the Bernoulli equation for ferrofluids can be derived by integrating the equations of motion (Rosensweig, 1985; Cowley and Rosensweig, 1967)

$$p_0 + \mu_0\int_0^H\left(\frac{\partial(\upsilon M)}{\partial\upsilon}\right)_{T,H}\mathrm{d}H + \frac{1}{2}\rho v^2 + \rho g_0 h_g - \mu_0\bar{M}H = 0 \tag{3.15}$$

where h_g is the height difference between the point of interest and the reference point in the direction opposite to gravity and $\bar{M}$ is the field-averaged magnetization denoted as

$$\bar{M} = \frac{1}{H}\int_0^H M\mathrm{d}H \tag{3.16}$$

The ferrofluid Bernoulli equation has numerous applications in steady-state flows where viscous drag is disregarded, encompassing diverse scenarios such as the contour of the meniscus in a stationary ferrofluid, ferrofluid jets in magnetic fields, seals, bearings, static supports, and sink-float separation devices. Notably, in ferrofluid seals, the pressure resistance of the seal can be estimated by utilizing the ferrofluid Bernoulli equation when the shaft's rotational speed is relatively low.

3.3 THE KELVIN FORCE IN QUASI-EQUILIBRIUM FERROFLUIDS

By substituting Eqs. (3.7) and (3.8) into Eq. (3.5), the Kelvin force can be expressed under constant temperature conditions as

$$\boldsymbol{F}_{\text{Kelvin}} = Nk_{\text{B}}TL(\alpha)\nabla\alpha \tag{3.17}$$

With the aid of the Langevin function,

$$L(\alpha) = \frac{\mathrm{d}}{\mathrm{d}\alpha}\ln\frac{\sinh\alpha}{\alpha} \tag{3.18}$$

the Kelvin force can be further expressed as

$$\boldsymbol{F}_{\text{Kelvin}} = Nk_{\text{B}}T\nabla\left(\ln\frac{\sinh\alpha}{\alpha}\right) \tag{3.19}$$

If the magnetic particles are uniformly distributed within ferrofluids, the concentration of particles N remains constant, and the Kelvin force becomes

$$\boldsymbol{F}_{\text{Kelvin}} = \nabla\left(Nk_{\text{B}}T\ln\frac{\sinh\alpha}{\alpha}\right) \tag{3.20}$$

It is evident that the Kelvin force is a potential force.

From Eq. (3.20), the average force per particle in ferrofluids is given as

$$\frac{\boldsymbol{F}_{\text{Kelvin}}}{N} = -\nabla U \tag{3.21}$$

where U is the energy potential denoted as

$$U = -k_{\text{B}}T\ln\frac{\sinh\alpha}{\alpha} \tag{3.22}$$

It is assumed that the particles within a ferrofluid at equilibrium adhere to a Boltzmann distribution, the number concentration of particles can be expressed as (Odenbach, 2002)

$$N = N_0\exp\left(-\frac{U}{k_{\text{B}}T}\right) \tag{3.23}$$

Substituting Eq. (3.22) into Eq. (3.23) yields the expression for particle number concentration

$$N = N_0 \frac{\sinh \alpha}{\alpha} \tag{3.24}$$

Substitution of Eq. (3.24) into Eq. (3.20) gives

$$\begin{aligned} \boldsymbol{F}_{\text{Kelvin}} &= N_0 k_{\text{B}} T \frac{\sinh \alpha}{\alpha} \nabla \left(\ln \frac{\sinh \alpha}{\alpha} \right) \\ &= N_0 k_{\text{B}} T \nabla \left(\frac{\sinh \alpha}{\alpha} \right) \\ &= \nabla \left(N k_{\text{B}} T \right) \end{aligned} \tag{3.25}$$

Substituting

$$p_m = N k_{\text{B}} T \tag{3.26}$$

in Eq. (3.25) indeed gives

$$\boldsymbol{F}_{\text{Kelvin}} = \nabla p_m \tag{3.27}$$

Equation (3.27) demonstrates that the Kelvin force can be represented as the gradient of a specific magnetic pressure.

The Kelvin force can be regarded as a simplified form of the magnetic body force in ferrofluids, although its validity conditions remain a topic of debate. Liu (2000) argued that the Kelvin force is derived from the Helmholtz force and is only effective when the ferrofluid's susceptibility is directly proportional to its density. Aquino et al. (2003) suggested that the Helmholtz force is suitable for linear ferrofluids. Liu et al. (2008) discussed the applicable conditions of these two forces in ferrofluids and noted that when the susceptibility of a ferrofluid is directly proportional to its density, the Helmholtz force can be transformed into the Kelvin force.

3.4 MAGNETIC LEVITATION FORCES IN FERROFLUIDS

Another important application of quasi-equilibrium ferrohydrodynamics is the stable levitation properties of magnetic or non-magnetic objects within a ferrofluid (Rosensweig, 1966a, 1966b). The magnetic levitation force exerted on a levitating body can be derived from the expression of the equivalent magnetic stress tensor of a ferrofluid $\mathbf{T}_m'$ under quasi-equilibrium conditions and ferrohydrodynamic Bernoulli Eq. (3.15) (Rosensweig, 1985)

$$\boldsymbol{F}_m = \oint_{S_m} \left[\left(H_n B_n - \int_0^B B \mathrm{d}H \right) \boldsymbol{n} + H_\tau B_n \boldsymbol{\tau} \right] \mathrm{d}S_m \tag{3.28}$$

The surface integral is calculated over the suspension's surface S_m, with the normal and tangential unit vectors designated as $\boldsymbol{n}$ and $\boldsymbol{\tau}$, respectively.

Based on Eq. (3.28), researchers have studied the magnetic levitation forces acting on various shapes of permanent magnets in ferrofluids, such as the force variations on a spherical permanent magnet placed at the centre of a ferrofluid under small perturbations (Kvitantsev et al., 2002; Ivanov and Pshenichnikov, 2023), the trajectories of a spherical permanent magnet moving within a spherical container filled with a ferrofluid under the assumption of no induction (Naletova et al., 2003), and the influence of various parameters on the magnetic levitation force in such scenarios (Liu et al., 2009). In this section, drawing from the methods of Blums (1997) and Rosensweig (1977), the radial and axial magnetic levitation forces acting on cylindrical permanent magnets and composite magnet structures in ferrofluids with cylindrical boundaries are derived.

3.4.1 Radial Magnetic Levitation Force on a Cylindrical Permanent Magnet in a Ferrofluid

3.4.1.1 Equations Satisfied by a Scalar Magnetic Potential in a Bipolar Coordinate System

A permanent magnet is presumed to be magnetized along its axis and submerged within a ferrofluid enclosed by a cylindrical, non-magnetic shell. Given that the length of the permanent magnet significantly exceeds its diameter, Figure 3.1 illustrates its cross-section, where the inner circle demarcates the perimeter of the magnet and the outer circle outlines the inner boundary of the shell. Utilizing the plane bipolar coordinate system, one can establish a relationship between this system and the orthogonal coordinate system

$$x = a\frac{\operatorname{sh}\gamma}{\operatorname{ch}\gamma - \cos\theta} \tag{3.29}$$

$$y = a\frac{\sin\theta}{\operatorname{ch}\gamma - \cos\theta} \tag{3.30}$$

where a is the coordinate of the foci located on the right side of the y-axis in the bipolar coordinate system, θ represents the angle in radians between the line segments connecting

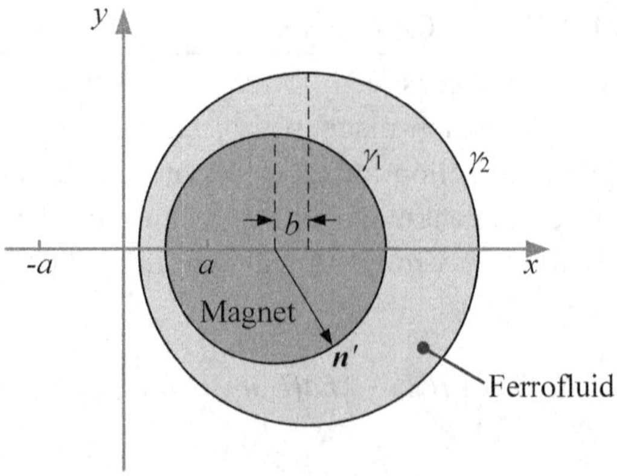

FIGURE 3.1 The model for calculating radial magnetic levitation forces in the bipolar coordinate.

any point $P(x, y)$ in the plane to the two foci, and γ denotes the logarithmic value of the ratio between the lengths of the line segments from P to the two foci. Since $\mathrm{ch}\gamma \geq 1$, the point P lies on the y-axis when $\mathrm{ch}\gamma = 1$. Because θ does not adopt the values of 0 or 2π for points situated within the finite plane, it follows $\cos\theta < 1$, which subsequently implies that $\mathrm{ch}\gamma \neq \cos\theta$.

From Eq. (3.29), one has

$$\cos\theta = \mathrm{ch}\gamma - \frac{a}{x}\mathrm{sh}\gamma \tag{3.31}$$

Substituting Eq. (3.31) into Eq. (3.30) gives

$$\sin\theta = \frac{y}{x}\mathrm{sh}\gamma \tag{3.32}$$

Putting Eqs. (3.31) and (3.32) into the equality

$$\sin^2\theta + \cos^2\theta = 1$$

one gets

$$y^2\mathrm{sh}^2\gamma + \left(x\mathrm{sh}\gamma - a\mathrm{ch}\gamma\right)^2 = a^2 \tag{3.33}$$

which leads to isolines for different values of γ

$$\left(x - a\coth\gamma\right)^2 + y^2 = \frac{a^2}{\mathrm{sh}^2\gamma} \tag{3.34}$$

Evidently, the isolines of γ are a set of non-intersecting circles with varying radii centred on the x-axis, surrounding a focal point. As γ increases, the radii of these circles decrease. Therefore, it is reasonable to assume that the contour lines corresponding to $\gamma = \gamma_1$ and $\gamma = \gamma_2$ align with the outer circumference of the permanent magnet and the inner circumference of the shell, respectively, with a distance of b between their axes.

Assuming that the outer radius of the permanent magnet and the inner radius of the shell are R_1 and R_2, respectively, it follows from the equation of the isolines of γ that $R_1 = a/\mathrm{sh}\gamma_1$ and $R_2 = a/\mathrm{sh}\gamma_2$, which leads to

$$a = R_1\mathrm{sh}\gamma_1 = R_2\mathrm{sh}\gamma_2 \tag{3.35}$$

From this equation, one can get

$$R_1^2\mathrm{ch}^2\gamma_1 = a^2 + R_1^2 \tag{3.36}$$

$$R_2^2\mathrm{ch}^2\gamma_2 = a^2 + R_2^2 \tag{3.37}$$

Based on the isoline equation (Eq. (3.34)) for γ and the spacing b between the permanent magnet and the outer shell axis, one obtains

$$a\coth\gamma_2 - a\coth\gamma_1 = b$$

It can be converted into

$$a\frac{\mathrm{ch}\gamma_2}{\mathrm{sh}\gamma_2} - a\frac{\mathrm{ch}\gamma_1}{\mathrm{sh}\gamma_1} = b \tag{3.38}$$

Substituting Eq. (3.35) into Eq. (3.38) yields

$$R_2\mathrm{ch}\gamma_2 - R_1\mathrm{ch}\gamma_1 = b \tag{3.39}$$

After rearranging the terms in Eq. (3.39) and squaring both sides, the substitution of Eq. (3.36) and Eq. (3.37) into the resultant equation yields

$$\mathrm{ch}\gamma_2 = \frac{R_2^2 - R_1^2 + b^2}{2R_2 b} \tag{3.40}$$

Introducing Eq. (3.40) into Eq. (3.39) yields the following equation:

$$\mathrm{ch}\gamma_1 = \frac{R_2^2 - R_1^2 - b^2}{2R_1 b} \tag{3.41}$$

Based on Figure 3.1, it can be observed that $R_2 > R_1 + b$, which implies that $R_2^2 - R_1^2 - b^2 > 2R_1 b$. Consequently, according to Eq. (3.41), one has $\mathrm{ch}\gamma_1 > 1$.

The coordinate plane depicted in Figure 3.1 divides the space into three distinct regions: The permanent magnet region (denoted by 1), the ferrofluid region (denoted by 2), and the non-magnetic medium region outside the shell (denoted by 3). Within these regions, as there is no current present and displacement current is neglected, the magnetic field intensity satisfies Eq. (2.207) according to Ampere's law. This relationship allows one to derive the scalar magnetic potential in relation to the magnetic field intensity

$$\boldsymbol{H} = -\nabla\varphi_m \tag{3.42}$$

The volume density of magnetic charges in magnetically permeable media can be represented as $-\mu_0 \nabla \cdot \boldsymbol{M}$, while for permanent magnets, magnetic charges are distributed exclusively on their surfaces, thus resulting in

$$\nabla\cdot\boldsymbol{M} = 0 \tag{3.43}$$

Thus, from Eqs. (2.82) and (2.108), one can deduce

$$\nabla\cdot\boldsymbol{H} = 0 \tag{3.44}$$

Assuming a constant permeability μ for the ferrofluid, one can derive Eq. (3.44) from $\boldsymbol{B} = \mu \boldsymbol{H}$ and Eq. (2.82). Consequently, by combining Eqs. (3.42) and (3.44), it follows that the scalar magnetic potential satisfies the Laplace equation in all three regions. Letting φ_m represent the scalar magnetic potential for these regions, one gets

$$\nabla^2 \varphi_m = 0 \tag{3.45}$$

Based on the definition of Lamé coefficients, the Lamé coefficients in a bipolar coordinate system are calculated as follows:

$$h_\theta = \sqrt{\left(\frac{\partial x}{\partial \theta}\right)^2 + \left(\frac{\partial y}{\partial \theta}\right)^2} \tag{3.46}$$

$$h_\gamma = \sqrt{\left(\frac{\partial x}{\partial \gamma}\right)^2 + \left(\frac{\partial y}{\partial \gamma}\right)^2} \tag{3.47}$$

From Eqs. (3.29) and (3.30), one gets

$$\frac{\partial x}{\partial \theta} = \frac{-a\mathrm{sh}\gamma \sin\theta}{(\mathrm{ch}\gamma - \cos\theta)^2}, \frac{\partial x}{\partial \gamma} = \frac{a(1 - \mathrm{ch}\gamma\cos\theta)}{(\mathrm{ch}\gamma - \cos\theta)^2}, \frac{\partial y}{\partial \theta} = \frac{a(\mathrm{ch}\gamma\cos\theta - 1)}{(\mathrm{ch}\gamma - \cos\theta)^2}, \frac{\partial y}{\partial \gamma} = \frac{-a\mathrm{sh}\gamma \sin\theta}{(\mathrm{ch}\gamma - \cos\theta)^2} \tag{3.48}$$

Introducing them into Eqs. (3.46) and (3.47) yields

$$h_\theta = \frac{a}{\mathrm{ch}\gamma - \cos\theta} \tag{3.49}$$

$$h_\gamma = \frac{a}{\mathrm{ch}\gamma - \cos\theta} \tag{3.50}$$

From the above equations, one obtains the Laplace equation satisfied by φ_m in the bipolar coordinate system

$$\begin{aligned} \nabla^2 \varphi_m &= \frac{a}{h_\theta h_\gamma}\left[\frac{\partial}{\partial \theta}\left(\frac{h_\gamma}{h_\theta}\frac{\partial \varphi_m}{\partial \theta}\right) + \frac{\partial}{\partial \gamma}\left(\frac{h_\theta}{h_\gamma}\frac{\partial \varphi_m}{\partial \gamma}\right)\right] \\ &= \left(\frac{\mathrm{ch}\gamma - \cos\theta}{a}\right)^2 \left(\frac{\partial^2 \varphi_m}{\partial \theta^2} + \frac{\partial^2 \varphi_m}{\partial \gamma^2}\right) = 0 \end{aligned}$$

As $\mathrm{ch}\gamma \neq \cos\theta$, this equation becomes

$$\frac{\partial^2 \varphi_m}{\partial \theta^2} + \frac{\partial^2 \varphi_m}{\partial \gamma^2} = 0 \tag{3.51}$$

3.4.1.2 Boundary Conditions in the Bipolar Coordinate System

The boundary conditions corresponding to Eq. (3.51) within the solution domains are as follows:

$$\varphi_{m,1} \text{ maintains a finite value, as } \gamma \to +\infty \tag{3.52}$$

$$\varphi_{m,3} \text{ maintains a finite value, as } \gamma \to -\infty \tag{3.53}$$

$$\varphi_{m,1} = \varphi_{m,2}, \text{ when } \gamma = \gamma_1 \tag{3.54}$$

$$\varphi_{m,2} = \varphi_{m,3}, \text{ when } \gamma = \gamma_2 \tag{3.55}$$

$$B_{n,1} = B_{n,2}, \text{ when } \gamma = \gamma_1 \tag{3.56}$$

$$B_{n,2} = B_{n,3}, \text{ when } \gamma = \gamma_2 \tag{3.57}$$

Meanwhile, in the regions being considered, one has

$$\boldsymbol{B}_1 = \mu_0\left(\boldsymbol{H}_1 + \boldsymbol{M}\right), \boldsymbol{B}_1 = \mu \boldsymbol{H}_2, \boldsymbol{B}_3 = \mu_0 \boldsymbol{H}_3 \tag{3.58}$$

Based on the relationships between the unit vectors in the bipolar coordinate system and the Cartesian coordinate system, the unit vectors along the coordinate directions in the bipolar coordinate system can be represented as

$$\boldsymbol{e}_\gamma = \frac{1}{h_\gamma}\frac{\partial x}{\partial \gamma}\boldsymbol{e}_x + \frac{1}{h_\gamma}\frac{\partial y}{\partial \gamma}\boldsymbol{e}_y, \quad \boldsymbol{e}_\theta = \frac{1}{h_\theta}\frac{\partial x}{\partial \theta}\boldsymbol{e}_x + \frac{1}{h_\theta}\frac{\partial y}{\partial \theta}\boldsymbol{e}_y \tag{3.59}$$

Substitution of the results from Eqs. (3.48), (3.49), and (3.50) into Eq. (3.59) yields

$$\boldsymbol{e}_\gamma = \frac{1}{\operatorname{ch}\gamma - \cos\theta}\left[\left(1 - \operatorname{ch}\gamma\cos\theta\right)\boldsymbol{e}_x - \operatorname{sh}\gamma\sin\theta\boldsymbol{e}_y\right] \tag{3.60}$$

$$\boldsymbol{e}_\theta = \frac{1}{\operatorname{ch}\gamma - \cos\theta}\left[-\operatorname{sh}\gamma\sin\theta\boldsymbol{e}_x + \left(\operatorname{ch}\gamma\cos\theta - 1\right)\boldsymbol{e}_y\right] \tag{3.61}$$

According to the equation of the isoline of γ (Eq. 3.34), it is known that the unit vector along the direction of the outer normal at a point (x, y) on the contour line of γ is given by

$$\boldsymbol{n}' = \left(x - a\coth\gamma\right)\boldsymbol{e}_x + y\boldsymbol{e}_y \tag{3.62}$$

Substituting Eqs. (3.29) and (3.30) into (3.62) and simplifying yields

$$\boldsymbol{n}' = -\frac{a}{\operatorname{sh}\gamma}\boldsymbol{e}_\gamma \tag{3.63}$$

The term shγ in Eq. (3.63) is a constant for the γ-isoline corresponding to a specific γ value. Notably, $\boldsymbol{n}'$ and $\boldsymbol{e}_\gamma$ are collinear but have opposite directions, indicating that the unit vectors at the outer normal directions of the γ_1 and γ_2 isolines are collinear but opposite to the unit vector at the γ direction. Therefore, on the boundaries represented by these two isolines, the relationship between the magnetic field intensity in the two coordinate systems is given by

$$H_n = -H_\gamma \tag{3.64}$$

While in the bipolar coordinate system, one has

$$\boldsymbol{H} = -\nabla \varphi_m = -\boldsymbol{e}_\theta \frac{1}{h_\theta}\frac{\partial \varphi_m}{\partial \theta} - \boldsymbol{e}_\gamma \frac{1}{h_\gamma}\frac{\partial \varphi_m}{\partial \gamma} \tag{3.65}$$

namely,

$$H_\gamma = -\frac{1}{h_\gamma}\frac{\partial \varphi_m}{\partial \gamma} \tag{3.66}$$

Using Eqs. (3.64) and (3.66), and substituting Eq. (3.50) into them, one can obtain the normal components of the magnetic field intensity on the interfaces $\gamma = \gamma_1$ and $\gamma = \gamma_2$ among the three regions

$$H_n = \frac{1}{h_\gamma}\frac{\partial \varphi_m}{\partial \gamma} = \frac{\mathrm{ch}\gamma - \cos\theta}{a}\frac{\partial \varphi_m}{\partial \gamma} \tag{3.67}$$

Additionally, the component of $\boldsymbol{M}$ along the outer normal direction on the interface $\gamma = \gamma_1$ is

$$M_n = \boldsymbol{M}\cdot(-\boldsymbol{e}_\gamma) = (M_x\boldsymbol{e}_x + M_y\boldsymbol{e}_y)\cdot(-\boldsymbol{e}_\gamma) \tag{3.68}$$

where M_x and M_y represent the components of $\boldsymbol{M}$ along the x and y axes in a Cartesian coordinate system, respectively. Substituting the expression for $\boldsymbol{e}_\gamma$ from Eq. (3.60) into Eq. (3.68), one obtains the following expression:

$$M_n = -\frac{\mathrm{ch}\gamma - \cos\theta}{a}\left[\frac{M_x a(1 - \mathrm{ch}\gamma\cos\theta)}{(\mathrm{ch}\gamma - \cos\theta)^2} - \frac{M_y a\,\mathrm{sh}\gamma\sin\theta}{(\mathrm{ch}\gamma - \cos\theta)^2}\right](M_x\boldsymbol{e}_x + M_y\boldsymbol{e}_y)\cdot(-\boldsymbol{e}_\gamma) \tag{3.69}$$

Using the results from Eqs. (3.67) and (3.69), one can get the following equations based on Eq. (3.58):

$$B_{n,1} = \mu_0(H_{n,1} + M_n) = \mu_0\frac{\mathrm{ch}\gamma - \cos\theta}{a}\left[\frac{\partial \varphi_{m,1}}{\partial \gamma} - \frac{M_x a(1 - \mathrm{ch}\gamma\cos\theta)}{(\mathrm{ch}\gamma - \cos\theta)^2} + \frac{M_y a\,\mathrm{sh}\gamma\sin\theta}{(\mathrm{ch}\gamma - \cos\theta)^2}\right] \tag{3.70}$$

$$B_{n,2} = \mu H_{n,2} = \mu\frac{\mathrm{ch}\gamma - \cos\theta}{a}\frac{\partial \varphi_{m,2}}{\partial \gamma} \tag{3.71}$$

$$B_{n,3} = \mu_0 H_{n,3} = \mu_0 \frac{\text{ch}\gamma - \cos\theta}{a} \frac{\partial \varphi_{m,3}}{\partial \gamma} \tag{3.72}$$

Thus, the boundary conditions (3.56) and (3.57) become

$$\mu_0 \left.\frac{\partial \varphi_{m,1}}{\partial \gamma}\right|_{\gamma_1} - \mu_0 \frac{M_x a(1 - \text{ch}\gamma_1 \cos\theta)}{(\text{ch}\gamma_1 - \cos\theta)^2} + \mu_0 \frac{M_y a \text{sh}\gamma_1 \sin\theta}{(\text{ch}\gamma_1 - \cos\theta)^2} = \mu \left.\frac{\partial \varphi_{m,2}}{\partial \gamma}\right|_{\gamma_1} \tag{3.73}$$

and

$$\mu \left.\frac{\partial \varphi_{m,2}}{\partial \gamma}\right|_{\gamma_2} = \mu_0 \left.\frac{\partial \varphi_{m,3}}{\partial \gamma}\right|_{\gamma_2}, \tag{3.74}$$

respectively.

Now it is need to expand the two functions of θ in Eq. (3.73) into Fourier series within the interval $[0, 2\pi]$. For the first function, $((1 - \text{ch}\gamma_1 \cos\theta))/(\text{ch}\gamma_1 - \cos\theta)^2$, since it is an even function of θ, the expansion will be a cosine series. When n is a positive integer greater than 2, the coefficients are given by

$$a_n = \frac{2}{\pi} \int_0^{\pi} \frac{1 - \text{ch}\gamma_1 \cos\theta}{(\text{ch}\gamma_1 - \cos\theta)^2} \cos(n\theta) \mathrm{d}\theta \tag{3.75}$$

Expressing the term $\cos(n\theta)$ in Eq. (3.75) as

$$\cos(n\theta) = \cos[(n-2)\theta] - 2\sin[(n-1)\theta]\sin\theta$$

yields

$$\begin{aligned} a_n = a_{n-2} &- \frac{4}{\pi} \int_0^{\pi} \frac{\sin[(n-1)\theta]}{(\text{ch}\gamma_1 - \cos\theta)^2} \sin\theta \mathrm{d}\theta + \frac{2\text{ch}\gamma_1}{\pi} \int_0^{\pi} \frac{\sin(n\theta)\sin\theta}{(\text{ch}\gamma_1 - \cos\theta)^2} \mathrm{d}\theta \\ &+ \frac{2\text{ch}\gamma_1}{\pi} \int_0^{\pi} \frac{\sin[(n-2)\theta]\sin\theta}{(\text{ch}\gamma_1 - \cos\theta)^2} \mathrm{d}\theta \end{aligned} \tag{3.76}$$

where

$$\begin{aligned} \int_0^{\pi} \frac{\sin[(n-1)\theta]}{(\text{ch}\gamma_1 - \cos\theta)^2} \sin\theta \mathrm{d}\theta &= \int_0^{\pi} \sin[(n-1)\theta] \mathrm{d}\left(\frac{1}{\text{ch}\gamma_1 - \cos\theta}\right) \\ &= \left[\frac{\sin[(n-1)\theta]}{\text{ch}\gamma_1 - \cos\theta}\right]_0^{\pi} - \int_0^{\pi} \frac{(n-1)\cos[(n-1)\theta]}{\text{ch}\gamma_1 - \cos\theta} \mathrm{d}\theta \\ &= -\frac{n-1}{\text{ch}\gamma_1} \int_0^{\pi} \frac{\cos[(n-1)\theta]}{1 - \dfrac{\cos\theta}{\text{ch}\gamma_1}} \mathrm{d}\theta \end{aligned} \tag{3.77}$$

Since $\mathrm{ch}\gamma_1 > 1$, it follows that $(-1/\mathrm{ch}\gamma_1)^2 < 1$, and consequently, the result of Eq. (3.77) can be obtained

$$\begin{aligned}\int_0^{\pi} \frac{\sin\left[(n-1)\theta\right]}{(\mathrm{ch}\gamma_1-\cos\theta)^2}\sin\theta\mathrm{d}\theta &= -\frac{n-1}{\mathrm{ch}\gamma_1}\frac{\pi}{\sqrt{1-\dfrac{1}{\mathrm{ch}^2\gamma_1}}}\left(\frac{\sqrt{1-\dfrac{1}{\mathrm{ch}^2\gamma_1}}-1}{-\dfrac{1}{\mathrm{ch}\gamma_1}}\right)^{n-1}\\ &= -\frac{(n-1)\pi}{\mathrm{sh}\gamma_1}(\mathrm{ch}\gamma_1-\mathrm{sh}\gamma_1)^{n-1}\\ &= -\frac{(n-1)\pi}{\mathrm{sh}\gamma_1}e^{-(n-1)\gamma_1}\end{aligned} \tag{3.78}$$

With a similar method, one gets

$$\int_0^{\pi}\frac{\sin(n\theta)\sin\theta}{(\mathrm{ch}\gamma_1-\cos\theta)^2}\mathrm{d}\theta = -\frac{n\pi}{\mathrm{sh}\gamma_1}e^{-n\gamma_1} \tag{3.79}$$

$$\int_0^{\pi}\frac{\sin\left[(n-2)\theta\right]\sin\theta}{(\mathrm{ch}\gamma_1-\cos\theta)^2}\mathrm{d}\theta = -\frac{(n-2)\pi}{\mathrm{sh}\gamma_1}e^{-(n-2)\gamma_1} \tag{3.80}$$

Substituting the integration results from Eqs. (3.77) to (3.79) into Eq. (3.76) and simplifying, one obtains

$$\begin{aligned}a_n &= a_{n-2}+\frac{2}{\mathrm{sh}\gamma_1}e^{-(n-2)\gamma_1}\left[2(n-1)e^{-\gamma_1}-n\mathrm{ch}\gamma_1 e^{-2\gamma_1}-(n-2)\mathrm{ch}\gamma_1\right]\\ &= a_{n-2}+\frac{2}{\mathrm{sh}\gamma_1}e^{-(n-2)\gamma_1}\left\{2(n-1)e^{-\gamma_1}-\frac{1}{2}\left[ne^{-2\gamma_1}+(n-2)\right](e^{\gamma_1}+e^{-\gamma_1})\right\}\\ &= a_{n-2}+\frac{2}{\mathrm{sh}\gamma_1}e^{-(n-2)\gamma_1}\left[(n-1)e^{-\gamma_1}-ne^{-\gamma_1}\mathrm{ch}2\gamma_1+e^{\gamma_1}\right]\\ &= a_{n-2}+\frac{2}{\mathrm{sh}\gamma_1}e^{-(n-2)\gamma_1}\left[2\mathrm{sh}\gamma_1+ne^{-\gamma_1}(1-\mathrm{ch}^2\gamma_1-\mathrm{sh}^2\gamma_1)\right]\\ &= a_{n-2}+\frac{2}{\mathrm{sh}\gamma_1}e^{-(n-2)\gamma_1}\left[\mathrm{sh}\gamma_1(2-n+ne^{-2\gamma_1})\right]\\ &= a_{n-2}-2(n-2)e^{-(n-2)\gamma_1}+2ne^{-n\gamma_1}\end{aligned} \tag{3.81}$$

namely,

$$a_n-2ne^{-n\gamma_1} = a_{n-2}-2(n-2)e^{-(n-2)\gamma_1} \tag{3.82}$$

where $n \geq 2$. When $n = 0$, one has

$$a_0 = \frac{2}{\pi}\int_0^{\pi}\frac{1-\mathrm{ch}\gamma_1\cos\theta}{(\mathrm{ch}\gamma_1-\cos\theta)^2}\mathrm{d}\theta = \frac{2}{\pi}\int_0^{\pi}\left[\frac{\mathrm{ch}\gamma_1}{\mathrm{ch}\gamma_1-\cos\theta}-\frac{\mathrm{sh}^2\gamma_1}{(\mathrm{ch}\gamma_1-\cos\theta)^2}\right]\mathrm{d}\theta$$
$$= \frac{2}{\pi}\int_0^{\pi}\frac{1}{1-\dfrac{\cos\theta}{\mathrm{ch}\gamma_1}}\mathrm{d}\theta - \frac{2}{\pi}\frac{\mathrm{sh}^2\gamma_1}{\mathrm{ch}^2\gamma_1}\int_0^{\pi}\frac{1}{\left(1-\dfrac{\cos\theta}{\mathrm{ch}\gamma_1}\right)^2}\mathrm{d}\theta \tag{3.83}$$

Since $1/\mathrm{ch}\gamma_1 < 1$, the following equations hold

$$\int_0^{\pi}\frac{1}{1-\dfrac{\cos\theta}{\mathrm{ch}\gamma_1}}\mathrm{d}\theta = \frac{\pi}{\sqrt{1-\dfrac{1}{\mathrm{ch}^2\gamma_1}}} = \frac{\pi\mathrm{sh}\gamma_1}{\mathrm{ch}\gamma_1},\quad \int_0^{\pi}\frac{1}{\left(1-\dfrac{\cos\theta}{\mathrm{ch}\gamma_1}\right)^2}\mathrm{d}\theta = \frac{\pi}{\left(1-\dfrac{1}{\mathrm{ch}^2\gamma_1}\right)^{\frac{3}{2}}} = \frac{\pi\mathrm{ch}^3\gamma_1}{\mathrm{sh}^3\gamma_1}$$

Substitution of them into Eq. (3.83) yields

$$a_0 = \frac{2\mathrm{ch}\gamma_1}{\mathrm{sh}\gamma_1} - \frac{2\mathrm{ch}^3\gamma_1}{\mathrm{sh}^3\gamma_1}\frac{\mathrm{sh}^2\gamma_1}{\mathrm{ch}^2\gamma_1} = 0 \tag{3.84}$$

The same derivation leads to

$$a_1 = 2e^{-\gamma_1} \tag{3.85}$$

Upon combining the results from Eqs. (3.82), (3.84), and (3.85), the coefficient a_n can be uniformly expressed as

$$a_n = 2ne^{-n\gamma_1}, (n = 0,1,2\cdots\cdots) \tag{3.86}$$

Thus, the Fourier series expansion of the function $((1 - \mathrm{ch}\gamma_1 \cos\theta))/(\mathrm{ch}\gamma_1 - \cos\theta)^2$ is

$$\frac{1-\mathrm{ch}\gamma_1\cos\theta}{(\mathrm{ch}\gamma_1-\cos\theta)^2} = \frac{a_0}{2} + \sum_{n=1}^{\infty}a_n\cos(n\theta) = \sum_{n=1}^{\infty}2ne^{-n\gamma_1}\cos(n\theta) \tag{3.87}$$

Similarly, for the second function in Eq. (3.73), which is $\mathrm{sh}\gamma_1 \sin\theta/(\mathrm{ch}\gamma_1 - \cos\theta)^2$, as it is an odd function of θ, its Fourier series expansion is a sine series. When n is a positive integer, the coefficients of this series are given as

$$b_n = \frac{2}{\pi}\int_0^{\pi}\frac{\mathrm{sh}\gamma_1\sin\theta}{(\mathrm{ch}\gamma_1-\cos\theta)^2}\sin(n\theta)\mathrm{d}\theta = \frac{2\mathrm{sh}\gamma_1}{\pi}\int_0^{\pi}\sin(n\theta)\mathrm{d}\left(\frac{1}{\mathrm{ch}\gamma_1-\cos\theta}\right)$$
$$= \frac{2\mathrm{sh}\gamma_1}{\pi}\left[\frac{\sin(n\theta)}{\mathrm{ch}\gamma_1-\cos\theta}\right]_0^{\pi} - \frac{2n\mathrm{sh}\gamma_1}{\pi}\int_0^{\pi}\frac{\cos(n\theta)}{\mathrm{ch}\gamma_1-\cos\theta}\mathrm{d}\theta = -\frac{2n\mathrm{sh}\gamma_1}{\pi\mathrm{ch}\gamma_1}\int_0^{\pi}\frac{\cos(n\theta)}{1-\dfrac{\cos\theta}{\mathrm{ch}\gamma_1}}\mathrm{d}\theta \tag{3.88}$$

Since $1/\text{ch}\gamma_1 < 1$, the expression of b_n given in (3.88) is transformed into

$$b_n = -\frac{2n\text{sh}\gamma_1}{\pi\text{ch}\gamma_1}\frac{\pi}{\sqrt{1-\frac{1}{\text{ch}^2\gamma_1}}}\left(\frac{\sqrt{1-\frac{1}{\text{ch}^2\gamma_1}}-1}{-\frac{1}{\text{ch}\gamma_1}}\right) = -2ne^{-n\gamma_1} \tag{3.89}$$

Then, one can get

$$\frac{\text{sh}\gamma_1 \sin\theta}{\left(\text{ch}\gamma_1 - \cos\theta\right)^2} = \sum_{n=1}^{\infty} b_n \sin(n\theta) = -\sum_{n=1}^{\infty} 2ne^{-n\gamma_1}\sin(n\theta) \tag{3.90}$$

Substituting the results from Eq. (3.87) and (3.90) into Eq. (3.73), the boundary condition becomes

$$\mu_0 \frac{\partial \varphi_{m,1}}{\partial \gamma}\bigg|_{\gamma_1} - 2\mu_0 a\left[\sum_{n=1}^{\infty}\left[nM_x e^{-n\gamma_1}\cos(n\theta)\right] - \sum_{n=1}^{\infty}\left[nM_y e^{-n\gamma_1}\sin(n\theta)\right]\right] = \mu \frac{\partial \varphi_{m,2}}{\partial \gamma}\bigg|_{\gamma_1} \tag{3.91}$$

Until now, the boundary conditions for this physical problem are composed of Eqs. (3.52) to (3.55) and Eqs. (3.74) and (3.91).

3.4.1.3 Solving for Scalar Magnetic Potential

The method of separation of variables is used to solve Eq. (3.51). The general solutions of scalar magnetic potential in the three regions are expressed by

$$\varphi_{m,1} = \left(C_1 e^{k\gamma} + C_2 e^{-k\gamma}\right)\cos(k\theta) + \left(C_3 e^{k\gamma} + C_4 e^{-k\gamma}\right)\sin(k\theta) \tag{3.92}$$

$$\varphi_{m,2} = \left(C_5 e^{k\gamma} + C_6 e^{-k\gamma}\right)\cos(k\theta) + \left(C_7 e^{k\gamma} + C_8 e^{-k\gamma}\right)\sin(k\theta) \tag{3.93}$$

$$\varphi_{m,3} = \left(C_9 e^{k\gamma} + C_{10} e^{-k\gamma}\right)\cos(k\theta) + \left(C_{11} e^{k\gamma} + C_{12} e^{-k\gamma}\right)\sin(k\theta) \tag{3.94}$$

where $C_1 \sim C_{12}$ are constants determined by boundary conditions and k is a coefficient to be determined. Based on the boundary conditions (3.52) and (3.53), one can deduce that $C_1 = C_3 = 0$ and $C_{10} = C_{12} = 0$. Consequently, the general solutions (3.92) and (3.94) reduce to

$$\varphi_{m,1} = C_2 e^{-k\gamma}\cos(k\theta) + C_4 e^{-k\gamma}\sin(k\theta) \tag{3.95}$$

$$\varphi_{m,3} = C_{10} e^{-k\gamma}\cos(k\theta) + C_{12} e^{-k\gamma}\sin(k\theta) \tag{3.96}$$

Letting $k = 1, 2, \cdots, n$ and based on the superposition principle of solutions to Laplace's equation, the general solution for the scalar magnetic potential can be expressed as

$$\varphi_{m,1} = \sum_{n=1}^{\infty} A_n e^{-n\gamma} \cos(n\theta) + \sum_{n=1}^{\infty} E_n e^{-n\gamma} \sin(n\theta) \tag{3.97}$$

$$\varphi_{m,2} = \sum_{n=1}^{\infty} \left(O_n e^{-n\gamma} + C_n e^{n\gamma}\right)\cos(n\theta) + \sum_{n=1}^{\infty} \left(F_n e^{-n\gamma} + G_n e^{n\gamma}\right)\sin(n\theta) \tag{3.98}$$

$$\varphi_{m,3} = \sum_{n=1}^{\infty} D_n e^{n\gamma} \cos(n\theta) + \sum_{n=1}^{\infty} Q_n e^{n\gamma} \sin(n\theta) \tag{3.99}$$

where A_n~Q_n are the coefficients to be determined. Substituting Eqs. (3.97) and (3.98) into the boundary condition (3.54), and utilizing the fact that the corresponding terms of $\cos(n\theta)$ and $\sin(n\theta)$ are equal when n is the same, one obtains

$$\begin{cases} A_n = O_n + C_n e^{2n\gamma_1} \\ E_n = F_n + G_n e^{2n\gamma_1} \end{cases} \tag{3.100}$$

Substituting Eqs. (3.98) and (3.99) into the boundary conditions (3.55) and (3.74) yields

$$\begin{cases} D_n = C_n + O_n e^{-2n\gamma_2} \\ Q_n = G_n + F_n e^{-2n\gamma_2} \end{cases} \tag{3.101}$$

$$\begin{cases} -\mu O_n e^{-2n\gamma_2} + \mu C_n = \mu_0 D_n \\ -\mu F_n e^{-2n\gamma_2} + \mu G_n = \mu_0 Q_n \end{cases} \tag{3.102}$$

Putting Eq. (3.101) into (3.102) gives

$$\begin{cases} O_n = \dfrac{\mu - \mu_0}{\mu + \mu_0} e^{2n\gamma_2} C_n \\ F_n = \dfrac{\mu - \mu_0}{\mu + \mu_0} e^{2n\gamma_2} G_n \end{cases} \tag{3.103}$$

Substituting Eqs. (3.97) and (3.98) into the boundary condition (3.91) yields

$$\begin{cases} \mu_0 A_n + 2\mu_0 M_x a = \mu O_n - \mu C_n e^{2n\gamma_1} \\ \mu_0 E_n + 2\mu_0 M_y a = \mu F_n - \mu G_n e^{2n\gamma_1} \end{cases} \tag{3.104}$$

With the aid of Eq. (3.100), one gets

$$\begin{cases} (\mu - \mu_0) O_n - (\mu + \mu_0) C_n e^{2n\gamma_1} = 2\mu_0 M_x a \\ (\mu - \mu_0) F_n - (\mu + \mu_0) G_n e^{2n\gamma_1} = 2\mu_0 M_y a \end{cases} \tag{3.105}$$

By introducing Eq. (3.103) into Eq. (3.105) and simplifying, one has

$$C_n = \frac{2\mu_0(\mu+\mu_0)aM_x}{(\mu-\mu_0)^2 e^{2n\gamma_2} - (\mu+\mu_0)^2 e^{2n\gamma_1}} \tag{3.106}$$

$$G_n = \frac{2\mu_0(\mu+\mu_0)aM_y}{(\mu-\mu_0)^2 e^{2n\gamma_2} - (\mu+\mu_0)^2 e^{2n\gamma_1}} \tag{3.107}$$

Substitution of Eqs. (3.106) and (3.107) into Eq. (3.103) yields

$$O_n = \frac{2\mu_0(\mu-\mu_0)aM_x e^{2n\gamma_2}}{(\mu-\mu_0)^2 e^{2n\gamma_2} - (\mu+\mu_0)^2 e^{2n\gamma_1}} \tag{3.108}$$

$$F_n = \frac{2\mu_0(\mu-\mu_0)aM_y e^{2n\gamma_2}}{(\mu-\mu_0)^2 e^{2n\gamma_2} - (\mu+\mu_0)^2 e^{2n\gamma_1}} \tag{3.109}$$

Now putting Eqs. (3.106)~(3.109) into Eq. (3.98) yields the scalar magnetic potential $\varphi_{m,2}$ within the ferrofluid region.

3.4.1.4 The Radial Magnetic Levitation Force Experienced by the Permanent Magnet

The integral term in the equivalent magnetic stress tensor from Eq. (3.12) can be calculated as

$$\mu_0 \int_0^H \left(\frac{\partial(\upsilon M)}{\partial \upsilon}\right)_{T,H} \mathrm{d}H = \mu_0 \int_0^H M\mathrm{d}H + \mu_0 \int_0^H \upsilon \left(\frac{\partial M}{\partial \upsilon}\right)_{T,H} \mathrm{d}H \tag{3.110}$$

Assuming the ferrofluid is incompressible, its specific volume remains constant, rendering the second integral on the right-hand side of the above equation zero. Furthermore, it is supposed that ***B***//***H***//***M*** when the ferrofluid is placed in a magnetic field, and its permeability is constant under equilibrium conditions, then one can derive that $B = \mu_0(M + H)$, $B = \mu H$, and

$$\mu_0 M = (\mu - \mu_0)H \tag{3.111}$$

Substituting Eq. (3.111) into the integral of Eq. (3.110) yields

$$\mu_0 \int_0^H \upsilon \left(\frac{\partial M}{\partial \upsilon}\right)_{T,H} \mathrm{d}H = \int_0^H (\mu-\mu_0)H\mathrm{d}H = \frac{1}{2}(\mu-\mu_0)H^2 \tag{3.112}$$

With the aid of Eq. (3.112), the equivalent magnetic stress tensor can be written as

$$\mathbf{T}_m' = -\frac{1}{2}\mu H^2 \delta_{ij} + \mu H_i H_j \tag{3.113}$$

and consequently,

$$\begin{aligned} \boldsymbol{n}\cdot\mathbf{T}_m' &= n_k\boldsymbol{e}_k\cdot\left(-\frac{1}{2}\mu H^2\delta_{ij}+\mu H_iH_j\right)\boldsymbol{e}_i\boldsymbol{e}_j \\ &= \left(n_k\mu H_kH_j-\frac{1}{2}\mu H^2 n_j\right)\boldsymbol{e}_j \end{aligned} \tag{3.114}$$

Replacing the force acting on the permanent magnet with equivalent surface pressure, the magnetic levitation force acting on the permanent magnet immersed in the ferrofluid can be expressed according to Eq. (3.114) as

$$\begin{aligned} \boldsymbol{F}_m &= \oint\!\!\!\oint_{S_m}\boldsymbol{n}\cdot\mathbf{T}_m'\mathrm{d}S_m \\ &= \boldsymbol{e}_j\mu\oint\!\!\!\oint_{S_m}\left(n_kH_kH_j-\frac{1}{2}H^2n_j\right)\mathrm{d}S_m \end{aligned} \tag{3.115}$$

Assuming that the permanent magnet is infinitely long along its axial direction, it experiences a radial magnetic levitation force with equal magnitude per unit length. It follows that $\mathrm{d}S_m = \mathrm{d}l_m$ for the unit length of the permanent magnet, where l_m is the cross-sectional contour of the permanent magnet. Consequently, the radial magnetic levitation force per unit length can be expressed as

$$\boldsymbol{F}_m = \boldsymbol{e}_j\mu\oint\!\!\!\oint_{l_m}\left(n_kH_kH_j-\frac{1}{2}H^2n_j\right)\mathrm{d}l_m \tag{3.116}$$

Based on the symmetry of the model depicted in Figure 3.1, it can be inferred that the radial magnetic levitation force acts along the x-direction. According to Eq. (3.116), this force is given by

$$F_{m,x} = \mu\oint\!\!\!\oint_{l_m}\left(n_yH_yH_x-\frac{1}{2}H^2n_x\right)\mathrm{d}l_m \tag{3.117}$$

In the neighbourhood of a point on the circle γ_1 representing the contour of a permanent magnet, one has

$$\mathrm{d}l_m = \sqrt{(\mathrm{d}x)^2+(\mathrm{d}y)^2} \tag{3.118}$$

According to Eqs. (3.29) and (3.30), it can be derived that

$$\mathrm{d}x = \frac{a\,\mathrm{sh}\gamma_1\sin\theta}{(\mathrm{ch}\gamma_1-\cos\theta)^2}\mathrm{d}\theta,\ \mathrm{d}y = \frac{a(\mathrm{ch}\gamma_1\cos\theta-1)}{(\mathrm{ch}\gamma_1-\cos\theta)^2}\mathrm{d}\theta \tag{3.119}$$

By substituting them into Eq. (3.118), one can obtain

$$dl_m = \frac{a}{\mathrm{ch}\gamma_1 - \cos\theta} d\theta \tag{3.120}$$

As l_m traverses around the perimeter of the permanent magnet, the variation of θ ranges from 0 to 2π.

In a planar bipolar coordinate system, the magnetic field intensity $\boldsymbol{H}$ in a ferrofluid can be calculated using the scalar magnetic potential as

$$\boldsymbol{H} = -\nabla\varphi_{m,2} = -\boldsymbol{e}_\theta \frac{1}{h_\theta}\frac{\partial\varphi_{m,2}}{\partial\theta} - \boldsymbol{e}_\gamma \frac{1}{h_\gamma}\frac{\partial\varphi_{m,2}}{\partial\gamma} \tag{3.121}$$

Putting Eqs. (3.49), (3.50), (3.60), and (3.61) into this equation gives

$$\begin{aligned}\boldsymbol{H} = &\frac{1}{a}\left[\mathrm{sh}\gamma\sin\theta\frac{\partial\varphi_{m,2}}{\partial\theta} - (1-\mathrm{ch}\gamma\cos\theta)\frac{\partial\varphi_{m,2}}{\partial\gamma}\right]\boldsymbol{e}_x \\ &+\frac{1}{a}\left[(1-\mathrm{ch}\gamma\cos\theta)\frac{\partial\varphi_{m,2}}{\partial\theta} + \mathrm{sh}\gamma\sin\theta\frac{\partial\varphi_{m,2}}{\partial\gamma}\right]\boldsymbol{e}_y\end{aligned} \tag{3.122}$$

Due to the fact that the unit vector $\boldsymbol{n}$ in the direction of the outer normal on the surface of the permanent magnet is collinear but opposite in direction to the unit vector $\boldsymbol{e}_{\gamma 1}$ in the direction of γ, it follows that

$$\boldsymbol{n} = -\boldsymbol{e}_{\gamma_1} \tag{3.123}$$

Based on this, one can deduce that the components of $\boldsymbol{n}$ along the two coordinate directions in the Cartesian coordinate system are

$$n_x = -\boldsymbol{e}_{\gamma_1}\cdot\boldsymbol{e}_x = -\frac{1-\mathrm{ch}\gamma_1\cos\theta}{\mathrm{ch}\gamma_1 - \cos\theta} \tag{3.124}$$

$$n_y = -\boldsymbol{e}_{\gamma_1}\cdot\boldsymbol{e}_y = \frac{\mathrm{sh}\gamma_1\sin\theta}{\mathrm{ch}\gamma_1 - \cos\theta} \tag{3.125}$$

By substituting Eqs. (3.122), (3.123), and (3.125) into the term $n_y H_y H_x$ in Eq. (3.117), one obtains the following result:

$$\begin{aligned}n_y H_y H_x = &\frac{1}{a^2}\frac{\mathrm{sh}\gamma_1\sin\theta}{\mathrm{ch}\gamma_1 - \cos\theta}\left\{\mathrm{sh}\gamma_1\sin\theta(1-\mathrm{ch}\gamma_1\cos\theta)\left[\left(\frac{\partial\varphi_{m,2}}{\partial\theta}\right)^2 - \left(\frac{\partial\varphi_{m,2}}{\partial\gamma}\right)^2\right]\right. \\ &\left.+\left[(\mathrm{sh}\gamma_1\sin\theta)^2 - (1-\mathrm{ch}\gamma_1\cos\theta)^2\right]\frac{\partial\varphi_{m,2}}{\partial\theta}\frac{\partial\varphi_{m,2}}{\partial\gamma}\right\}\end{aligned} \tag{3.126}$$

Introducing Eq. (3.122) to (3.124) into the term $\frac{1}{2}H^2 n_x$ in Eq. (3.117) gives

$$\frac{1}{2}H^2 n_x = -\frac{1}{2a^2}\frac{1-\text{ch}\gamma_1\cos\theta}{\text{ch}\gamma_1-\cos\theta}(\text{ch}\gamma_1-\cos\theta)^2\left[\left(\frac{\partial\varphi_{m,2}}{\partial\theta}\right)^2+\left(\frac{\partial\varphi_{m,2}}{\partial\gamma}\right)^2\right] \tag{3.127}$$

Substituting the results from Eqs. (3.126) and (3.127) into Eq. (3.117), and then applying Eq. (3.120), one obtains

$$F_{m,x} = \frac{\mu}{2a}\int_0^{2\pi}\left\{\left[\left(\frac{\partial\varphi_{m,2}}{\partial\theta}\right)^2-\left(\frac{\partial\varphi_{m,2}}{\partial\gamma}\right)^2\right](1-\text{ch}\gamma_1\cos\theta)-2\text{sh}\gamma_1\sin\theta\frac{\partial\varphi_{m,2}}{\partial\theta}\frac{\partial\varphi_{m,2}}{\partial\gamma}\right\}\text{d}\theta \tag{3.128}$$

From Eq. (3.98), it can be deduced that

$$\frac{\partial\varphi_{m,2}}{\partial\theta} = \sum_{n=1}^{\infty}\left[-\left(O_n e^{-n\gamma}+C_n e^{n\gamma}\right)n\sin(n\theta)+\left(F_n e^{-n\gamma}+G_n e^{n\gamma}\right)n\cos(n\theta)\right] \tag{3.129}$$

$$\frac{\partial\varphi_{m,2}}{\partial\gamma} = \sum_{n=1}^{\infty}\left[-\left(O_n e^{-n\gamma}-C_n e^{n\gamma}\right)n\cos(n\theta)-\left(F_n e^{-n\gamma}-G_n e^{n\gamma}\right)n\sin(n\theta)\right] \tag{3.130}$$

Letting n' be a positive integer, and by applying Eqs. (3.129) and (3.130), one obtains

$$\begin{aligned}
\left(\frac{\partial\varphi_{m,2}}{\partial\theta}\right)^2 &= \sum_{n=1}^{\infty}\sum_{n'=1}^{\infty}\left\{\begin{matrix}\left[-\left(O_n e^{-n\gamma}+C_n e^{n\gamma}\right)n\sin(n\theta)+\left(F_n e^{-n\gamma}+G_n e^{n\gamma}\right)n\cos(n\theta)\right]\cdot\\ \left[-\left(O_{n'} e^{-n'\gamma}+C_{n'} e^{n'\gamma}\right)n'\sin(n'\theta)+\left(F_{n'} e^{-n'\gamma}+G_{n'} e^{n'\gamma}\right)n'\cos(n'\theta)\right]\end{matrix}\right\}\\
&= \sum_{n=1}^{\infty}\sum_{n'=1}^{\infty}\left\{\begin{matrix}\left[O_nO_{n'}e^{-(n+n')\gamma}+C_nC_{n'}e^{(n+n')\gamma}+O_nC_{n'}e^{(n'-n)\gamma}+C_nO_{n'}e^{(n-n')\gamma}\right]nn'\sin(n\theta)\sin(n'\theta)\\ +\left[F_nF_{n'}e^{-(n+n')\gamma}+G_nG_{n'}e^{(n+n')\gamma}+F_nG_{n'}e^{(n'-n)\gamma}+G_nF_{n'}e^{(n-n')\gamma}\right]nn'\cos(n\theta)\cos(n'\theta)\\ -\left[O_nF_{n'}e^{-(n+n')\gamma}+C_nG_{n'}e^{(n+n')\gamma}+O_nG_{n'}e^{(n'-n)\gamma}+C_nF_{n'}e^{(n-n')\gamma}\right]nn'\sin(n\theta)\cos(n'\theta)\\ -\left[F_nO_{n'}e^{-(n+n')\gamma}+G_nC_{n'}e^{(n+n')\gamma}+F_nC_{n'}e^{(n'-n)\gamma}+G_nO_{n'}e^{(n-n')\gamma}\right]nn'\cos(n\theta)\sin(n'\theta)\end{matrix}\right\}
\end{aligned} \tag{3.131}$$

$$\begin{aligned}
\left(\frac{\partial\varphi_{m,2}}{\partial\gamma}\right)^2 &= \sum_{n=1}^{\infty}\sum_{n'=1}^{\infty}\left\{\begin{matrix}\left[-\left(O_n e^{-n\gamma}-C_n e^{n\gamma}\right)n\cos(n\theta)-\left(F_n e^{-n\gamma}-G_n e^{n\gamma}\right)n\sin(n\theta)\right]\cdot\\ \left[-\left(O_{n'} e^{-n'\gamma}-C_{n'} e^{n'\gamma}\right)n'\cos(n'\theta)-\left(F_{n'} e^{-n'\gamma}-G_{n'} e^{n'\gamma}\right)n'\sin(n'\theta)\right]\end{matrix}\right\}\\
&= \sum_{n=1}^{\infty}\sum_{n'=1}^{\infty}\left\{\begin{matrix}\left[O_nO_{n'}e^{-(n+n')\gamma}+C_nC_{n'}e^{(n+n')\gamma}-O_nC_{n'}e^{(n'-n)\gamma}-C_nO_{n'}e^{(n-n')\gamma}\right]nn'\cos(n\theta)\cos(n'\theta)\\ +\left[F_nF_{n'}e^{-(n+n')\gamma}+G_nG_{n'}e^{(n+n')\gamma}-F_nG_{n'}e^{(n'-n)\gamma}-G_nF_{n'}e^{(n-n')\gamma}\right]nn'\sin(n\theta)\sin(n'\theta)\\ +\left[O_nF_{n'}e^{-(n+n')\gamma}+C_nG_{n'}e^{(n+n')\gamma}-O_nG_{n'}e^{(n'-n)\gamma}-C_nF_{n'}e^{(n-n')\gamma}\right]nn'\cos(n\theta)\sin(n'\theta)\\ +\left[F_nO_{n'}e^{-(n+n')\gamma}+G_nC_{n'}e^{(n+n')\gamma}-F_nC_{n'}e^{(n'-n)\gamma}-G_nO_{n'}e^{(n-n')\gamma}\right]nn'\sin(n\theta)\cos(n'\theta)\end{matrix}\right\}
\end{aligned} \tag{3.132}$$

$$\frac{\partial\varphi_{m,2}}{\partial\theta}\frac{\partial\varphi_{m,2}}{\partial\gamma}=\sum_{n=1}^{\infty}\sum_{n'=1}^{\infty}\left\{\begin{array}{l}\left[-\left(O_ne^{-n\gamma}+C_ne^{n\gamma}\right)n\sin(n\theta)+\left(F_ne^{-n\gamma}+G_ne^{n\gamma}\right)n\cos(n\theta)\right]\cdot\\ \left[-\left(O_{n'}e^{-n'\gamma}-C_{n'}e^{n'\gamma}\right)n'\cos(n'\theta)-\left(F_{n'}e^{-n'\gamma}-G_{n'}e^{n'\gamma}\right)n'\sin(n'\theta)\right]\end{array}\right\}$$

$$=\sum_{n=1}^{\infty}\sum_{n'=1}^{\infty}\left\{\begin{array}{l}\left[O_nO_{n'}e^{-(n+n')\gamma}-C_nC_{n'}e^{(n+n')\gamma}-O_nC_{n'}e^{(n'-n)\gamma}+C_nO_{n'}e^{(n-n')\gamma}\right]nn'\sin(n\theta)\cos(n'\theta)\\ -\left[F_nF_{n'}e^{-(n+n')\gamma}-G_nG_{n'}e^{(n+n')\gamma}-F_nG_{n'}e^{(n'-n)\gamma}+G_nF_{n'}e^{(n-n')\gamma}\right]nn'\cos(n\theta)\sin(n'\theta)\\ +\left[O_nF_{n'}e^{-(n+n')\gamma}-C_nG_{n'}e^{(n+n')\gamma}-O_nG_{n'}e^{(n'-n)\gamma}+C_nF_{n'}e^{(n-n')\gamma}\right]nn'\sin(n\theta)\sin(n'\theta)\\ -\left[F_nO_{n'}e^{-(n+n')\gamma}-G_nC_{n'}e^{(n+n')\gamma}-F_nC_{n'}e^{(n'-n)\gamma}+G_nO_{n'}e^{(n-n')\gamma}\right]nn'\cos(n\theta)\cos(n'\theta)\end{array}\right\}$$

(3.133)

If the values of Eqs. (3.131) to (3.133) at $\gamma=\gamma_1$ are substituted into Eq. (3.128), one obtains

$$F_{m,x}=\frac{\mu}{2a}\sum_{n=1}^{\infty}\sum_{n'=1}^{\infty}nn'\left\{\begin{array}{l}\left[\begin{array}{l}\left(O_nO_{n'}-F_nF_{n'}\right)e^{-(n+n')\gamma_1}+\left(C_nC_{n'}-G_nG_{n'}\right)e^{(n+n')\gamma_1}+\\ \left(O_nC_{n'}+F_nG_{n'}\right)e^{(n'-n)\gamma_1}+\left(C_nO_{n'}+G_nF_{n'}\right)e^{(n-n')\gamma_1}\end{array}\right]\int_0^{2\pi}\sin(n\theta)\sin(n'\theta)(1-\mathrm{ch}\gamma_1\cos\theta)\mathrm{d}\theta\\ +\left[\begin{array}{l}\left(F_nF_{n'}-O_nO_{n'}\right)e^{-(n+n')\gamma_1}+\left(G_nG_{n'}-C_nC_{n'}\right)e^{(n+n')\gamma_1}+\\ \left(F_nG_{n'}+O_nC_{n'}\right)e^{(n'-n)\gamma_1}+\left(G_nF_{n'}+C_nO_{n'}\right)e^{(n-n')\gamma_1}\end{array}\right]\int_0^{2\pi}\cos(n\theta)\cos(n'\theta)(1-\mathrm{ch}\gamma_1\cos\theta)\mathrm{d}\theta\\ -\left[\begin{array}{l}\left(O_nF_{n'}+F_nO_{n'}\right)e^{-(n+n')\gamma_1}+\left(C_nG_{n'}+G_nC_{n'}\right)e^{(n+n')\gamma_1}+\\ \left(O_nG_{n'}-F_nC_{n'}\right)e^{(n'-n)\gamma_1}+\left(C_nF_{n'}-G_nO_{n'}\right)e^{(n-n')\gamma_1}\end{array}\right]\int_0^{2\pi}\sin(n\theta)\cos(n'\theta)(1-\mathrm{ch}\gamma_1\cos\theta)\mathrm{d}\theta\\ -\left[\begin{array}{l}\left(F_nO_{n'}+O_nF_{n'}\right)e^{-(n+n')\gamma_1}+\left(G_nC_{n'}+C_nG_{n'}\right)e^{(n+n')\gamma_1}+\\ \left(F_nC_{n'}-O_nG_{n'}\right)e^{(n'-n)\gamma_1}+\left(G_nO_{n'}-C_nF_{n'}\right)e^{(n-n')\gamma_1}\end{array}\right]\int_0^{2\pi}\cos(n\theta)\sin(n'\theta)(1-\mathrm{ch}\gamma_1\cos\theta)\mathrm{d}\theta\end{array}\right\}$$

$$-\frac{\mu\mathrm{sh}\gamma_1}{a}\sum_{n=1}^{\infty}\sum_{n'=1}^{\infty}nn'\left\{\begin{array}{l}\left[O_nO_{n'}e^{-(n+n')\gamma_1}-C_nC_{n'}e^{(n+n')\gamma_1}-O_nC_{n'}e^{(n'-n)\gamma_1}+C_nO_{n'}e^{(n-n')\gamma_1}\right]\int_0^{2\pi}\sin(n\theta)\cos(n'\theta)\sin\theta\mathrm{d}\theta\\ -\left[F_nF_{n'}e^{-(n+n')\gamma_1}-G_nG_{n'}e^{(n+n')\gamma_1}-F_nG_{n'}e^{(n'-n)\gamma_1}+G_nF_{n'}e^{(n-n')\gamma_1}\right]\int_0^{2\pi}\cos(n\theta)\sin(n'\theta)\sin\theta\mathrm{d}\theta\\ +\left[O_nF_{n'}e^{-(n+n')\gamma_1}-C_nG_{n'}e^{(n+n')\gamma_1}-O_nG_{n'}e^{(n'-n)\gamma_1}+C_nF_{n'}e^{(n-n')\gamma_1}\right]\int_0^{2\pi}\sin(n\theta)\sin(n'\theta)\sin\theta\mathrm{d}\theta\\ -\left[F_nO_{n'}e^{-(n+n')\gamma_1}-G_nC_{n'}e^{(n+n')\gamma_1}-F_nC_{n'}e^{(n'-n)\gamma_1}+G_nO_{n'}e^{(n-n')\gamma_1}\right]\int_0^{2\pi}\cos(n\theta)\cos(n'\theta)\sin\theta\mathrm{d}\theta\end{array}\right\}$$

(3.134)

When calculating the integral terms in Eq. (3.134), it is necessary to further simplify their integrand functions, for instance,

$$\begin{array}{l}\sin(n\theta)\sin(n'\theta)(1-\mathrm{ch}\gamma_1\cos\theta)=\frac{1}{2}\left\{\cos\left[(n'-n)\theta\right]-\cos\left[(n'+n)\theta\right]\right\}\\ -\frac{1}{4}\mathrm{ch}\gamma_1\left\{\cos\left[(n'-n+1)\theta\right]+\cos\left[(n'-n-1)\theta\right]-\cos\left[(n'+n+1)\theta\right]-\cos\left[(n'+n-1)\theta\right]\right\}\end{array}\qquad(3.135)$$

The integral in Eq. (3.134) can be calculated according to Eq. (3.135) as

$$\int_0^{2\pi}\sin(n\theta)\sin(n'\theta)(1-\mathrm{ch}\gamma_1\cos\theta)\mathrm{d}\theta=\begin{cases}\pi,n'=n\\ -\frac{1}{2}\pi\mathrm{ch}\gamma_1,n'=n+1\\ -\frac{1}{2}\pi\mathrm{ch}\gamma_1,n=n'+1\end{cases}\qquad(3.136)$$

The same derivation gives the following equations:

$$\int_0^{2\pi}\cos(n\theta)\cos(n'\theta)(1-\text{ch}\gamma_1\cos\theta)\text{d}\theta=\begin{cases}\pi, n'=n\\ -\frac{1}{2}\pi\text{ch}\gamma_1, n'=n+1\\ -\frac{1}{2}\pi\text{ch}\gamma_1, n=n'+1\end{cases} \tag{3.137}$$

$$\int_0^{2\pi}\sin(n\theta)\cos(n'\theta)(1-\text{ch}\gamma_1\cos\theta)\text{d}\theta=0 \tag{3.138}$$

$$\int_0^{2\pi}\cos(n\theta)\sin(n'\theta)(1-\text{ch}\gamma_1\cos\theta)\text{d}\theta=0 \tag{3.139}$$

$$\int_0^{2\pi}\sin(n\theta)\cos(n'\theta)\sin\theta\text{d}\theta=\begin{cases}\frac{\pi}{2}, n=n'+1\\ -\frac{\pi}{2}, n'=n+1\end{cases} \tag{3.140}$$

$$\int_0^{2\pi}\cos(n\theta)\sin(n'\theta)\sin\theta\text{d}\theta=\begin{cases}\frac{\pi}{2}, n'=n+1\\ -\frac{\pi}{2}, n=n'+1\end{cases} \tag{3.141}$$

$$\int_0^{2\pi}\sin(n\theta)\sin(n'\theta)\sin\theta\text{d}\theta=0 \tag{3.142}$$

$$\int_0^{2\pi}\cos(n\theta)\cos(n'\theta)\sin\theta\text{d}\theta=0 \tag{3.143}$$

After substituting Eqs. (3.136)–(3.143) into Eq. (3.134) and simplifying, one gets

$$F_{m,x}=\frac{\pi\mu}{a}\sum_{n=1}^{\infty}\begin{Bmatrix}2n^2(O_nC_n+F_nG_n)-\\ n(n+1)\left[(O_nC_{n+1}+F_nG_{n+1})e^{2\gamma_1}+(C_nO_{n+1}+G_nF_{n+1})e^{-2\gamma_1}\right]\end{Bmatrix} \tag{3.144}$$

Expanding the terms in Eq. (3.144) based on Eqs. (3.106)–(3.109)

$$O_nC_n+F_nG_n=\frac{4\mu_0^2(\mu-\mu_0)a^2M^2}{(\mu+\mu_0)^3}\frac{e^{-2n(\gamma_1-\gamma_2)}e^{-2n\gamma_1}}{\left[1-\frac{(\mu-\mu_0)^2}{(\mu+\mu_0)^2}e^{-2n(\gamma_1-\gamma_2)}\right]^2} \tag{3.145}$$

$$O_nC_{n+1}+F_nG_{n+1}=\frac{4\mu_0^2\left(\mu-\mu_0\right)a^2M^2}{\left(\mu+\mu_0\right)^3}\frac{e^{-2n(\gamma_1-\gamma_2)}}{1-\frac{\left(\mu-\mu_0\right)^2}{\left(\mu+\mu_0\right)^2}e^{-2n(\gamma_1-\gamma_2)}}\frac{e^{-2(n+1)\gamma_1}}{1-\frac{\left(\mu-\mu_0\right)^2}{\left(\mu+\mu_0\right)^2}e^{-2(n+1)(\gamma_1-\gamma_2)}} \tag{3.146}$$

$$C_nO_{n+1}+G_nF_{n+1}=\frac{4\mu_0^2\left(\mu-\mu_0\right)a^2M^2}{\left(\mu+\mu_0\right)^3}\frac{e^{-2n\gamma_1}}{1-\frac{\left(\mu-\mu_0\right)^2}{\left(\mu+\mu_0\right)^2}e^{-2n(\gamma_1-\gamma_2)}}\frac{e^{-2(n+1)(\gamma_1-\gamma_2)}}{1-\frac{\left(\mu-\mu_0\right)^2}{\left(\mu+\mu_0\right)^2}e^{-2(n+1)(\gamma_1-\gamma_2)}} \tag{3.147}$$

By substituting Eqs. (3.145)–(3.147) into Eq. (3.144), the radial magnetic levitation force experienced by the permanent magnet per unit length from the ferrofluid can be obtained as

$$F_{m,x}=\frac{4\pi a\mu_0^2\mu\left(\mu-\mu_0\right)M^2}{\left(\mu+\mu_0\right)^3}\sum_{n=1}^{\infty}\left[2n^2d_nc_n-n\left(n+1\right)\left(d_nc_{n+1}e^{2\gamma_1}+c_nd_{n+1}e^{-2\gamma_1}\right)\right], \tag{3.148}$$

where

$$d_n=\frac{e^{-2n(\gamma_1-\gamma_2)}}{1-\frac{\left(\mu-\mu_0\right)^2}{\left(\mu+\mu_0\right)^2}e^{-2n(\gamma_1-\gamma_2)}} \tag{3.149}$$

$$c_n=\frac{e^{-2n\gamma_1}}{1-\frac{\left(\mu-\mu_0\right)^2}{\left(\mu+\mu_0\right)^2}e^{-2n(\gamma_1-\gamma_2)}} \tag{3.150}$$

To investigate the influence of various parameters on the radial magnetic levitation force, dimensionless parameters are defined as $w_1 = b/R_1$, $w_2 = R_2/R_1$, and $w_3 = b/(R_2 - R_1)$, with w_3 referred to as eccentricity and satisfying the relation $w_1 = w_3(w_2 - 1)$. Based on Eqs. (3.35), (3.40), and (3.41) and the identity $\mathrm{ch}^2\gamma - \mathrm{sh}^2\gamma = 1$,

$$\left(\frac{R_2^2-R_1^2-b^2}{2R_1b}\right)^2-\left(\frac{a}{R_1}\right)^2=1 \tag{3.151}$$

Thus, one has

$$a=\frac{R_1}{2w_1}\sqrt{\left(w_2^2-w_1^2-1\right)^2-4w_1^2} \tag{3.152}$$

and the following equations can be derived from the identity $e^\gamma = \mathrm{ch}\gamma + \mathrm{sh}\gamma$:

$$\gamma_1=\ln\left(\frac{w_2^2-w_1^2-1}{2w_1}+\frac{a}{R_1}\right) \tag{3.153}$$

$$\gamma_2 = \ln\left(\frac{w_2^2 + w_1^2 - 1}{2w_1w_2} + \frac{a}{w_2R_1}\right) \tag{3.154}$$

Assuming the relative permeability of the ferrofluid is μ_r, then

$$\mu = \mu_0\mu_r \tag{3.155}$$

The substitution of this representation into Eq. (3.148) and the incorporation of the representations of the parameters from Eqs. (3.152) to (3.154) give the dimensionless form of the force function

$$\frac{F_{m,x}}{\mu_0 M^2} = \frac{4\pi a\mu_r(\mu_r - 1)}{(\mu_r + 1)^3}\sum_{n=1}^{\infty}\left[2n^2 d_n c_n - n(n+1)\left(d_n c_{n+1} e^{2\gamma_1} + c_n d_{n+1} e^{-2\gamma_1}\right)\right] \tag{3.156}$$

where d_n and c_n are given by Eqs. (3.149) and (3.150), respectively.

The relationship between the dimensionless radial magnetic levitation force per unit length of the permanent magnet, $\frac{F_{m,x}}{\mu_0 M^2}$, obtained from Eq. (3.156) and the eccentricity and radius ratio is shown in Figure 3.2. It can be observed that for a given type of permanent magnet, the radial magnetic levitation force monotonically increases with increasing eccentricity, a trend that has been experimentally confirmed (Yang, 2017, 2021; Yang et al., 2017). When R_2/R_1 is relatively small, indicating a small difference between the radii of the permanent magnet and the outer shell, the radial magnetic levitation force changes almost linearly with eccentricity. Within this range, a larger value of R_2/R_1 corresponds to a faster rate of linear change. For larger values of R_2/R_1 (as shown in Figure 3.2 for R_2/R_1 = 1.4, 2,

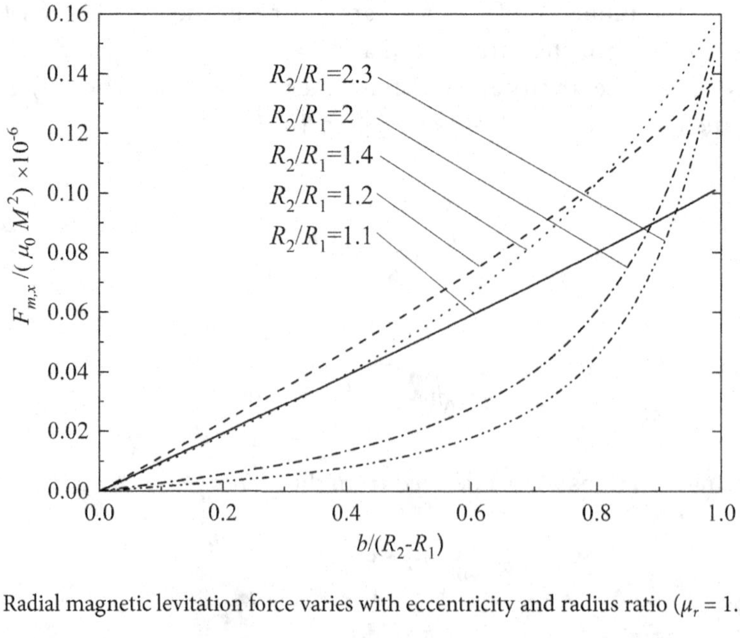

FIGURE 3.2 Radial magnetic levitation force varies with eccentricity and radius ratio (μ_r = 1.1, R_1 = 5 mm).

and 2.3), the force changes slowly at lower eccentricities but gradually accelerates at higher eccentricities. It can be anticipated that as R_2/R_1 increases further, the radial magnetic levitation force can reach significantly high values when the permanent magnet experiences its maximum eccentricity.

Additional results indicate that, under similar conditions, as the radius of the permanent magnet R_1 increases, the radial magnetic levitation force also increases. In other words, the greater the gap between the centred permanent magnet and the outer shell, the stronger the levitation force acting on the permanent magnet with the same eccentricity. Furthermore, the radial magnetic levitation force increases with an increase in the relative permeability μ_r of a ferrofluid.

In a spatial environment where gravity is neglected, a permanent magnet remains radially centred within a ferrofluid when undisturbed by external factors. The stability of this centred position, commonly referred to as the initial magnetic levitation force, is of particular interest. Assuming a slight displacement of the permanent magnet, $b \ll (R_2 - R_1)$, Figure 3.3 illustrates the relationship between the initial radial magnetic levitation force and the radius ratio R_2/R_1 under different permeabilities when the permanent magnet experiences unit eccentricity at $R_1 = 5$ mm and $\frac{b}{R_1} = 0.02$ based on Eq. (3.151). It is evident that as the radius ratio decreases, indicating a closer proximity between the permanent magnet's radius and the outer shell's radius, the magnetic levitation force increases. Furthermore, a higher permeability of the ferrofluid leads to a greater radial levitation force experienced by the magnet during unit eccentricity. Therefore, it can be concluded that by increasing the permeability of the ferrofluid and decreasing the radius ratio between the magnet and the outer shell, a larger initial radial magnetic levitation force can be achieved.

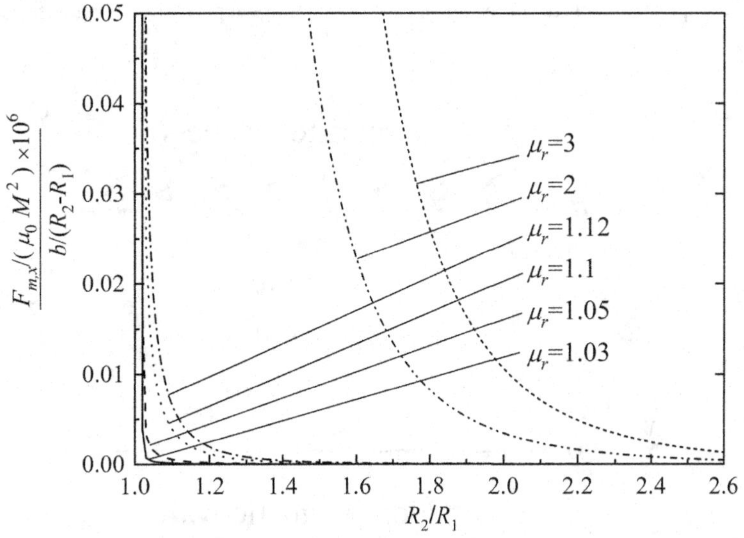

FIGURE 3.3 Relationship between the initial radial magnetic levitation force and the ratio of the outer shell to the permanent magnet radius ($b/R_1 = 0.02$, $R_1 = 5$ mm).

3.4.2 Radial Magnetic Levitation Force Exerted on Cylindrical Composite Magnets Immersed in Ferrofluids

3.4.2.1 Magnetic Field Distribution of Cylindrical Composite Magnets

Sometimes devices utilize cylindrical composite structures made up of multiple cylindrical permanent magnets to enhance the magnetic levitation force while minimizing the adsorption of ferrofluid onto the magnets as depicted in Figure 3.4. In this context, it is assumed that the gap between adjacent permanent magnets is significantly smaller than the overall structural dimensions. Consequently, when calculating the magnetic field distribution of this structure, it can be simplified to a two-dimensional plane as shown in Figure 3.5. This plane consists of one surface of the composite structure and the surface of the non-magnetic wall, maintaining a distance of h_c between them.

The magnetization within the composite structure can be represented as (Rosensweig, 1977)

$$\boldsymbol{M} = \tilde{M}_0 e^{-k_m(y-h_c)}\left[\cos\left(k_m x\right)\boldsymbol{e}_x - \sin\left(k_m x\right)\boldsymbol{e}_y\right] \tag{3.157}$$

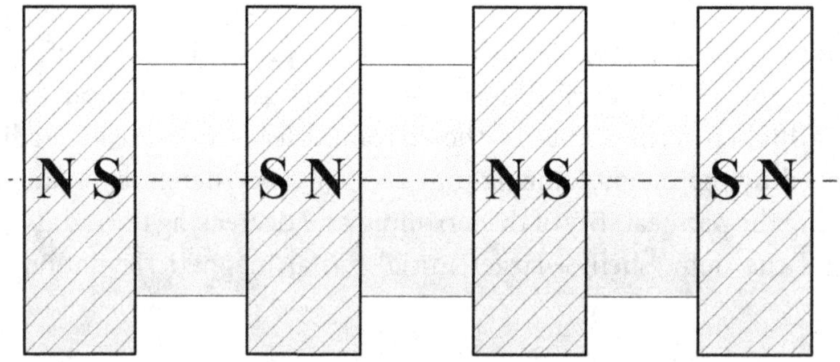

FIGURE 3.4 A composite structure composed of multiple cylindrical permanent magnets.

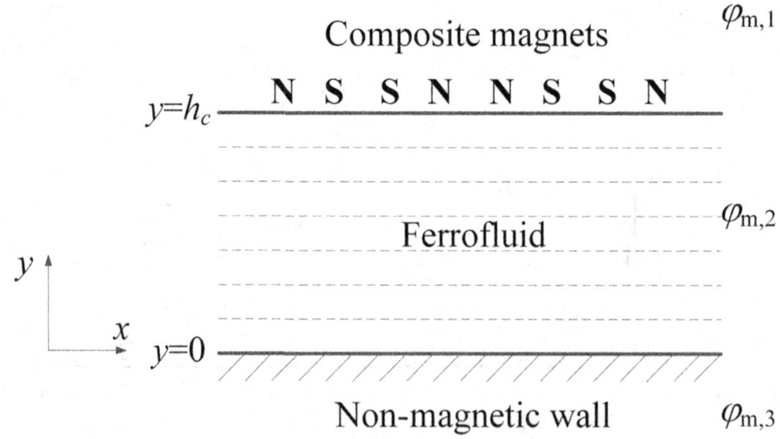

FIGURE 3.5 A model for calculating magnetic field distribution. (Reproduced from Yang and Liu (2018) with permission from IOP Publishing.)

In the absence of ferrofluid, the peak magnetization on the surface of the composite structure attains a value of $\tilde{M}_0 / 2$. The wave number of the periodically arranged permanent magnets within the composite structure is denoted as k_m, which is related to the spacing of the magnetic pole faces of the permanent magnets, $\lambda_m/2$, by the equation $k_m = 2\pi/\lambda_m$.

In the absence of electric current and neglecting displacement current within a ferrofluid, considering a uniform permeability μ that is independent of the magnetic field, the magnetic scalar potential within the fluid still satisfies the Laplace equation (3.45). Furthermore, this equation remains valid in the other two regions, thus leading to a general solution for the scalar magnetic potential in all three regions depicted in Figure 3.5,

$$\varphi_{m,1} = \left(C_1 e^{k_m y} + C_2 e^{-k_m y}\right)\sin\left(k_m x\right) \tag{3.158}$$

$$\varphi_{m,2} = \left(C_3 e^{k_m y} + C_4 e^{-k_m y}\right)\sin\left(k_m x\right) \tag{3.159}$$

$$\varphi_{m,3} = \left(C_5 e^{k_m y} + C_6 e^{-k_m y}\right)\sin\left(k_m x\right) \tag{3.160}$$

Here, one of the undetermined coefficients is set equal to the wave number of the permanent magnet, k_m, while C_1 to C_6 represent constants determined by the boundary conditions. Within the computational domain, the boundary conditions are

$$\varphi_{m,1} \to 0, y \to +\infty \tag{3.161}$$

$$\varphi_{m,3} \to 0, y \to -\infty \tag{3.162}$$

$$\varphi_{m,1} = \varphi_{m,2}, y = h_c \tag{3.163}$$

$$\left(\boldsymbol{B}_1 - \boldsymbol{B}_2\right)\cdot\boldsymbol{n} = 0, y = h_c \tag{3.164}$$

$$\varphi_{m,2} = \varphi_{m,3}, y = 0 \tag{3.165}$$

$$\left(\boldsymbol{B}_2 - \boldsymbol{B}_3\right)\cdot\boldsymbol{n} = 0, y = 0 \tag{3.166}$$

where $\boldsymbol{n} = -\boldsymbol{e}_y$. On the interface where $y = h_c$, there exists a side where $y > h_c$ and on this side, one has

$$H_{n,1} = \frac{\partial \varphi_{m,1}}{\partial y}, M_{n,1} = -\boldsymbol{M}\cdot\boldsymbol{e}_y \,|_{y=h_c} = \tilde{M}_0 \sin\left(k_m x\right) \tag{3.167}$$

Therefore, it can be inferred that

$$B_{n,1} = \mu_0\left(H_{n,1} + M_{n,1}\right) = \mu_0\left[\frac{\partial \varphi_{m,1}}{\partial y} + \tilde{M}_0 \sin\left(k_m x\right)\right] \tag{3.168}$$

Within the region where $0 < y < h_c$, one gets

$$B_{n,1} = \mu H_{n,2} = \mu \frac{\partial \varphi_{m,2}}{\partial y} \tag{3.169}$$

and in the region where $y < 0$,

$$B_{n,3} = \mu_0 H_{n,3} = \mu_0 \frac{\partial \varphi_{m,3}}{\partial y} \tag{3.170}$$

Substitution of Eqs. (3.168)–(3.170) into the boundary conditions (3.164) and (3.166) gets

$$\mu_0 \left[\left(\frac{\partial \varphi_{m,1}}{\partial y} \right)_{y=h} + \tilde{M}_0 \sin(k_m x) \right] = \left(\mu \frac{\partial \varphi_{m,2}}{\partial y} \right)_{y=h_c}, y = h_c \tag{3.171}$$

$$\left(\mu \frac{\partial \varphi_{m,2}}{\partial y} \right)_{y=0} = \mu_0 \left(\frac{\partial \varphi_{m,3}}{\partial y} \right)_{y=0}, y = 0 \tag{3.172}$$

From Eqs. (3.161) and (3.162), it can be observed that $C_1 = 0$ and $C_6 = 0$. Furthermore, from Eq. (3.163) one yields

$$C_2 - e^{2k_m h_c} C_3 - C_4 = 0 \tag{3.173}$$

Introducing Eqs. (3.158) and (3.159) into Eq. (3.171) and simplifying, one gets

$$e^{-k_m h_c} C_2 - \mu_r e^{k_m h_c} C_3 - \mu_r e^{-k_m h_c} C_4 = \frac{\tilde{M}_0}{k_m} \tag{3.174}$$

Then it follows from Eq. (3.165) that

$$C_3 + C_4 - C_5 = 0 \tag{3.175}$$

Substituting Eqs. (3.159) and (3.160) into Eq. (3.172) and simplifying, one obtains

$$\mu_r (C_3 - C_4) - C_5 = 0 \tag{3.176}$$

The system of equations consisting of Eqs. (3.173)–(3.176) is solved to obtain expressions for the unknown coefficients

$$\begin{cases} C_2 = -\dfrac{\tilde{M}_0 \left[e^{2k_m h_c} (\mu_r - 1) + (\mu_r + 1) \right]}{k_m D_m} \\ C_3 = -\dfrac{\tilde{M}_0 (\mu_r + 1)}{k_m D_m} \\ C_4 = -\dfrac{\tilde{M}_0 (\mu_r - 1)}{k_m D_m} \\ C_5 = -\dfrac{2\tilde{M}_0 \mu_r}{k_m D_m} \end{cases} \tag{3.177}$$

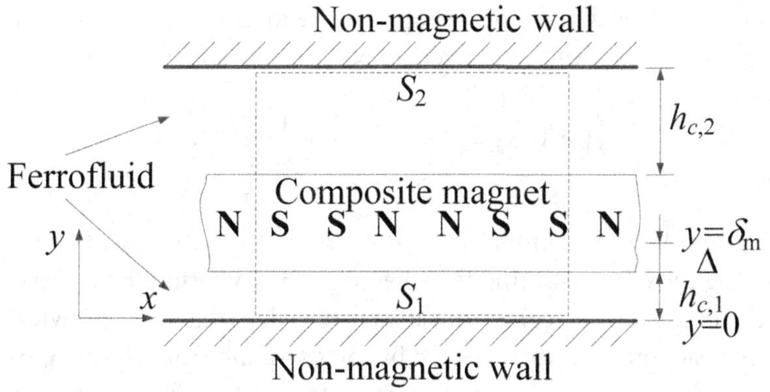

FIGURE 3.6 A model for calculating the magnetic levitation force of a composite magnet in a ferrofluid. (Reproduced from Yang and Liu (2018) with permission from IOP Publishing.)

where

$$D_m = (\mu_r - 1)^2 e^{-k_m h_c} - (\mu_r + 1)^2 e^{k_m h_c}$$

3.4.2.2 Radial Magnetic Levitation Force Exerted on the Cylindrical Composite Magnet

In calculating the levitation force acting on a composite magnet, it is assumed that the structure extends infinitely long in its axial direction as illustrated in Figure 3.6. When the composite magnet is centred between non-magnetic walls, there is a gap on each side, designated as δ_m. If the composite magnet deviates from the central position by Δ, the distance to the nearest wall, denoted as $y = 0$, is expressed as $y = h_{c,1}$. The surfaces S_1 and S_2 are considered to have a unit length in the direction perpendicular to the plane of the paper and are approximately planar.

It is supposed that the ferrofluid behaves as a linearly magnetizable medium, its stress tensor can be expressed according to Eq. (2.222) as

$$\mathbf{T} = -\left(p_0 + p_s + \frac{1}{2}\mu H^2\right)\mathbf{I} + \boldsymbol{BH} \tag{3.178}$$

where p_s is the magnetostrictive pressure. The body force density of ferrofluids can be represented by its stress tensor as

$$\boldsymbol{f} = \nabla \cdot \mathbf{T} = -\nabla(p_0 + p_s) - \mu H \nabla H + \frac{B}{H}(\boldsymbol{H} \cdot \nabla)\boldsymbol{H} = -\nabla(p_0 + p_s) \tag{3.179}$$

When the system depicted in Figure 3.6 reaches a state of equilibrium, the net force acting on a unit volume within the ferrofluid becomes zero. Neglecting the gravitational force of the ferrofluid, one has $\boldsymbol{f} = \boldsymbol{0}$, implying that the term $(p_0 + p_s)$ remains constant regardless of the position.

Taking the CV enclosed by the dashed lines in Figure 3.6 as the object of study, given the composite structure extends infinitely in the x-direction, only the upper and lower base surfaces S_1 and S_2 are considered. The average surface force acting on these two surfaces of the ferrofluid per unit area can be represented based on its stress tensor as

$$\frac{1}{S_m}\oiint_{S_m} \boldsymbol{n}\cdot\mathbf{T}\mathrm{d}S_m = \frac{1}{S_1}\iint_{S_1} \boldsymbol{n}_1\cdot\mathbf{T}\mathrm{d}S_1 + \frac{1}{S_2}\iint_{S_2} \boldsymbol{n}_2\cdot\mathbf{T}\mathrm{d}S_2 \tag{3.180}$$

where $\boldsymbol{n}_1$ and $\boldsymbol{n}_2$ are the unit vectors in the direction of the outer normal to the respective surfaces enclosing the CV. Assuming that the magnetic levitation force exerted by the ferrofluid on the unit area of the composite structure is denoted as $\boldsymbol{F}_m$, Newton's Third Law dictates that the reaction force experienced by the ferrofluid from the composite structure is $-\boldsymbol{F}_m$. Based on this, the force balance equation for the ferrofluid within the CV can be established

$$-\boldsymbol{F}_m + \frac{1}{S_1}\iint_{S_1} \boldsymbol{n}_1\cdot\mathbf{T}\mathrm{d}S_1 + \frac{1}{S_2}\iint_{S_2} \boldsymbol{n}_2\cdot\mathbf{T}\mathrm{d}S_2 = \mathbf{0}$$

Then, the magnetic levitation force acting on the composite structure can be derived as

$$\boldsymbol{F}_m = \frac{1}{S_1}\iint_{S_1} \boldsymbol{n}_1\cdot\mathbf{T}\mathrm{d}S_1 + \frac{1}{S_2}\iint_{S_2} \boldsymbol{n}_2\cdot\mathbf{T}\mathrm{d}S_2 \tag{3.181}$$

To simplify calculations, the stress tensor in Eq. (3.178) is expressed in the following form:

$$\begin{aligned}\mathbf{T} = &\left[-(p_0+p_s)+\frac{1}{2}\mu\left(H_x^2-H_y^2\right)\right]\boldsymbol{e}_x\boldsymbol{e}_y + \mu H_x H_y \boldsymbol{e}_x\boldsymbol{e}_y + \mu H_x H_y \boldsymbol{e}_y\boldsymbol{e}_x \\ &+\left[-(p_0+p_s)+\frac{1}{2}\mu\left(H_y^2-H_x^2\right)\right]\boldsymbol{e}_y\boldsymbol{e}_y + \left[-(p_0+p_s)-\frac{1}{2}\mu H^2\right]\boldsymbol{e}_z\boldsymbol{e}_z\end{aligned} \tag{3.182}$$

For surface S_1, with its unit normal vector $\boldsymbol{n}_1 = -\boldsymbol{e}_y$, the application of Eq. (3.182) results in the integration of the first term on the right-hand side of Eq. (3.181)

$$\frac{1}{S_1}\iint_{S_1} \boldsymbol{n}_1\cdot\mathbf{T}\mathrm{d}S_1 = -\frac{1}{S_1}\left\{\boldsymbol{e}_x\iint_{S_1}\mu H_x H_y \mathrm{d}S_1 + \boldsymbol{e}_y\iint_{S_1}\left[-(p_0+p_s)+\frac{1}{2}\mu\left(H_y^2-H_x^2\right)\right]\mathrm{d}S_1\right\} \tag{3.183}$$

Since the term $(p_0 + p_s)$ is independent of position, and considering the periodic arrangement of permanent magnets in the x-direction within the composite structure, the integration is limited to a length of λ_m, representing one wavelength in the x-direction. Consequently, $S_1 = S_2 = 2\pi/k_m$ and $dS_2 = dx$. Therefore, Eq. (3.183) becomes

$$\frac{1}{S_1}\iint_{S_1} \boldsymbol{n}_1\cdot\mathbf{T}\mathrm{d}S_1 = -\frac{k_m}{2\pi}\left[\boldsymbol{e}_x\iint_0^{\frac{2\pi}{k_m}}\mu H_x H_y \mathrm{d}x + \boldsymbol{e}_y\frac{\mu}{2}\iint_0^{\frac{2\pi}{k_m}}\left(H_y^2-H_x^2\right)\mathrm{d}x\right] + \boldsymbol{e}_y\left(p_0+p_s\right) \tag{3.184}$$

Calculating the magnetic field components using the relations between the scalar magnetic potential and the magnetic field intensity, as given by Eqs. (3.42) and (3.159), yields

$$H_x = -\frac{\partial \varphi_{m,2}}{\partial x} = -k_m\left(C_3 e^{k_m y} + C_4 e^{-k_m y}\right)\cos\left(k_m x\right) \tag{3.185}$$

$$H_y = -\frac{\partial \varphi_{m,2}}{\partial y} = -k_m\left(C_3 e^{k_m y} + C_4 e^{-k_m y}\right)\sin\left(k_m x\right) \tag{3.186}$$

After substituting Eqs. (3.185) and (3.186) into Eq. (3.184) and simplifying, one obtains

$$\frac{1}{S_1}\iint_{S_1} \boldsymbol{n}_1 \cdot \mathbf{T} dS_1 = \boldsymbol{e}_y \mu k_m^2 C_3 C_4 + \boldsymbol{e}_y\left(p_0 + p_s\right) \tag{3.187}$$

Putting the results of C_3 and C_4 in Eq. (3.177) into Eq. (3.187) gives

$$\frac{1}{S_1}\iint_{S_1} \boldsymbol{n}_1 \cdot \mathbf{T} dS_1 = \boldsymbol{e}_y \frac{\mu \tilde{M}_0^2\left(\mu_r^2 - 1\right)}{\left[\left(\mu_r - 1\right)^2 e^{-k_m h_{c,1}} - \left(\mu_r + 1\right)^2 e^{k_m h_{c,1}}\right]^2} + \boldsymbol{e}_y\left(p_0 + p_s\right) \tag{3.188}$$

For the surface S_2, its unit normal vector is $\boldsymbol{n}_2 = \boldsymbol{e}_y$. By the same token, one obtains

$$\frac{1}{S_2}\iint_{S_2} \boldsymbol{n}_1 \cdot \mathbf{T} dS_2 = -\boldsymbol{e}_y \frac{\mu \tilde{M}_0^2\left(\mu_r^2 - 1\right)}{\left[\left(\mu_r - 1\right)^2 e^{-k_m h_{c,2}} - \left(\mu_r + 1\right)^2 e^{k_m h_{c,2}}\right]^2} - \boldsymbol{e}_y\left(p_0 + p_s\right) \tag{3.189}$$

Substituting Eqs. (3.188) and (3.189) into Eq. (3.181), and further introducing $h_{c,1} = \delta_m - \Delta$ and $h_{c,2} = \delta_m + \Delta$, yields

$$\begin{aligned} \boldsymbol{F}_m &= \boldsymbol{e}_y \mu \tilde{M}_0^2\left(\mu_r^2 - 1\right)\left\{\frac{1}{\left[\left(\mu_r - 1\right)^2 e^{-k_m h_{c,1}} - \left(\mu_r + 1\right)^2 e^{k_m h_{c,1}}\right]^2} - \frac{1}{\left[\left(\mu_r - 1\right)^2 e^{-k_m h_{c,2}} - \left(\mu_r + 1\right)^2 e^{k_m h_{c,2}}\right]^2}\right\} \\ &= \boldsymbol{e}_y \mu \tilde{M}_0^2\left(\mu_r^2 - 1\right)\frac{2\left(\xi_1^2 - \xi_2^2\right)\mathrm{sh}2k_m\Delta}{\left(\xi_1^2 - 2\xi_1\xi_2\mathrm{ch}2k_m\Delta + \xi_2^2\right)^2} \end{aligned} \tag{3.190}$$

where

$$\xi_1 = \left(\mu_r + 1\right)^2 e^{k_m \delta_m}, \quad \xi_2 = \left(\mu_r - 1\right)^2 e^{-k_m \delta_m}$$

Equation (3.190) represents the radial magnetic levitation force per unit area experienced by a composite magnet when it deviates from its central position by Δ.

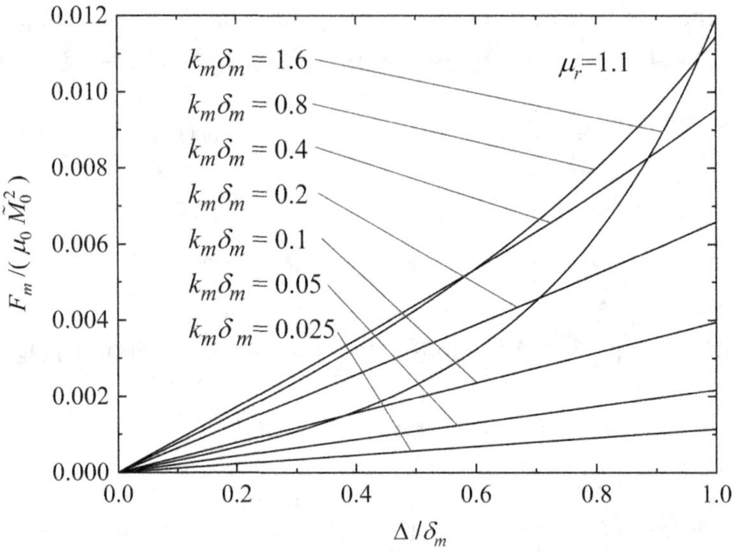

FIGURE 3.7 The magnetic levitation force exerted on composite magnets in a ferrofluid.

Figure 3.7 illustrates the dimensionless magnetic levitation force per unit area, $F/\mu_0 \tilde{M}_0^2$, calculated using Eq. (3.190) for various values of $k_m\delta_m$ when $\mu_r = 1.1$. The curves represent the dependence of force on eccentricity for composite structures composed of a specific type of permanent magnet. It is evident that the magnetic levitation force exhibits a monotonic increase with increasing eccentricity for this composite structure. When $k_m\delta_m$ is relatively small, indicating a larger spacing between the pole faces of the permanent magnet, the magnetic levitation force changes almost linearly with eccentricity. As the spacing decreases, the rate of linear change increases. Within this range, for a given eccentricity, a larger value of $k_m\delta_m$ (smaller spacing between the pole faces) leads to a larger magnetic levitation force. On the other hand, when $k_m\delta_m$ is larger (as shown in Figure 3.7 for $k_m\delta_m = 0.8$ and 1.6), indicating a smaller spacing between the magnetic pole faces, the magnetic levitation force exhibits a slower change at lower eccentricities but gradually accelerates at higher eccentricities. Additional findings suggest that, for a constant value of $k_m\delta_m$, an increased relative permeability of the ferrofluid leads to an augmented magnetic levitation force on the composite magnet.

When the composite magnet comes into contact with a non-magnetic wall, and the gap size is relatively large compared to the spacing of the magnetic pole faces, it follows that $\Delta = \delta_m$ and $k_m\delta_m \to \infty$. Substituting these values into Eq. (3.190) yields the maximum value of the magnetic levitation force

$$\boldsymbol{F}_{\max} = \boldsymbol{e}_y \mu \tilde{M}_0^2 \left(\mu_r^2 - 1\right) \frac{\left[\left(\mu_r + 1\right)^4 e^{2k_m\delta_m} - \left(\mu_r - 1\right)^4 e^{-2k_m\delta_m}\right]\left(e^{2k_m\Delta} - e^{-2k_m\Delta}\right)}{\left[\left(\mu_r + 1\right)^4 e^{2k_m\delta_m} - \left(\mu_r + 1\right)^2 \left(\mu_r - 1\right)^2 \left(e^{2k_m\Delta} + e^{-2k_m\Delta}\right) + \left(\mu_r - 1\right)^4 e^{-2k_m\delta_m}\right]^2}$$

$$= \boldsymbol{e}_y \tilde{M}_0^2 \frac{\mu_0\left(\mu_r^2 - 1\right)}{16\mu_r} \tag{3.191}$$

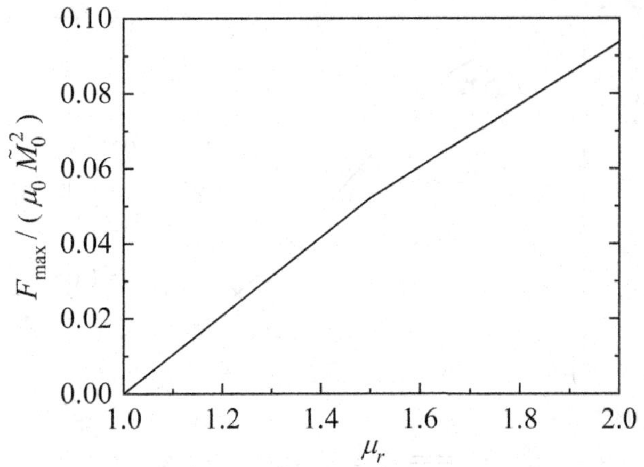

FIGURE 3.8 Variation of the maximum magnetic levitation force experienced by the composite structure with the relative permeability of the ferrofluid.

Figure 3.8 depicts the variation of the maximum magnetic levitation force experienced by the composite magnet with respect to the relative permeability of the ferrofluid. As the relative permeability increases, the maximum magnetic levitation force also rises, demonstrating a nearly linear trend.

In a spatial environment where gravity is negligible, the composite magnet radially centres itself within the ferrofluid when undisturbed by external factors. Of primary concern is the stability of maintaining this centred position, which is determined by the magnitude of the initial magnetic levitation force. Assuming a minute displacement of the composite magnet, denoted as $\Delta \to 0$, where $k_m\Delta \ll 1$, one has

$$e^{2k_m\Delta} + e^{-2k_m\Delta} \approx 2$$

$$e^{2k_m\Delta} - e^{-2k_m\Delta} \approx (1 + k_m\Delta)^2 - (1 - k_m\Delta)^2 = 4k_m\Delta$$

With the aid of these two equations and drawing upon Eq. (3.190), the dimensionless initial magnetic levitation force per unit area for a composite magnet with unit eccentricity can be obtained as

$$\frac{F_{\text{ini}}}{\mu_0 M_0^2} / \frac{\Delta}{\delta_m} = 4k_m\delta_m\mu_r\left(\mu_r^2 - 1\right)\frac{(\mu_r + 1)^2 e^{k_m\delta_m} + (\mu_r - 1)^2 e^{-k_m\delta_m}}{\left[(\mu_r + 1)^2 e^{k_m\delta_m} - (\mu_r - 1)^2 e^{-k_m\delta_m}\right]^3} \tag{3.192}$$

Figure 3.9 illustrates the variation of the initial magnetic levitation force experienced by a composite magnet with unit eccentricity as a function of the permanent magnets' pole face spacing when ferrofluids with multiple relative permeabilities are employed. It is evident that for a given ferrofluid, there exists an optimal value of $k_m\delta_m$, representing the most favourable distance between the pole faces of permanent magnets, that maximizes the initial magnetic levitation force at unit eccentricity. Notably, this maximum value increases with the relative permeability of ferrofluids, but the optimal value shifts towards lower

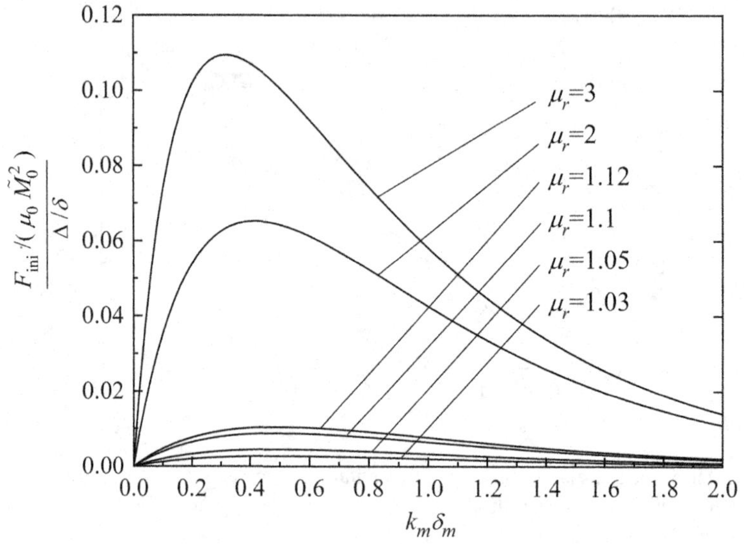

FIGURE 3.9 Variation of the initial magnetic levitation force with the spacing between the magnetic pole faces of permanent magnets at unit eccentricity.

$k_m\delta_m$ values. Therefore, to achieve the maximum initial magnetic levitation force by enhancing the ferrofluid permeability, it is imperative to increase the pole face spacing of the permanent magnet in the composite structure.

3.4.3 Calculation of Axial Magnetic Levitation Force on Permanent Magnets in Ferrofluids Using the Magnetic Charge Image Method

This section employs the classical method of image charges from the perspective of magnetic charges in electromagnetic fields to directly compute the axial magnetic levitation force exerted on a cylindrical permanent magnet immersed in a ferrofluid.

3.4.3.1 Equivalent Magnetic Charge Model for Cylindrical Permanent Magnets

The equivalent magnetic charge perspective views magnetic poles as regions where magnetic charges accumulate, with a concentration of positive charges at the N-pole and negative charges at the S-pole. Analogous to the corresponding expressions in electricity, the static magnetic field, from the perspective of magnetic charges, also follows a magnetic Coulomb's law, which describes the interaction force between the two magnetic charges. In the case of a cylindrical permanent magnet, the magnetic charges are exclusively distributed on its outer surface, with a surface density of magnetic charge

$$\sigma_m = \boldsymbol{J}\cdot\boldsymbol{n} \tag{3.193}$$

where $\boldsymbol{J}$ is the magnetic polarization of the permanent magnet and $\boldsymbol{n}$ is the outward normal unit vector on the surface of the permanent magnet. In accord with the relationship between $\boldsymbol{J}$ and the magnetization for the permanent magnet $\boldsymbol{M}$,

$$\boldsymbol{J} = \mu_0\boldsymbol{M} \tag{3.194}$$

one obtains the charge density on the outer surface of the permanent magnet

$$\sigma_m = \mu_0 \boldsymbol{M} \cdot \boldsymbol{n} \tag{3.195}$$

It can be inferred from Eq. (3.195) that for a cylindrical permanent magnet uniformly magnetized along its axis, the magnetic charge is solely localized on its two planar end faces. Consequently, the cylindrical permanent magnet can be regarded as a collection of uniformly distributed magnetic charges on its planar end faces.

For the uniformly magnetized permanent magnet shown in Figure 3.10, there exists an approximate relationship between its magnetization and its residual magnetic induction B_r

$$\boldsymbol{M} = \frac{B_r}{\mu_0} \boldsymbol{e}_z \tag{3.196}$$

Then the magnitude of the magnetic charge surface density on the end face of a uniformly magnetized cylindrical permanent magnet can be obtained

$$\sigma_m = B_r \tag{3.197}$$

The $\boldsymbol{M}$-pointing end is distributed with positive magnetic charges, while the other end is distributed with negative magnetic charges.

3.4.3.2 Magnetic Field Distribution

A cylindrical permanent magnet with a radius of R_1 is vertically immersed in a ferrofluid exhibiting a constant permeability of μ. The ferrofluid is separated from a non-magnetic medium by an infinite interface, with the magnet's axis aligned perpendicularly to the

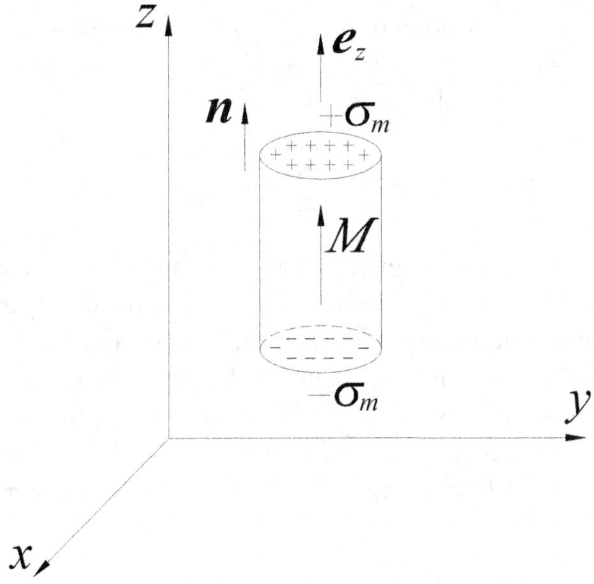

FIGURE 3.10 A cylindrical permanent magnet and equivalent magnetic charges on its two end faces.

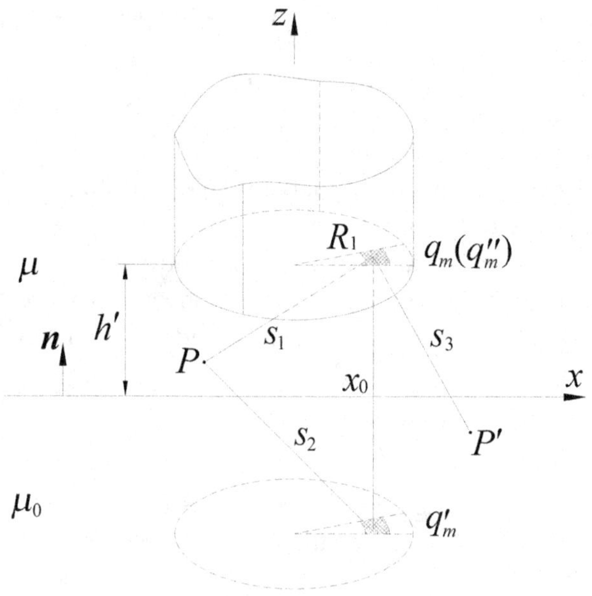

FIGURE 3.11 Model of calculating the image magnetic charge intensity.

interface. The distance from the end face of the permanent magnet, adjacent to the interface, is denoted as h'. A coordinate system is established with the interface serving as one of the coordinate planes, as shown in Figure 3.11. At any position on the end face of the permanent magnet near the interface, a surface element is chosen. The magnetic charges distributed on this surface element can be regarded as point magnetic charges with an intensity of q_m.

Based on the uniqueness theorem of magnetic fields, regardless of the distribution of magnetic charges in two different media, the fields they generate must adhere to the following boundary conditions on the plane $z = 0$.

$$\varphi_{m,1} = \varphi_{m,2}, \mu \frac{\partial \varphi_{m,1}}{\partial z} = \mu_0 \frac{\partial \varphi_{m,2}}{\partial z} \tag{3.198}$$

where $\varphi_{m,1}$ and $\varphi_{m,2}$ are the scalar magnetic potentials on both sides of the interface, respectively. Assuming that the scalar magnetic potential $\varphi_{m,1}$ is generated by the actual magnetic charge q_m, the interface's effect is represented by an image magnetic charge with intensity q'_m. This image magnetic charge must be situated at the geometric mirror point of the actual magnetic charge to satisfy the boundary conditions. Furthermore, it is supposed that the image magnetic charge generates a scalar magnetic potential $\varphi_{m,2}$. Then, the total scalar magnetic potential at point P within the ferrofluid is given by the sum of the two scalar magnetic potentials

$$\varphi_{m,1} + \varphi_{m,2} = \frac{q_m}{4\pi\mu s_1} + \frac{q'_m}{4\pi\mu s_2} \tag{3.199}$$

Here, s_1 and s_2 represent the distances from point $P(x, z)$ to the actual and image magnetic charges, respectively. Assuming that the scalar magnetic potential $\varphi_{m,3}$ in a non-magnetic medium is generated by a magnetic charge with intensity q_m'' located at the original magnetic charge position, the scalar magnetic potential at any point $P'(x', z')$ in the non-magnetic medium is given by

$$\varphi_{m,3} = \frac{q_m''}{4\pi\mu s_3} \tag{3.200}$$

where s_3 is the distance from point P' to the assumed location of the magnetic charge.

According to the boundary conditions of the original magnetic field problem, at a specific point $(x, 0)$ on the plane $z = 0$, one has

$$\varphi_{m,1} + \varphi_{m,2} = \varphi_{m,3}, \mu\frac{\partial\left(\varphi_{m,1} + \varphi_{m,2}\right)}{\partial z} = \mu_0\frac{\partial\varphi_{m,3}}{\partial z} \tag{3.201}$$

Substitution of Eqs. (3.199) and (3.200) into the boundary condition (3.201), and with the aid of

$$s_1 = \sqrt{\left(x - x_0\right)^2 + \left(h' - z\right)^2}, s_2 = \sqrt{\left(x - x_0\right)^2 + \left(h' + z\right)^2}, s_3 = \sqrt{\left(x' - x_0\right)^2 + \left(h' - z\right)^2}$$

and

$$s_1 = s_2 = s_3, \text{when } z = 0$$

one gets

$$\frac{q_m + q_m'}{\mu} = \frac{q_m''}{\mu_0} \tag{3.202}$$

$$\left.\left(\frac{q_m}{4\pi}\frac{h' - z}{s_1^3} - \frac{q_m'}{4\pi}\frac{h' + z}{s_2^3}\right)\right|_{z=0, s_1=s_2=s_3} = \left.\frac{q_m''}{4\pi}\frac{h' - z}{s_3^3}\right|_{z=0, s_1=s_2=s_3} \tag{3.203}$$

Consequently, a system of equations is obtained

$$\begin{cases} \mu_0\left(q_m + q_m'\right) = \mu q_m'' \\ q_m - q_m' = q_m'' \end{cases} \tag{3.204}$$

Solving this system of equations leads to

$$q_m' = \frac{\mu_r - 1}{\mu_r + 1} q_m \tag{3.205}$$

Based on the arbitrary position of the original magnetic charge on the end face of the permanent magnet, it can be inferred that the image magnetic charges of all the end-face magnetic charges of a cylindrical permanent magnet occupy geometrically mirrored positions with respect to the interface. The area occupied by the image magnetic charges also forms a circular shape with a radius of R_1, and the surface density of the image magnetic charge intensity is

$$\sigma_m' = \frac{\mu_r - 1}{\mu_r + 1}\sigma_m \tag{3.206}$$

Correspondingly, the magnetic field distribution generated by a cylindrical permanent magnet within a ferrofluid, particularly near the boundary of its end face, is equivalent to the magnetic field distribution produced by the magnetic charges on the end face of the permanent magnet and their corresponding image charges. Likewise, the magnetic levitation force exerted by the ferrofluid on the permanent magnet is equivalent to the interaction force between the magnetic charges on its end face and their corresponding image charges.

3.4.3.3 Calculation of the Magnetic Levitation Force

Utilizing the equivalence between the magnetic levitation force exerted on a cylindrical permanent magnet within a ferrofluid and the interaction force between the end-face magnetic charges of the permanent magnet and its image charges near the interface of the medium, now one can determine the magnetic levitation force acting on the permanent magnet when it is situated a distance of h' away from the interface.

As shown in Figure 3.12, a permanent magnet is immersed in a ferrofluid, with one end located near the interface of two media and the other end extending to infinity. It is assumed that the ferrofluid surrounding the permanent magnet extends infinitely above the interface, while the region beneath the interface is non-magnetic. A surface element is chosen on the end face of the permanent magnet that is close to the media interface. The radius of the surface on which this element lies is r, and the angle formed between the line

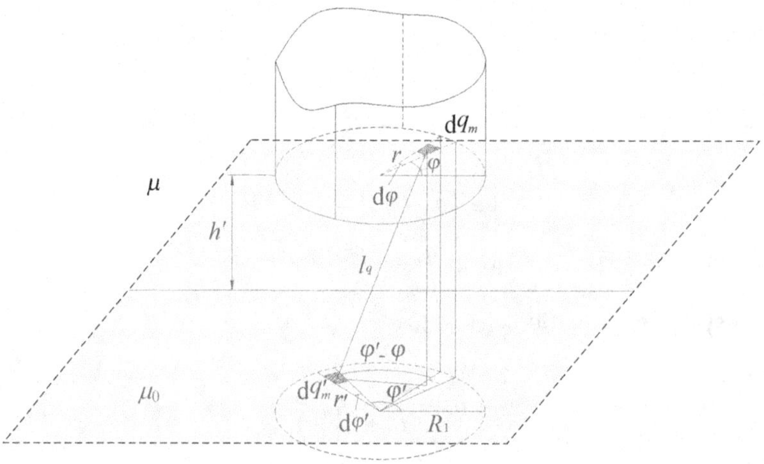

FIGURE 3.12 Model of calculating the magnetic levitation force using the magnetic charge image method.

connecting it to the centre of the end face and a plane passing through the axis of the permanent magnet is designated as φ. The magnetic charge intensity on this surface element is given by $dq_m = \sigma_m r d\varphi dr$. Similarly, a surface element is chosen on the face where the image magnetic charge resides. The radius of the surface on which this element lies is denoted as r', and the angle formed between the line connecting it to the centre of the end face and the zero-angle plane is designated as φ'. The magnetic charge intensity on this surface element is given by $dq'_m = \sigma'_m r' d\varphi' dr'$.

According to Coulomb's law, the magnitude of the interactive force between the magnetic charges on the two surface elements is given by

$$dF_m = \frac{dq_m dq'_m}{4\pi\mu l_q^2} = \frac{\sigma_m \sigma'_m}{4\pi\mu} \frac{rr' d\varphi d\varphi' dr dr'}{l_q^2} \tag{3.207}$$

where l_q is the distance between the two elements, its magnitude is

$$l_q = \sqrt{4h'^2 + r^2 + r'^2 - 2rr'\cos(\varphi' - \varphi)}$$

Then the magnitude of the interactive force between all magnetic charges located on the two end faces is given by $F_m = \int dF_m$. Due to the axial symmetry of the magnetic charge distribution, only the vertical component of the total interactive force is non-zero. Therefore, the interactive force between the magnetic charges reads

$$F_m = \int_0^{2\pi}\int_0^{2\pi}\int_0^{R_1}\int_0^{R_1} \frac{h'\sigma_m \sigma'_m}{2\pi\mu} \frac{rr' d\varphi d\varphi' dr dr'}{\left[4h'^2 + r^2 + r'^2 - 2rr'\cos(\varphi' - \varphi)\right]^{3/2}} \tag{3.208}$$

Letting $\xi = r/R_1$, $\xi' = r'/R_1$, and $\xi'' = 2h'/R_1$, Then, one has $dr = R_1 d\xi$ and $dr' = R_1 d\xi'$. Consequently, Eq. (3.208) transforms into

$$F_m = \frac{h' R_1 \sigma_m \sigma'_m}{2\pi\mu} \int_0^{2\pi}\int_0^{2\pi}\int_0^{1}\int_0^{1} \frac{h'\sigma_m \sigma'_m}{2\pi\mu} \frac{\xi\xi' d\xi d\xi' d\varphi d\varphi'}{\left[\xi''^2 + \xi^2 + \xi'^2 - 2\xi\xi'\cos(\varphi' - \varphi)\right]^{3/2}} \tag{3.209}$$

Because the integral is calculated as

$$\begin{aligned}
&\int_0^1 \frac{\xi d\xi}{\left[\xi''^2 + \xi^2 + \xi'^2 - 2\xi\xi'\cos(\varphi' - \varphi)\right]^{3/2}} \\
&= \left. \frac{\xi\xi'\cos(\varphi' - \varphi) - \xi''^2 - \xi'^2}{\left[\xi''^2 + \xi'^2 \sin^2(\varphi' - \varphi)\right]\left[\xi''^2 + \xi^2 + \xi'^2 - 2\xi\xi'\cos(\varphi' - \varphi)\right]^{1/2}} \right|_0^1 \\
&= \frac{\xi'\cos(\varphi' - \varphi) - \xi''^2 - \xi'^2}{\left[\xi''^2 + \xi'^2 \sin^2(\varphi' - \varphi)\right]\left[\xi''^2 + 1 + \xi'^2 - 2\xi'\cos(\varphi' - \varphi)\right]^{1/2}} + \frac{\left(\xi''^2 + \xi'^2\right)^{1/2}}{\xi''^2 + \xi'^2 \sin^2(\varphi' - \varphi)}
\end{aligned}$$

then Eq. (3.209) becomes

$$F_m = \frac{R_1(\mu_r - 1)\sigma_m^2 h'}{2\pi\mu(\mu_r + 1)} \int_0^{2\pi}\int_0^{2\pi}\int_0^{1} \left\{ \frac{\xi'\left[\xi'\cos(\varphi' - \varphi) - \xi''^2 - \xi'^2\right]}{\left[\xi''^2 + \xi'^2 \sin^2(\varphi' - \varphi)\right]\left[\xi''^2 + 1 + \xi'^2 - 2\xi'\cos(\varphi' - \varphi)\right]^{1/2}} + \frac{\xi'\left(\xi''^2 + \xi'^2\right)^{1/2}}{\xi''^2 + \xi'^2 \sin^2(\varphi' - \varphi)} \right\} \mathrm{d}\xi' \mathrm{d}\varphi \mathrm{d}\varphi' \tag{3.210}$$

Since the image magnetic charges represented by Eq. (3.205) share the same sign as the original magnetic charges, the interaction force between them is repulsive. Equation (3.210) represents the interaction force between the magnetic charges on the end face of a cylindrical permanent magnet and its corresponding image magnetic charges, which is also referred to as the axial magnetic levitation force experienced by the permanent magnet within the ferrofluid.

The Romberg integration method is employed to compute the integral in Eq. (3.210). Figure 3.13 illustrates the relationship between the axial magnetic levitation force exerted by ferrofluids on the permanent magnet and its levitation height. This analysis considers two distinct permanent magnet radii: 5 mm and 12 mm, with corresponding relative permeabilities of the ferrofluid being 1.07 and 1.12, respectively. The cylindrical container housing the ferrofluid has a radius of 40 mm.

Figure 3.13 illustrates that the axial magnetic levitation force experienced by the permanent magnet decreases significantly with increasing levitation height. For ferrofluids with the same permeability, a larger radius of the permanent magnet results in a greater magnetic levitation force at a given levitation position, as evident from Eq. (3.210).

The results depicted in Figure 3.13 further reveal that, for permanent magnets with identical radii, a higher permeability of the ferrofluid leads to a greater magnetic levitation force when suspended at the same position. However, this correlation is only valid within a specific range of permeabilities. Given a fixed radius of the permanent magnet and a constant suspension height, the integral in Eq. (3.210) remains unchanged, denoted as C. Taking the derivative of Eq. (3.210) with respect to μ_r yields

$$\frac{\partial F_m}{\partial \mu_r} = \frac{R_1 \sigma_m^2 h' C}{2\pi\mu(\mu_r + 1)} \frac{2\mu_r - \mu_r^2 + 1}{\left[\mu_r(\mu_r + 1)\right]^2} \tag{3.211}$$

From Eq. (3.211) it can be inferred that when $\mu_r < 1 + \sqrt{2}$, one has $\partial F_m / \partial \mu_r > 0$, indicating that the magnetic levitation force acting on the permanent magnet increases as the permeability of the surrounding ferrofluid increases. However, once the relative permeability exceeds this value, the magnetic levitation force declines with further increases in permeability. In practical applications, stability constraints often limit the relative permeability of ferrofluids to values well below $1 + \sqrt{2}$. Therefore, it is generally concluded that the axial magnetic levitation force increases as permeability increases.

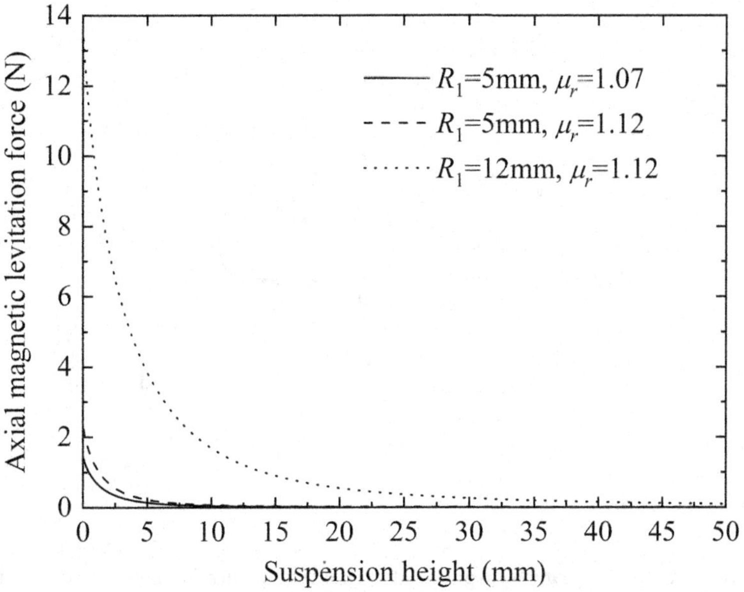

FIGURE 3.13 The axial magnetic levitation force calculated using the magnetic charge image method.

3.4.4 Calculation of Axial Magnetic Levitation Force on Permanent Magnets in Ferrofluids Using the Current Image Method

3.4.4.1 A Magnetization Current Model for Cylindrical Permanent Magnets

For a given axially magnetized cylindrical permanent magnet, the magnetic field it generates is equivalent to that produced by the magnetization current distributed on its cylindrical surface. The magnetization current per unit length, denoted as the magnetization current surface density $\boldsymbol{i}_m$, is

$$\boldsymbol{i}_m = \boldsymbol{M} \times \boldsymbol{n} \tag{3.212}$$

where $\boldsymbol{n}$ represents the outward-pointing unit normal vector for the cylindrical surface of the permanent magnet.

For a uniformly magnetized cylindrical permanent magnet as shown in Figure 3.14, the relationship between the magnetization $\boldsymbol{M}$ and its residual magnetic induction B_r can be derived from Eq. (3.196) as

$$\boldsymbol{i}_m = \frac{B_r}{\mu_0}\boldsymbol{e}_z \times \boldsymbol{n} = \frac{B_r}{\mu_0}\boldsymbol{e}_\theta \tag{3.213}$$

where the unit vector $\boldsymbol{e}_\theta$ is tangent to the circumferential direction of the cylindrical surface of the permanent magnet. Consequently, the magnetization current is tangent to the cylindrical surface and perpendicular to its magnetization. This magnetization current is uniformly distributed along the axial direction of the permanent magnet, with a density approximately equal to B_r/μ_0, as illustrated in Figure 3.14, where I_m represents the magnitude of the magnetization current.

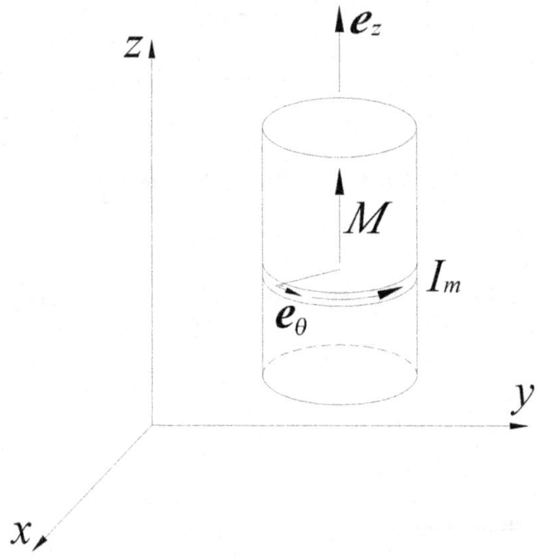

FIGURE 3.14 A cylindrical permanent magnet and magnetization current on its cylindrical surface.

3.4.4.2 Magnetic Field Distribution

Based on the uniqueness theorem of magnetic fields, the boundary's influence on the field can be replaced by equivalent currents outside the field domain provided that the current distribution and boundary conditions within the magnetic field region remain unchanged. In the case of a cylindrical permanent magnet immersed in a ferrofluid occupying an infinite region, as shown in Figure 3.15, the boundary conditions at the interface $z = 0$ are

$$H_{\tau,1} = H_{\tau,2}, B_{n,1} = B_{n,2} \tag{3.214}$$

where $H_{\tau,1}$ and $H_{\tau,2}$ are the tangential components of the magnetic field intensity on the ferrofluid and non-magnetic medium sides of the interface, respectively, $B_{n,1}$ and $B_{n,2}$ are the normal components of the magnetic flux density on these two sides.

Under the same assumptions as in Section 3.4.3, a coordinate system is established with the interface as one of the coordinate planes as shown in Figure 3.15. A current $\boldsymbol{I}_m$ is selected at any position on the cylindrical surface of the permanent magnet, flowing in a tangential direction aligned with the current's flow. Due to the vertical placement of the magnet, the orientation of $\boldsymbol{I}_m$ aligns with the horizontal plane interface.

When calculating the magnetic field in the region above the interface (ferrofluid region), the effect of the interface is replaced by an image current $\boldsymbol{I}_m'$, which must be located at the geometric mirror point of the magnetization current $\boldsymbol{I}_m$ relative to the interface, to satisfy the boundary conditions. Assuming that the direction of $\boldsymbol{I}_m'$ aligns with that of $\boldsymbol{I}_m$, the magnetic induction and magnetic field intensity, generated by both the magnetization current $\boldsymbol{I}_m$ and its corresponding image current $\boldsymbol{I}_m'$ at any point P on the intersection line between the plane perpendicular to the current $\boldsymbol{I}_m$ and the interface are calculated. This calculation

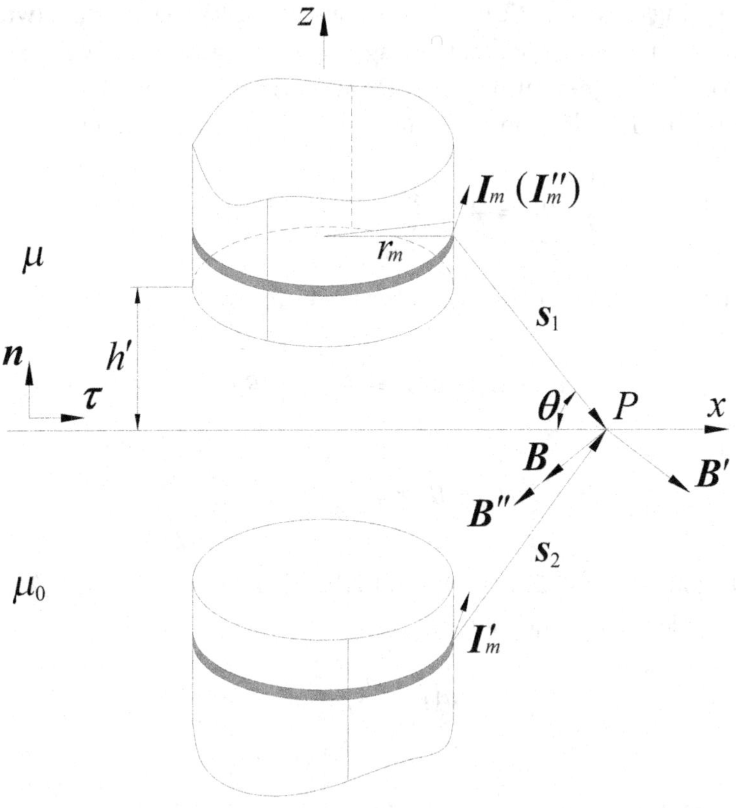

FIGURE 3.15 Model of calculating the magnetization current.

is performed using the Biot–Savart law, with the angle between the line connecting the point P to the location of the magnetization current and the interface designated as θ.

$$B = \frac{\mu}{4\pi}\frac{I_m \times s_1}{s_1^3}, B' = \frac{\mu}{4\pi}\frac{I'_m \times s_2}{s_2^3}, H = \frac{1}{4\pi}\frac{I_m \times s_1}{s_1^3}, H' = \frac{1}{4\pi}\frac{I'_m \times s_2}{s_2^3} \tag{3.215}$$

where $\boldsymbol{s}_1$ and $\boldsymbol{s}_2$ are the vectors originating from the magnetization current $\boldsymbol{I}_m$ and the image current $\boldsymbol{I}'_m$, respectively, and extending towards point P. Then, at point P on the ferrofluid side, one has

$$B_{n,1} = \boldsymbol{B}\cdot\boldsymbol{n} + \boldsymbol{B}'\cdot\boldsymbol{n}, H_{\tau,1} = \boldsymbol{H}\cdot\boldsymbol{\tau} + \boldsymbol{H}'\cdot\boldsymbol{\tau} \tag{3.216}$$

Here, $\boldsymbol{n}$ and $\boldsymbol{\tau}$ represent the normal and tangential unit vectors of the interface, respectively. Given the condition $s_1 = s_2$ at point P, substituting Eq. (3.215) into Eq. (3.216) yields

$$B_{n,1} = \frac{\mu}{4\pi s_1^3}\left[\left(\boldsymbol{I}_m \times \boldsymbol{s}_1\right)\cdot\boldsymbol{n} + \left(\boldsymbol{I}'_m \times \boldsymbol{s}_2\right)\cdot\boldsymbol{n}\right] = -\frac{\mu}{4\pi s_1^3}\left(I_m + I'_m\right)\cos\theta \tag{3.217}$$

$$H_{\tau,1} = \frac{1}{4\pi s_1^3}\left[\left(\boldsymbol{I}_m \times \boldsymbol{s}_1\right)\cdot\boldsymbol{\tau} + \left(\boldsymbol{I}'_m \times \boldsymbol{s}_2\right)\cdot\boldsymbol{\tau}\right] = \frac{1}{4\pi s_1^3}\left(-I_m + I'_m\right)\sin\theta \tag{3.218}$$

When calculating the magnetic field in the region beneath the interface (within the non-magnetic medium), it is assumed that the magnetic field is generated by a current $\boldsymbol{I}_m''$ located at the position of the magnetization current. This current flows in the same direction as $\boldsymbol{I}_m$ and produces magnetic induction and magnetic field intensity at point P

$$\boldsymbol{B}'' = \frac{\mu_0}{4\pi}\frac{\boldsymbol{I}_m'' \times \boldsymbol{s}_1}{s_1^3}, \boldsymbol{H}'' = \frac{1}{4\pi}\frac{\boldsymbol{I}_m'' \times \boldsymbol{s}_1}{s_1^3} \tag{3.219}$$

At point P on the non-magnetic side of the medium, one has

$$B_{n,2} = \boldsymbol{B}''\cdot\boldsymbol{n} = -\frac{\mu_0 I_m''}{4\pi s_1^3}\cos\theta \tag{3.220}$$

$$H_{\tau,2} = \boldsymbol{H}''\cdot\boldsymbol{\tau} = -\frac{I_m''}{4\pi s_1^3}\sin\theta \tag{3.221}$$

The substitution of Eqs. (3.218)–(3.221) into the boundary condition (3.214) and simplification yields the system of equations

$$\begin{cases} -\mu\left(I_m + I_m'\right) = \mu_0 I_m'' \\ -I_m + I_m' = -I_m'' \end{cases} \tag{3.222}$$

The solution of this system of equations is

$$I_m' = -\frac{\mu - \mu_0}{\mu + \mu_0} I_m \tag{3.223}$$

Equation (3.223) represents the magnitude of the magnetization current's image current when determining the magnetic field within a ferrofluid region.

Based on the arbitrariness of $\boldsymbol{I}_m$ selected on the cylindrical surface of the permanent magnet, the image currents of all magnetization current elements are located at positions mirrored about the interface. It means that the cylindrical surface, on which the image currents are distributed, has a length of l_m and a radius of R_1. Additionally, both types of currents have a relationship given by Eq. (3.223). Consequently, the relationship between the image current density and the surface density of the magnetization current is

$$i_m' = \frac{\mu - \mu_0}{\mu + \mu_0} i_m \tag{3.224}$$

Both the magnetization current and its image align on the same line, and from Eq. (3.223), it can be deduced that they have opposite directions, resulting in a repulsive force between them. Consequently, the magnetic field distribution generated by a cylindrical permanent magnet within a ferrofluid is analogous to that produced by the combined effects of the magnetization current and its image. The axial magnetic levitation force experienced by the permanent magnet within a ferrofluid is equivalent to the interactive force between these two current sources.

3.4.4.3 Calculation of the Magnetic Levitation Force

Based on the equivalence between the magnetic levitation force acting on a cylindrical permanent magnet within a ferrofluid and the interaction force between the magnet's magnetization current and its corresponding image current, now one can determine the axial magnetic levitation force experienced by the permanent magnet as it departs from the interface by a distance h'. As shown in Figure 3.16, a cylindrical permanent magnet, with a length l_m and radius R_1, is vertically immersed in the ferrofluid. It is assumed that the ferrofluid surrounding the permanent magnet extends infinitely far above the interface, while the region beneath the interface is non-magnetic.

On the cylindrical surface of the permanent magnet, at a distance of z_1 from its base, let us consider an elemental surface with an axial width of $\mathrm{d}z_1$ and a circumferential length of $\mathrm{d}l$. The vector $\mathrm{d}\boldsymbol{l}$ is defined to be tangent to the elemental surface and aligned along the magnetization current. Therefore, the magnetization current element residing on this surface element is given by $i_m \mathrm{d}z_1 \mathrm{d}\boldsymbol{l}$. Additionally, assuming that within a plane parallel to the horizontal plane, the angle between the line connecting the elemental surface and the axis of the permanent magnet with the $x0z$ plane is θ, then it follows that $\mathrm{d}l = R_1 \mathrm{d}\theta$.

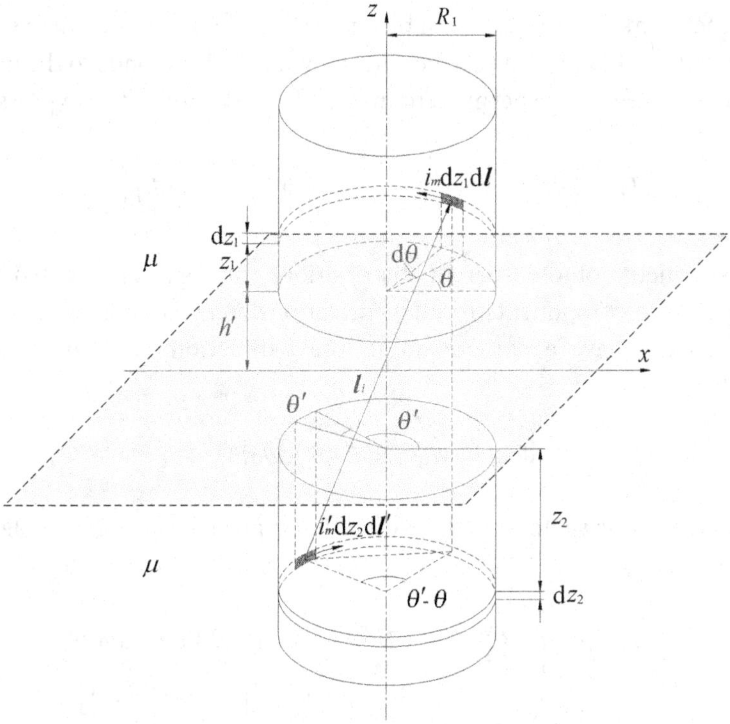

FIGURE 3.16 Model of calculating the magnetic levitation force using the current image method.

An elemental surface is considered as the cylindrical surface where the image current resides, situated at a distance of z_2 from the nearest end face to the x-axis. This surface has an axial width of dz_2 and a circumferential length of dl'. It is tangent to the vector $d\boldsymbol{l}'$, which points along the direction of the image current on the surface. Therefore, the image current element on this surface is given by $i'_m dz_2 d\boldsymbol{l}'$. In a plane parallel to the horizontal plane, the angle between the line connecting this elemental surface and the axis of the permanent magnet with the $x0z$ plane is denoted as θ', which implies that $dl' = R_1 d\theta'$.

The distance between the two current elements is

$$l_i = \sqrt{(z_1 + 2h' + z_2)^2 + 2R_1^2\left[1 - \cos(\theta - \theta')\right]}$$

The direction of the vector $\boldsymbol{l}_i$ is determined from the image current element towards the magnetization current element. Therefore, according to Ampere's law, the magnetization current element experiences a magnetic force exerted by the image current element

$$d\boldsymbol{F}_m = \frac{\mu}{4\pi} \frac{i_m i'_m dz_1 dz_2 d\boldsymbol{l} \times (d\boldsymbol{l}' \times \boldsymbol{l}_i)}{l_i^3} \tag{3.225}$$

Using the vector identity

$$d\boldsymbol{l} \times (d\boldsymbol{l}' \times \boldsymbol{l}_i) = d\boldsymbol{l}'(d\boldsymbol{l} \cdot \boldsymbol{l}_i) - \boldsymbol{l}_i(d\boldsymbol{l} \cdot d\boldsymbol{l}')$$

the cross-product term in Eq. (3.225) can be represented, and the total force exerted by the entire magnetization current on its image current, which corresponds to the magnetic levitation force experienced by the permanent magnet in the ferrofluid, is expressed as

$$\boldsymbol{F}_m = \int d\boldsymbol{F}_m = \int \frac{\mu i_m i'_m dz_1 dz_2}{4\pi l_i^3}\left[d\boldsymbol{l}'(d\boldsymbol{l} \cdot \boldsymbol{l}_i) - \boldsymbol{l}_i(d\boldsymbol{l} \cdot d\boldsymbol{l}')\right] \tag{3.226}$$

From the symmetry of the current distribution, the force is oriented towards the z-direction, with zero components in other directions. Yet, the first vector expression in Eq. (3.226) does not have a component in the z-direction and can be disregarded. Additionally, because of

$$d\boldsymbol{l} \cdot d\boldsymbol{l}' = -R_1^2 \cos(\theta - \theta') d\theta d\theta'$$

and the component of the vector $\boldsymbol{l}_i$ in the z-direction is given by $(z_1 + 2h' + z_2)\boldsymbol{e}_z$, Eq. (3.226) becomes

$$\boldsymbol{F}_m = \boldsymbol{e}_z \frac{\mu i_m i'_m R_1^2}{4\pi} \int_0^{2\pi}\int_0^{2\pi}\int_0^{l_m}\int_0^{l_m} \frac{(z_1 + 2h' + z_2)\cos(\theta - \theta') dz_1 dz_2 d\theta d\theta'}{\left[(z_1 + 2h' + z_2)^2 + 2R_1^2\left[1 - \cos(\theta - \theta')\right]\right]^{3/2}} \tag{3.227}$$

Letting $\varpi = z_1/l_m$, $\varpi' = z_2/l_m$, $k = R_1/l_m$, and $c_m = 2h'/l_m$, then one has $dz_1 = l_m d\varpi$ and $dz_2 = l_m d\varpi'$. Consequently, Eq. (3.227) becomes

$$\boldsymbol{F}_m = \boldsymbol{e}_z \frac{\mu i_m i'_m R_1^2}{4\pi} \int_0^{2\pi}\int_0^{2\pi}\int_0^1\int_0^1 \frac{(\varpi + c_m + \varpi')\cos(\theta - \theta')\mathrm{d}\varpi \mathrm{d}\varpi' \mathrm{d}\theta \mathrm{d}\theta'}{\left[(\varpi + c_m + \varpi')^2 + 2k^2\left[1 - \cos(\theta - \theta')\right]\right]^{3/2}} \tag{3.228}$$

Equation (3.228) can be analytically integrated twice, with the integrand

$$\int_0^1 \frac{(\varpi + c_m + \varpi')\mathrm{d}\varpi}{\left[(\varpi + c_m + \varpi')^2 + 2k^2\left[1 - \cos(\theta - \theta')\right]\right]^{3/2}} = \left\{\frac{-1}{\left[(\varpi + c_m + \varpi')^2 + 2k^2\left[1 - \cos(\theta - \theta')\right]\right]^{1/2}}\right\}\Bigg|_0^1$$

$$= \frac{1}{\left[(c_m + \varpi')^2 + 2k^2\left[1 - \cos(\theta - \theta')\right]\right]^{1/2}} - \frac{1}{\left[(c_m + \varpi' + 1)^2 + 2k^2\left[1 - \cos(\theta - \theta')\right]\right]^{1/2}} \tag{3.229}$$

Furthermore, the two terms in the result of Eq. (3.229) are integrated with respect to $\mathrm{d}\varpi'$, namely,

$$\int_0^1 \frac{\mathrm{d}\varpi'}{\left[(c_m + \varpi')^2 + 2k^2\left[1 - \cos(\theta - \theta')\right]\right]^{1/2}} = \left\{\ln\frac{\left[\frac{1}{2}(c_m + \varpi')^2 + k^2\left[1 - \cos(\theta - \theta')\right]\right]^{1/2} + c_m + \varpi'}{\left[2k^2\left[1 - \cos(\theta - \theta')\right]\right]^{1/2}}\right\}\Bigg|_0^1$$

$$= \ln\frac{\left[\frac{1}{2}(c_m + 1)^2 + k^2\left[1 - \cos(\theta - \theta')\right]\right]^{1/2} + c_m + 1}{\left[\frac{1}{2}c_m^2 + k^2\left[1 - \cos(\theta - \theta')\right]\right]^{1/2} + c_m} \tag{3.230}$$

and

$$\int_0^1 \frac{\mathrm{d}\varpi'}{\left[(c_m + \varpi' + 1)^2 + 2k^2\left[1 - \cos(\theta - \theta')\right]\right]^{1/2}} = \ln\frac{\left[\frac{1}{2}(c_m + 2)^2 + k^2\left[1 - \cos(\theta - \theta')\right]\right]^{1/2} + c_m + 2}{\left[\frac{1}{2}(c_m + 1)^2 + k^2\left[1 - \cos(\theta - \theta')\right]\right]^{1/2} + c_m + 1} \tag{3.231}$$

By substituting the results of Eqs. (3.229)–(3.231) into Eq. (3.228), the expression for the magnetic levitation force is obtained as

$$\boldsymbol{F}_m = \boldsymbol{e}_z \frac{\mu i_m i'_m R_1^2}{4\pi} \int_0^{2\pi}\int_0^{2\pi} \frac{\left\{\left[\frac{1}{2}(c_m + 1)^2 + k^2\left[1 - \cos(\theta - \theta')\right]\right]^{1/2} + c_m + 1\right\}^2 \cos(\theta - \theta')\mathrm{d}\theta \mathrm{d}\theta'}{\left\{\left[\frac{1}{2}c_m^2 + k^2\left[1 - \cos(\theta - \theta')\right]\right]^{1/2} + c_m\right\}\left\{\left[\frac{1}{2}(c_m + 2)^2 + k^2\left[1 - \cos(\theta - \theta')\right]\right]^{1/2} + c_m + 2\right\}} \tag{3.232}$$

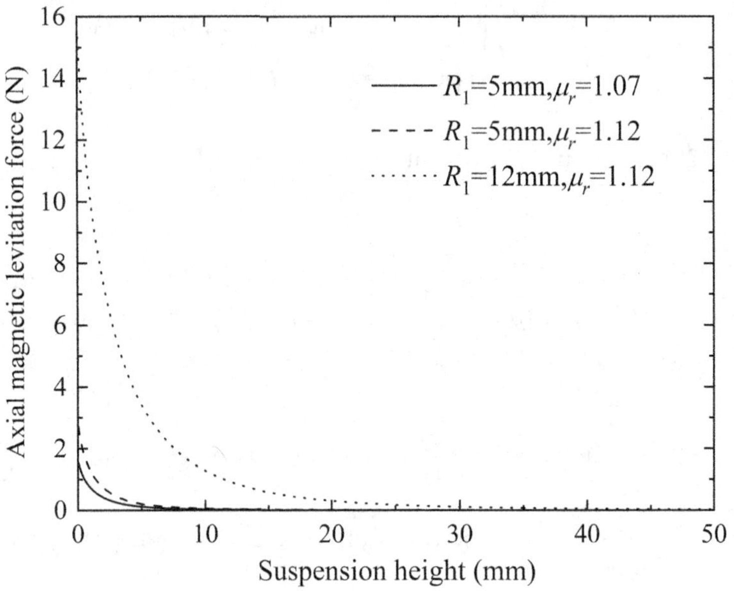

FIGURE 3.17 Axial magnetic levitation force calculated by the current image method.

Equation (3.232) represents the axial magnetic levitation force exerted on a cylindrical permanent magnet within a ferrofluid, where i_m and i'_m are calculated using Eqs. (3.213) and (3.224), respectively.

Figure 3.17 illustrates the relationships between the axial magnetic levitation force acting on the permanent magnets and its levitation height. These relationships are calculated using the Romberg integration method from Eq. (3.232) with the same parameters for the permanent magnets and ferrofluids as those used in Section 3.4.3. One can observe the same variation pattern of magnetic levitation force with respect to various parameters as obtained in Section 3.4.3. However, in this case, the magnetic levitation force does not exhibit a critical value of permeability when considering the variations in the permeability of the ferrofluid. Upon deriving Eq. (3.232) with respect to μ_r, it is found that the derivative is always positive, indicating a consistent increase in magnetic levitation force as permeability rises across its entire range.

REFERENCES

Aquino M. d', Miano G., Serpico C., Zamboni W., Coppola G., Forces in magnetic fluids subject to stationary magnetic fields, *IEEE Transactions on Magnetics*, 39(5): 2657–2659, 2003.

Blums E., Cebers A., Maiorov M. M., *Magnetic fluids*. New York: Walter de Gruyter. 1997.

Cowley M. D., Rosensweig R. E., The interfacial stability of a ferromagnetic fluid. *Journal of Fluid Mechanics*, 30(4): 671–688, 1967.

Ivanov A. S., Pshenichnikov A. F., Magnetostatic buoyancy force acting on a non-magnetic sphere immersed in a ferrofluid magnetized by a gradient field, *Journal of Magnetism and Magnetic Materials*, 565: 170294, 2023.

Kvitantsev A. S., Naletova V. A., Turkov V. A., Levitation of magnets and paramagnetic bodies in vessels filled with magnetic fluid, *Fluid Dynamics*, 37(3): 361–368, 2002.

Liu G., Pu Y., Xu C., Definition of Helmholtz and Kelvin forces in magnetic fluids, *ACTA Physica Sinica*, 57(4): 2500–2503, 2008.

Liu G., Xu C., Zhang P., Wu T., Magnetomechanical modeling of magnet immersed in magnetic fluid and controllability of self-suspension. *ACTA Physical Sinica*, 58(3): 2005–2010, 2009.

Liu M., Range of validity for the Kelvin force, *Physical Review Letters*, 84(12): 2762, 2000.

Naletova V. A., Kvitantsev A. S., Turkov V. A., Movement of a magnet and a paramagnetic body inside a vessel with a magnetic fluid, *Journal of Magnetism and Magnetic Materials*, 258–259: 439–442, 2003.

Odenbach S., *Ferrofluids, magnetically controllable fluids and their applications*, Heidelberg: Berlin: Springer, 2002.

Odenbach S., Liu M., Invalidation of the Kelvin force in ferrofluids, *Physical Review Letters*, 86(2): 328–331, 2001.

Rosensweig R. E., Fluidmagnetic buoyancy, *AIAA Journal*, 4(10): 1751–1758, 1966a.

Rosensweig R. E., Buoyancy and stable levitation of a magnetic body immersed in a magnetizable fluid, *Nature*, 210(5036): 613–614, 1966b.

Rosensweig R. E., Phenomena and relationships of magnetic fluid bearings, *Thermomechanics of magnetic fluids*. Washington & London: Hemisphere Publishing Corporation, 231–253. 1977.

Rosensweig R. E., *Ferrohydrodynamics*. Cambridge: Cambridge University Press. 1985.

Shliomis M. I., Lyubimova T. P., and Lyubimov D. V., Ferrohydrodynamics: An essay on the progress of ideas, *Chemical Engineering Communications*, 67: 275–290, 1988.

Yang W. M., Magnetic levitation force exerted on the cylindrical magnet in a ferrofluid damper, *Journal of Vibration and Control*, 23(14): 2345–2354, 2017.

Yang W. M., On the boundary conditions of magnetic field in OpenFOAM and a magnetic field solver for multi-region applications, *Computer Physics Communications*, 263: 107883, 2021.

Yang W. M., Liu B. Y., Magnetic levitation force of composite magnets in a ferrofluid damper, *Smart Materials and Structures*, 27: 115009, 2018.

Yang W. M., Wang P. K., Hao R. C., and Ma B. C., Experimental verification of radial magnetic levitation force on the cylindrical magnets in ferrofluid dampers, *Journal of Magnetism and Magnetic Materials*, 426: 334–339, 2017.

CHAPTER 4

Magnetic Properties of Ferrofluids

EACH MAGNETIC PARTICLE WITHIN a ferrofluid possesses a magnetic moment, which, in the absence of a magnetic field, is randomly distributed, resulting in an overall absence of net magnetization. However, upon the application of an external magnetic field, the particle magnetic moments tend to align with the field. The stronger the magnetic field, the greater the number of particle magnetic moments that align towards the external field. Once the magnetic field intensity exceeds a certain threshold, all particle magnetic moments align completely with the magnetic field direction, resulting in saturation magnetization of the ferrofluid. This chapter delves into the mathematical description of the magnetization process of ferrofluids.

4.1 EQUILIBRIUM MAGNETIZATION EQUATION

When the concentration of magnetic particles in a ferrofluid is relatively low, typically below 10% (Odenbach, 2002), the interactions between particles can be neglected. Under these conditions, the magnetic particles can be treated as tiny permanent magnets suspended within the carrier liquid, subject to thermal fluctuations. This results in the ferrofluid exhibiting overall paramagnetic behaviour. For superparamagnetic ferrofluids containing monodisperse particles, under the assumption of equilibrium magnetization, the magnetization always aligns with the magnetic field. Neglecting magnetic interactions among particles, the magnitude of the equilibrium magnetization is described by the Langevin equation (Rosensweig, 1985),

$$M_0 = M_S\left(\coth\alpha - \frac{1}{\alpha}\right) = M_S L(\alpha), \alpha = \mu_0 m_p H / k_B T \tag{4.1}$$

where M_S is the saturation magnetization of ferrofluids. It can be expressed as

$$M_S = \phi M_d = N m_p \tag{4.2}$$

DOI: 10.1201/9781003540342-4

Here, M_d is the saturation magnetization of the particulate material and ϕ is the volume fraction of the particulate phase in ferrofluids which is given by

$$\phi = V_p N \tag{4.3}$$

The quantity of magnetic particles within a unit volume of ferrofluid is denoted as N. The Langevin parameter, denoted as α, represents the ratio between the energy of a particle's magnetic moment m_p in a magnetic field H and the thermal motion energy. The Langevin function is denoted as $L(\alpha)$. The magnetic moment m_p of a single particle is calculated as

$$m_p = M_d V_p \tag{4.4}$$

where V_p is the volume of a single particle and it is equal to

$$V_p = \frac{\pi}{6} d_p^3 \tag{4.5}$$

Here d_p is the diameter of magnetic particles.

Regarding the ferrofluid with typical ferrite magnetic particles, where $M_d = 4.46 \times 10^5$ A/m, the typical volume fraction is $\phi = 4\%$, and the particle size is $d_p = 10$ nm ($V_p = 5.25 \times 10^{-25}$ m^3), one can calculate the saturation magnetization of the ferrofluid to be $\mu_0 M_S = 0.0244$ T (Rindldi et al., 2005). Additionally, the number concentration of magnetic particles is $N = 7.6 \times 10^{22}$ m^{-3} (Rindldi et al., 2005), and the dimensionless magnetic field intensity is approximately $\alpha \approx 7 \times 10^{-5} H$. Therefore, when the external magnetic field is relatively weak ($H <$ ~10^3 A/m), one has $\alpha \ll 1$.

One of the direct methods to obtain the equilibrium magnetization properties of ferrofluids is by measuring their magnetization curves, as exemplified in Figure 4.1, which depicts the magnetization curve of a typical ferrofluid (APG513A, Ferrotec). Once the

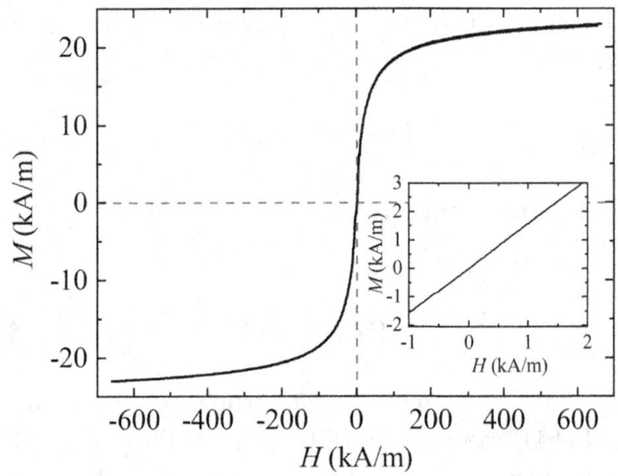

FIGURE 4.1 Magnetization curve of a typical ferrofluid (APG513A, Ferrotec).

saturation magnetization is determined from the magnetization curve, the volume fraction of particles in a ferrofluid can be further obtained using the formula $\phi = M_S/M_d$.

Researchers have also employed the arctan function to fit the magnetization curves of ferrofluids obtained through experiments, expressing the relationship between magnetization and magnetic field intensity as (Yang, 2012)

$$M_0 = a_1 \arctan(b_1 H) \tag{4.6}$$

The coefficients a_1 and b_1, measured in units of A/m and m/A, determine the saturation magnetization and initial susceptibility of ferrofluids, respectively. However, due to the lack of direct physical significance, they are only utilized in mathematical calculations of the ferrofluid equilibrium magnetization. In contrast, the Langevin function representation of equilibrium magnetization is more widely applied.

When the external magnetic field is weak, specifically when the parameter α is small ($\alpha \ll 1$), it is well-established from the Taylor expansion of the Langevin function that

$$L(\alpha) \approx \alpha / 3 \tag{4.7}$$

In this case, the initial susceptibility of ferrofluids is

$$\chi_0 = \frac{M_0}{H} = \frac{\mu_0 m_p^2 N}{3k_B T} = \frac{\pi\phi\mu_0 M_d^2 d_p^3}{18k_B T} \tag{4.8}$$

Given the initial susceptibility χ_0 and the volume fraction of particles ϕ, the average particle size can be estimated using (Chantrell et al., 1978)

$$\bar{d}_p = \left(\frac{18k_B T \chi_0}{\pi\phi\mu_0 M_d^2} \right)^{1/3} \tag{4.9}$$

When the external magnetic field is strong, specifically when the magnitude of α is large ($\alpha \gg 1$), it follows that

$$L(\alpha) \approx 1 - 1/\alpha \tag{4.10}$$

Therefore, it can be inferred from this that

$$\frac{M_0}{M_S} \approx 1 - \frac{1}{\alpha} = 1 - \frac{6k_B T}{\pi\mu_0 M_d H d_p^3} \tag{4.11}$$

Evidently, when the external magnetic field is strong, $1/\alpha \sim 0$, $M_0 \sim M_S$.

Equations (4.8) and (4.11) can be utilized to estimate the particle size in ferrofluids in terms of magnetization curves, assuming the particles are spherical. In weak magnetic fields, the primary contribution to magnetization originates from larger particles, as their larger size facilitates the alignment of their magnetic moments with the magnetic field. Conversely, as a ferrofluid approaches saturation, magnetization becomes increasingly dependent on smaller particles, as the alignment of these particles requires stronger magnetic field actions.

If the initial susceptibility is significant, the interaction among the particle magnetic moments within ferrofluids must be accounted for. Using the Debye–Onsager theory for polar fluids and assuming monodisperse particles, Shliomis (1974) and Morozov et al. (1987) derived the following relationship for estimating the initial magnetic susceptibility:

$$\frac{\chi_0(2\chi_0+3)}{\chi_0+1}=\frac{\mu_0 m_p^2 N}{k_B T}=\frac{\pi\phi\mu_0 M_d^2 d_p^3}{6k_B T} \tag{4.12}$$

4.2 MAGNETIC RELAXATION

For ferrofluids, when the direction of the external magnetic field changes, the particles undergo a magnetic torque $\mu_0 \boldsymbol{m}_p \times \boldsymbol{H}$, which tends to align $\boldsymbol{m}_p$ with the direction of $\boldsymbol{H}$. Subsequently, the magnetization of the ferrofluid reaches a new equilibrium through two mechanisms: Brownian relaxation and Néel relaxation. The Brownian relaxation process occurs through the rotation of particles, with the magnetic moment remaining fixed relative to the particles. During this rotation, the particles experience viscous drag torque from the surrounding carrier fluid. In contrast, during the Néel relaxation process, the magnetic moment within a particle rotates relative to the crystal axis towards the direction of $\boldsymbol{H}$, while the particle itself remains stationary. Figure 4.2 compares these two relaxation mechanisms.

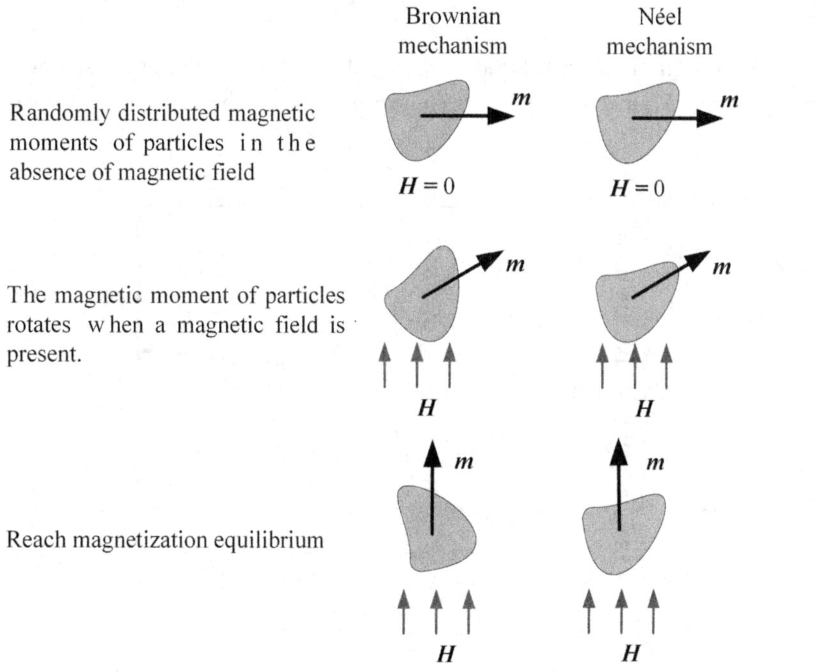

FIGURE 4.2 Brownian mechanism and Néel mechanism in the magnetization process of ferrofluids.

The speed of both relaxation processes is characterized by their respective relaxation times, with the Brownian relaxation mechanism represented by the Brownian diffusion time τ_B (Rosensweig, 1985)

$$\tau_B = \frac{3V_{\mathrm{h}}\eta_0}{k_{\mathrm{B}}T} \tag{4.13}$$

where V_{h} is the volume of particles containing the surfactant layer and η_0 is the dynamic viscosity of the carrier liquid. The Néel relaxation process is characterized by the Néel relaxation time τ_N, which represents the energy barrier for the magnetic moment to overcome the jump with respect to the lattice structure (Rosensweig, 1985).

$$\tau_N = \tau_0 \exp\left(\frac{KV_{\mathrm{p}}}{k_{\mathrm{B}}T}\right) \tag{4.14}$$

Here K is the anisotropy constant for granular materials. For ferrites, its value is dependent on the particle size, typically ranging from 2.3×10^4 to 10^5 J/m^3. The inverse of τ_0 corresponds to the Larmour frequency of the particle's magnetic moment in the anisotropic field. When the particle size is 12.6 nm, the value of K is 7.8×10^4 J/m^3, resulting in a τ_0 value of approximately 10^{-9} s (Lehlooh et al., 2002).

When both relaxation mechanisms coexist, the total magnetization relaxation time of ferrofluids is given by (Rosensweig, 1985; Berkovsky et al., 1993)

$$\tau = \frac{\tau_B \tau_N}{\tau_B + \tau_N} \tag{4.15}$$

Clearly, the magnetization relaxation time of a ferrofluid depends on the smaller value between τ_B and τ_N. Figure 4.3 illustrates a comparison of the two magnetization relaxation

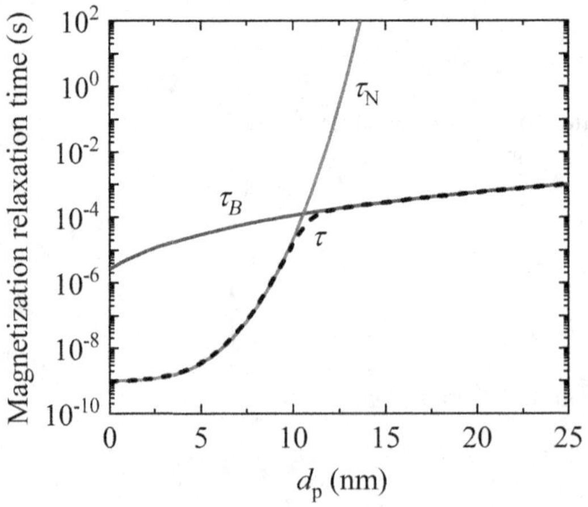

FIGURE 4.3 Comparison of the magnetization relaxation time of ferrofluids. (Surfactant thickness 2 nm, $\eta_0 = 0.108$ Pa·s, $K = 7.8 \times 10^4$ J/m^3, $T = 298$ K, and $\tau_0 = 10^{-9}$ s.)

times. If $\tau_N \ll \tau_B$, the Néel mechanism dominates, and the corresponding material is referred to as having intrinsic superparamagnetism. Conversely, if $\tau_B \ll \tau_N$, the Brownian mechanism prevails, and the material is described as exhibiting extrinsic superparamagnetism.

Upon comparing Eqs. (4.13) and (4.14), it is evident that τ_N exhibits an exponential relationship with particle volume, and is more sensitive to particle size compared to its proportional relationship with τ_B. Consequently, as the particle size decreases, the Néel mechanism increasingly dominates, resulting in shorter relaxation times, typically less than 10^{-8} s.

The ratio of the anisotropic energy KV_p to the thermal motion energy $k_\mathrm{B}T$ of magnetic particles in a ferrofluid determines the degree of solidification between the particles' magnetic moment and the particles themselves. If the particle diameter $d_\mathrm{p} \geq 10$ nm, then $KV_\mathrm{p} \gg k_\mathrm{B}T$, resulting in a complete solidification of the magnetic moments with the corresponding particles. Under these conditions, the direction of the magnetic moment can only be changed through particle rotation. This scenario is primarily dominated by the Brownian mechanism, with a characteristic time $\tau \approx \tau_B \approx 10^{-4}\sim10^{-5}$ s. Furthermore, when dealing with dynamic flow issues of ferrofluids under these conditions, one cannot assume that the magnetization $\boldsymbol{M}$ is collinear with the applied magnetic field $\boldsymbol{H}$. In this case, in addition to the force density $\mu_0\boldsymbol{M} \cdot \nabla \boldsymbol{H}$, ferrofluids also experience a magnetic torque density $\mu_0\boldsymbol{M} \times \boldsymbol{H}$.

The formulas for calculating the relaxation times corresponding to the two magnetization relaxation mechanisms are both based on the assumption of monodisperse particles. For a comprehensive understanding of the polydispersity of particles present in genuine ferrofluids, please refer to the study by Ivanov and Camp (2023).

In a flowing ferrofluid, if the magnetization relaxation time is significantly shorter than the characteristic flow time, the magnetization relaxation process can be neglected ($\tau \approx 0$). This means that the magnetization $\boldsymbol{M}$ instantaneously aligns with the direction of the magnetic field $\boldsymbol{H}$ at any given moment. In this case, the magnetization is represented by the equilibrium magnetization formula (4.1), and the corresponding theory of ferrohydrodynamics is referred to as quasistationary ferrofluid dynamics.

4.3 PHENOMENOLOGICAL MAGNETIZATION EQUATION I

The non-negativity of the dissipation function Φ in the constitutive relation implies that the final term on the right-hand side of Eq. (2.153) must be non-negative. To satisfy this condition, it is assumed that the term enclosed within the bracket is proportional to $(\boldsymbol{H} - \boldsymbol{H}^{(\mathrm{e})})$. Let $\boldsymbol{M}_0$ represent the equilibrium magnetization of a ferrofluid when a magnetic field $\boldsymbol{H}$ is present and $\boldsymbol{M}$ denote the actual magnetization of the ferrofluid. Furthermore, an assumption is made that $\boldsymbol{M}$ corresponds to an equilibrium value at a magnetic field intensity $\boldsymbol{H}_\mathrm{e}$, which is also referred to as the effective field (Martsenyuk et al., 1974). During the initial magnetization stage, when the magnetization curve approximates a straight line, or when $\boldsymbol{M}$ deviates slightly from $\boldsymbol{M}_0$, $(\boldsymbol{M}_0 - \boldsymbol{M})$ is directly proportional to $(\boldsymbol{H} - \boldsymbol{H}^{(\mathrm{e})})$. Based on this, the magnetization equation can be derived (Odenbach, 2002; Felderhof and Kroh, 1999) as

$$\frac{\mathrm{D}\boldsymbol{M}}{\mathrm{D}t} + \boldsymbol{M}(\nabla \cdot \boldsymbol{v}) - \boldsymbol{\omega}_\mathrm{p} \times \boldsymbol{M} = \frac{1}{\tau}(\boldsymbol{M}_0 - \boldsymbol{M}) \tag{4.16}$$

or

$$\frac{\mathrm{D}\boldsymbol{M}}{\mathrm{D}t} = \boldsymbol{\omega}_{\mathrm{p}} \times \boldsymbol{M} - \boldsymbol{M}(\nabla \cdot \boldsymbol{v}) + \frac{1}{\tau}(\boldsymbol{M}_0 - \boldsymbol{M}) \tag{4.17}$$

where the first term on the right-hand side represents the change in magnetization caused by the rotation of magnetized particles, while the last term denotes the process of magnetization relaxing towards the equilibrium magnetization.

Substituting the simplified form of the angular momentum Eq. (2.217) into Eq. (4.17), assuming the ferrofluid is incompressible, and applying $\nabla \cdot \boldsymbol{v} = 0$, one obtains

$$\frac{\mathrm{D}\boldsymbol{M}}{\mathrm{D}t} = \boldsymbol{\Omega} \times \boldsymbol{M} - \frac{1}{\tau}(\boldsymbol{M} - \boldsymbol{M}_0) - \frac{\mu_0}{4\zeta}\boldsymbol{M} \times (\boldsymbol{M} \times \boldsymbol{H}) \tag{4.18}$$

For dilute ferrofluids, the coefficient ζ is calculated using Eq. (2.213). For real ferrofluids, τ_B is often used as a substitute for τ during calculation.

For ferrofluids with high initial susceptibility, it can be derived from Eq. (4.12) that

$$M_{\mathrm{S}}^2 = M_d^2\phi^2 = \frac{6k_{\mathrm{B}}T\phi}{\pi\mu_0 d_{\mathrm{p}}^3}\frac{\chi_0(2\chi_0+3)}{\chi_0+1}$$

Then one has

$$\frac{3\chi_0}{2\tau_B M_{\mathrm{S}}^2} = \frac{3\chi_0 k_{\mathrm{B}}T\pi\mu_0 d_{\mathrm{p}}^3(\chi_0+1)}{2\cdot 3\eta V_h 6k_{\mathrm{B}}T\phi\chi_0(2\chi_0+3)} = \frac{\mu_0(\chi_0+1)}{2\cdot\eta\phi(2\chi_0+3)} \tag{4.19}$$

Due to the fact that $\chi_0 \ll 1$, it is therefore

$$\frac{3\chi_0}{2\tau_B M_{\mathrm{S}}^2} \approx \frac{\mu_0}{6\eta\phi} \tag{4.20}$$

The term at the right-hand side of Eq. (4.20) is identical to the coefficient of the last term of Eq. (4.18), leading to an alternative expression for the magnetization equation

$$\frac{\partial \boldsymbol{M}}{\partial t} + (\boldsymbol{v} \cdot \nabla)\boldsymbol{M} = \boldsymbol{\Omega} \times \boldsymbol{M} - \frac{1}{\tau}(\boldsymbol{M} - \boldsymbol{M}_0) - \frac{3\chi_0}{2\tau_B M_{\mathrm{S}}^2}\boldsymbol{M} \times (\boldsymbol{M} \times \boldsymbol{H}) \tag{4.21}$$

Shliomis (1972) extended the Debye relaxation equation to derive a magnetization relaxation equation identical to Eq. (4.18). To accomplish this, he introduced a local reference frame Σ' that is stationary relative to the magnetic particles within ferrofluids. In this reference frame, the average rotational velocity of the magnetic particles is zero, denoted as $\boldsymbol{\omega}_{\mathrm{p}}' = 0$. He assumed that the deviation of the magnetization of the ferrofluid from its equilibrium value decays exponentially

$$(\boldsymbol{M} - \boldsymbol{M}_0) \sim \exp\left(-\frac{t}{\tau_B}\right) \tag{4.22}$$

Then, the magnetization equation in this reference frame possesses a similar form to the Debye equation

$$\frac{\mathrm{d}'\boldsymbol{M}}{\mathrm{d}t} = -\frac{1}{\tau_B}(\boldsymbol{M} - \boldsymbol{M}_0) \tag{4.23}$$

The vectorial form of the equilibrium magnetization can be expressed according to Eq. (4.1) as

$$\boldsymbol{M}_0 = M_S L(\alpha)\boldsymbol{\alpha}/\alpha, \ \boldsymbol{\alpha} = \mu_0 m_p \boldsymbol{H}/k_B T \tag{4.24}$$

In this fixed reference frame, the reference frame Σ′ rotates with an angular velocity $\boldsymbol{\omega}_p$, allowing one to derive a relationship for any vector, taking $\boldsymbol{M}$ as an illustrative example.

$$\frac{\mathrm{d}\boldsymbol{M}}{\mathrm{d}t} = \boldsymbol{\omega}_p \times \boldsymbol{M} + \frac{\mathrm{d}'\boldsymbol{M}}{\mathrm{d}t} \tag{4.25}$$

Substitution of Eq. (2.217) into Eq. (4.25) yields an equation with the same form as the magnetization equation (4.18), with the sole difference being that the relaxation time is now τ_B.

The last term on the right-hand side of Eq. (4.18) represents the process of magnetization approaching its equilibrium direction, with the magnitude of $\boldsymbol{M}$ remaining constant throughout. Consequently, the relaxation rates of the two components of $\boldsymbol{M}$, parallel and perpendicular to the magnetic field direction, are not identical. To calculate these components, the term $(\boldsymbol{M} - \boldsymbol{M}_0)$ is expressed as

$$\begin{aligned}\boldsymbol{M} - \boldsymbol{M}_0 &= \frac{\boldsymbol{M}(\boldsymbol{H}\cdot\boldsymbol{H})}{H^2} - \frac{\boldsymbol{H}(\boldsymbol{H}\cdot\boldsymbol{M}_0)}{H^2} \\ &= \frac{\boldsymbol{M}(\boldsymbol{H}\cdot\boldsymbol{H})}{H^2} - \frac{\boldsymbol{H}(\boldsymbol{H}\cdot\boldsymbol{M})}{H^2} + \frac{\boldsymbol{H}(\boldsymbol{H}\cdot\boldsymbol{M})}{H^2} - \frac{\boldsymbol{H}(\boldsymbol{H}\cdot\boldsymbol{M}_0)}{H^2} \\ &= \frac{\boldsymbol{H}\times(\boldsymbol{M}\times\boldsymbol{H})}{H^2} + \frac{\boldsymbol{H}\left[\boldsymbol{H}\cdot(\boldsymbol{M}-\boldsymbol{M}_0)\right]}{H^2}\end{aligned} \tag{4.26}$$

wherein obtaining the first equality use is made of the condition that $\boldsymbol{H}$ and $\boldsymbol{M}_0$ are collinear. The first term and the second term on the right-hand side of the last equality can be regarded as the perpendicular component and the parallel component relative to the magnetic field direction, respectively. Substituting these components into the magnetization equation (4.18) yields

$$\frac{\mathrm{D}\boldsymbol{M}}{\mathrm{D}t} = \boldsymbol{\Omega}\times\boldsymbol{M} - \frac{\boldsymbol{H}\left[\boldsymbol{H}\cdot(\boldsymbol{M}-\boldsymbol{M}_0)\right]}{\tau_B H^2} - \left(\frac{\boldsymbol{H}}{\tau_B H^2} + \frac{\mu_0}{4\zeta}\boldsymbol{M}\right)\times(\boldsymbol{M}\times\boldsymbol{H}) \tag{4.27}$$

To linearize Eq. (4.27) with respect to $\boldsymbol{M}$, the first occurrence of $\boldsymbol{M}$ in the third term on the right-hand side is replaced with its equilibrium value, $\boldsymbol{M}_0$. By applying Eqs. (2.213), (4.13), and (4.24), the linearized magnetization equation is obtained

$$\frac{\mathrm{D}\boldsymbol{M}}{\mathrm{D}t}=\boldsymbol{\Omega}\times\boldsymbol{M}-\frac{\boldsymbol{H}\left[\boldsymbol{H}\cdot\left(\boldsymbol{M}-\boldsymbol{M}_0\right)\right]}{\tau_{\|}H^2}-\frac{\boldsymbol{H}\times\left(\boldsymbol{M}\times\boldsymbol{H}\right)}{\tau_{\perp}H^2}, \tag{4.28}$$

where

$$\tau_{\|}=\tau_B,$$

$$\frac{1}{\tau_{\perp}}=\frac{1}{\tau_B}+\frac{\mu_0}{6\eta\phi}HNm_{\mathrm{p}}L(\alpha)=\frac{1}{\tau_B}\left[1+\frac{1}{2}\alpha L(\alpha)\right]$$

and

$$\tau_{\perp}=\frac{2\tau_B}{2+\alpha L(\alpha)}.$$

The linearized magnetization equation (4.28) is applicable when the deviation of $\boldsymbol{M}$ from $\boldsymbol{M}_0$ is relatively small.

The phenomenological magnetization equation (4.18) can also be derived from the irreversible thermodynamic theory put forward by Landau (Landau and Lifshitz, 1960). By applying this theory, if magnetization $\boldsymbol{M}$ is regarded as an order parameter, it reaches its equilibrium value $\boldsymbol{M}_0$ when the thermodynamic potential ψ (typically the Gibbs or Helmholtz free energy) attains its minimum. The thermodynamic potential ψ is dependent on magnetization and other thermodynamic variables. In other words, at equilibrium states, one gives

$$\frac{\partial\psi}{\partial\boldsymbol{M}}=0 \tag{4.29}$$

When the magnetization deviates from its equilibrium state, this equation is no longer satisfied. The relaxation process can also be perceived as a gradual approach of $\boldsymbol{M}$ towards $\boldsymbol{M}_0$ over time. In cases where the deviation from the equilibrium value is minimal, both $\frac{\partial\psi}{\partial\boldsymbol{M}}$ and $\frac{\mathrm{d}\boldsymbol{M}}{\mathrm{d}t}$ exhibit relatively small values. Landau's theory suggests a straightforward proportional relationship between these two derivatives

$$\frac{\mathrm{d}\boldsymbol{M}}{\mathrm{d}t}=-\gamma_M\frac{\partial\psi}{\partial\boldsymbol{M}} \tag{4.30}$$

where the constant coefficient $\gamma_M > 0$, thereby resulting in

$$\frac{\mathrm{d}\psi}{\mathrm{d}t} = \frac{\partial\psi}{\partial \boldsymbol{M}} \cdot \frac{\mathrm{d}\boldsymbol{M}}{\mathrm{d}t} = -\gamma_M \left(\frac{\mathrm{d}\boldsymbol{M}}{\mathrm{d}t}\right)^2 < 0 \tag{4.31}$$

Evidently, as the system approaches its equilibrium state, its free energy decreases. In cases of weak non-equilibrium states, deviating slightly from the equilibrium state, one has

$$\frac{\partial\psi}{\partial \boldsymbol{M}} = \left(\frac{\partial\psi}{\partial \boldsymbol{M}}\right)_0 + \left(\frac{\partial^2\psi}{\partial \boldsymbol{M}^2}\right)_0 \left(\boldsymbol{M} - \boldsymbol{M}_0\right) + \cdots \tag{4.32}$$

where the subscript 0 denotes the equilibrium point, at which the first-order derivative is zero and the second-order derivative is positive. Setting

$$\gamma_M \left(\frac{\partial^2\psi}{\partial \boldsymbol{M}^2}\right)_0 = \frac{1}{\tau_B} \tag{4.33}$$

indeed deduces Eq. (4.23) from Eq. (4.30). It can be shown that Eq. (4.18) is validated and

$$\frac{\mathrm{d}\psi}{\mathrm{d}t} = -\frac{\left(\boldsymbol{M} - \boldsymbol{M}_0\right)^2}{\gamma_M \tau_B^2} \tag{4.34}$$

Equation (4.18) predicts the rotational viscosity of ferrofluids in static magnetic fields, agreeing well with experimental results under the condition $\Omega\tau_B \ll 1$, which is generally satisfied in most practical situations. However, for highly disequilibrium states where magnetization deviates significantly from its equilibrium value, the condition $\Omega\tau_B \ll 1$ is no longer satisfied, or when an alternating magnetic field is applied, Eq. (4.18) is no longer valid.

4.4 MAGNETIZATION EQUATION DERIVED MICROSCOPICALLY

When ferrofluids move in a magnetic field, the interaction between fluid flow and the magnetic field gives rise to remarkably intricate phenomena. This interaction arises from the fact that, under the influence of a magnetic field $\boldsymbol{H}$, magnetic particles possessing a magnetic moment of $\boldsymbol{m}_\mathrm{p}$ experience a magnetic force $(\boldsymbol{m}_\mathrm{p} \cdot \nabla)\boldsymbol{H}$ and a magnetic torque $\boldsymbol{m}_\mathrm{p} \times \boldsymbol{H}$. Here,

$$\boldsymbol{m}_\mathrm{p} \times \boldsymbol{H} = 6\eta V_\mathrm{p} \left(\boldsymbol{\omega}_\mathrm{p} - \boldsymbol{\Omega}\right) \tag{4.35}$$

The linkage between the magnetic and rotational degrees of freedom of a particle arises from its magnetic anisotropy. The ratio of anisotropic energy KV_p to thermal motion energy $k_\mathrm{B}T$ determines the degree of "pinning" of the magnetic moment within the particle, also known as the pinning stiffness between the magnetic moment vector $\boldsymbol{m}_\mathrm{p}$ and the crystal axis. For a particle with a diameter of $d_\mathrm{p} \geq 100$, $KV_\mathrm{p} \gg k_\mathrm{B}T$, the magnetic moment is completely pinned to the particle. Under these conditions, changes in the direction of the magnetic moment can only occur through the rotation of the particle leading to

$$\frac{d\boldsymbol{m}_{\text{p}}}{dt} = \boldsymbol{\omega}_{\text{p}} \times \boldsymbol{m}_{\text{p}} \tag{4.36}$$

Use of Eq. (4.35) to eliminate $\boldsymbol{\omega}_{\text{p}}$ in Eq. (4.36) then obtains

$$\frac{d\boldsymbol{m}_{\text{p}}}{dt} = \boldsymbol{\Omega} \times \boldsymbol{m}_{\text{p}} - \frac{1}{6\eta V_{\text{p}}} \boldsymbol{m}_{\text{p}} \times \left(\boldsymbol{m}_{\text{p}} \times \boldsymbol{H}\right) \tag{4.37}$$

Although Eq. (4.37) is able to describe the evolution process of particle magnetic moment, it does not consider the Brownian rotational diffusion of particles. To account for particle orientational fluctuations, it is necessary to replace $\boldsymbol{H}$ in this equation with the sum of the external field and the fluctuation field, namely,

$$\boldsymbol{H} \to \boldsymbol{H} - \left(k_{\text{B}} T / m_{\text{p}}\right) \nabla \ln W \tag{4.38}$$

Here, $W(\boldsymbol{e}, t)$ is the orientation distribution function of particle magnetic moments, ∇ represents the derivative relative to the unit vector angular variable, specifically given by $\nabla = \partial/\partial \boldsymbol{e}$, and $\boldsymbol{e} = \boldsymbol{m}_{\text{p}}/m_{\text{p}}$ is the unit vector along the direction of the particle magnetic moment. Using the replacement Eq. (4.38) in Eq. (4.37), one obtains

$$\frac{d\boldsymbol{e}}{dt} = \boldsymbol{\Omega} \times \boldsymbol{e} - \frac{1}{2\tau_B} \boldsymbol{e} \times \left[\boldsymbol{e} \times \left(\boldsymbol{\alpha} - \nabla \ln W\right)\right] \tag{4.39}$$

where the non-equilibrium distribution function satisfies the law of probability conservation

$$\frac{dW}{dt} + \nabla \cdot \left(\frac{d\boldsymbol{e}}{dt} W\right) = 0 \tag{4.40}$$

Substituting the expression for $\frac{d\boldsymbol{e}}{dt}$ from Eq. (4.39) into Eq. (4.40) yields the Fokker–Planck equation for a flowing ferrofluid under the influence of a magnetic field, neglecting interactions between particles (Martsenyuk et al., 1974).

$$2\tau_B \frac{\partial W}{\partial t} = \hat{\boldsymbol{R}} \cdot \left(\hat{\boldsymbol{R}} - 2\tau_B \boldsymbol{\Omega} - \boldsymbol{e} \times \boldsymbol{\alpha}\right) W \tag{4.41}$$

where $\hat{\boldsymbol{R}} = \boldsymbol{e} \times \partial / \partial \boldsymbol{e}$ is the infinitesimal rotation operator. The Fokker–Planck equation is utilized to depict the diffusion of particles within a colloid. The macroscopic magnetization is determined by the relationship

$$\boldsymbol{M}(t) = N m_{\text{p}} \langle \boldsymbol{e} \rangle \tag{4.42}$$

where $\langle \boldsymbol{e} \rangle$ represents the statistical average of the distribution function. Performing a cross-product with $\boldsymbol{e}$ on both sides of Eq. (4.41) and integrating over the entire angular range, one obtains

$$\tau_B \frac{\partial \langle \boldsymbol{e} \rangle}{\partial t} = \tau_B \boldsymbol{\Omega} \times \langle \boldsymbol{e} \rangle - \langle \boldsymbol{e} \rangle - \langle \boldsymbol{e} \times (\boldsymbol{e} \times \boldsymbol{\alpha}) \rangle \tag{4.43}$$

Using the effective field method (EFM; Martsenyuk et al., 1974), Eq. (4.43) is closed and then solved. Under the influence of a constant magnetic field, at the equilibrium state ($\boldsymbol{\Omega} = 0$), the static value of Eq. (4.43) corresponds to the Gibbs distribution function

$$W_0(\boldsymbol{e}) = \frac{\alpha}{4\pi \sinh \alpha} \exp(\boldsymbol{\alpha} \cdot \boldsymbol{e}) \tag{4.44}$$

The expression for the equilibrium magnetization Eq. (4.24) can be obtained by taking the average of the vector $\boldsymbol{e}$ within this function. Furthermore, any magnetization $\boldsymbol{M}$ deviating from the equilibrium state can be regarded as an equilibrium magnetization value under a specific magnetic field intensity $\boldsymbol{H}_e$, which exhibits a relationship with the non-equilibrium magnetization,

$$\boldsymbol{M} = M_S L(\varsigma) \boldsymbol{\varsigma} / \varsigma, \boldsymbol{\varsigma} = \mu_0 m_p \boldsymbol{H}_e / k_B T \tag{4.45}$$

Here, ς represents the dimensionless effective field. As the effective field $\boldsymbol{H}_e$ approaches the true field $\boldsymbol{H}$ during the relaxation process towards the equilibrium state, the magnetization $\boldsymbol{M}$ relaxes to its equilibrium value $\boldsymbol{M}_0$. The functional relationship Eq. (4.45) can be obtained by averaging the distribution function

$$W_e(\boldsymbol{e}) = \frac{\varsigma}{4\pi \sinh \varsigma} \exp(\boldsymbol{\varsigma} \cdot \boldsymbol{e}) \tag{4.46}$$

with respect to $\boldsymbol{e}$. The function can be obtained by replacing the actual field α with the effective field ς in the Gibbs distribution function. Taking the average of Eq. (4.43) with respect to $W_e(\boldsymbol{e})$ yields its solution

$$\frac{\mathrm{D}\boldsymbol{M}}{\mathrm{D}t} = \boldsymbol{\Omega} \times \boldsymbol{M} - \frac{1}{\tau_B}\left(\boldsymbol{M} - M_S \frac{L(\varsigma)}{\varsigma} \boldsymbol{\alpha} \right) - \frac{\mu_0}{L(\varsigma)} \left(\frac{1}{L(\varsigma)} - \frac{3}{\varsigma} \right) \frac{\boldsymbol{M} \times (\boldsymbol{M} \times \boldsymbol{H})}{6\eta\phi} \tag{4.47}$$

Equation (4.47) and relation (4.45) implicitly determine $\boldsymbol{M}(t; \boldsymbol{H}, \boldsymbol{\Omega})$ together.

Sometimes, in order to solve for $\boldsymbol{\varsigma}$, one has to utilize Eq. (4.45) to derive an equation related to $\boldsymbol{\varsigma}$. Substitution of $\boldsymbol{M} = M_S L(\varsigma)\boldsymbol{\varsigma}/\varsigma$ into Eq. (4.47) gives a term

$$\frac{\mu_0}{L(\varsigma)}\left(\frac{1}{L(\varsigma)}-\frac{3}{\varsigma}\right)\frac{\boldsymbol{M}\times(\boldsymbol{M}\times\boldsymbol{H})}{6\eta\phi}=\frac{M_S^2}{6\eta\phi}\frac{\mu_0}{L(\varsigma)}\left(\frac{1}{L(\varsigma)}-\frac{3}{\varsigma}\right)\left(\frac{L(\varsigma)}{\varsigma}\right)^2\frac{k_BT}{\mu_0 m_p}\boldsymbol{\varsigma}\times(\boldsymbol{\varsigma}\times\boldsymbol{\alpha})$$
$$=\frac{k_BT}{3\eta V}\frac{M_S^2V}{2m_p\phi}\frac{1}{\varsigma^2}\left(1-\frac{3L(\varsigma)}{\varsigma}\right)\boldsymbol{\varsigma}\times(\boldsymbol{\varsigma}\times\boldsymbol{\alpha})$$
$$=\frac{Nm_dV}{m_p\phi}\frac{M_S}{2\tau_B\varsigma^2}\left(1-\frac{3L(\varsigma)}{\varsigma}\right)\boldsymbol{\varsigma}\times(\boldsymbol{\varsigma}\times\boldsymbol{\alpha})$$
$$=\frac{M_S}{2\tau_B\varsigma^2}\left(1-\frac{3L(\varsigma)}{\varsigma}\right)\boldsymbol{\varsigma}\times(\boldsymbol{\varsigma}\times\boldsymbol{\alpha}) \tag{4.48}$$

Substituting the result from Eq. (4.48) into Eq. (4.47) yields an alternative form of the microscopic magnetization equation

$$\frac{\mathrm{D}}{\mathrm{D}t}\left(\frac{L(\varsigma)}{\varsigma}\boldsymbol{\varsigma}\right)=\boldsymbol{\Omega}\times\left(\frac{L(\varsigma)}{\varsigma}\boldsymbol{\varsigma}\right)-\frac{1}{\tau_B}\frac{L(\varsigma)}{\varsigma}(\boldsymbol{\varsigma}-\boldsymbol{\alpha})-\frac{1}{2\tau_B\varsigma^2}\left(1-\frac{3L(\varsigma)}{\varsigma}\right)\boldsymbol{\varsigma}\times(\boldsymbol{\varsigma}\times\boldsymbol{\alpha}) \tag{4.49}$$

If the external magnetic field is weak, that is when the parameter α is small ($\alpha \ll 1$), it can be derived from Eqs. (4.8) and (4.13) as

$$M_S\frac{L(\varsigma)}{\varsigma}\boldsymbol{\alpha}=M_S\frac{L(\varsigma)}{\varsigma}\frac{\mu_0 m_d\boldsymbol{H}}{k_BT}=\frac{L(\varsigma)}{\varsigma}\frac{\mu_0 Nm_d^2}{k_BT}\boldsymbol{H}=3\frac{L(\varsigma)}{\varsigma}\chi_0\boldsymbol{H} \tag{4.50}$$

$$\frac{\mu_0}{6\eta\phi L(\varsigma)}\left(\frac{1}{L(\varsigma)}-\frac{3}{\varsigma}\right)=\frac{\mu_0}{6\eta\phi L^2(\varsigma)}\left(1-\frac{3L(\varsigma)}{\varsigma}\right)=\frac{\mu_0 M_S^2}{6\eta\phi M^2}\left(1-\frac{3L(\varsigma)}{\varsigma}\right)$$
$$=\frac{\mu_0}{6\eta\phi M^2}\frac{3Nk_BT\chi_0}{\mu_0}\left(1-\frac{3L(\varsigma)}{\varsigma}\right)=\frac{k_BT}{3V_h\eta}\frac{3\chi_0}{2M^2}\left(1-\frac{3L(\varsigma)}{\varsigma}\right) \tag{4.51}$$
$$=\frac{3\chi_0}{2\tau_BM^2}\left(1-\frac{3L(\varsigma)}{\varsigma}\right)$$

Introducing Eqs. (4.50) and (4.51) into Eq. (4.47) casts into the following microscopic magnetization equation that holds for weak magnetic fields:

$$\frac{\mathrm{D}\boldsymbol{M}}{\mathrm{D}t}=\boldsymbol{\Omega}\times\boldsymbol{M}-\frac{1}{\tau_B}\left(\boldsymbol{M}-\chi_0\frac{3L(\varsigma)}{\varsigma}\boldsymbol{H}\right)-\frac{3\chi_0}{2\tau_BM^2}\left(1-\frac{3L(\varsigma)}{\varsigma}\right)\boldsymbol{M}\times(\boldsymbol{M}\times\boldsymbol{H}) \tag{4.52}$$

If the deviation of magnetization from its equilibrium state is minor, the effective field can be expressed as the sum of the true field and a small correction field: $\boldsymbol{\varsigma}=\boldsymbol{\alpha}+\boldsymbol{\nu}$.

By utilizing a Taylor series expansion for the Langevin function, inserting the derived results into Eq. (4.45), and linearizing with respect to $\boldsymbol{\nu}$, one obtains

$$\boldsymbol{M}-\boldsymbol{M}_0 = M_S\left[\frac{\mathrm{d}L(\alpha)}{\mathrm{d}\alpha}\boldsymbol{\nu}_{\parallel}+\frac{L(\alpha)}{\alpha}\boldsymbol{\nu}_{\perp}\right] \tag{4.53}$$

where

$$\boldsymbol{\nu}_{\parallel}=\frac{\boldsymbol{\alpha}(\boldsymbol{\nu}\cdot\boldsymbol{\alpha})}{\alpha^2},\boldsymbol{\nu}_{\perp}=\frac{\boldsymbol{\alpha}\times(\boldsymbol{\nu}\times\boldsymbol{\alpha})}{\alpha^2} \tag{4.54}$$

are correction fields in the direction parallel and perpendicular to the true magnetic field, respectively. Substituting Eq. (4.53) into Eq. (4.47) yields the linearized magnetization Eq. (4.28), with the magnetization relaxation time components now represented as

$$\tau_{\parallel}=\frac{\mathrm{d}\ln L(\alpha)}{\mathrm{d}\ln\alpha}\tau_B, \tau_{\perp}=\frac{2L(\alpha)}{\alpha-L(\alpha)}\tau_B \tag{4.55}$$

The magnetization Eq. (4.47), derived from the EFM, is capable of accurately describing the magnetization process of actual ferrofluids, even in scenarios where the magnetization deviates significantly from its equilibrium value (i.e., when $\boldsymbol{\Omega}\tau_B \gg 1$) (Shliomis et al., 1988; Bacri et al., 1995; Zeuner et al., 1998; Embs et al., 2000).

4.5 PHENOMENOLOGICAL MAGNETIZATION EQUATION II

The derivation of the phenomenological magnetization equation II follows a similar procedure to that of the phenomenological magnetization equation I in Section 4.3. However, in this case, the effective field $\boldsymbol{H}_e$ is chosen as the independent variable instead of the magnetization $\boldsymbol{M}$. Correspondingly, the thermodynamic potential is denoted as $\tilde{\psi}(\boldsymbol{H}_e)$. This leads to an equation analogous to Eq. (4.30) (Shliomis, 2001)

$$\frac{\mathrm{d}\boldsymbol{H}_e}{\mathrm{d}t}=-\tilde{\gamma}\frac{\partial\tilde{\psi}}{\partial\boldsymbol{H}_e} \tag{4.56}$$

where $\tilde{\gamma}$ is a constant coefficient. Using the same method applied in Section 4.3, the following equation can be derived:

$$\frac{\mathrm{d}'\boldsymbol{H}_e}{\mathrm{d}t}=-\frac{1}{\tau_B}(\boldsymbol{H}_e-\boldsymbol{H}) \tag{4.57}$$

where in obtaining this equation use was made of

$$\frac{1}{\tilde{\gamma}}=\frac{\partial^2\tilde{\psi}}{\partial\boldsymbol{H}_e^2}\tau_B \tag{4.58}$$

Under weak field conditions, where the actual magnetization and its corresponding equilibrium value are given by $\boldsymbol{M} = \chi_0 \boldsymbol{H}_e$ and $\boldsymbol{M}_0 = \chi_0 \boldsymbol{H}$, respectively, Eq. (4.57) can be reduced to Eq. (4.23). By transforming Eq. (4.57) into a form in a stationary coordinate system and applying Eq. (2.217), the phenomenological magnetization equation II is obtained

$$\frac{\mathrm{D}\boldsymbol{H}_e}{\mathrm{D}t} = \boldsymbol{\Omega} \times \boldsymbol{H}_e - \frac{1}{\tau_B}(\boldsymbol{H}_e - \boldsymbol{H}) - \frac{\mu_0}{4\zeta}\boldsymbol{H}_e \times (\boldsymbol{M} \times \boldsymbol{H}) \tag{4.59}$$

If the spin viscosity of ferrofluids can be neglected, substituting Eq. (2.217) into Eq. (4.59) leads to an alternative form of the phenomenological magnetization equation II,

$$\frac{\mathrm{D}\boldsymbol{H}_e}{\mathrm{D}t} = \boldsymbol{\omega}_p \times \boldsymbol{H}_e - \frac{1}{\tau_B}(\boldsymbol{H}_e - \boldsymbol{H}) \tag{4.60}$$

Equations (4.18) and (4.59) are consistent in weak field limits, but when the value of α is relatively large, there is a significant difference in the predicted values of $\boldsymbol{M}$ between the two equations.

Additionally, for practical purposes, the vector identity

$$\nabla \cdot (\boldsymbol{v}\boldsymbol{M}) = \boldsymbol{M}(\nabla \cdot \boldsymbol{v}) + (\boldsymbol{v} \cdot \nabla)\boldsymbol{M}$$

and the continuity equation are occasionally employed in the magnetization Eqs. (4.18), (4.47), and (4.59) to replace the term $(\boldsymbol{v} \cdot \nabla)\boldsymbol{M}$ in each equation with $\nabla \cdot (\boldsymbol{v}\boldsymbol{M})$.

The phenomenological magnetization equation I is the most frequently utilized among the aforementioned three magnetization equations (Rindldi et al., 2005), being suitable for ferrofluids in weak non-equilibrium states with relatively small shear rates (Felderhof, 2000). The microscopic magnetization equation is applicable across the entire magnetic field range (Martsenyuk et al., 1974; Patel et al., 2003), while the phenomenological magnetization equation II is suitable in far-from-equilibrium states (Shliomis, 2001). Müller and Liu (2001) derived an alternative form of magnetization equation based on non-equilibrium thermodynamic theory, encompassing the three magnetization equations discussed in this chapter as particular instances of this generalized framework.

REFERENCES

Bacri J.-C., Perzynski R., Shliomis M. I., Burde G. I., "Negative-viscosity" effect in a magnetic fluid, *Physical Review Letters*, 75(11): 2128–2131, 1995.

Berkovsky B. M., Medvedev V. F., Krakov M. S., *Magnetic fluids engineering applications*. Oxford: Oxford University Press. 1993.

Chantrell R. W., Popplewell J., Charles S. W., Measurements of particle size distribution parameters in ferrofluids, *IEEE Transactions on Magnetics*, 14: 975–977, 1978.

Embs J., Müller H. W., Wagner C., Knorr K., Lücke M., Measuring the rotational viscosity of ferrofluids without shear flow, *Physical Review E*, 61: R2196, 2000.

Felderhof B. U., Magnetoviscosity and relaxation in ferrofluids, *Physical Review E*, 62: 3848–3854, 2000.

Felderhof B. U., Kroh H. J., Hydrodynamics of magnetic and dielectric fluids in interaction with the electromagnetic field, *The Journal of Chemical Physics*, 110: 7403, 1999.

Ivanov A. O., Camp P. J., Magnetization relaxation dynamics in polydisperse ferrofluids, *Physical Review E*, 107: 034604, 2023.

Landau L. D., Lifshitz E. M., *Electrodynamics of continuous media*. New York: Pergamon Press. 1960.

Lehlooh A. F., Mahmood S. H., Williams J. M., On the particle size dependence of the magnetic anisotropy energy, *Physica B*, 321: 159–162, 2002.

Martsenyuk M. A., Raikher Y. L., Shliomis M. I., On the kinetics of magnetization of suspensions of ferromagnetic particles, *Soviet Physics JETP*, 28(2): 413–416, 1974.

Morozov K. I., Pshenichnikov A. F., Raikher Y. L., Shliomis M. I., Magnetic properties of ferrocolloids: The effect of interparticle interactions, *Journal of Magnetism and Magnetic Materials*, 65: 269–272, 1987.

Müller H. W., Liu M., Structure of ferrofluid dynamics, *Physical Review E*, 64: 061405, 2001.

Odenbach S., *Magnetoviscous effects in ferrofluids*. Berlin: Springer. 2002.

Patel R., Upadhyay R. V., Mehta R. V., Viscosity measurements of a ferrofluid: Comparison with various hydrodynamic equations, *Journal of Colloid and Interface Science*, 263: 661–664, 2003.

Rindldi C., Chaves A., Elborai S., He X., Zahn M., Magnetic fluid rheology and flows, *Current Opinion in Colloid and Interface Science*, 10: 141–157, 2005.

Rosensweig R. E., *Ferrohydrodynamics*, Cambridge: Cambridge University Press, 1985.

Shliomis M. I., Effective viscosity of magnetic suspensions, *Soviet Physics JETP*, 34(6): 1291–1294, 1972.

Shliomis M. I., Magnetic fluids, *Soviet Phys Uspekhi*, 17(2): 153–169, 1974.

Shliomis M. I., Ferrohydrodynamics: Testing a third magnetization equation, *Physical Review E*, 64: 060501, 2001.

Shliomis M. I., Lyubimova T. P., and Lyubimov D. V., Ferrohydrodynamics: an essay on the progress of ideas, *Chemical Engineering Communications*, 67: 275–290, 1988.

Yang W. M., *Basic theory of a ferrofluid damper and its experiments*. Beijing: Beijing Jiaotong University. 2012.

Zeuner A., Richter R., Rehberg I., Experiments on negative and positive magnetoviscosity in an alternating magnetic field, *Physical Review E*, 58(5): 6287–6293, 1998.

CHAPTER 5

Viscosity and Magnetoviscous Effects of Ferrofluids

VISCOSITY IS A FUNDAMENTAL property of ferrofluids, and interestingly, the apparent viscosity of ferrofluids can vary with applied magnetic fields. Essentially, the magnetic field introduces a magnetic stress component into the shear stress within the fluid. This results in an additional viscosity that overlays the intrinsic viscosity of the ferrofluid (Odenbach, 2002; Zubarev and Chirikov, 2010; Soto-Aquino and Rinaldi, 2010; Fang, 2019). The magnitude of this additional viscosity varies depending on the type of magnetic field (steady, alternating, rotating, etc.). To investigate the effect of a magnetic field on the viscosity of ferrofluids, it is typically necessary to solve the magnetization equation and use the resultant magnetization to calculate the magnetic stress component of the shear stress in Eq. (2.229). In other words, the contribution of the magnetic field to the shear stress determines the magnitude of the additional viscosity.

5.1 INTRINSIC VISCOSITY OF FERROFLUIDS

Ferrofluids maintain their fluidity even after being magnetized; yet, the application of a magnetic field can alter their apparent viscosity. In this context, the viscosity of ferrofluids without an external magnetic field is referred to as intrinsic viscosity. Due to the presence of magnetic particles, the intrinsic viscosity of a ferrofluid surpasses that of its carrier liquid. In the absence of an external magnetic field, the viscosity of ferrofluids shares the same expression as that of non-magnetic particle suspensions, with the earliest theoretical model describing this viscosity being the Einstein formula (Rosensweig, 1985)

$$\eta = \eta_0 \left(1 + \frac{5}{2}\phi \right) \tag{5.1}$$

where the assumption has been made that there are no surfactants or other coatings on the particles, η_0 is the viscosity of the carrier liquid, and ϕ is the volume fraction of solid

DOI: 10.1201/9781003540342-5

particles in ferrofluids. Generally speaking, the Einstein formula is considered valid only for ferrofluids with low concentrations.

Rosensweig (1985) proposed another theoretical model specifically tailored for high-concentration ferrofluids

$$\frac{\eta-\eta_0}{\phi\eta}=\frac{5}{2}\left(1+\frac{\delta_0}{r_0}\right)^3-\left(\frac{\frac{5}{2}\phi_c-1}{\phi_c^2}\right)\left(1+\frac{\delta}{r_0}\right)^6\phi \tag{5.2}$$

Here, ϕ_c represents the volume fraction corresponding to the close-packed state of particles in a ferrofluid, specifically $\phi_c = 0.74$. r_0 denotes the radius of the particles without a surfactant coating, while δ_0 represents the thickness of the surfactant layer coating the particles. Over time, ferrofluids may undergo aggregation phenomena such as particle chaining (Weis and Levesque, 1993, Morozov and Shliomis, 2004; Zubarev et al., 2011). The viscosity model expressed in Eq. (5.2) does not incorporate these aggregation effects. Zubarev et al. (2002) introduced a viscosity model for particles in an aggregated state.

For a ferrofluid moving at a velocity $\boldsymbol{v}$, it is assumed that each fluid element rotates with a local angular velocity $\boldsymbol{\Omega}$, known as vorticity. In the absence of an external magnetic field, the magnetic particles within the fluid element rotate with the same angular velocity, that is $\boldsymbol{\omega}_\mathrm{p} = \boldsymbol{\Omega}$. Herein, the macroscopic angular velocity of particle rotation is defined as the average over a small physical volume of the fluid element. Any deviation of $\boldsymbol{\omega}_\mathrm{p}$ from $\boldsymbol{\Omega}$ will decay to zero within a very short time (Martsenyuk et al., 1974) which is equal to

$$\tau_s=\frac{\rho_s d_\mathrm{p}^2}{60\eta_0} \tag{5.3}$$

For particles with a diameter of $d_\mathrm{p} = 10$ nm, a density of $\rho_s = 6 \times 10^3$ kg/m^3, and a carrier liquid with a viscosity of $\eta_0 = 10^{-3}$ Pa·s, the decay time is $\tau_s = 10^{-11}$ s. In the presence of a static magnetic field, the magnetic moment of the particles aligns with the field direction, hindering their free rotation in the vorticity field. This results in a deviation between the particles' angular velocity and flow vorticity, denoted as $\boldsymbol{\omega}_\mathrm{p} - \boldsymbol{\Omega}$. To maintain this angular velocity difference, ferrofluids require additional kinetic energy, manifesting as increased rotational viscosity (or vortex viscosity) (Odenbach, 2002)

$$\eta_r=\frac{3}{2}\eta\phi\frac{\alpha-\tanh\alpha}{\alpha+\tanh\alpha}\sin^2\beta \tag{5.4}$$

where β represents the angle between the vectors $\boldsymbol{H}$ and $\boldsymbol{\Omega}$. Notably, when $\boldsymbol{H}$ and $\boldsymbol{\Omega}$ are parallel, the viscosity becomes independent of the magnetic field, allowing the particle magnetic moment to align with the magnetic field and rotate with the same angular velocity as the flow vorticity. Conversely, when $\beta=\frac{\pi}{2}$, indicating that $\boldsymbol{H} \perp \boldsymbol{\Omega}$, such as in the cases of planar Couette or Poiseuille flow with a magnetic field applied perpendicular to the flow direction, or in a Poiseuille pipe flow with a capillary tube aligned along the axis of an electromagnetic coil, Eq. (5.4) indicates that in a weak static field, η_r increases proportionally with α^2

$$\eta_r = \frac{1}{4}\eta\phi\alpha^2, (\alpha \ll 1) \tag{5.5}$$

In a strong magnetic field, η_r approaches a saturated value, that is

$$\eta_r = \frac{3}{2}\eta\phi, (\alpha \gg 1) \tag{5.6}$$

The rotational viscosity η_r is comparable to the viscosity increase $\frac{5}{2}\eta\phi$ in Einstein's formula Eq. (5.1) due to the fact that both models target dilute suspensions, neglecting hydrodynamic and magnetic interactions among particles.

In a linearly magnetized alternating magnetic field, if the ferrofluid remains stationary (i.e. $\mathbf{\Omega} = \mathbf{0}$), the magnetic field induces rotational oscillations of magnetic particles, but these rotations lack directional bias, resulting in an overall zero angular velocity within a small physical volume, $\boldsymbol{\omega}_\mathrm{p} = \mathbf{0}$. However, any vortical flow ($\mathbf{\Omega} \neq \mathbf{0}$) is sufficient to disrupt this directional symmetry and induce a non-zero macroscopic angular velocity $\boldsymbol{\omega}_\mathrm{p}$. In this case, a slowly oscillating magnetic field opposes the free rotation of particles, resulting in $\eta_r > 0$. Conversely, a rapidly oscillating magnetic field causes the particles to rotate faster than the vorticity of the flow, that is $\boldsymbol{\omega}_\mathrm{p} > \mathbf{\Omega}$. In this scenario, particle rotations accelerate fluid rotation, converting a portion of the alternating magnetic field's energy into fluid kinetic energy. This conversion manifests as a decrease in ferrofluid apparent viscosity, a phenomenon known as the "negative viscosity" effect (Shliomis and Morozov, 1994; Zahn and Pioch, 1998, 1999).

Both the rotational viscosity in a static magnetic field and the "negative viscosity" in an alternating magnetic field are influenced by the Brownian magnetization relaxation mechanism as both phenomena originate from the difference in angular velocity between the particles and the carrier liquid (Pop and Odenbach, 2006). This mechanism is particularly significant in ferrofluids where the magnetic particles are relatively large.

5.2 ROTATIONAL VISCOSITY IN STATIC MAGNETIC FIELD

For dilute ferrofluids, the Einstein formula (Eq. 5.1) can be used to represent the viscosity of the ferrofluid when neglecting the rotational motion of solid particles relative to the carrier liquid. However, when a magnetic field is applied to the ferrofluid, it hinders the free rotation of particles. Any deviation between the angular velocity of particles, $\boldsymbol{\omega}_\mathrm{p}$, and the angular velocity of fluid motion, $\mathbf{\Omega}$, leads to additional energy dissipation, which manifests as an increase in rotational viscosity η_r.

To calculate the rotational viscosity in planar shear flows such as Poiseuille or Couette flow, as depicted in Figure 5.1, consider the upper wall moving with a velocity v_0 along the x-direction. The velocity of the ferrofluid is given by $\boldsymbol{v} = (v(y), 0, 0)$. Under the influence of

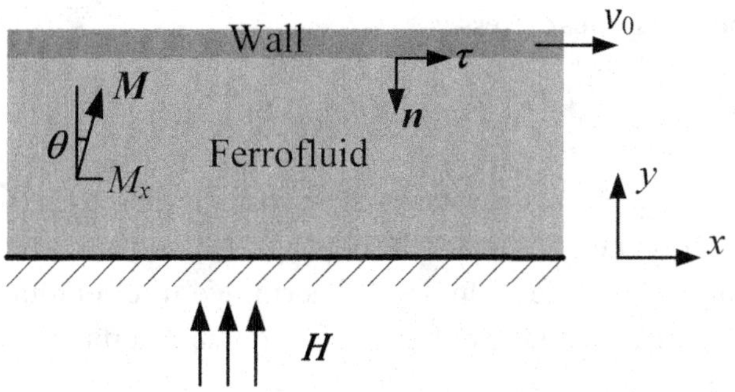

FIGURE 5.1 Couette flow of a ferrofluid.

a longitudinal static magnetic field $\boldsymbol{H} = (0, H, 0)$, the magnetization within the fluid possesses two components, $\boldsymbol{M} = (M_x, M_y, 0)$. According to Eq. (2.229), the tangential force exerted by the ferrofluid on the upper wall is given by

$$f_\tau = \eta \frac{\partial v}{\partial y} + \frac{\mu_0}{2} M_x H \tag{5.7}$$

It can be reformulated in the form of the shear stress of a general fluid

$$f_\tau = \eta_{\text{eff}} \frac{\partial v}{\partial y} \tag{5.8}$$

where η_{eff} represents the effective or apparent viscosity of ferrofluids when a magnetic field is applied. Comparing Eqs. (5.7) and (5.8), it can be observed that

$$\eta_{\text{eff}} = \eta + \frac{\mu_0}{2\dfrac{\partial v}{\partial y}} M_x H \tag{5.9}$$

where the second term on the right-hand side represents the rotational viscosity. It can be expressed in terms of the vorticity magnitude Ω

$$2\Omega = \frac{\partial v}{\partial y} \tag{5.10}$$

as

$$\eta_r = \frac{\mu_0 M_x H}{4\Omega} \tag{5.11}$$

Now the corresponding shear stress is

$$f_\tau = 2(\eta + \eta_r)\Omega \tag{5.12}$$

As demonstrated in Eq. (5.11), the rotational viscosity is directly related to the magnetization M_x in the direction perpendicular to the magnetic field. Once the velocity distribution and magnetization of the ferrofluid are determined, the rotational viscosity can be determined. Additionally, due to the presence of a non-zero component M_x perpendicular to the magnetic field, the shear stress is no longer proportional to the strain rate $\frac{\partial v}{\partial y}$. This indicates that ferrofluids exhibit non-Newtonian behaviour when subjected to a magnetic field. Next, the three magnetization equations introduced in Chapter 4 will be applied to derive the corresponding rotational viscosities.

To determine the rotational viscosity, the phenomenological magnetization equation I is first utilized. If $\Omega\tau_B \ll 1$, this equation can be linearized to Eq. (4.28). By considering only the component equation in the x-direction, replacing $\boldsymbol{M}$ in the first term on the right-hand side of Eq. (4.28) with $\boldsymbol{M}_0$, setting the second term to zero, and simplifying the third term to $M_x/\tau_\perp$, one obtains

$$\boldsymbol{M}_\perp = \tau_\perp \boldsymbol{\Omega} \times \boldsymbol{M}_0, M_x = \tau_\perp M_0 \Omega \tag{5.13}$$

Substitution of Eq. (5.13) into Eq. (5.11) yields

$$\eta_r = \frac{1}{4}\mu_0 \tau_\perp M_0 H \tag{5.14}$$

Introducing the expressions for $\tau_\perp$ in Eq. (4.28) and M_0 from Eq. (4.1) into Eq. (5.14) gives

$$\eta_r = \frac{\mu_0}{2}\frac{L(\alpha)}{2+\alpha L(\alpha)}\frac{3V_h\eta_0}{k_B T}\phi M_d H$$

where use has been made of Eqs. (4.13) and (4.2) to represent τ_B and M_S, respectively. Then Eq. (4.4) and Eq. (4.1) are subsequently utilized to express M_d and H, respectively, and gives

$$\eta_r = \frac{3}{2}\phi\frac{\alpha L(\alpha)}{2+\alpha L(\alpha)}\frac{V_h\eta_0}{V_p} \tag{5.15}$$

If the differences between V_h and V_p are neglected, and under the assumption of a dilute ferrofluid, the viscosity of the carrier liquid can be approximately replaced by the ferrofluid's viscosity. Consequently, the rotational viscosity can be expressed using Eq. (5.15) as

$$\eta_r(\alpha) = \frac{3}{2}\eta\phi\frac{\alpha L(\alpha)}{2+\alpha L(\alpha)} = \frac{3}{2}\eta\phi\frac{\alpha - \tanh\alpha}{\alpha + \tanh\alpha} \tag{5.16}$$

Equation (5.16) represents the rotational viscosity derived from the phenomenological magnetization equation I. Under weak field conditions, where $\alpha \ll 1$, using the approximation from Eq. (4.7), the asymptotic value of the rotational viscosity can be obtained,

$$\eta_r(\alpha) \approx \frac{3}{2}\eta\phi\frac{\alpha^2/3L(\alpha)}{2+\alpha^2/3} \approx \frac{1}{4}\eta\phi\alpha^2 \tag{5.17}$$

In the absence of an external magnetic field, particles rotate freely around the shear plane with an angular velocity $\boldsymbol{\omega}_\mathrm{p}$, where $\boldsymbol{\omega}_\mathrm{p} = \boldsymbol{\Omega}$ in this case, resulting in $\eta_r = 0$. However, when a sufficiently strong magnetic field is applied, the particles' direction is fixed by the field, and the action of the carrier liquid cannot alter its orientation. Consequently, relative sliding occurs between the particles and the carrier liquid, causing η_r to reach its limiting value $\eta_r(\infty) = \frac{3}{2}\eta\phi$, also known as the saturation value. Notably, this saturation value remains invariant regardless of the specific magnetization equation employed and can be directly derived from the motion equation (2.206). Since η_r reaches its saturation value when $\boldsymbol{\omega}_\mathrm{p} = \mathbf{0}$, substituting this value into the angular momentum equation (2.209) leads to

$$\mu_0 \boldsymbol{M} \times \boldsymbol{H} = -6\eta\phi\boldsymbol{\Omega} \tag{5.18}$$

Substituting Eq. (5.18) into Eq. (2.208) and expanding it using vector identities, then combining the second and fifth terms on the right-hand side obtains

$$\eta\nabla^2\boldsymbol{v} + \frac{\mu_0}{2}\nabla\times(\boldsymbol{M}\times\boldsymbol{H}) = \left(\eta + \frac{3}{2}\eta\phi\right)\nabla^2\boldsymbol{v} \tag{5.19}$$

Therefore, the rotational viscosity $\eta_r(\infty) = \frac{3}{2}\eta\phi$ caused by magnetization relaxation can be directly obtained by comparing to the intrinsic viscosity η.

Next the expression for rotational viscosity based on the microscopic magnetization equation will be derived. Specifically, substitution of $\tau_\perp$ in Eq. (4.55) into Eq. (5.14) gives

$$\eta_r(\alpha) = \frac{3}{2}\eta\phi\frac{\alpha L^2(\alpha)}{\alpha - L(\alpha)} \tag{5.20}$$

This result differs slightly from the one derived from the phenomenological magnetization equation I (Eq. 5.16), but their outcomes are relatively close throughout the entire range of α, and both tend to the same saturation value of $\frac{3}{2}\eta\phi$ when $\alpha \gg 1$. As shown in Figure 5.2, which compares the rotational viscosity calculated using Eqs. (5.16) and (5.20), it can be observed that they approach each other in both the very weak and very strong magnetic field limits, with minor deviations observed in the intermediate range. However, the deviation between them remains less than 15% throughout the entire range of α. Additionally, according to Eq. (5.14), η_r is independent of vorticity Ω under the condition $\Omega\tau_B \ll 1$. However, for finite $\Omega\tau_B$ values, η_r becomes dependent on Ω, indicating rheological properties of the ferrofluid.

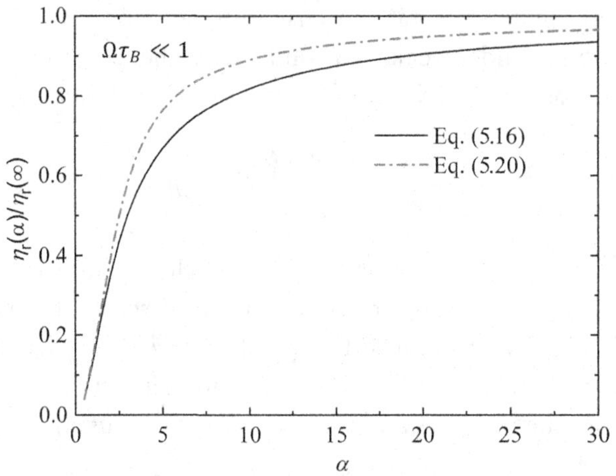

FIGURE 5.2 Rotational viscosity of a ferrofluid under the condition of $\Omega\tau_B \ll 1$.

Utilizing Eqs. (5.13), (5.11), and (5.20), the magnetization component perpendicular to the external magnetic field in a simple shear flow can be obtained

$$M_x = \tau_\perp M_0 \Omega = \frac{4\eta_r \Omega}{\mu_0 H} = \frac{6\eta\phi\Omega}{\mu_0 H}\frac{\alpha L^2(\alpha)}{\alpha - L(\alpha)} \tag{5.21}$$

Then, one has

$$\boldsymbol{M}\times\boldsymbol{H} = -M_x H = -\frac{6\eta\phi\Omega}{\mu_0}\frac{\alpha L^2(\alpha)}{\alpha - L(\alpha)} \tag{5.22}$$

After substituting it in Eq. (2.217), the magnitude of the particle's angular velocity of rotation is obtained

$$\omega_\text{p} = \left(1 - \frac{\alpha L^2(\alpha)}{\alpha - L(\alpha)}\right)\Omega \tag{5.23}$$

Evidently, under the influence of a static magnetic field, the rotational angular velocity of the particles is slower than the local angular velocity of the carrier liquid. Consequently, the rotation of these particles hinders the flow of the carrier liquid, resulting in an increase in tangential stress within the fluid.

Now compare the non-Newtonian viscosities in ferrofluids with limited $\Omega\tau_B$ values in static magnetic fields when the magnetization Eqs. (4.18), (4.47), and (4.59) are used. For Eq. (4.59), by substituting $\boldsymbol{H}$ and $\boldsymbol{H}_e$ with dimensionless fields $\boldsymbol{\alpha}$ (defined in Eq. (4.24)) and $\boldsymbol{\varsigma}$ (defined in Eq. (4.45)), respectively, and utilizing Eqs. (4.13), (4.45), and (4.3), one obtains

$$\frac{\text{D}\boldsymbol{\varsigma}}{\text{D}t} = \boldsymbol{\Omega}\times\boldsymbol{\varsigma} - \frac{1}{\tau_B}(\boldsymbol{\varsigma} - \boldsymbol{\alpha}) - \frac{L(\varsigma)}{2\tau_B\varsigma}\boldsymbol{\varsigma}\times(\boldsymbol{\varsigma}\times\boldsymbol{\alpha}) \tag{5.24}$$

Under steady-state conditions, the left-hand side of Eq. (5.24) equals zero. Let θ_m be the angle between the effective field $\boldsymbol{\varsigma}$ and the actual field $\boldsymbol{\alpha}$ vectors, and one can formulate the component equation along the direction parallel to $\boldsymbol{\varsigma}$,

$$\cos\theta_m = \frac{\varsigma}{\alpha} \tag{5.25}$$

and the component equation in the direction perpendicular to $\boldsymbol{\varsigma}$,

$$\Omega\varsigma - \frac{\alpha\sin\theta_m}{\tau_B} - \frac{L(\varsigma)\varsigma\alpha\sin\theta_m}{2\tau_B} = 0 \tag{5.26}$$

Combining Eqs. (5.25) and (5.26) yields

$$\alpha\sin\theta_m = \sqrt{\alpha^2 - \varsigma^2} = \frac{2\tau_B\Omega\varsigma}{L(\varsigma)\varsigma + 2} \tag{5.27}$$

The use of the expression

$$M_x = M_S L(\varsigma)\sin\theta_m \tag{5.28}$$

and Eq. (5.27) gives the rotational viscosity based on Eq. (4.59)

$$\eta_r = \frac{3}{2}\eta\phi\frac{\varsigma L(\varsigma)}{L(\varsigma)\varsigma + 2}, \cos\theta_m = \frac{\varsigma}{\alpha} \tag{5.29}$$

Using the dimensionless magnetic field represented by Eqs. (4.24) and (4.45) in Eq. (4.47), the terms on the right-hand side can be expressed as

$$\begin{aligned}
&\frac{M_S L(\varsigma)}{\varsigma}\boldsymbol{\Omega}\times\boldsymbol{\varsigma} - \left(1 - \frac{\boldsymbol{\alpha}\cdot\boldsymbol{\varsigma}}{\varsigma^2}\right)\frac{M_S L(\varsigma)\boldsymbol{\varsigma}}{\varsigma\tau_B} - \frac{\mu_0}{L(\varsigma)}\left(\frac{1}{L(\varsigma)} - \frac{1}{\varsigma}\right)\left(\frac{M_S L(\varsigma)}{\varsigma}\right)^2 \frac{k_B T}{\mu_0 m_p}\frac{\boldsymbol{\varsigma}\times(\boldsymbol{\varsigma}\times\boldsymbol{\alpha})}{6\eta\phi} \\
&= \boldsymbol{\Omega}\times\boldsymbol{\varsigma} - \left(1 - \frac{\boldsymbol{\alpha}\cdot\boldsymbol{\varsigma}}{\varsigma^2}\right)\frac{\boldsymbol{\varsigma}}{\tau_B} - \left(\frac{1}{L(\varsigma)} - \frac{1}{\varsigma}\right)\frac{\boldsymbol{\varsigma}\times(\boldsymbol{\varsigma}\times\boldsymbol{\alpha})}{2\tau_B\varsigma} = 0
\end{aligned} \tag{5.30}$$

As given by Eq. (5.30), in the direction parallel to $\boldsymbol{\varsigma}$, one has

$$1 - \frac{\boldsymbol{\alpha}\cdot\boldsymbol{\varsigma}}{\varsigma^2} = 0 \tag{5.31}$$

and

$$\cos\theta_m = \frac{\varsigma}{\alpha}$$

Then in the direction perpendicular to $\boldsymbol{\varsigma}$, one has

$$\Omega\varsigma - \left(\frac{1}{L(\varsigma)} - \frac{1}{\varsigma}\right)\frac{\varsigma^2 \alpha \sin\theta_m}{2\tau_B \varsigma} = 0 \tag{5.32}$$

and

$$\alpha \sin\theta_m = \sqrt{\alpha^2 - \varsigma^2} = \frac{2\tau_B \varsigma \Omega L(\varsigma)}{\varsigma - L(\varsigma)} \tag{5.33}$$

The corresponding rotational viscosity is

$$\eta_r = \frac{3}{2}\eta\phi \frac{\varsigma L^2(\varsigma)}{\varsigma - L(\varsigma)}, \cos\theta_m = \frac{\varsigma}{\alpha} \tag{5.34}$$

For Eq. (4.18), let

$$\boldsymbol{M} = M_0 \boldsymbol{\varsigma} / \alpha, M_0 = M_S L(\alpha) \tag{5.35}$$

and it is no longer necessary to indicate the effective field at this point. Substituting Eq. (5.35) into the right-hand side of Eq. (4.18) gives

$$\begin{aligned} &\frac{M_S L(\alpha)}{\alpha}\boldsymbol{\Omega} \times \boldsymbol{\varsigma} - \frac{1}{\tau}\frac{M_S L(\alpha)}{\alpha}(\boldsymbol{\varsigma} - \boldsymbol{\alpha}) - \frac{\mu_0}{6\eta\phi}\left(\frac{M_S L(\alpha)}{\alpha}\right)^2 \frac{k_B T}{\mu_0 m_p}\boldsymbol{\varsigma} \times (\boldsymbol{\varsigma} \times \boldsymbol{\alpha}) \\ &= \boldsymbol{\Omega} \times \boldsymbol{\varsigma} - \frac{1}{\tau_B}(\boldsymbol{\varsigma} - \boldsymbol{\alpha}) - \frac{L(\alpha)}{2\tau_B \alpha}\boldsymbol{\varsigma} \times (\boldsymbol{\varsigma} \times \boldsymbol{\alpha}) = 0 \end{aligned} \tag{5.36}$$

Based on Eq. (5.36), the component equation parallel to the direction of $\boldsymbol{\varsigma}$ yields $\cos\theta_m = \frac{\varsigma}{\alpha}$, while the component equation perpendicular to the direction of $\boldsymbol{\varsigma}$ leads to

$$\Omega\varsigma - \frac{\alpha \sin\theta_m}{\tau_B} - \frac{\varsigma^2 L(\alpha) \alpha \sin\theta_m}{2\tau_B \alpha} = 0 \tag{5.37}$$

After a rearrangement, one gets

$$\alpha \sin\theta_m = \sqrt{\alpha^2 - \varsigma^2} = \frac{2\tau_B \alpha \varsigma \Omega}{2\alpha + \varsigma^2 L(\alpha)} \tag{5.38}$$

Combining with

$$M_x = M_0 (\varsigma / \alpha) \sin\theta_m \tag{5.39}$$

one obtains

$$\eta_r = \frac{3}{2}\eta\phi\frac{\varsigma^2 L(\alpha)}{2\alpha + \varsigma^2 L(\alpha)}, \cos\theta_m = \frac{\varsigma}{\alpha} \tag{5.40}$$

Given the condition $\tau_B \ll 1$, the final term in Eqs. (5.27), (5.33), and (5.38) can be considered to approach zero, resulting in $\varsigma = \alpha$ from all three equations. Consequently, Eqs. (5.29) and (5.40) simplify to Eq. (5.5), while Eq. (5.34) simplifies to Eq. (5.20). However, in the case of higher shear rates, that is $\Omega\tau_B \geq 1$, viscous shear causes the rotation of magnetic particles to deviate from the magnetic field direction, leading to demagnetization. This effect is attributed to the decrease in the parameter ς.

Regarding Eqs. (5.27), (5.33), and (5.38), it is evident that, with a constant α, an increase in $\Omega\tau_B$ leads to a decrease in ς. The decrease in magnetization, in turn, leads to a further reduction in rotational viscosity. However, in practice, the decrease in rotational viscosity is minimal during the initial stage of $\Omega\tau_B$ increasing up to $\Omega\tau_B$~1. Significant reductions in rotational viscosity only occur when $\Omega\tau_B$ is much greater than 1. Figures 5.3–5.5 illustrate the relationship between rotational viscosity and magnetic field intensity for $\Omega\tau_B$ values of 2, 4, and 6, respectively. It can be observed that, when the shear rate is finite, the predictions of Eqs. (5.29) and (5.40) for rotational viscosity diverge, and this discrepancy increases as the shear rate rises. In scenarios with high shear rates and strong magnetic field intensities (as shown in Figure 5.5), the rotational viscosity obtained from the phenomenological magnetization equation I exhibits relaxation characteristics. This observation remains unconfirmed through experiments or comparisons with numerical solutions of microscopic equations (Shliomis et al., 1988; Ambacher et al., 1992). In contrast, the phenomenological magnetization equation II provides results that are more consistent with those of microscopic equation across a broader range of α and $\Omega\tau_B$ values. Consequently, for static magnetic fields, the predictions of the phenomenological equation II align closely with

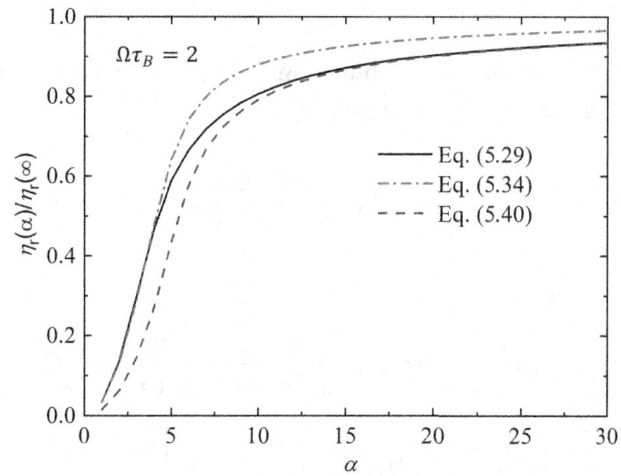

FIGURE 5.3 Rotational viscosity varies with the external magnetic field ($\Omega\tau_B = 2$).

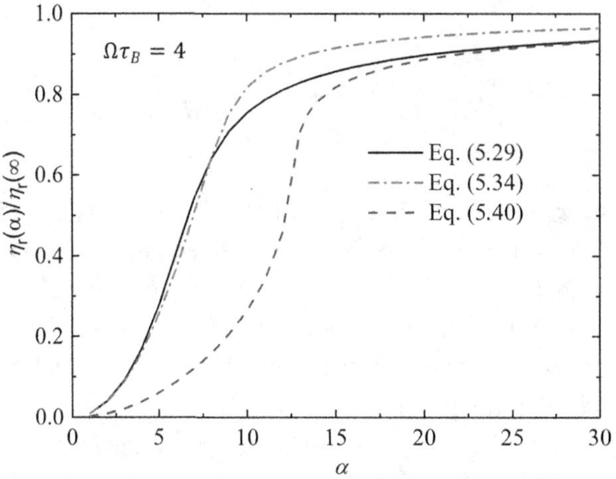

FIGURE 5.4 Rotational viscosity varies with the external magnetic field ($\Omega\tau_B = 4$).

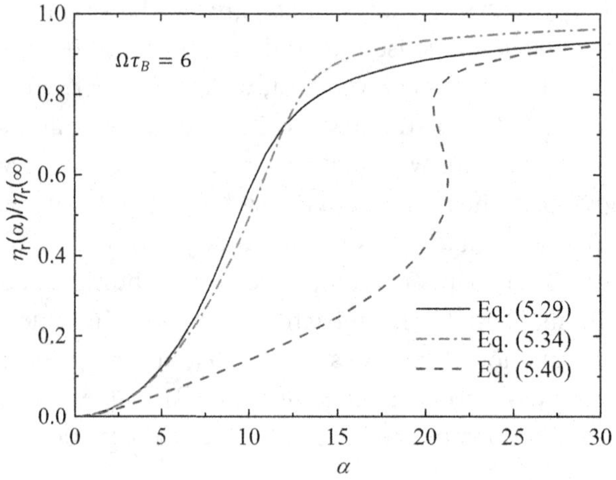

FIGURE 5.5 Rotational viscosity varies with the external magnetic field ($\Omega\tau_B = 6$).

those of the microscopic equation. In obtaining the curve corresponding to Eq. (5.40) in Figure 5.5, the relationship was transformed into

$$\frac{\eta_r}{\eta_r(\infty)} / \sin\theta_m = \frac{\varsigma L(\alpha)}{2\tau_B\Omega} \tag{5.41}$$

$$\sin\theta_m = \frac{2\tau_B\Omega}{2\dfrac{\alpha}{\varsigma} + \varsigma L(\alpha)} = \frac{1}{\dfrac{1}{\tau_B\Omega\cos\theta_m} + \dfrac{\eta_r}{\eta_r(\infty)} / \sin\theta_m}$$
$$\Rightarrow \theta_m = \arctan\left[\left(1 - \frac{\eta_r}{\eta_r(\infty)}\right)\tau_B\Omega\right] \tag{5.42}$$

After obtaining θ_m from Eq. (5.42), one can replace ς in Eq. (5.41) with $\alpha \cos \theta_m$, and then numerically solve the resultant equation to determine the corresponding α value for a given $\eta_r/\eta_r(\infty)$ value.

It is worth noting that the calculation of shear flow described above is also applicable to rigidly rotating ferrofluids under the influence of a transverse magnetic field, where the angular velocity is perpendicular to the magnetic field ($\boldsymbol{\Omega} \perp \boldsymbol{H}$), as well as to stationary ferrofluids in a rotating uniform magnetic field ($\boldsymbol{H} = (H \cos(wt), H \sin(wt), 0)$). Furthermore, for a comprehensive examination of how the apparent viscosity of ferrofluids varies with shear rate, kindly refer to studies conducted by Odenbach (2002), Odenbach and Störk (1998), Cunha (2022), Müller et al. (2006), Pinho et al. (2011), and Zubarev and Iskakova (2008). For insights into how it varies with the suspending medium, please refer to studies by Borin et al. (2016), Gerth-Noritzsch et al. (2011), and Ilg et al. (2005).

5.3 "NEGATIVE" VISCOSITY IN ALTERNATING MAGNETIC FIELDS

5.3.1 Rotational Viscosity under the Action of a Low-amplitude Alternating Magnetic Field at Low Shear Rates

The alternating magnetic field refers to a magnetic field with a fixed direction and a periodically varying magnitude over time. Assuming the magnetic field is along the x-direction, its magnetic field intensity can be represented as

$$\boldsymbol{H} = \left(H_0 \cos(wt), 0, 0\right) \tag{5.43}$$

where H_0 is the amplitude of the alternating magnetic field and w represents the frequency of the magnetic field multiplied by 2π. Under the assumption of small amplitude, one has

$$\mu_0 m_p H_0 \ll k_B T \tag{5.44}$$

Based on the definition provided in Eq. (4.1), the parameter α_0 can be expressed by

$$\alpha_0 = \mu_0 m_p H_0 / k_B T \tag{5.45}$$

From Eq. (5.44), it is evident that $\alpha_0 \ll 1$. Utilizing Eqs. (4.7) and (4.8), the Langevin function can be linearized and the equilibrium magnetization can be expressed as

$$\boldsymbol{M}_0(t) = \chi_0 \boldsymbol{H}(t), \chi_0 = \frac{\mu_0 m_p^2 N}{3 k_B T} \tag{5.46}$$

The existence of the linearized approximation given by Eq. (5.46) under the condition of a small-amplitude alternating magnetic field allows for the derivation of an analytical solution for rotational viscosity.

Under the assumption of linear magnetization, the magnetic field represented by Eq. (5.43) can be regarded as the superposition of two rotating magnetic fields (Shliomis and Morozov, 1994),

$$\boldsymbol{H}_{+} = \left(H_0 \cos(wt), H_0 \sin(wt), 0\right) \tag{5.47}$$

$$\boldsymbol{H}_{-} = \left(H_0 \cos(wt), -H_0 \sin(wt), 0\right) \tag{5.48}$$

Now the magnetic field intensity can be represented as

$$\boldsymbol{H} = \left(\boldsymbol{H}_{+} + \boldsymbol{H}_{-}\right)/2 \tag{5.49}$$

When the ferrofluid remains stationary ($\boldsymbol{\Omega} = 0$), under the separate actions of these two rotating magnetic fields, the magnetization will rotate with an angular frequency identical to that of the field, but will lag by a certain angle ϑ. Additionally, the particles rotate stably at a specific angular velocity, perpendicular to the plane of magnetic field rotation. Assuming that the magnetization intensity can be represented as

$$\boldsymbol{M}_{\pm} = \left(M \cos(wt - \vartheta), \pm M \sin(wt - \vartheta), 0\right) \tag{5.50}$$

To derive an expression for M by solving the magnetization Eq. (4.18) in the Cartesian coordinate system, considering solely the effect of the magnetic field $\boldsymbol{H}_{+}$, one has to expand the terms in the equation individually as follows:

$$\frac{\mathrm{D}\boldsymbol{M}}{\mathrm{D}t} = -Mw \sin(wt - \vartheta)\boldsymbol{e}_x + Mw \cos(wt - \vartheta)\boldsymbol{e}_y \tag{5.51}$$

$$\boldsymbol{M} - \boldsymbol{M}_0 = \left[M \cos(wt - \vartheta) - \chi_0 H_0 \cos(wt)\right]\boldsymbol{e}_x + \left[M \sin(wt - \vartheta) - \chi_0 H_0 \sin(wt)\right]\boldsymbol{e}_y \tag{5.52}$$

$$\boldsymbol{M} \times \boldsymbol{H} = M H_0 \sin \vartheta \boldsymbol{e}_z \tag{5.53}$$

$$\boldsymbol{M} \times \left(\boldsymbol{M} \times \boldsymbol{H}\right) = M^2 H_0 \sin \vartheta \left[\sin(wt - \vartheta)\boldsymbol{e}_x - \cos(wt - \vartheta)\boldsymbol{e}_y\right] \tag{5.54}$$

The component equations of Eq. (4.18) in the x and y directions are now given by:

$$-Mw \sin(wt - \vartheta) = -\frac{1}{\tau_B}\left[M \cos(wt - \vartheta) - \chi_0 H_0 \cos(wt)\right] - \frac{\mu_0}{6\eta\phi} M^2 H_0 \sin \vartheta \sin(wt - \vartheta) \tag{5.55}$$

$$Mw \cos(wt - \vartheta) = -\frac{1}{\tau_B}\left[M \sin(wt - \vartheta) - \chi_0 H_0 \sin(wt)\right] + \frac{\mu_0}{6\eta\phi} M^2 H_0 \sin \vartheta \cos(wt - \vartheta) \tag{5.56}$$

Rearranging Eqs. (5.55) and (5.56), respectively, and dividing the resultant results gives

$$\tan\left(wt-\vartheta\right)=\frac{Mw\sin\left(wt-\vartheta\right)-\dfrac{1}{\tau_B}\left[M\cos\left(wt-\vartheta\right)-\chi_0 H_0\cos\left(wt\right)\right]}{Mw\cos\left(wt-\vartheta\right)+\dfrac{1}{\tau_B}\left[M\sin\left(wt-\vartheta\right)-\chi_0 H_0\sin\left(wt\right)\right]} \tag{5.57}$$

which can be simplified as

$$\begin{aligned}&M\left\{\tan\left(wt-\vartheta\right)\left[w\cos\left(wt-\vartheta\right)+\frac{1}{\tau_B}\sin\left(wt-\vartheta\right)\right]-w\sin\left(wt-\vartheta\right)+\frac{1}{\tau_B}\cos\left(wt-\vartheta\right)\right\}\\&\quad=\frac{1}{\tau_B}\chi_0 H_0\cos\left(wt\right)+\frac{1}{\tau_B}\chi_0 H_0\sin\left(wt\right)\tan\left(wt-\vartheta\right)\end{aligned} \tag{5.58}$$

namely,

$$\begin{aligned}M&=\frac{\dfrac{1}{\tau_B}\chi_0 H_0\left[\cos\left(wt\right)+\sin\left(wt\right)\tan\left(wt-\vartheta\right)\right]}{\dfrac{1}{\tau_B}\left[\tan\left(wt-\vartheta\right)\sin\left(wt-\vartheta\right)+\cos\left(wt-\vartheta\right)\right]}\\&=\frac{\chi_0 H_0\left[\cos\left(wt\right)\cos\left(wt-\vartheta\right)+\sin\left(wt\right)\sin\left(wt-\vartheta\right)\right]}{\left[\sin^2\left(wt-\vartheta\right)+\cos^2\left(wt-\vartheta\right)\right]}\\&=\chi_0 H_0\cos\vartheta\end{aligned} \tag{5.59}$$

The following calculation determines the average angular velocity of particles in a stationary ferrofluid under the influence of the magnetic field represented by Eq. (5.47). Using Eqs. (2.217), (2.213), (5.53), and (5.59), the average angular velocity of the particles is expressed as

$$\boldsymbol{\omega}_{\mathrm{p}}^{+}=\frac{\mu_0}{6\eta\phi}MH_0\sin\vartheta\boldsymbol{e}_z=\frac{\mu_0}{6\eta\phi}\chi_0 H_0^2\cos\vartheta\sin\vartheta\boldsymbol{e}_z \tag{5.60}$$

To solve for ϑ, substituting the calculated value of M from Eq. (5.59) into the first component Eq. (5.55), one has

$$\begin{aligned}&w\cos\vartheta\sin\left(wt-\vartheta\right)\\&=\frac{1}{\tau_B}\left[\cos\vartheta\cos\left(wt-\vartheta\right)-\cos wt\right]+\frac{\mu_0}{6\eta\phi}\chi_0 H_0^2\cos^2\vartheta\sin\vartheta\sin\left(wt-\vartheta\right)\\&=\frac{1}{\tau_B}\left[\cos\vartheta\cos\left(wt-\vartheta\right)-\cos\left(\vartheta-\left(\vartheta-wt\right)\right)\right]+\frac{\mu_0}{6\eta\phi}\chi_0 H_0^2\cos^2\vartheta\sin\vartheta\sin\left(wt-\vartheta\right)\\&=\frac{1}{\tau_B}\sin\vartheta\sin\left(wt-\vartheta\right)+\frac{\mu_0}{6\eta\phi}\chi_0 H_0^2\cos^2\vartheta\sin\vartheta\sin\left(wt-\vartheta\right)\end{aligned}$$

which can be simplified as

$$w\cos\vartheta = \frac{1}{\tau_B}\sin\vartheta + \frac{\mu_0}{6\eta\phi}\chi_0 H_0^2\cos^2\vartheta\sin\vartheta \tag{5.61}$$

By applying Eq. (5.60) to the right-hand side of Eq. (5.61), one obtains

$$\tan\vartheta = \tau_B\left(w - \frac{\mu_0}{6\eta\phi}\chi_0 H_0^2\cos\vartheta\sin\vartheta\right) = \tau_B\left(w - \omega_p^+\right) \tag{5.62}$$

Similarly, when considering a situation where only the magnetic field $\boldsymbol{H}_-$ is acting alone, it holds that

$$\boldsymbol{\omega}_p^- = -\frac{\mu_0}{6\eta\phi}MH_0\sin\vartheta\boldsymbol{e}_z = -\frac{\mu_0}{6\eta\phi}\chi_0 H_0^2\cos\vartheta\sin\vartheta\boldsymbol{e}_z \tag{5.63}$$

From the principle of superposition, it is understood that under the influence of the composite magnetic field $\boldsymbol{H}$ represented by Eq. (5.49), both the average rotational speed $\boldsymbol{\omega}_p$ of the particles and the magnetization along the y-direction vanish. This occurs because, in a stationary fluid, a linearly magnetizing alternating magnetic field does not favour a preferential direction. Alternatively, at any given moment, half of the particles rotate clockwise, while the other half rotate counterclockwise, ultimately resulting in a macroscopic average rotational speed of zero (Shliomis et al., 1988).

When a ferrofluid exhibits flow and possesses vorticity $\boldsymbol{\Omega}$, such as in boundary-driven Couette flow or pressure gradient-driven Poiseuille flow, the symmetry of the aforementioned positive and negative solutions for particle angular velocity is disrupted by the presence of flow vorticity. Now representing the flow vorticity as

$$\boldsymbol{\Omega} = (0,0,\Omega) \tag{5.64}$$

The magnetization resulting from the individual actions of the two rotating magnetic fields, $\boldsymbol{H}_+$ and $\boldsymbol{H}_-$, can be represented in accordance with Eq. (5.59) in the context of stationary ferrofluids as

$$M_+ = \chi_0 H_0\cos\vartheta_+, M_- = \chi_0 H_0\cos\vartheta_- \tag{5.65}$$

The corresponding angular velocity of particle rotation and the lag angle are respectively

$$\omega_p^+ = \frac{\mu_0}{6\eta\phi}\chi_0 H_0^2\cos\vartheta\sin\vartheta + \Omega, \omega_p^- = \frac{\mu_0}{6\eta\phi}\chi_0 H_0^2\cos\vartheta\sin\vartheta - \Omega \tag{5.66}$$

$$\tan\vartheta_+ = \tau_B\left(w - \omega_p^+\right), \tan\vartheta_- = \tau_B\left(w - \omega_p^-\right) \tag{5.67}$$

Evidently, when the angular velocity of the particle aligns with the flow vorticity, the rotational speed of the particle increases. Under the weak-field approximation, the first term in the expressions for ω_p^+ and ω_p^- can be neglected, leading to

$$\tan\vartheta_+ = \tau_B(w-\Omega), \tan\vartheta_- = \tau_B(w+\Omega) \tag{5.68}$$

and

$$\cos^2\vartheta_+ = \frac{1}{1+(w-\Omega)^2\tau_B^2}, \cos^2\vartheta_- = \frac{1}{1+(w+\Omega)^2\tau_B^2} \tag{5.69}$$

Based on Eqs. (5.50), (5.65), (5.68), and (5.69), the magnetization components in the y-direction resulting from the two types of rotating magnetic fields can be derived as follows:

$$\begin{aligned} M_y^+ &= M_+\sin(wt-\vartheta_+) \\ &= \chi_0 H_0\cos^2\vartheta_+\left[\sin(wt)-\tan\vartheta_+\cos(wt)\right] \\ &= \frac{\chi_0 H_0}{1+(w-\Omega)^2\tau_B^2}\left[\sin(wt)-\tau_B(w-\Omega)\cos(wt)\right] \end{aligned} \tag{5.70}$$

$$\begin{aligned} M_y^- &= -M_-\sin(wt-\vartheta_-) \\ &= -\chi_0 H_0\cos\vartheta_-\cos^2\vartheta_-\left[\sin(wt)-\tan\vartheta_-\cos(wt)\right] \\ &= -\frac{\chi_0 H_0}{1+(w+\Omega)^2\tau_B^2}\left[\sin(wt)-\tau_B(w+\Omega)\cos(wt)\right] \end{aligned} \tag{5.71}$$

Assuming $\Omega\tau_B \ll 1$, the utilization of Eqs. (5.70) and (5.71) leads to the magnetization component in the y-direction when subjected to the combined magnetic field $\boldsymbol{H}$

$$M_y = \chi_0 H_0 \tau_B \Omega\cos^2\vartheta\cos(wt-2\vartheta) \tag{5.72}$$

The proof of Eq. (5.72) is as follows:

$$\begin{aligned} M_y &= \frac{1}{2}\left(M_y^+ + M_y^-\right) \\ &= \frac{1}{2}\chi_0 H_0\left\{\begin{aligned} &\frac{1}{1+(w-\Omega)^2\tau_B^2}\left[\sin(wt)-\tau_B(w-\Omega)\cos(wt)\right] \\ &-\frac{1}{1+(w+\Omega)^2\tau_B^2}\left[\sin(wt)-\tau_B(w+\Omega)\cos(wt)\right] \end{aligned}\right\} \end{aligned}$$

$$
=\frac{1}{2}\chi_0H_0\left\{\begin{array}{l}\sin(wt)\left[\dfrac{1}{1+(w-\Omega)^2\tau_B^2}-\dfrac{1}{1+(w+\Omega)^2\tau_B^2}\right]\\+\tau_B\cos(wt)\left[\dfrac{w+\Omega}{1+(w+\Omega)^2\tau_B^2}-\dfrac{w-\Omega}{1+(w-\Omega)^2\tau_B^2}\right]\end{array}\right\}
$$

$$
=\frac{1}{2}\chi_0H_0\left\{\begin{array}{l}\tau_B^2\sin(wt)\dfrac{(w+\Omega)^2-(w-\Omega)^2}{\left[1+(w-\Omega)^2\tau_B^2\right]\left[1+(w+\Omega)^2\tau_B^2\right]}\\+\tau_B\cos(wt)\dfrac{2\Omega+w\tau_B^2\left[(w-\Omega)^2-(w+\Omega)^2\right]+\Omega\tau_B^2\left[(w-\Omega)^2+(w+\Omega)^2\right]}{\left[1+(w-\Omega)^2\tau_B^2\right]\left[1+(w+\Omega)^2\tau_B^2\right]}\end{array}\right\}
$$

$$
=\frac{\frac{1}{2}\chi_0H_0}{\left[1+(w-\Omega)^2\tau_B^2\right]\left[1+(w+\Omega)^2\tau_B^2\right]}\left\{\begin{array}{l}4\Omega w\tau_B^2\sin(wt)\\+\tau_B\cos(wt)\left[2\Omega-4w^2\tau_B^2\Omega+2\Omega\tau_B^2\left(w^2+\Omega^2\right)\right]\end{array}\right\}
$$

$$
=\frac{\chi_0H_0\tau_B\Omega}{\left[1+(w-\Omega)^2\tau_B^2\right]\left[1+(w+\Omega)^2\tau_B^2\right]}\left\{2w\tau_B\sin(wt)+\cos(wt)\left[1-\tau_B^2\left(w^2-\Omega^2\right)\right]\right\}
$$

$$
\approx\frac{\chi_0H_0\tau_B\Omega}{\left(1+w^2\tau_B^2\right)^2}\left[2w\tau_B\sin(wt)+\left(1-w^2\tau_B^2\right)\cos(wt)\right]
$$

$$
=\chi_0H_0\tau_B\Omega\cos^2\vartheta\left[\frac{2\tan\vartheta}{1+\tan^2\vartheta}\sin(wt)+\frac{1-\tan^2\vartheta}{1+\tan^2\vartheta}\cos(wt)\right]
$$

$$
=\chi_0H_0\tau_B\Omega\cos^2\vartheta\left[\sin(2\vartheta)\sin(wt)+\cos(2\vartheta)\cos(wt)\right]
$$

$$
=\chi_0H_0\tau_B\Omega\cos^2\vartheta\cos(wt-2\vartheta)
$$

where use has been made of $\Omega\tau_B \ll 1$ and $\tan\vartheta = w\tau_B$.

The non-zero value of M_y in Eq. (5.72) indicates that the ferrofluid is subject to the action of magnetic moment. The corresponding magnetic torque density can be determined using the result of Eqs. (5.72) and (5.43) as

$$
\begin{aligned}
\mu_0\mathbf{M}\times\mathbf{H}&=-\mu_0M_yH_x\mathbf{k}\\
&=-\mu_0\chi_0H_0^2\tau_B\mathbf{\Omega}\cos^2\vartheta\cos(wt-2\vartheta)\cos(wt)\\
&=-\mu_0\chi_0H_0^2\tau_B\mathbf{\Omega}\cos^2\vartheta\left[\cos^2(wt)\cos(2\vartheta)+\sin(wt)\cos(wt)\sin(2\vartheta)\right]
\end{aligned}\tag{5.73}
$$

From Eqs. (5.45), (4.13) and (5.46), one obtains

$$
\mu_0\mathbf{M}\times\mathbf{H}=-\frac{1}{2}\eta\phi\alpha_0^2\mathbf{\Omega}\cos^2\vartheta\left\{\cos2\vartheta\left[1+\cos(2wt)\right]+\sin(2\vartheta)\sin(2wt)\right\}\tag{5.74}
$$

There are two distinct time scales in Eq. (5.74). One is the explicit time variable t, multiplied by w and characterized by a scale of $\frac{1}{\tau_B}$. The other is implicitly integrated into the vorticity $\boldsymbol{\Omega}$, exhibiting a characteristic time scale $\tau_h \sim l^2/\nu$, where l represents the characteristic length scale of the flow and ν denotes the kinematic viscosity of the ferrofluid. Notably, τ_h is typically much longer than the magnetization relaxation time, satisfying the condition $\tau_h \gg \tau_B$. Consequently, when solving the momentum Eq. (2.219) under the influence of an alternating magnetic field, the solution can be expressed as a superposition of a slowly varying component and a rapidly varying component

$$\boldsymbol{v} = \boldsymbol{v}_1 + \boldsymbol{v}_2, p = p_1 + p_2 \tag{5.75}$$

For $\boldsymbol{M}$ and $\boldsymbol{H}$, there is no distinction between slow and rapid variations, as both exhibit rapid oscillatory over time. By substituting Eq. (5.75) into the momentum Eq. (2.219), one obtains

$$\rho\left(\frac{\mathrm{D}\boldsymbol{v}_1}{\mathrm{D}t} + \frac{\mathrm{D}\boldsymbol{v}_2}{\mathrm{D}t} + (\boldsymbol{v}_2\cdot\nabla)\boldsymbol{v}_1\right) = -\nabla(p_1 + p_2) + \eta\nabla^2(\boldsymbol{v}_1 + \boldsymbol{v}_2) + \mu_0\boldsymbol{M}\cdot\nabla\boldsymbol{H} + \frac{\mu_0}{2}\nabla\times(\boldsymbol{M}\times\boldsymbol{H}) \tag{5.76}$$

where use has been made of

$$(\boldsymbol{v}\cdot\nabla)\boldsymbol{v} = \boldsymbol{v}_1\cdot\nabla\boldsymbol{v}_1 + \boldsymbol{v}_1\cdot\nabla\boldsymbol{v}_2 + \boldsymbol{v}_2\cdot\nabla\boldsymbol{v}_1 + \boldsymbol{v}_2\cdot\nabla\boldsymbol{v}_2$$

and the higher order small quantity $\boldsymbol{v}_2 \cdot \nabla\, \boldsymbol{v}_2$ has been ignored.

By averaging both sides of Eq. (5.76) over a rapidly varying time period, specifically the period of magnetic field variation $2\pi/w$, the slowly varying components can be approximated as constants at this scale. The averaging operation results in the elimination of terms containing $\boldsymbol{v}_2$ and p_2, as their periods are the same as the magnetic field's variation period, ultimately yielding

$$\rho\frac{\mathrm{D}\boldsymbol{v}_1}{\mathrm{D}t} = -\nabla p_1 + \eta\nabla^2\boldsymbol{v}_1 + \mu_0\langle\boldsymbol{M}\cdot\nabla\boldsymbol{H}\rangle + \frac{\mu_0}{2}\nabla\times\langle\boldsymbol{M}\times\boldsymbol{H}\rangle \tag{5.77}$$

where the pointed brackets "⟨⟩" signify the computation of the average value over the duration $2\pi/w$. In addition, from Eq. (5.74), it is known that

$$\langle\boldsymbol{M}\times\boldsymbol{H}\rangle = -\frac{1}{2}\eta\phi\alpha_0^2\boldsymbol{\Omega}\cos^2\vartheta\cos(2\vartheta) \tag{5.78}$$

Substituting Eq. (5.78) into Eq. (5.77) and consolidating the terms containing the intrinsic viscosity gives

$$\eta\nabla^2\boldsymbol{v}_1 - \frac{1}{4}\eta\phi\alpha_0^2\cos^2\vartheta\cos(2\vartheta)\nabla\times\boldsymbol{\Omega} = -2\left[\eta + \frac{1}{8}\eta\phi\alpha_0^2\cos^2\vartheta\cos(2\vartheta)\right]\nabla\times\boldsymbol{\Omega} \tag{5.79}$$

where use has been made of the incompressibility condition for ferrofluids and the vector equality

$$\nabla^2 \boldsymbol{v}_1 = \nabla(\nabla \cdot \boldsymbol{v}_1) - \nabla \times (\nabla \times \boldsymbol{v}_1) = -2\nabla \times \boldsymbol{\Omega} \tag{5.80}$$

As can be seen from Eq. (5.79), the term added to the intrinsic viscosity represents the rotational viscosity, namely,

$$\eta_r = \frac{1}{8}\eta\phi\alpha_0^2 \cos^2\vartheta \cos(2\vartheta) \tag{5.81}$$

Substituting $\tan\vartheta = w\tau_B$ into Eq. (5.81) yields

$$\begin{aligned} \eta_r &= \frac{1}{16}\eta\phi\alpha_0^2\left[1+\cos(2\vartheta)\right]\cos(2\vartheta) \\ &= \frac{1}{16}\eta\phi\alpha_0^2\left(1+\frac{1-\tan^2\vartheta}{1+\tan^2\vartheta}\right)\frac{1-\tan^2\vartheta}{1+\tan^2\vartheta} \\ &= \frac{1}{8}\eta\phi\alpha_0^2\frac{1-w^2\tau_B^2}{\left(1+w^2\tau_B^2\right)^2} \end{aligned} \tag{5.82}$$

Figure 5.6 illustrates the relationship between the relative rotational viscosity represented by Eq. (5.82) and $w\tau_B$. It is evident that when $w = 0$, corresponding to a static magnetic field, $\alpha_0 = \alpha$, and the result of Eq. (5.82) aligns with that of Eq. (5.17). As $w\tau_B$ approaches 1, the sign of rotational viscosity reverses, and once $w\tau_B$ reaches $\sqrt{3}$, the rotational viscosity attains its minimum value. For higher frequencies of alternating magnetic fields, $\eta_r < 0$, indicating a decrease in the ferrofluid's apparent viscosity compared to its

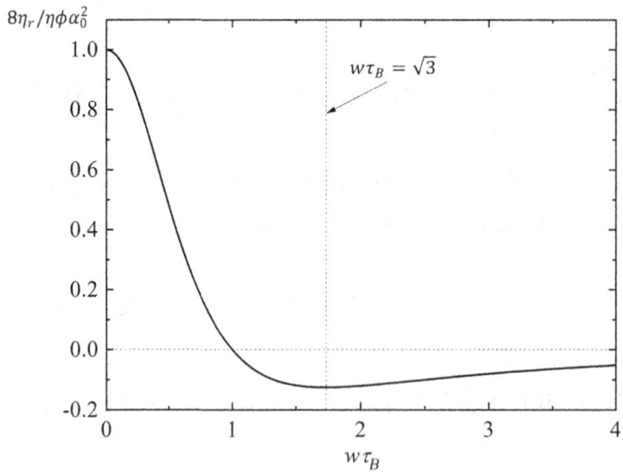

FIGURE 5.6 Variation of relative rotational viscosity with magnetic field frequency under weak field conditions.

intrinsic viscosity, a phenomenon known as the "negative viscosity" effect. In the scenario where $w\tau_B \gg 1$, one has

$$\eta_r = \frac{1}{2}\frac{\phi}{\eta}\left(\frac{M_d H_0}{w}\right)^2, w\tau_B \gg 1 \tag{5.83}$$

At this point, the rotational viscosity is inversely proportional to w^2. The tendency of ferrofluid viscosity to vary with the field frequency, under the influence of an alternating magnetic field, has been experimentally confirmed.

Utilizing the angular momentum Eq. (2.217) and the time-averaged Eq. (5.78), an expression for the time-averaged particle angular velocity over the magnetic field oscillation period can be derived,

$$\langle \boldsymbol{\omega}_\text{p} \rangle = \boldsymbol{\Omega} + \frac{\mu_0}{6\eta\phi}\boldsymbol{M}\times\boldsymbol{H} = \boldsymbol{\Omega}\left[1 - \frac{1}{12}\alpha_0^2 \cos^2\vartheta \cos(2\vartheta)\right] = \boldsymbol{\Omega}\left[1 - \frac{\alpha_0^2}{12}\frac{1 - w^2\tau_B^2}{\left(1 + w^2\tau_B^2\right)^2}\right] \tag{5.84}$$

The relationship between the effect of a magnetic field on viscosity can be derived from Eq. (5.84). A steady or slowly varying magnetic field ($w\tau_B < 1$) impedes the rotation of particles in a ferrofluid, resulting in $\boldsymbol{\omega}_\text{p} < \boldsymbol{\Omega}$, where the particle rotation speed is slower than the surrounding carrier liquid. This situation is analogous to the carrier liquid flowing over the particle surface, causing additional kinetic energy dissipation in the fluid, which manifests as $\eta_r > 0$.

By combining Eqs. (5.82) and (5.84), one can obtain

$$\eta_r = \frac{3}{2}\eta\phi\frac{\boldsymbol{\Omega} - \boldsymbol{\omega}_\text{p}}{\boldsymbol{\Omega}} \tag{5.85}$$

Examining Eq. (5.85) reveals a clear relationship between the rotational speed of particles and the rotational viscosity within a ferrofluid. In the absence of a magnetic field, the particles rotate synchronously with the fluid ($\boldsymbol{\Omega} = \boldsymbol{\omega}_\text{p}$), resulting in no additional frictional energy loss and $\eta_r = 0$. Conversely, in the presence of a strong magnetic field, the particle magnetic moments are fully aligned, causing the particles to cease rotation ($\boldsymbol{\omega}_\text{p} = \boldsymbol{0}$) and the rotational viscosity to saturate at $\eta_r = \frac{3}{2}\eta\phi$. In alternating magnetic fields with $w\tau_B > 1$, the particles rotate faster than the fluid ($\boldsymbol{\omega}_\text{p} > \boldsymbol{\Omega}$), leading to a tendency for the particles to enhance the flow of the ferrofluid. This phenomenon results in the ferrofluid exhibiting "negative viscosity", where $\eta_r < 0$, indicating that a portion of the energy from the alternating magnetic field is converted into kinetic energy of the fluid. The "negative viscosity" effect of ferrofluids has been experimentally verified (Bacri et al., 1995; Zeuner et al., 1998).

Equation (5.85) can also be derived from the phenomenological magnetization Eq. (4.28) under the conditions of a steady magnetic field and $\Omega\tau_B \ll 1$, specifically by considering component equation of Eq. (4.28) that is perpendicular to the magnetic field,

$$\boldsymbol{M} \times \boldsymbol{H} = -\tau_{\perp} M_0 H \boldsymbol{\Omega} \tag{5.86}$$

By comparing this equation with the angular momentum Eq. (2.217) and applying the rotational viscosity in a static magnetic field Eq. (5.4), one can arrive at Eq. (5.85).

5.3.2 Rotational Viscosity under the Action of a Limited-Amplitude Alternating Magnetic Field at Low Shear Rates

To analyze the flow behaviour of ferrofluids under the influence of a finite-amplitude alternating magnetic field, it is necessary to solve the microscopic magnetization equation to determine the rotational viscosity. When the value of $\Omega\tau_B$ is small or satisfies the condition $\Omega\tau_B \ll 1$, the regular perturbation theory can be applied to solve the dimensionless microscopic magnetization equation (Odenbach, 2002).

First, the various physical quantities present in the microscopic magnetization equation are transformed into dimensionless ones based on the following expressions:

$$t^* = \frac{t}{\tau_B}, \boldsymbol{\Omega}^* = \boldsymbol{\Omega}\tau_B, \boldsymbol{M}^* = \frac{\boldsymbol{M}}{M_S} = L(\varsigma)\boldsymbol{\varsigma}/\varsigma,\ \boldsymbol{H}^* = \frac{\boldsymbol{H}}{M_S} = \frac{k_B T}{\mu_0 m_p M_S}\boldsymbol{\alpha} = \frac{k_B T N}{\mu_0 M_S^2}\boldsymbol{\alpha} \tag{5.87}$$

where τ_B was used to measure time, M_S was used to measure magnetic field intensity and magnetization, and the asterisks "*" represent dimensionless quantities. Substituting the dimensionless quantities into the microscopic magnetization equation (4.47) and applying Eq. (4.13), while omitting the asterisks (except for $\boldsymbol{\Omega}^*$ which is represented as $\boldsymbol{\Omega}\tau_B$), one gets

$$\frac{\mathrm{D}}{\mathrm{D}t}\left(\frac{L(\varsigma)\boldsymbol{\varsigma}}{\varsigma}\right) = \tau_B\boldsymbol{\Omega} \times \left(\frac{L(\varsigma)\boldsymbol{\varsigma}}{\varsigma}\right) - \frac{L(\varsigma)}{\varsigma}(\boldsymbol{\varsigma} - \boldsymbol{\alpha}) - \frac{1}{2\varsigma^2}\left(1 - \frac{3L(\varsigma)}{\varsigma}\right)\boldsymbol{\varsigma} \times (\boldsymbol{\varsigma} \times \boldsymbol{\alpha}) \tag{5.88}$$

Equation (5.88) contains the small parameter $\boldsymbol{\Omega}\tau_B$, and its zeroth-order approximation corresponds to $\boldsymbol{\Omega}\tau_B = \boldsymbol{0}$, indicating τ_B is infinitely small. In this case, the effective field $\boldsymbol{\varsigma}$ aligns with the actual field $\boldsymbol{\alpha}$, and Eq. (5.88) becomes

$$\frac{\mathrm{D}}{\mathrm{D}t}\left(\frac{L(\varsigma)\boldsymbol{\varsigma}}{\varsigma}\right) = -\frac{L(\varsigma)}{\varsigma}(\boldsymbol{\varsigma} - \boldsymbol{\alpha}) \tag{5.89}$$

Since all the vectors in this equation are in the same direction, it can be further simplified into a scalar form

$$\frac{\mathrm{D}}{\mathrm{D}t}\left(L(\varsigma)\right) = -L(\varsigma)\left(1 - \frac{\alpha}{\varsigma}\right) \tag{5.90}$$

Expanding the terms on the left-hand side of Eq. (5.90) yields

$$\frac{\mathrm{D}\varsigma}{\mathrm{D}t} = -\frac{L(\varsigma)}{\varsigma L'(\varsigma)}(\varsigma - \alpha) \tag{5.91}$$

where

$$\frac{L(\varsigma)}{\varsigma L'(\varsigma)}=\frac{\frac{1}{\varsigma}}{\frac{L'(\varsigma)}{L(\varsigma)}}=\frac{\frac{\mathrm{d}\ln\varsigma}{\mathrm{d}\varsigma}}{\frac{\mathrm{d}\ln L(\varsigma)}{\mathrm{d}\varsigma}}=\frac{\mathrm{d}\ln\varsigma}{\mathrm{d}\ln L(\varsigma)} \tag{5.92}$$

Hence, the zeroth-order approximate solution of the magnitude of the dimensionless effective field ς satisfies

$$\frac{\mathrm{D}\varsigma}{\mathrm{D}t}=-\left[\frac{\mathrm{d}\ln L(\varsigma)}{\mathrm{d}\ln\varsigma}\right]^{-1}\left[\varsigma-\alpha_0\cos(wt)\right] \tag{5.93}$$

where α_0 is given in Eq. (5.45).

Under the first-order linear approximation of the small parameter $\boldsymbol{\Omega}\tau_B$, the magnetization can be considered as the composition of a significant component aligned with the magnetic field and a minor component perpendicular to the magnetic field direction as shown in Figure 5.7. That is, the solution of Eq. (5.89) can be expressed as

$$\boldsymbol{M}^*=\frac{L(\varsigma)\boldsymbol{\varsigma}}{\varsigma}=M^{(0)}\boldsymbol{e}_H+M^{(1)}\tau_B\boldsymbol{\Omega}\times\boldsymbol{e}_H \tag{5.94}$$

where

$$M^{(0)}=L(\varsigma),M^{(1)}=M^{(0)}\Psi(\varsigma) \tag{5.95}$$

and $\boldsymbol{e}_H$ is the unit vector along the magnetic field direction, $\boldsymbol{e}_H=\boldsymbol{H}/H$, and $\Psi(\varsigma)$ is the undetermined function.

For the sake of simplification, Eq. (5.88) is written as

$$\frac{\mathrm{D}}{\mathrm{D}t}\left(\frac{L(\varsigma)\boldsymbol{\varsigma}}{\varsigma}\right)=\tau_B\boldsymbol{\Omega}\times\left(\frac{L(\varsigma)\boldsymbol{\varsigma}}{\varsigma}\right)-\left(\frac{L(\varsigma)}{\varsigma}\boldsymbol{\varsigma}-\frac{L(\varsigma)\boldsymbol{\varsigma}\cdot\boldsymbol{h}_e}{\varsigma}\frac{\boldsymbol{\alpha}}{\varsigma}\right)-\frac{\alpha}{2\varsigma}\left(\boldsymbol{\varsigma}-\frac{3L(\varsigma)\boldsymbol{\varsigma}}{\varsigma}\right)\times\left(\boldsymbol{h}_e\times\boldsymbol{e}_H\right) \tag{5.96}$$

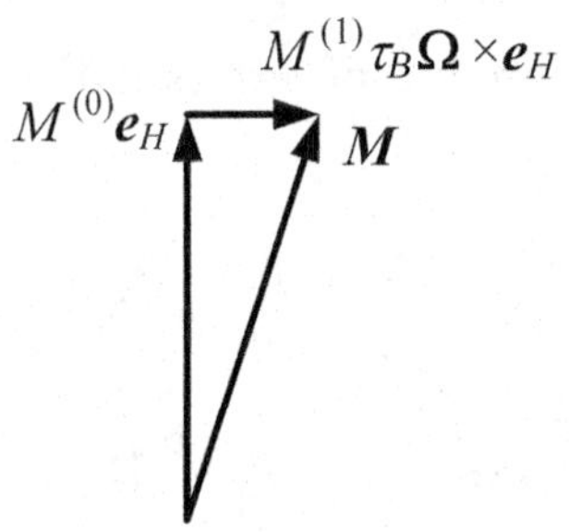

FIGURE 5.7 Decomposition of the magnetization vector.

where $\boldsymbol{h}_e$ represents the unit vector in the same direction as ς. Introducing Eqs. (5.94) and (5.95) into Eq. (5.96) gives

$$\begin{aligned}
&\boldsymbol{e}_H \frac{\mathrm{D}M^{(0)}}{\mathrm{D}t} + \tau_B \boldsymbol{\Omega} \times \boldsymbol{e}_H \frac{\mathrm{D}M^{(1)}}{\mathrm{D}t} = \tau_B \boldsymbol{\Omega} \times \boldsymbol{e}_H M^{(0)} + \tau_B \boldsymbol{\Omega} \times \left(\tau_B \boldsymbol{\Omega} \times \boldsymbol{e}_H\right) M^{(1)} \\
&-\left[\boldsymbol{e}_H M^{(0)} + \left(\tau_B \boldsymbol{\Omega} \times \boldsymbol{e}_H\right) M^{(1)} - M^{(0)} \boldsymbol{e}_H \cdot \boldsymbol{h}_e \frac{\boldsymbol{\alpha}}{\varsigma} - M^{(1)} \left(\tau_B \boldsymbol{\Omega} \times \boldsymbol{e}_H\right) \cdot \boldsymbol{h}_e \frac{\boldsymbol{\alpha}}{\varsigma}\right] \\
&-\frac{\alpha}{2\varsigma}\left[\varsigma - 3\boldsymbol{e}_H M^{(0)} - 3\left(\tau_B \boldsymbol{\Omega} \times \boldsymbol{e}_H\right) M^{(1)}\right] \times \left(\boldsymbol{h}_e \times \boldsymbol{e}_H\right)
\end{aligned} \tag{5.97}$$

Based on the arbitrariness of $\boldsymbol{\Omega}\tau_B$, it follows from Eq. (5.97) that

$$\boldsymbol{e}_H \frac{\mathrm{D}M^{(0)}}{\mathrm{D}t} = -\boldsymbol{e}_H M^{(0)} + M^{(0)} \boldsymbol{e}_H \cdot \boldsymbol{h}_e \frac{\boldsymbol{\alpha}}{\varsigma} - \frac{\alpha}{2\varsigma} \varsigma \times \left(\boldsymbol{h}_e \times \boldsymbol{e}_H\right) + \frac{3\alpha}{2\varsigma} M^{(0)} \boldsymbol{e}_H \times \left(\boldsymbol{h}_e \times \boldsymbol{e}_H\right) \tag{5.98}$$

and

$$\begin{aligned}
\tau_B \boldsymbol{\Omega} \times \boldsymbol{e}_H \frac{\mathrm{D}M^{(1)}}{\mathrm{D}t} &= \tau_B \boldsymbol{\Omega} \times \boldsymbol{e}_H M^{(0)} - \left(\tau_B \boldsymbol{\Omega} \times \boldsymbol{e}_H\right) M^{(1)} \\
&+ \left(\tau_B \boldsymbol{\Omega} \times \boldsymbol{e}_H\right) \cdot \boldsymbol{h}_e M^{(1)} \frac{\boldsymbol{\alpha}}{\varsigma} + \frac{3\alpha}{2\varsigma} M^{(1)} \left(\tau_B \boldsymbol{\Omega} \times \boldsymbol{e}_H\right) \times \left(\boldsymbol{h}_e \times \boldsymbol{e}_H\right)
\end{aligned} \tag{5.99}$$

Taking into account the directionality of the expressions on both sides of Eqs. (5.98) and (5.99), they can be simplified to the following forms, respectively:

$$\frac{\mathrm{D}M^{(0)}}{\mathrm{D}t} = -M^{(0)} + M^{(0)} \boldsymbol{e}_H \cdot \boldsymbol{h}_e \frac{\alpha}{\varsigma} - \frac{\alpha}{2} \left(\boldsymbol{h}_e \cdot \boldsymbol{e}_H\right)^2 + \frac{\alpha}{2} \tag{5.100}$$

$$\frac{\mathrm{D}M^{(1)}}{\mathrm{D}t} = M^{(0)} - M^{(1)} \tag{5.101}$$

where use has been made of

$$\boldsymbol{h}_e \times \left(\boldsymbol{h}_e \times \boldsymbol{e}_H\right) = \boldsymbol{h}_e \left(\boldsymbol{h}_e \cdot \boldsymbol{e}_H\right) - \boldsymbol{e}_H$$

The terms that involve the dot product of vectors at the right-hand side of Eq. (5.100) can be simplified as

$$\begin{aligned}
M^{(0)} \boldsymbol{e}_H \cdot \boldsymbol{h}_e \frac{\alpha}{\varsigma} - \frac{\alpha}{2} \left(\boldsymbol{h}_e \cdot \boldsymbol{e}_H\right)^2 &= M^{(0)} \boldsymbol{e}_H \cdot \boldsymbol{h}_e \frac{\alpha}{\varsigma} - \frac{\alpha}{2}\left[\left(\boldsymbol{h}_e \cdot \boldsymbol{e}_H\right)^2 - 2\left(\boldsymbol{h}_e \cdot \boldsymbol{e}_H\right) + 1\right] - \alpha\left(\boldsymbol{h}_e \cdot \boldsymbol{e}_H\right) + \frac{\alpha}{2} \\
&= \boldsymbol{e}_H \cdot \boldsymbol{h}_e \alpha \left(\frac{M^{(0)}}{\varsigma} - 1\right) - \frac{\alpha}{2} \left(\boldsymbol{h}_e \cdot \boldsymbol{e}_H - 1\right)^2 + \frac{\alpha}{2}
\end{aligned} \tag{5.102}$$

From Figure 5.7 one has $\boldsymbol{h}_e \cdot \boldsymbol{e}_H \approx 1$, and then

$$\boldsymbol{h}_e \cdot \boldsymbol{e}_H = \frac{M^{(0)}}{|\boldsymbol{M}|} = \frac{L(\varsigma)}{\left[\left(M^{(0)}\right)^2 + \left(M^{(1)}\tau_B\Omega\right)^2\right]^{\frac{1}{2}}}$$
$$= \frac{L(\varsigma)}{\left\{\left(L(\varsigma)\right)^2 + \left[L(\varsigma)\Psi(\varsigma)\tau_B\Omega\right]^2\right\}^{\frac{1}{2}}} = \frac{1}{\left\{1 + \left[\Psi(\varsigma)\tau_B\Omega\right]^2\right\}^{\frac{1}{2}}} \tag{5.103}$$

Substituting the result from Eq. (5.103) into Eq. (5.102) and neglecting the term $(\boldsymbol{h}_e \cdot \boldsymbol{e}_H - 1)^2$, one gets

$$M^{(0)}\boldsymbol{e}_H \cdot \boldsymbol{h}_e \frac{\alpha}{\varsigma} - \frac{\alpha}{2}\left(\boldsymbol{h}_e \cdot \boldsymbol{e}_H\right)^2 = \left(\frac{L(\varsigma)}{\varsigma} - 1\right)\frac{\alpha}{\left\{1 + \left[\Psi(\varsigma)\tau_B\Omega\right]^2\right\}^{\frac{1}{2}}} + \frac{\alpha}{2} \tag{5.104}$$

Bringing the result of Eq. (5.104) back into Eq. (5.100) gives

$$\frac{\mathrm{D}M^{(0)}}{\mathrm{D}t} = -M^{(0)} + \left(\frac{L(\varsigma)}{\varsigma} - 1\right)\frac{\alpha}{\left\{1 + \left[\Psi(\varsigma)\tau_B\Omega\right]^2\right\}^{\frac{1}{2}}} + \alpha \tag{5.105}$$

Using Eq. (5.95) and substituting Eq. (5.105) into Eq. (5.101) gives

$$M^{(0)} - M^{(1)} = \frac{\mathrm{D}M^{(1)}}{\mathrm{D}t} = M^{(0)}\frac{d\Psi(\varsigma)}{\mathrm{d}t} + \Psi(\varsigma)\frac{\mathrm{D}M^{(0)}}{\mathrm{D}t}$$
$$= M^{(0)}\frac{d\Psi(\varsigma)}{\mathrm{d}t} + \Psi(\varsigma)\left[-M^{(0)} + \left(\frac{L(\varsigma)}{\varsigma} - 1\right)\frac{\alpha}{\left\{1 + \left[\Psi(\varsigma)\tau_B\Omega\right]^2\right\}^{\frac{1}{2}}} + \alpha\right] \tag{5.106}$$

which can be simplified as

$$\frac{d\Psi(\varsigma)}{\mathrm{d}t} = 1 - \alpha_0\Psi(\varsigma)\left[\frac{1}{\varsigma} - \frac{1}{L(\varsigma)}\left(\frac{1}{\sqrt{1 + \left(\Psi(\varsigma)\tau_B\Omega\right)^2}} - 1\right)\right]\cos(wt) \tag{5.107}$$

If the minor terms are neglected in this equation, the differential equation is satisfied by the function $\Psi(\varsigma)$ is obtained

$$\frac{d\Psi(\varsigma)}{dt} = 1 - \frac{\alpha_0}{\varsigma}\Psi(\varsigma)\cos(wt) \tag{5.108}$$

Furthermore, by utilizing Eqs. (5.94) and (5.95), one can derive an expression for magnetization:

$$\boldsymbol{M} = \boldsymbol{M}^* M_S = M_S L(\varsigma)\boldsymbol{e}_H + M_S L(\varsigma)\Psi(\varsigma)\tau_B \boldsymbol{\Omega} \times \boldsymbol{e}_H \tag{5.109}$$

However, the final result of magnetization can only be obtained after determining ς and $\Psi(\varsigma)$ by solving Eqs. (5.93), (5.107), and (5.109) simultaneously.

Assuming that the characteristic time scale of magnetic field, τ_B, is significantly shorter than the characteristic time scale of fluid flow, $\tau_h \sim l^2/\nu$, namely $\tau_h \gg \tau_B$, such that the same time-averaging approach employed in Section 5.3.1 can be applied to derive the equation of motion for ferrofluids given by Eq. (5.77). To determine the rotational viscosity of the ferrofluid corresponding to the flow depicted in Figure 5.1, it is necessary to integrate the magnetic torque, $\mu_0 \boldsymbol{M} \times \boldsymbol{H}$, over the magnetic field period $2\pi/w$, resulting in a time-averaged magnetic torque. To accomplish this, one has to substitute Eq. (5.109) into the expression for magnetic torque which gives

$$\mu_0 \langle \boldsymbol{M} \times \boldsymbol{H} \rangle = \mu_0 \langle M_S \left[L(\varsigma)\boldsymbol{e}_H \times \boldsymbol{H} + L(\varsigma)\Psi(\varsigma)\tau_B (\boldsymbol{\Omega} \times \boldsymbol{e}_H) \times \boldsymbol{H} \right] \rangle \tag{5.110}$$

Considering that $\boldsymbol{e}_H$ is aligned with $\boldsymbol{H}$, resulting in $\boldsymbol{e}_H \times \boldsymbol{H} = \boldsymbol{0}$, and in the flow depicted in Figure 5.1, $\boldsymbol{\Omega}$ is oriented along the z-direction, $\boldsymbol{\Omega} \times \boldsymbol{e}_H$ points towards the $-x$ direction, and $(\boldsymbol{\Omega} \times \boldsymbol{e}_H) \times \boldsymbol{H}$ points towards the $-z$ direction. Since all these quantities vary slowly, it follows that

$$\mu_0 \langle \boldsymbol{M} \times \boldsymbol{H} \rangle = -\mu_0 M_S \tau_B H_0 \langle L(\varsigma)\Psi(\varsigma)\cos(wt) \rangle \boldsymbol{\Omega} \tag{5.111}$$

Substitution of Eq. (4.13) into Eq. (5.111) with the aid of Eqs. (4.2) and (5.45) gives

$$\mu_0 \langle \boldsymbol{M} \times \boldsymbol{H} \rangle = -6\eta\phi\gamma\boldsymbol{\Omega} \tag{5.112}$$

where

$$\gamma = \frac{1}{2}\alpha_0 \langle L(\varsigma)\Psi(\varsigma)\cos(wt) \rangle \tag{5.113}$$

Using the results obtained from Eq. (5.80), the terms represented by Eq. (5.112) are combined with the terms containing the intrinsic viscosity in Eq. (5.77)

$$\eta\nabla^2 \boldsymbol{v}_1 + \frac{\mu_0}{2}\nabla \times \langle \boldsymbol{M} \times \boldsymbol{H} \rangle = -2\eta\nabla \times \boldsymbol{\Omega} + \frac{1}{2}\nabla \times (-6\eta\phi\gamma\boldsymbol{\Omega})$$

$$= -2\left(\eta + \frac{3}{2}\eta\phi\gamma\right)\nabla\times\mathbf{\Omega}$$

$$= \left(\eta + \frac{3}{2}\eta\phi\gamma\right)\nabla^2\mathbf{v}_0 \tag{5.114}$$

As shown in Eq. (5.114), the rotational viscosity of a ferrofluid under the action of a finite-amplitude alternating magnetic field at low shear rates is

$$\eta_r = \frac{3}{2}\eta\phi\gamma \tag{5.115}$$

The result of Eq. (5.112) can also be substituted into Eq. (2.217) to obtain

$$\boldsymbol{\omega}_\mathrm{p} = (1-\gamma)\mathbf{\Omega} \tag{5.116}$$

Thereby, the rotational viscosity in this case can also be expressed in the form of Eq. (5.85).

To determine the rotational viscosity at low shear rates under the influence of a finite-amplitude alternating magnetic field, it is necessary to solve Eqs. (5.93) and (5.107) to obtain the value of γ. Utilizing the fourth-order Runge–Kutta method, one can numerically solve these equations within a periodic interval of $t = [0, 2\pi/w\tau_B]$, subject to the initial conditions of $\varsigma = \alpha_0$ and $\Psi(\varsigma) = 0$ at $t = 0$. This numerical approach allows to obtain discrete values of ς and Ψ at various time points. Subsequently, the time-averaged values, represented by Eq. (5.113), are calculated. Figures 5.8 and 5.9 depict the variations of γ with α_0 and the γ contours in the $(w\tau_B, \alpha_0)$ plane. Notably, negative γ values emerge when $w\tau_B$ is relatively large. In Figure 5.9, the contour $\gamma = 0$ divides the region into two parts: $\gamma > 0$ when $w\tau_B \leq 1$ and $\gamma < 0$ when $w\tau_B \geq 1$.

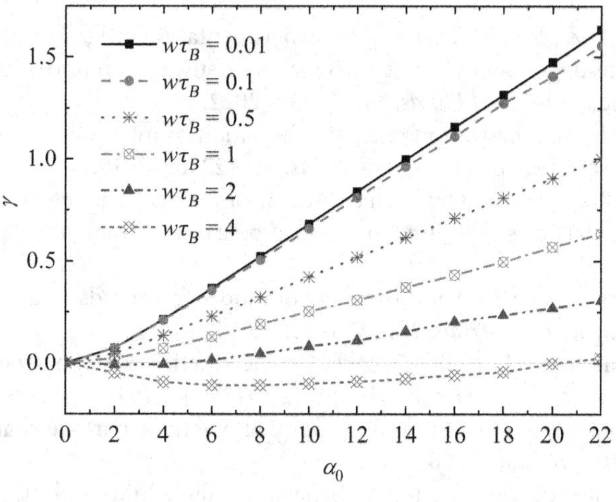

FIGURE 5.8 Variation of γ with α_0 for different values of $w\tau_B$.

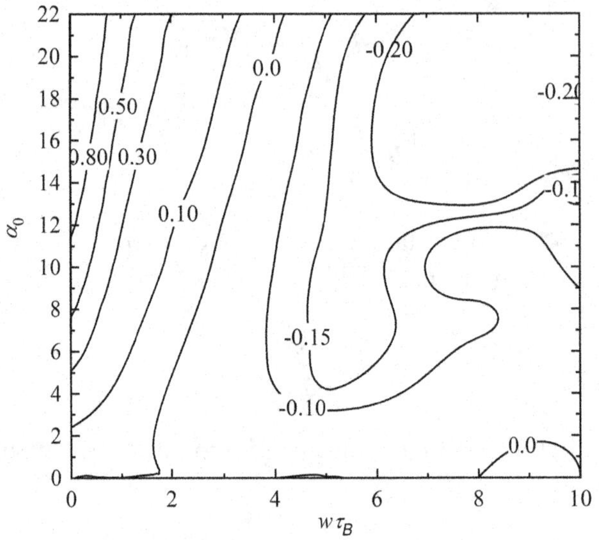

FIGURE 5.9 Iso-γ lines in the ($w\tau_B$, α_0) plane.

Researchers have experimentally observed the "negative viscosity" effect at $w\tau_B \geq 1$ in a cobalt-ferrite water-based ferrofluid, with a particle volume fraction of 20%, an external magnetic field amplitude of 6.37×10^4 A/m, and a frequency of 1480 Hz, resulting in a maximum viscosity reduction by 25% (Bacri et al., 1995).

REFERENCES

Ambacher O., Odenbach S., Stierstadt K., Rotational viscosity in ferrofluids, *Zeitschrift für Physics B Condensed Matters*, 86(1): 29–32, 1992.

Bacri J.-C., Perzynski R., Shliomis M. I., Burde G. I., "Negative-viscosity" effect in a magnetic fluid, *Physical Review Letters*, 75(11): 2128–2131, 1995.

Borin D. Y., Korolev V. V., Ramazanova A. G., Odenbach S., Balmasova O. V., Yashkova V. I., Korolev D. V., Magnetoviscous effect in ferrofluids with different dispersion media, *Journal of Magnetism and Magnetic Materials*, 416: 110–116, 2016.

Cunha F. R., Sinzato Y. Z., Pereira I. D. O., An experimental investigation on the magnetoviscous effect and shear rate-dependent viscosity of a magnetic suspension under longitudinal and transverse magnetic fields, *Physics of Fluids*, 34: 093314, 2022.

Fang A., First-principles magnetization relaxation equation of interacting ferrofluids with applications to magnetoviscous effects, *Physics of Fluids*, 31: 122002, 2019.

Gerth-Noritzsch M., Borin D. Y., Odenbach S., Anisotropy of the magnetoviscous effect in ferrofluids containing nanoparticles exhibiting magnetic dipole interaction, *Journal of Physics: Condensed Matter*, 23: 346002, 2011.

Ilg P., Kröger M., Hess S., Structure and rheology of model-ferrofluids under shear flow, *Journal of Magnetism and Magnetic Materials*, 289: 325–327, 2005.

Martsenyuk M. A., Raikher Y. L., Shliomis M. I., On the kinetics of magnetization of suspensions of ferromagnetic particles, *Soviet Physics JETP*, 28(2): 413–416, 1974.

Morozov K. I., Shliomis M. I., Ferrofluids: Flexibility of magnetic particle chains, *Journal of Physics D: Condensed Matter*, 16: 3807, 2004.

Müller O., Hahn D., Liu M., Non-Newtonian behavior in ferrofluids and magnetization relaxation, *Journal of Physics: Condensed Matter*, 18: S2623–S2632, 2006.

Odenbach S., *Magnetoviscous effects in ferrofluids*, Berlin: Springer, 2002.
Odenbach S., Störk H., Shear dependence of field-induced contributions to the viscosity of magnetic fluids at low shear rates, *Journal of Magnetism and Magnetic Materials*, 183: 188–194, 1998.
Pinho M., Brouard B., Génevaux J. M., Dauchez N., Volkova O., Mézière H., Collas P., Investigation into ferrofluid magnetoviscous effects under an oscillating shear flow, *Journal of Magnetism and Magnetic Materials*, 323: 2386–2390, 2011.
Pop L. M., Odenbach S., Investigation of the microscopic reason for the magnetoviscous effect in ferrofluids studied by small angle neutron scattering, *Journal of Physics: Condensed Matter*, 18: S2785–2802, 2006.
Rosensweig R. E., *Ferrohydrodynamics*, Cambridge: Cambridge University Press. 1985.
Shliomis M. I., Lyubimova T. P., Lyubimov D. V., Ferrohydrodynamics: An essay on the progress of ideas, *Chemical Engineering Communications*, 67: 275–290, 1988.
Shliomis M. I., Morozov K. I., Negative viscosity of ferrofluid under alternating magnetic field, *Physics of Fluids*, 6(8): 2855–2861, 1994.
Soto-Aquino D., Rinaldi C., Magnetoviscosity in dilute ferrofluids from rotational Brownian dynamics simulations, *Physical Review E*, 82: 046310, 2010.
Weis J. J., Levesque D., Chain formation in low density dipolar hard spheres: A Monte Carlo study, *Physical Review Letters*, 71: 2729, 1993.
Zahn M., Pioch L. L., Magnetizable fluid behaviour with effective positive, zero or negative dynamic viscosity, *Indian Journal of Engineering & Materials Science*, 5: 400–410, 1998.
Zahn M., Pioch L. L., Ferrofluid flows in AC and traveling wave magnetic fields with effective positive, zero or negative dynamic viscosity, *Journal of Magnetism and Magnetic Materials*, 201: 144–148, 1999.
Zeuner A., Richter R., Rehberg I., Experiments on negative and positive magnetoviscosity in an alternating magnetic field, *Physical Review E*, 58(5): 6287–6293, 1998.
Zubarev A. Y., Chirikov D. N., On the theory of the magnetoviscous effect in ferrofluids, *Journal of Experimental and Theoretical Physics*, 110(6): 995–1004, 2010.
Zubarev A. Y., Iskakova L., Rheological properties of magnetic suspensions, *Journal of Physics: Condensed Matter*, 20: 204138, 2008.
Zubarev A. Y., Iskakova L. Y., Chirikov D. N., On the nonlinear rheology of magnetic fluids, *Colloid Journal*, 73(3): 327–339, 2011.
Zubarev A. Y., Odenbach S., Fleischer J., Rheological properties of dense ferrofluids. Effect of chain-like aggregates, *Journal of Magnetism and Magnetic Materials*, 252(1–3): 241–243, 2002.

CHAPTER 6

Ferrofluid Planar Couette–Poiseuille Flows

PLANAR COUETTE–POISEUILLE FLOWS ARE parallel flows, that is only one of the velocity components is non-zero and all the masses within the fluid flow in the same direction. The Couette–Poiseuille flow of ferrofluids serves as a fluid flow model for many of its applications (Matsuno et al., 1993; Paz et al., 2022). A typical instance of such a flow occurs between two infinitely long parallel flat plates, as shown in Figure 6.1. Herein, it is assumed that the distance between the two plates is h_0, the fluid is subjected to a pressure gradient $\partial p / \partial x$ along the x-direction, and the translational upper plate is moving along the x-direction with velocity v_0. If there is no pressure gradient acting in the Couette–Poiseuille flow, resulting in a pure shear flow, it is called Couette flow. If only a pressure gradient acts and the upper plate is at rest, it is called Poiseuille flow.

Under the influence of a magnetic field, the planar Couette–Poiseuille flow of ferrofluids exhibits distinct flow characteristics compared to those of regular fluids, such as different velocity and vorticity distributions, different shear stress on the upper plate, and magnetization relaxation effects. To provide a contrast, this chapter initially presents the flow characteristics of the Couette–Poiseuille flow of regular fluids.

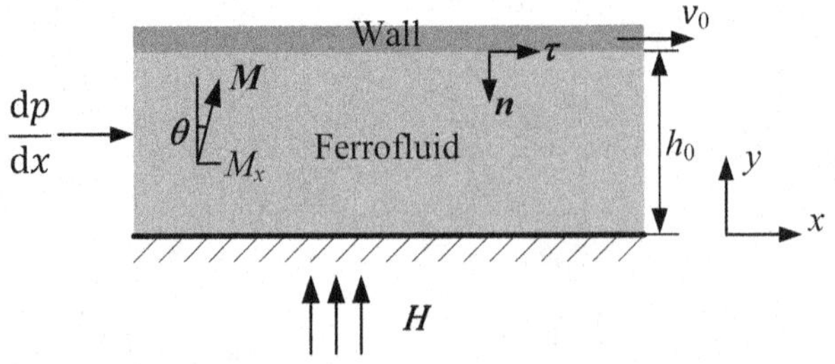

FIGURE 6.1 Ferrofluid planar Couette–Poiseuille flow.

DOI: 10.1201/9781003540342-6

6.1 COUETTE–POISEUILLE FLOW OF REGULAR FLUIDS

For the Couette–Poiseuille flow of a Newtonian fluid, an exact analytical solution exists for its velocity profile. Referring to the model depicted in Figure 6.1, the fluid's flow velocity in the absence of a magnetic field is given by (Landau and Lifshitz, 1987)

$$v_x = \frac{y}{h_0} v_0 - \frac{h_0^2}{2\eta} \frac{\mathrm{d}p}{\mathrm{d}x} \frac{y}{h_0}\left(1 - \frac{y}{h_0}\right) \tag{6.1}$$

Letting the dimensionless pressure gradient P^* be

$$P^* = -\frac{h_0^2}{2\eta v_0} \frac{\mathrm{d}p}{\mathrm{d}x} \tag{6.2}$$

which gives

$$\frac{v_x}{v_0} = \frac{y}{h_0} + P^* \frac{y}{h_0}\left(1 - \frac{y}{h_0}\right) \tag{6.3}$$

It is evident that the Couette–Poiseuille flow of a regular fluid exhibits a parabolic velocity profile as depicted in Figure 6.2. When $P^* > 0$, indicating that the pressure drop aligns with the motion of the upper plate, the velocity remains positive throughout the entire flow cross-section. Conversely, when $P^* \leq -1$, backflow (negative velocity) occurs in a portion of the cross sections, indicating that the drag exerted by the upper plate on the fluid layer cannot overcome the influence of the opposing pressure gradient.

The maximum velocity of the Couette–Poiseuille flow is

$$v_{x,\max} = \frac{v_0\left(1 + P^*\right)^2}{4P^*} \tag{6.4}$$

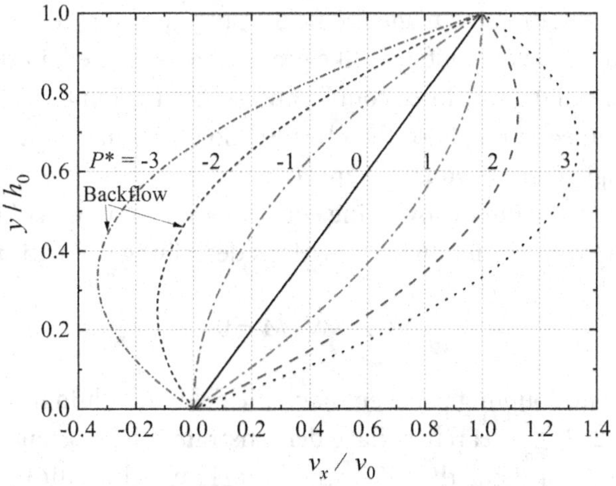

FIGURE 6.2 Velocity distributions of Couette–Poiseuille flow for a Newtonian fluid.

and the location of the maximum flow velocity is

$$y_{\max} = \frac{h_0\left(1+P^*\right)}{2P^*} \tag{6.5}$$

From Eq. (6.3), the vorticity of the flow is found to be

$$\boldsymbol{\Omega} = \frac{1}{2}\nabla\times\boldsymbol{v} = \left(\frac{P^*}{h_0^2}y - \frac{1+P^*}{2h_0}\right)v_0\boldsymbol{e}_z \tag{6.6}$$

The volume of fluid passing through a unit width cross-section per unit time, known as single-width flow rate, is

$$Q = \int_0^{h_0} v_x \mathrm{d}y = \frac{1}{6}v_0 h_0\left(3+P^*\right) \tag{6.7}$$

The viscous drag force acting on the upper plate per unit length is

$$\Gamma = \frac{1}{L_0}\int_0^{L_0} \eta\frac{\mathrm{d}v_x}{\mathrm{d}y}\mathrm{d}x = \eta\frac{v_0}{h_0}\left(1-P^*\right) \tag{6.8}$$

where L_0 is the length of the upper plate.

6.2 FLOW CHARACTERISTICS IN CONSTANT UNIFORM MAGNETIC FIELDS

6.2.1 Normalized Ferrohydrodynamic Equations

Assuming a constant fluid temperature, the equations governing the motion of ferrofluids include the conservation of momentum and angular momentum, the magnetization equation, and Maxwell's equation for magnetic field (Yang and Liu, 2020). Specifically, within the momentum equation, volume forces like gravity are neglected; in the angular momentum equation, the inertial, body force couple, and spin diffusion terms are disregarded. As for the magnetization equation, both the phenomenological magnetization equation I and the microscopic magnetization equation are considered. Furthermore, assuming the ferrofluid is incompressible, the fluid's governing equations are composed of Eqs. (2.219), (2.217), (4.18) or (4.47), along with the following equations describing the static magnetic field.

$$\nabla^2\varphi_m - \nabla\cdot\boldsymbol{M} = 0 \tag{6.9}$$

Despite the intricate coupling between magnetic fields and fluid flow within these equations (Ghosh et al., 2022), researchers have demonstrated the existence of solutions to the system of equations comprising them (Amirat and Hamdache, 2016).

To simplify the solution process and gain insights into the influence of various parameters on flow characteristics, it is necessary to convert the equation set into its dimensionless form. This can be achieved by introducing dimensionless variables with the aid of characteristic length h_0 (the spacing between the two plates), characteristic velocity v_0, characteristic time h_0/v_0, characteristic pressure $v_0\eta/h_0$, characteristic magnetic field intensity H_0, and characteristic angular velocity $\omega_0 = v_0/h_0$. These dimensionless quantities are listed as follows:

$$\begin{gathered}\nabla^* = \nabla h_0, x^* = \frac{x}{h_0}, y^* = \frac{y}{h_0}, z^* = \frac{z}{h_0}, t^* = \frac{t}{h_0/v_0}, \boldsymbol{v}^* = \frac{\boldsymbol{v}}{v_0}, p^* = \frac{p}{v_0\eta/h_0} \\ \boldsymbol{H}^* = \frac{\boldsymbol{H}}{H_0}, \boldsymbol{\omega}_\mathrm{p}^* = \frac{\boldsymbol{\omega}_\mathrm{p}}{v_0/h_0}, \tau^* = \frac{\tau}{h_0/v_0}, \boldsymbol{M}^* = \frac{\boldsymbol{M}}{H_0}, \boldsymbol{M}_0^* = \frac{\boldsymbol{M}_0}{H_0}, \varphi_m^* = \frac{\varphi_m}{H_0 h_0}\end{gathered} \tag{6.10}$$

Representing each physical quantity as a product of characteristic quantities and dimensionless numbers, and then substituting them into each equation obtains their dimensionless form:

$$\boldsymbol{\omega}_\mathrm{p}^* = \frac{1}{2}\nabla^* \times \boldsymbol{v}^* + \frac{\mu_0 H_0^2}{4\zeta v_0/h_0}\boldsymbol{M}^* \times \boldsymbol{H}^*$$

$$\frac{\partial \boldsymbol{v}^*}{\partial t^*} + \left(\boldsymbol{v}^* \cdot \nabla^*\right)\boldsymbol{v}^* = -\frac{\eta}{\rho v_0 h_0}\nabla^* p^* + \frac{\eta}{\rho v_0 h_0}\nabla^2 \boldsymbol{v}^* + \frac{\mu_0 H_0^2}{\rho v_0^2}\boldsymbol{M}^* \cdot \nabla^* \boldsymbol{H}^* + \frac{\mu_0 H_0^2}{2\rho v_0^2}\nabla^* \times \left(\boldsymbol{M}^* \times \boldsymbol{H}^*\right)$$

$$\frac{\partial \boldsymbol{M}^*}{\partial t^*} + \left(\boldsymbol{v}^* \cdot \nabla^*\right)\boldsymbol{M}^* = \frac{1}{2}\left(\nabla^* \times \boldsymbol{v}^*\right) \times \boldsymbol{M}^* - \frac{1}{\tau^*}\left(\boldsymbol{M}^* - \boldsymbol{M}_0^*\right) - \frac{\mu_0 h_0 H_0^2}{6 v_0 \eta \phi}\boldsymbol{M}^* \times \left(\boldsymbol{M}^* \times \boldsymbol{H}^*\right)$$

$$\begin{aligned}\frac{\partial \boldsymbol{M}^*}{\partial t^*} + \left(\boldsymbol{v}^* \cdot \nabla^*\right)\boldsymbol{M}^* = {} & \frac{1}{2}\left(\nabla^* \times \boldsymbol{v}^*\right) \times \boldsymbol{M}^* - \frac{1}{\tau^*}\left(\boldsymbol{M}^* - \frac{M_S}{H_0}\frac{L(\varsigma)}{\varsigma}\boldsymbol{\alpha}\right) \\ & - \frac{\mu_0 h_0 H_0^2}{6 v_0 \eta \phi}\frac{1}{L(\varsigma)}\left(\frac{1}{L(\varsigma)} - \frac{3}{\varsigma}\right)\boldsymbol{M}^* \times \left(\boldsymbol{M}^* \times \boldsymbol{H}^*\right)\end{aligned}$$

Several dimensionless parameters are required to be introduced here

$$Re = \frac{\rho v_0 h_0}{\eta}, Mn_f = \frac{\mu_0 H_0^2}{\zeta v_0/h_0}, Pe = \frac{\tau}{h_0/v_0} \tag{6.11}$$

where the Reynolds number Re represents the ratio of inertial force to viscous force per unit mass of fluid, Mn_f denotes the ratio of magnetic torque to viscous torque, and the Peclet number Pe characterizes the ratio of magnetization relaxation time to macroscale flow characteristic time (Carvalho and Gontijo, 2020). These dimensionless parameters lead to the derivation of a dimensionless set of equations governing the hydrodynamics of ferrofluids.

$$\boldsymbol{\omega}_\mathrm{p}^* = \frac{1}{2}\nabla^* \times \boldsymbol{v}^* + \frac{1}{4}Mn_f \boldsymbol{M}^* \times \boldsymbol{H}^* \tag{6.12}$$

$$\frac{\partial \boldsymbol{v}^*}{\partial t^*} + \left(\boldsymbol{v}^* \cdot \nabla^*\right)\boldsymbol{v}^* = -\frac{1}{Re}\nabla^* p^* + \frac{1}{Re}\nabla^{*2}\boldsymbol{v}^* + \frac{1.5\phi Mn_f}{Re}\boldsymbol{M}^* \cdot \nabla^* \boldsymbol{H}^* + \frac{3\phi Mn_f}{4Re}\nabla^* \times \left(\boldsymbol{M}^* \times \boldsymbol{H}^*\right) \tag{6.13}$$

$$\frac{\partial \boldsymbol{M}^*}{\partial t^*}+\left(\boldsymbol{v}^*\cdot\nabla^*\right)\boldsymbol{M}^*=\frac{1}{2}\left(\nabla^*\times\boldsymbol{v}^*\right)\times\boldsymbol{M}^*-\frac{1}{Pe}\left(\boldsymbol{M}^*-\boldsymbol{M}_0^*\right)-\frac{Mn_f}{4}\boldsymbol{M}^*\times\left(\boldsymbol{M}^*\times\boldsymbol{H}^*\right) \tag{6.14}$$

or

$$\begin{aligned}\frac{\partial \boldsymbol{M}^*}{\partial t^*}+\left(\boldsymbol{v}^*\cdot\nabla^*\right)\boldsymbol{M}^*=\frac{1}{2}\left(\nabla^*\times\boldsymbol{v}^*\right)\times\boldsymbol{M}^*-\frac{1}{Pe}\left(\boldsymbol{M}^*-\frac{M_S}{H_0}\frac{L(\varsigma)}{\varsigma}\boldsymbol{\alpha}\right)\\-\frac{Mn_f}{4}\frac{1}{L(\varsigma)}\left(\frac{1}{L(\varsigma)}-\frac{3}{\varsigma}\right)\boldsymbol{M}^*\times\left(\boldsymbol{M}^*\times\boldsymbol{H}^*\right)\end{aligned} \tag{6.15}$$

$$\nabla^{*2}\varphi_m^*-\nabla\cdot\boldsymbol{M}^*=0 \tag{6.16}$$

It is evident that in examining the flow characteristics of ferrofluids, one can consider Re, Pe, and Mn_f as characteristic numbers to assess the similarity of flows.

6.2.2 Governing Equations of Ferrofluid Couette–Poiseuille Flows in Constant Uniform Magnetic Fields

For the Couette–Poiseuille flow of ferrofluid depicted in Figure 6.1, assuming that the field variables are independent of the coordinates x and z, except for the pressure gradient $\partial p^*/\partial x^*$, the velocity and spin angular velocity can be represented as $\boldsymbol{v}^*=\left(v_x^*,0,0\right)$ and $\boldsymbol{\omega}_\text{p}^*=\left(0,0,\omega_{\text{p},z}^*\right)$, respectively. Under steady-state flow conditions, the dimensionless equations ranging from (6.12) to (6.16) are simplified to

$$\frac{\partial^2 v_x^*}{\partial y^{*2}}-\frac{\partial p^*}{\partial x^*}+\frac{3\phi Mn_f}{4}\frac{\partial\left(M_x^*H_y^*\right)}{\partial y^*}=0 \tag{6.17}$$

$$\omega_{\text{p},z}^*=-\frac{1}{2}\frac{\partial v_x^*}{\partial y^*}+\frac{Mn_f}{4}M_x^*H_y^* \tag{6.18}$$

$$\frac{1}{2}M_y^*\frac{\partial v_x^*}{\partial y^*}-\frac{1}{Pe}M_x^*-\frac{Mn_f}{4}M_x^*M_y^*H_y^*=0 \tag{6.19}$$

$$-\frac{1}{2}M_x^*\frac{\partial v_x^*}{\partial y^*}-\frac{1}{Pe}\left(M_y^*-M_0^*\right)+\frac{Mn_f}{4}M_x^{*2}H_y^*=0 \tag{6.20}$$

$$\frac{1}{2}M_y^*\frac{\partial v_x^*}{\partial y^*}-\frac{1}{Pe}M_x^*-\frac{Mn_f}{4}\frac{1}{L(\varsigma)}\left(\frac{1}{L(\varsigma)}-\frac{3}{\varsigma}\right)M_x^*M_y^*H_y^*=0 \tag{6.21}$$

$$-\frac{1}{2}M_x^*\frac{\partial v_x^*}{\partial y^*}-\frac{1}{Pe}\left(M_y^*-\frac{M_S}{H_0}\frac{L(\varsigma)}{\varsigma}\alpha\right)+\frac{Mn_f}{4}\frac{1}{L(\varsigma)}\left(\frac{1}{L(\varsigma)}-\frac{3}{\varsigma}\right)M_x^{*2}H_y^*=0 \tag{6.22}$$

$$\frac{\partial^2\varphi_m^*}{\partial y^{*2}}-\frac{\partial M_y^*}{\partial y^*}=0 \tag{6.23}$$

where the magnetization equations are expressed as separate component equations in the x and y directions, respectively. With the aid of the definition provided in Eq. (6.2), Eq. (6.17) is transformed into

$$\frac{\partial^2 v_x^*}{\partial y^{*2}} + 2P^* + \frac{3\phi Mn_f}{4}\frac{\partial\left(M_x^* H_y^*\right)}{\partial y^*} = 0 \tag{6.24}$$

It is observed that the characteristic number Re is absent from the simplified momentum equation; thus, it has no impact on the steady-state velocity distribution. However, the plate driving velocity v_0 is incorporated in both parameters Mn_f and P^*; hence, it is still feasible to investigate the influence of plate driving velocity using Re.

Regarding the boundary conditions for the Couette–Poiseuille flow of ferrofluids as depicted in Figure 6.1, these conditions are analogous to those of ordinary fluids, where the translational slip effect is neglected (Korlie et al., 2008). Consequently, both the upper and lower boundaries of the fluid are subjected to no-penetration and no-slip boundary conditions.

$$v_x\left(x,h_0\right) = v_0, v_x\left(x,0\right) = 0 \tag{6.25}$$

A constant and uniform magnetic field boundary condition is applied to both the upper and lower plates that is denoted as

$$H_y\left(x,0\right) = \frac{\partial \varphi_m\left(x,0\right)}{\partial y} = H_0, H_y\left(x,h_0\right) = \frac{\partial \varphi_m\left(x,h_0\right)}{\partial y} = H_0 \tag{6.26}$$

Since the equation of angular momentum is explicit, there is no requirement to establish boundary conditions for $\boldsymbol{\omega}_{\mathrm{p}}$. The corresponding dimensionless boundary conditions corresponding to Eqs. (6.25) and (6.26) are as follows:

$$v_x^*\left(x^*,1\right) = 1, v_x^*\left(x^*,0\right) = 0 \tag{6.27}$$

$$H_y^*\left(x^*,0\right) = \frac{\partial \varphi_m^*\left(x^*,0\right)}{\partial y^*} = 1, H_y^*\left(x^*,1\right) = \frac{\partial \varphi_m^*\left(x^*,1\right)}{\partial y^*} = 1 \tag{6.28}$$

Furthermore, based on the expression for the stress tensor in ferrofluids, the tangential stress acting on the non-magnetic solid wall adjacent to the fluid is given in Eq. (2.229). For planar Couette–Poiseuille flow, as the magnetic field intensity along the x-direction is zero, the final result for the shear stress is

$$f_x = \eta \frac{\partial v_x}{\partial y} + \frac{\mu_0}{2} M_x H_y \tag{6.29}$$

6.2.3 Asymptotic Solutions for Flow Velocity Distributions in Weak Fields

When the applied constant magnetic field is weak and meets the condition $\Omega\tau_B \ll 1$, magnetization Eqs. (4.18) and (4.47) yield identical solutions. Given certain assumptions, magnetization Eq. (4.18) can be analytically solved. Since the ferrofluid is nearing a state of equilibrium under these conditions, the corresponding solutions are also referred to as asymptotic solutions (Yang, 2023).

Assuming that the magnetic field direction is perpendicular to the flow direction and under weak field conditions, where the magnetization relaxation effect is weak, it can be reasonably approximated that $\boldsymbol{M}//\boldsymbol{H}$ within the computational domain. This means that, in comparison to the characteristic time of the flow, the magnetization relaxation time is significantly shorter. When the magnetic field direction changes, the magnetic dipole moment will rapidly align with the new magnetic field direction, resulting in the disappearance of the magnetic torque $\mu_0\boldsymbol{M}\times\boldsymbol{H}$ within the fluid. This assumption is equivalent to treating the stress tensor of the ferrofluid as symmetric, as the term $\boldsymbol{BH}=\boldsymbol{HB}$ in the stress tensor expression. Therefore, under the assumptions of steady-state conditions and a weak magnetic field, the momentum and magnetization equations for the ferrofluid can be respectively simplified according to Eqs. (6.13) and (6.14) as

$$\left(\boldsymbol{v}^*\cdot\nabla^*\right)\boldsymbol{v}^* = -\frac{1}{Re}\nabla^* p^* + \frac{1}{Re}\nabla^2\boldsymbol{v}^* + \frac{1.5\phi Mn_f}{Re}\boldsymbol{M}^*\cdot\nabla^*\boldsymbol{H}^* \tag{6.30}$$

$$\left(\boldsymbol{v}^*\cdot\nabla^*\right)\boldsymbol{M}^* = \frac{1}{2}\left(\nabla^*\times\boldsymbol{v}^*\right)\times\boldsymbol{M}^* - \frac{1}{Pe}\left(\boldsymbol{M}^* - \boldsymbol{M}_0^*\right) \tag{6.31}$$

Another assumption is that the magnetization variations arising from convective transport are negligible compared to those induced by the vorticity of the flow, namely,

$$\left(\boldsymbol{v}^*\cdot\nabla^*\right)\boldsymbol{M}^* \ll \frac{1}{2}\left(\nabla^*\times\boldsymbol{v}^*\right)\times\boldsymbol{M}^*$$

Then magnetization Eq. (6.31) is simplified further as

$$\frac{1}{2}\left(\nabla^*\times\boldsymbol{v}^*\right)\times\boldsymbol{M}^* = \frac{1}{Pe}\left(\boldsymbol{M}^* - \boldsymbol{M}_0^*\right) \tag{6.32}$$

Rewriting Eq. (6.32) in component forms

$$\frac{Pe}{2}M_y^*\frac{dv_x^*}{dy^*} = M_x^* - M_{0x}^* \tag{6.33}$$

$$-\frac{Pe}{2}M_x^*\frac{\partial v_x^*}{\partial y^*} = M_y^* - M_{0y}^* \tag{6.34}$$

Under weak field conditions, equilibrium magnetization $\boldsymbol{M}_0^*$ only has a component along the direction of the external magnetic field, leading to $M_{0x}^* = 0$ and $M_{0y}^* = M_0^*$. From Eq. (6.33) one has

$$M_x^* = \frac{Pe}{2} M_y^* \frac{\mathrm{d}v_x^*}{\mathrm{d}y^*} \tag{6.35}$$

Substituting Eq. (6.35) into Eq. (6.34) and performing transformations gives

$$M_y^* = \frac{M_0^*}{1 + \dfrac{Pe^2}{4}\left(\dfrac{\mathrm{d}v_x^*}{\mathrm{d}y^*}\right)^2} \tag{6.36}$$

Using the binomial theorem, the denominator on the right-hand side of Eq. (6.36) can be expanded, and the result is approximated by considering the first two terms of the expansion

$$M_y^* \approx \left[1 - \frac{Pe^2}{4}\left(\frac{\mathrm{d}v_x^*}{\mathrm{d}y^*}\right)^2\right] M_0^* \tag{6.37}$$

The component equation of the momentum equation in the x-direction can be written according to Eq. (6.30) as

$$\frac{\mathrm{d}^2 v_x^*}{\mathrm{d}y^{*2}} + 1.5\phi Mn_f \left(M_x^* \frac{\partial H_x^*}{\partial x^*} + M_y^* \frac{\partial H_x^*}{\partial y^*} \right) = \frac{\partial p^*}{\partial x^*} \tag{6.38}$$

Ignoring the variation of magnetic field intensity along the x-direction and substituting Eq. (6.38) into Eq. (6.37), one obtains

$$\frac{\mathrm{d}^2 v_x^*}{\mathrm{d}y^{*2}} - \left(\frac{3\phi Mn_f M_0^* Pe^2}{8} \frac{\partial H_x^*}{\partial y^*} \right)\left(\frac{\mathrm{d}v_x^*}{\mathrm{d}y^*}\right)^2 = \frac{\partial p^*}{\partial x^*} - \frac{3\phi Mn_f M_0^*}{2} \frac{\partial H_x^*}{\partial y^*} \tag{6.39}$$

Equation (6.39) represents the governing equation that describes the ferrofluid flow between two parallel plates along the x-direction within a Couette–Poiseuille flow configuration. This equation incorporates the coupling between flow vorticity and magnetization. If two parameters are introduced, defined by

$$\varepsilon = \frac{3\phi Mn_f M_0^* Pe^2}{8} \frac{\partial H_x^*}{\partial y^*} \tag{6.40}$$

$$\gamma_0 = \frac{3\phi Mn_f M_0^*}{2} \frac{\partial H_x^*}{\partial y^*} - \frac{\partial p^*}{\partial x^*} \tag{6.41}$$

then Eq. (6.39) becomes

$$\frac{d^2 v_x^*}{dy^{*2}} - \varepsilon\left(\frac{dv_x^*}{dy^*}\right)^2 = -\gamma_0 \tag{6.42}$$

The inclusion of the nonlinear term $\left(\frac{dv_x^*}{dy^*}\right)^2$ in Eq. (6.39) renders its direct integration unfeasible. Typically, the magnetization relaxation time is much shorter than the flow characteristic time, that is $Pe \ll 1$, which implies $\varepsilon \ll 1$. Under these conditions, the regular perturbation method can be applied to solve Eq. (6.42) (Cunha and Sobral, 2004). The second-order accurate solution to Eq. (6.42) is expressed as

$$v_x^* = v_{x0}^*(y) + \varepsilon v_{x1}^*(y) + \varepsilon^2 v_{x2}^*(y) \tag{6.43}$$

Substituting Eq. (6.43) into Eq. (6.42) and neglecting the higher-order terms with respect to ε, one obtains

$$\left(\frac{d^2 v_{x0}^*}{dy^{*2}} + \gamma_0\right) + \varepsilon\left[\frac{d^2 v_{x1}^*}{dy^{*2}} - \left(\frac{dv_{x0}^*}{dy^*}\right)^2\right] + \varepsilon^2\left[\frac{d^2 v_{x2}^*}{dy^{*2}} - 2\left(\frac{dv_{x0}^*}{dy^*}\right)\left(\frac{dv_{x1}^*}{dy^*}\right)\right] = 0 \tag{6.44}$$

Solving this equation amounts to solving the following three linear differential equations:

$$\frac{d^2 v_{x0}^*}{dy^{*2}} + \gamma_0 = 0 \tag{6.45}$$

$$\frac{d^2 v_{x1}^*}{dy^{*2}} - \left(\frac{dv_{x0}^*}{dy^*}\right)^2 = 0 \tag{6.46}$$

$$\frac{d^2 v_{x2}^*}{dy^{*2}} - 2\left(\frac{dv_{x0}^*}{dy^*}\right)\left(\frac{dv_{x1}^*}{dy^*}\right) = 0 \tag{6.47}$$

Appling the boundaries

$$v_x^* = 0, \text{when } y^* = 0$$

$$v_x^* = 1, \text{when } y^* = 1$$

and solving equations ranging from Eq. (6.45) to (6.47) give

$$v_{x0}^* = -\frac{\gamma_0}{2} y^2 + \left(1 + \frac{\gamma_0}{2}\right) y \tag{6.48}$$

$$v_{x1}^{*} = \frac{\gamma_0^2}{12} y^4 - \frac{\gamma_0}{6}(2+\gamma_0) y^3 + \frac{1}{2}\left(1+\frac{\gamma_0}{2}\right)^2 y^2 + \left(-\frac{\gamma_0^2}{24} - \frac{\gamma_0}{6} + \frac{1}{2}\right) y \tag{6.49}$$

$$\begin{aligned} v_{x2}^{*} = & -\frac{\gamma_0^3}{180} y^6 + \frac{\gamma_0^2}{60}(2+\gamma_0) y^5 - \frac{\gamma_0}{48}(2+\gamma_0)^2 y^4 + \left(\frac{\gamma_0^3}{72} + \frac{11}{114}\gamma_0^2 + \frac{\gamma_0}{12} + \frac{1}{12}\right) y^3 \\ & + \left(-\frac{\gamma_0^3}{192} - \frac{\gamma_0^2}{32} + \frac{\gamma_0}{48} + \frac{1}{8}\right) y^2 - \left(\frac{7}{960}\gamma_0^3 - \frac{7}{1440}\gamma_0^2 + \frac{\gamma_0}{48} - \frac{9}{24}\right) y \end{aligned} \tag{6.50}$$

Substituting Eqs. (6.48) to (6.50) into Eq. (6.43), one can obtain the asymptotic solution for the flow velocity

$$\begin{aligned} v_x^{*} = & \left[-\frac{\gamma_0}{2} y^2 + \left(1+\frac{\gamma_0}{2}\right) y\right] + \varepsilon\left[\frac{\gamma_0^2}{12} y^4 - \frac{\gamma_0}{6}(2+\gamma_0) y^3 + \frac{1}{2}\left(1+\frac{\gamma_0}{2}\right)^2 y^2 + \left(-\frac{\gamma_0^2}{24} - \frac{\gamma_0}{6} + \frac{1}{2}\right) y\right] \\ & + \varepsilon^2 \left[\begin{array}{c} -\frac{\gamma_0^3}{180} y^6 + \frac{\gamma_0^2}{60}(2+\gamma_0) y^5 - \frac{\gamma_0}{48}(2+\gamma_0)^2 y^4 \\ + \left(\frac{\gamma_0^3}{72} + \frac{11}{114}\gamma_0^2 + \frac{\gamma_0}{12} + \frac{1}{12}\right) y^3 + \left(-\frac{\gamma_0^3}{192} - \frac{\gamma_0^2}{32} + \frac{\gamma_0}{48} + \frac{1}{8}\right) y^2 \\ - \left(\frac{7}{960}\gamma_0^3 - \frac{7}{1440}\gamma_0^2 + \frac{\gamma_0}{48} - \frac{9}{24}\right) y \end{array}\right] \end{aligned} \tag{6.51}$$

Equation (6.51) represents the velocity distribution in the absence of magnetization relaxation effects. Figure 6.3 compares the asymptotic solution results with numerical calculations when $Mn_f = 10$, $\frac{\partial p^*}{\partial x^*} = -4$, $Pe = 0.1$, and $\phi = 0.15$. It can be observed that under weak field conditions, the application of the regular perturbation method yields results that are in good agreement with the numerical solutions.

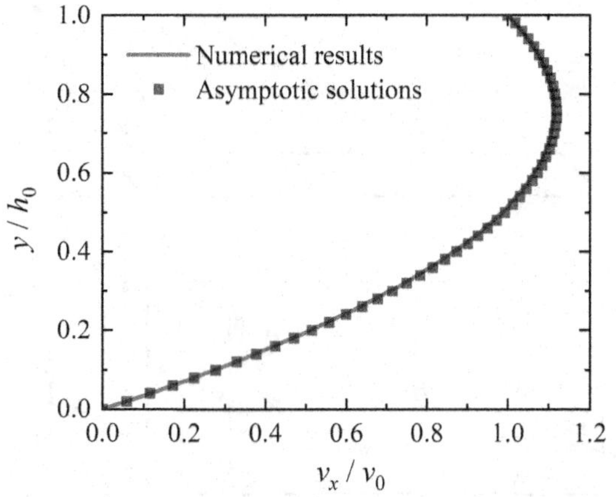

FIGURE 6.3 Comparison of the asymptotic solutions of Couette–Poiseuille flow with numerical results. (Reprinted from Yang et al., 2023 with permission from Elsevier.)

6.2.4 Methods and Parameters Applied in Numerically Solving the System of Ferrohydrodynamic Equations

When the applied constant magnetic field is not weak or the condition $\Omega\tau_B \ll 1$ is not met, the system of ferrohydrodynamic equations becomes analytically intractable, necessitating numerical methods to derive flow characteristics. During numerical computation, specific physical parameters must be defined. This section provides these parameters and outlines the numerical solution methods. Using the OpenFOAM library, which employs the finite volume method, a solver for the ferrohydrodynamic equations is developed and validated. This solver is then used to analyze the Couette–Poiseuille flow of ferrofluids under the influence of constant magnetic fields.

The discretized representation of the ferrofluid flow rate between two parallel plates, as shown in the meshing result depicted in Figure 6.4, can be expressed in numerical calculations as

$$Q = \int_0^{h_0} v_x \mathrm{d}y = \sum_{k=1}^{CN} v_{x,k} \left(\Delta y\right)_k \tag{6.52}$$

where CN represents the total number of elements within a longitudinal section, while $v_{x,k}$ and $\left(\Delta y\right)_k$ correspond to the flow velocity and the longitudinal length of the kth element, respectively.

When a magnetic field is applied, the force exerted by the ferrofluid on the upper plate consists of viscous drag and magnetic stress arising from the abrupt change in magnetization across the interface, which can be expressed as

$$\begin{aligned}\Gamma &= \frac{1}{L}\int_0^{L_0}\left(\eta\frac{\mathrm{d}v_x}{\mathrm{d}y} + \frac{\mu_0}{2}M_x H_y\right)\mathrm{d}x \\ &= \frac{1}{L}\sum_{k}^{CM}\left(\eta\frac{v_0 - v_{x,k}}{\left(\Delta y\right)_k/2} + \frac{\mu_0}{2}M_{x,k}H_{y,k}\right)\Delta x\end{aligned} \tag{6.53}$$

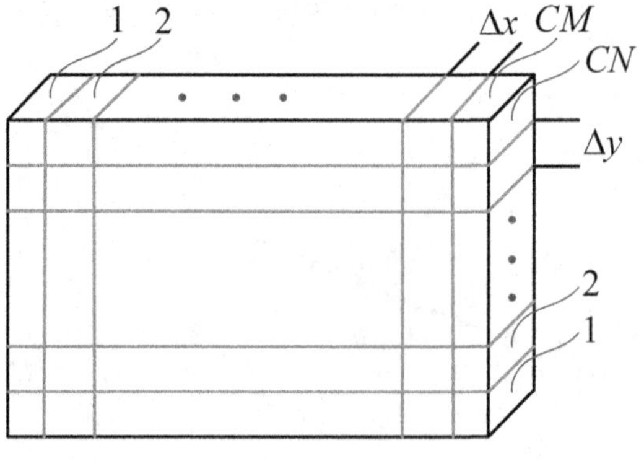

FIGURE 6.4 Schematic diagram of mesh generation in the fluid domain.

where CM represents the number of elements in the fluid layer adjacent to the upper plate and $(\Delta y)_k / 2$ signifies the vertical distance from the centroid of the kth element to the lower surface of the plate. Furthermore, $M_{x,k}$ and $H_{y,k}$ represent the magnetization in the x-direction and the magnetic field intensity in the y-direction, respectively, for the kth element.

To examine the variation of vortex intensity with various parameters of ferrofluids, the average vorticity (or root mean square vorticity) throughout the fluid domain is computed as

$$\Omega_{\text{rms}} = \left[\frac{1}{CN \times CM} \sum_{(i,j)=(1,1)}^{(CN,CM)} \left(\frac{1}{2} \nabla \times \boldsymbol{v} \right)_{i,j}^{2} \right]^{1/2} \tag{6.54}$$

The parameters used in the calculations are listed in Table 6.1, including typical physical parameters of a ferrofluid.

6.2.5 Magnetization Component in the Direction Perpendicular to the Magnetic Field

The ferrohydrodynamic equations, encompassing both the phenomenological magnetization equation I and the microscopic magnetization equation, are solved to determine the magnetization perpendicular to the external magnetic field. This magnetization component serves as a metric for characterizing the strength of the magnetization relaxation effect. A comparison of the results obtained for the two cases: $P^* = 0$, $Re = 1$ and $P = 2$, $Re = 20$ is shown in Figure 6.5.

In the case of $P^* = 0$, $Re = 1$, the calculated results of M_x using the phenomenological magnetization equation I agree well with those obtained from the microscopic magnetization equation in the weak magnetic field range. However, as the magnetic field becomes stronger, there is a deviation between them, with a maximum deviation of 19% within the calculated range. This case corresponds to a situation where $\Omega \tau_B = 5.12 \times 10^{-4} \ll 1$,

TABLE 6.1 Parameter Values Used in Numerical Calculations

Parameters	**Symbols**	**Values**	**Units**
Distance between two plates	h_0	2×10^{-4}	m
Density of ferrofluids	ρ	1600	kg/m^3
Dynamic viscosity of ferrofluids	η	0.01	Pa·s
Kinematic viscosity of ferrofluids	ν	6.25×10^{-6}	m^2/s
Saturation magnetization of ferrofluids	M_S	5.25×10^{4}	A/m
Magnetization relaxation time of ferrofluids	τ_B	4×10^{-4}	s
Saturation magnetization of bulk materials for particles	M_d	4.46×10^{5}	A/m
Volumetric fraction of magnetic particles in ferrofluids	ϕ	0.118	—
Average diameter of magnetic particles in ferrofluids	d_p	10×10^{-9}	m
Temperature	T	300	K

Source: Yang and Liu (2020).

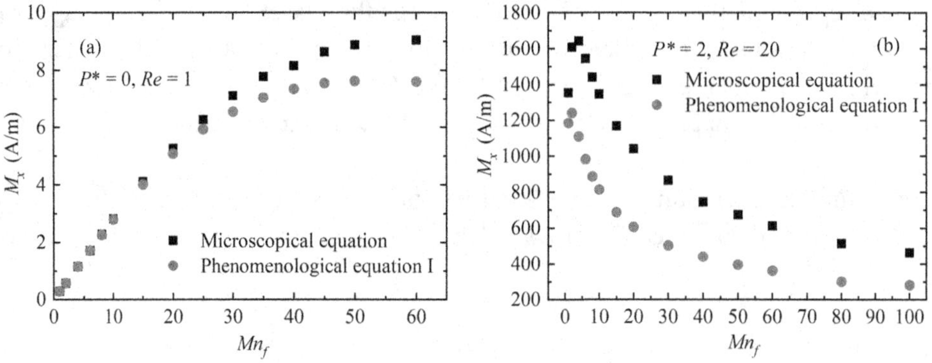

FIGURE 6.5 Comparison of the magnetization perpendicular to the magnetic field in the Couette–Poiseuille flows when different magnetization equations are applied. (a) $P^* = 0$, $Re = 1$ and (b) $P^* = 2$, $Re = 20$. (Reprinted from Yang et al., 2022 with permission from Elsevier.)

indicating that the ferrofluid is in a weakly non-equilibrium state. Under these conditions, the magnetization results obtained from both equations are similar, consistent with the observations made by Shliomis.

In the case of $P^* = 2$ and $\text{Re} = 20$, $\Omega\tau_B \sim 0.6$, and the phenomenological magnetization equation falls short of accurately describing the non-equilibrium magnetization process of the ferrofluid. In fact, the disparities between the results obtained using this equation and those derived from the microscopic magnetization equation reach a maximum of 69.6%. Therefore, all subsequent calculations will rely solely on the microscopic magnetization equation.

In Figure 6.5(b), it is observed that M_x exhibits an initial increase with Mn_f, followed by a subsequent decrease, which can be explained by Eq. (6.18). Under strong magnetic fields, the local vorticity of the fluid becomes inadequate for inducing particle rotation. Consequently, the self-rotation speed in Eq. (6.18) becomes very small, resulting in minimal changes in vorticity. Consequently, as Mn_f increases, M_x^* decreases in Eq. (6.18).

6.2.6 Magnetization Relaxation Effects in Couette–Poiseuille Flows in Constant Magnetic Fields

The most pronounced manifestation of the magnetization relaxation effects is the emergence of a magnetization component M_x, perpendicular to the external magnetic field within the ferrofluid. This distinction highlights a crucial difference between ferrofluids and regular fluids. This effect causes the magnetization vector lines to deviate from the magnetic field intensity vector lines as depicted in Figure 6.6. Consequently, a non-zero magnetic torque $\mu_0 \boldsymbol{M} \times \boldsymbol{H}$ arises within the fluid, ultimately leading to deviations in flow velocity and vorticity from those observed in regular fluids (Ghosh and Das, 2019). These deviations give rise to non-Newtonian flow characteristics within the ferrofluid (Weng et al., 2008). Figures 6.7 and 6.8 illustrate the distribution of flow velocity and vorticity, respectively, within a cross-section, both before and after the application of a magnetic field. Additionally, Figures 6.9 and 6.10 depict the distribution of the magnetization component perpendicular to the external magnetic field and particle spin velocity, respectively, within a cross-section.

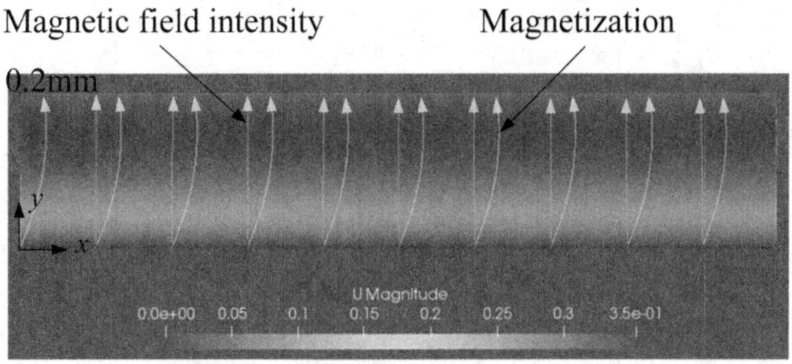

FIGURE 6.6 Magnetic field intensity and magnetization in a Couette–Poiseuille flow of a ferrofluid ($Re = 10$, $P^* = 2$, $Mn_f = 5$). (Reprinted from Yang et al., 2022 with permission from Elsevier.)

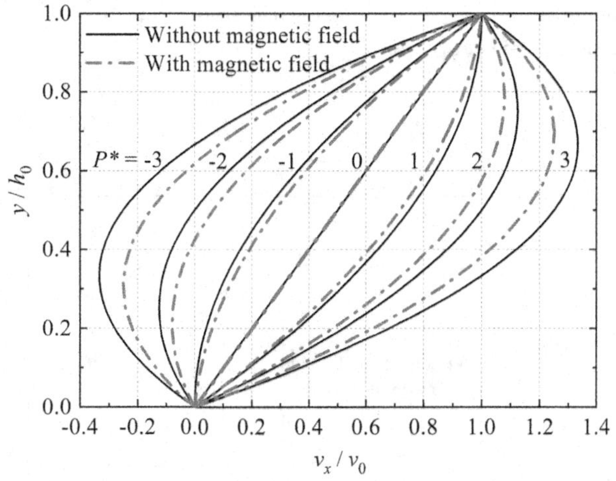

FIGURE 6.7 Deviation of flow velocity distribution caused by magnetization relaxation from that of ordinary fluids ($Re = 1$, $Mn_f = 100$). (Reprinted from Yang et al., 2022 with permission from Elsevier.)

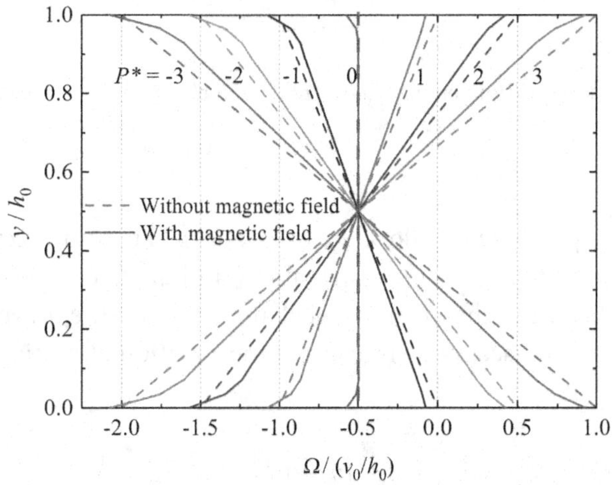

FIGURE 6.8 Deviation of flow vorticity caused by magnetization relaxation from that of ordinary fluids ($Re = 1$, $Mn_f = 100$). (Reprinted from Yang et al., 2022 with permission from Elsevier.)

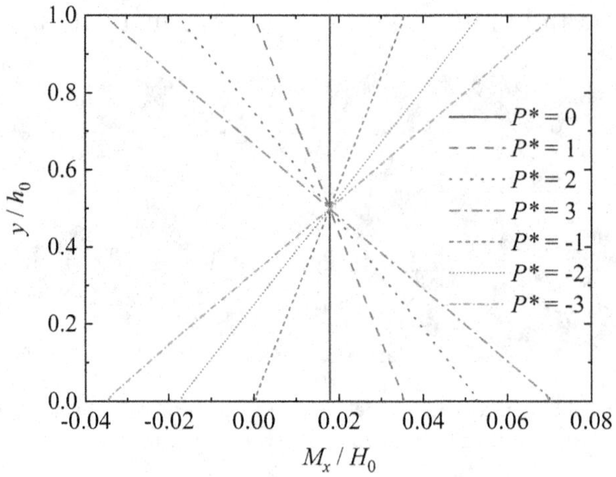

FIGURE 6.9 Distribution of magnetization perpendicular to the direction of the external magnetic field within the flow cross-section ($Re = 1$, $Mn_f = 100$).

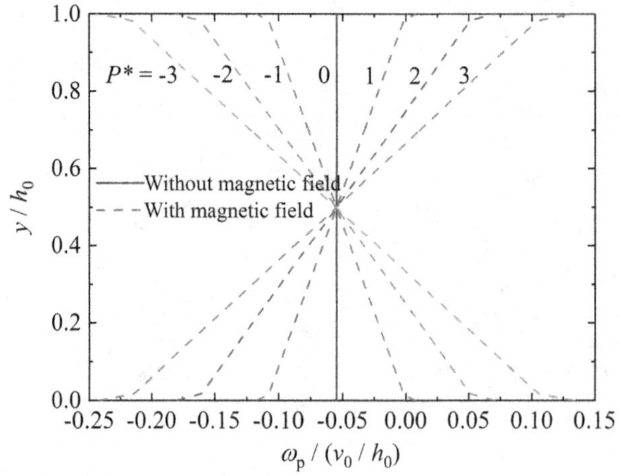

FIGURE 6.10 Spin velocity distribution in the flow cross-section ($Re = 1$, $Mn_f = 100$).

When the flow is a pure Couette flow, the magnetic field has no effect on either the flow velocity distribution or the vorticity distribution, and at this point, the magnetization component perpendicular to the magnetic field and the particle spin velocity remain constant throughout the cross-section. In this case, the equation of motion (Eq. 6.24) further simplifies to

$$\frac{\partial^2 v_x^*}{\partial y^{*2}} + \frac{3\phi Mn_f}{4} \frac{\partial \left(M_x^* H_y^* \right)}{\partial y^*} = 0 \tag{6.55}$$

From the velocity and M_x distributions shown in Figures 6.7 and 6.9, it is evident that the velocity gradient, spin velocity, and M_x remain constant for pure Couette flow. As stated in Eq. (6.18), the second term on the right-hand side of the equation is spatially invariant, leading to the elimination of the second term on the left-hand side of Eq. (6.55). Therefore, variations in the external magnetic field do not affect the velocity distribution characteristic of pure Couette flow.

The velocity profile of ferrofluids in a Couette–Poiseuille flow is modulated by the application of a magnetic field. When $P^* > 0$, indicating a pressure drop aligned with the motion of the upper plate, the magnetic field slows down the velocity at each location within the cross-section. Conversely, when $P^* < 0$, implying a pressure drop opposing the motion of the upper plate, the applied magnetic field enhances the velocity. Furthermore, examining the vorticity distribution, the magnetic field slows down velocity changes, resulting in an overall decrease in flow rate, as illustrated in Figure 6.7. This effect is attributed to the increase in apparent viscosity of the ferrofluid due to the influence of the applied constant magnetic field, which hinders the flow. This resistance is reflected in the distribution of particle rotation speeds. In the absence of a magnetic field, particles freely rotate along with the local vorticity. However, in the presence of a magnetic field, the free rotation of particles is restricted by the magnetic torque exerted by the field, leading to a significant decrease in $\boldsymbol{\omega}_\mathrm{p}$ values.

The strength of the magnetization relaxation effect can be gauged through the distribution of M_x when $P^* \neq 0$ as depicted in Figure 6.9. When a magnetic field is applied, the magnetization component M_x, perpendicular to the magnetic field at a given flow cross-section, varies linearly with position. The rate of this variation is related to the pressure gradient. For instance, when $P^* > 0$, M_x decreases with decreasing positional coordinates and when $P^* < 0$, M_x increases with increasing positional coordinates. Notably, higher P^* values lead to greater M_x values near the boundary.

By examining the distribution of spin velocity presented in Figure 6.10, it is evident that the introduction of a magnetic field results in an increase in the absolute value of $\boldsymbol{\omega}_\mathrm{p}$ as the pressure gradient rises. This observation indicates that the pressure gradient amplifies the rotation of particles.

6.2.7 Effects of Physical Quantities on Flow Characteristics

6.2.7.1 Effects of Pressure

The upper plate experiences a force from the ferrofluid, encompassing both viscous drag and magnetic stress. The magnetic field amplifies the viscous drag as shown in Figure 6.11. The magnetic stress arises from sudden changes in magnetization, adding to the viscous drag and increasing the total shear stress exerted on the plate. Compared to the case without a magnetic field, the flow rate of the ferrofluid decreases under an external magnetic field due to the augmented apparent viscosity caused by the constant magnetic field. Additionally, the volumetric flow rate increases linearly with the pressure gradient.

The magnetization component M_x, perpendicular to the external magnetic field within the ferrofluid near the upper plate, decreases linearly with the increase of P^*, as shown in Figure 6.12. Once P^* reaches a certain threshold value ($P^* \geq 2$), the direction of M_x reverses.

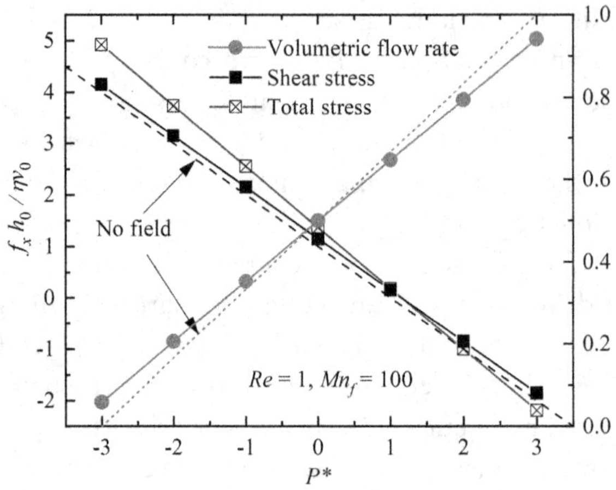

FIGURE 6.11 Variation of tangential stress on the upper plate and the overall flow rate with the dimensionless pressure gradient ($Re = 1, Mn_f = 100$). (Reprinted from Yang et al., 2022 with permission from Elsevier.)

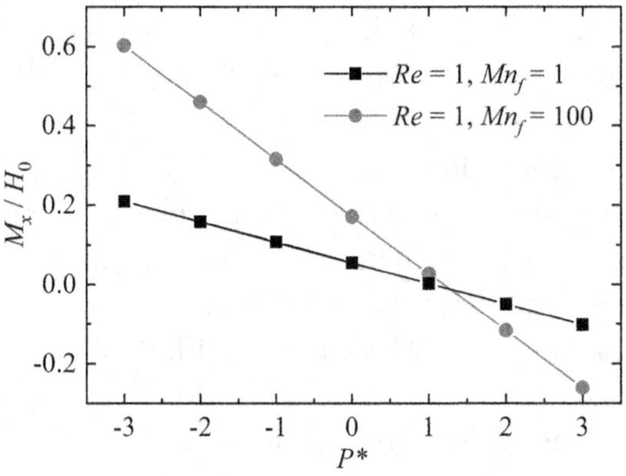

FIGURE 6.12 Variation of M_x with dimensionless pressure in a ferrofluid near the upper plate. (Reprinted from Yang et al., 2022 with permission from Elsevier.)

6.2.7.2 Effects of Magnetic Field

If the *Re* value remains constant, varying Mn_f is equivalent to altering the magnetic field gradient. Figure 6.13 illustrates the dependence of flow characteristics on Mn_f at $Re = 1$, $P^* = 2$, corresponding to changes in magnetic field intensity. Evidently, in a constant uniform magnetic field, the velocity of Couette–Poiseuille flow decreases as the magnetic field intensity increases. However, as the applied magnetic field approaches the saturation value of the ferrofluid, the flow velocity hardly changes anymore. As the magnetic field intensity rises, the rate of linear change in vorticity gradually diminishes, while the rate of linear change in particle rotation velocity increases. This change in particle rotation velocity is more significant than the change in vorticity, eventually reaching a negative slope, indicating that, at a certain magnetic field intensity, the particle rotation velocity decreases with position. Purely considering the magnitude of the vectors, a stronger applied magnetic field results in smaller vorticity and particle rotation rates at the same location. Once the

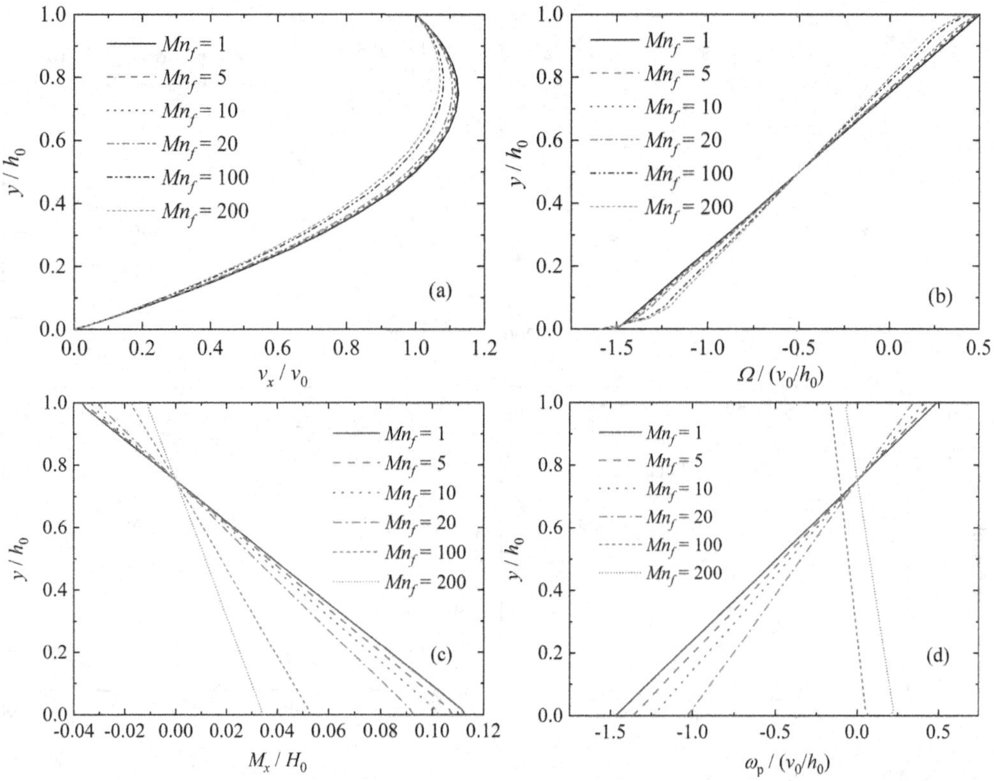

FIGURE 6.13 Distribution of flow properties within a cross-section at different M_{nf} ($Re = 1, P^* = 2$). (a) Flow velocity, (b) flow vorticity, (c) magnetization perpendicular to the direction of magnetic field, and (d) particle spin velocity. (Reprinted from Yang et al., 2022 with permission from Elsevier.)

magnetic field reaches a certain threshold, the direction of particle rotation velocity reverses, indicating that the magnetic torque's effect on particle rotation dominates over the influence of fluid vorticity. In weaker magnetic fields, particles rotate along with the local vorticity of the flow, but due to magnetization relaxation effect, particle rotation lags behind the flow vorticity, with both having the same direction. When the magnetic field intensity exceeds a certain value (as in the case of $M_{nf} = 100$), the magnitude of the second term on the right-hand side of Eq. (6.18) surpasses the first term. Since these two terms have opposite signs, the particle's self-rotation angular velocity reverses direction. Additionally, the magnetization component perpendicular to the magnetic field decreases linearly along the longitudinal axis of the channel cross-section, as shown in Figure 6.13(c), and the stronger the magnetic field, the smaller the magnitude of this magnetization component at a given location.

Figure 6.14 illustrates the variation in flow rate, average vorticity, stress, and M_x near the upper plate with magnetic field intensity. It can be observed that the shear stress experiences minimal fluctuations with the magnetic field, while the flow rate and the overall average vorticity of the fluid both decrease as the field intensity increases. This phenomenon is attributed to the increasing hindrance of particle rotation by the magnetic field,

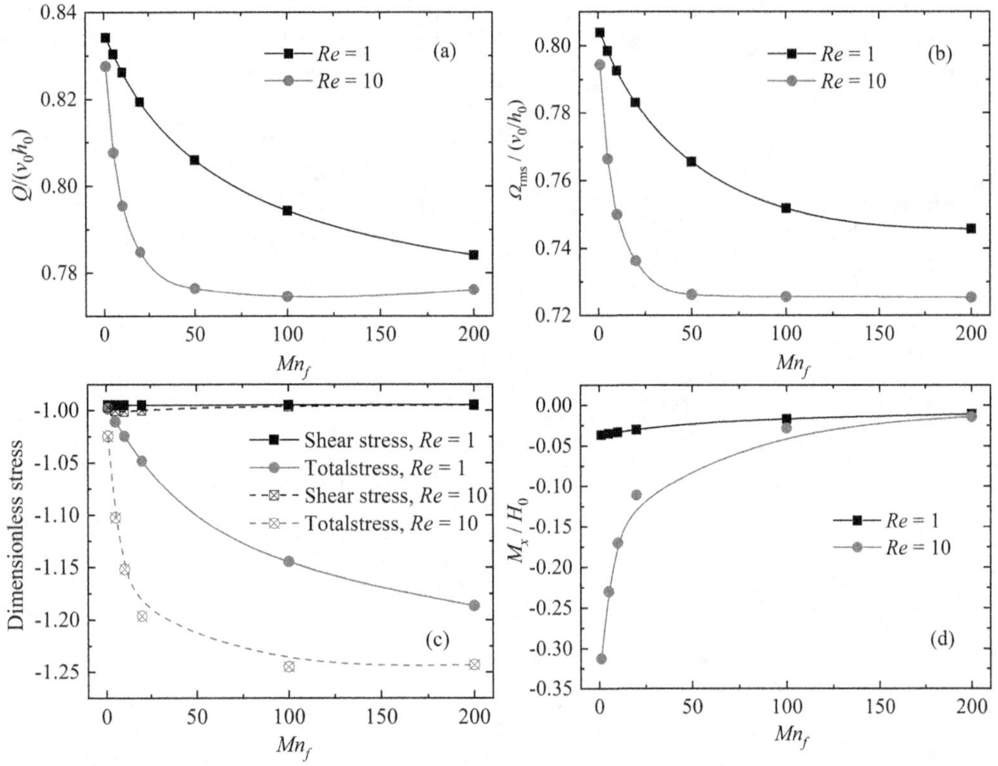

FIGURE 6.14 Variation of (a) flow rate, (b) mean vorticity, (c) dimensionless stress, and (d) M_x near the upper plate ($P^* = 2$). (Reprinted from Yang et al., 2020 with permission from Elsevier.)

subsequently leading to a reduction in fluid velocity. Furthermore, the magnitude of M_x near the upper plate increases with the field intensity. As the external magnetic field approaches the saturation value of the ferrofluid, the rates of change for all these variables gradually decrease, ultimately approaching constant values.

6.2.7.3 Effects of the Driving Speed by the Upper Plate

Figure 6.15 shows the distribution of particle spin velocity and magnetization component perpendicular to the external magnetic field within the flow cross-section at $P^* = 2$ and $Mn_f = 1$. Under the same pressure conditions, altering the Re value or varying the driving speed of the upper plate does not affect the distribution of dimensionless velocity and vorticity. Therefore, only the distribution of spin speed and M_x values near the upper plate are presented. As the driving speed increases, the linear rate of change in particle spin velocity along the cross-section gradually increases, while the rate of change in M_x along the cross-section gradually decreases.

Figure 6.16 gives the variations of M_x and dimensionless total stress within the ferrofluid near the upper plate with Re under the conditions of $P^* = 2$ and $Mn_f = 1$. As Re increases, both the magnitude of the dimensionless total stress and the magnitude of M_x decrease. Since there is no change in the dimensionless flow velocity and vorticity in this case, the average vorticity and the shear stress experienced by the upper plate also remain constant with respect to Re. At higher values of Re, the absolute velocity of the particles

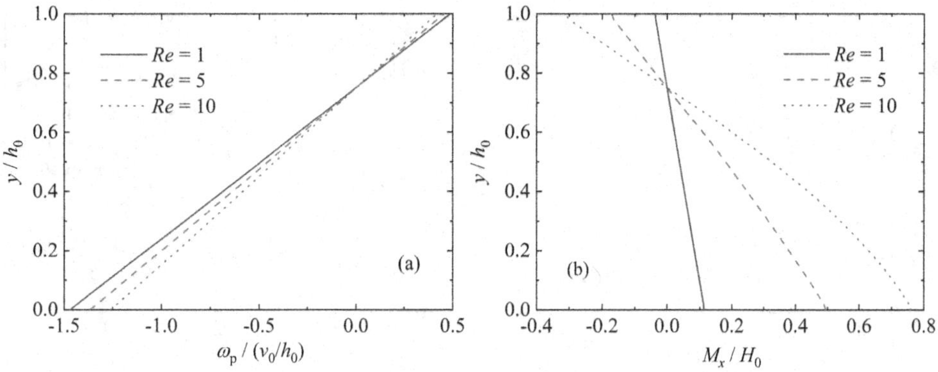

FIGURE 6.15 Flow characteristics at different Re ($P^* = 2$, $M_{nf} = 1$). (a) Particle spin velocity within the flow cross-section and (b) magnetization component perpendicular to the direction of the external magnetic field.

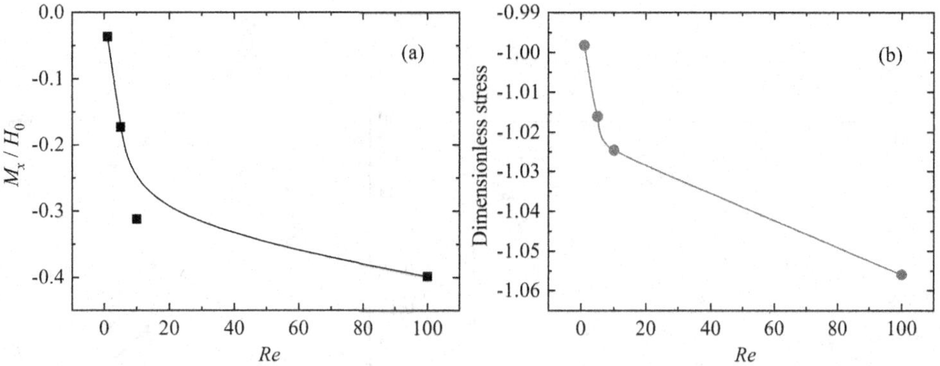

FIGURE 6.16 Variation in (a) M_x and (b) dimensionless total stress with *Re*.

moving with the fluid is increased, causing them to traverse the magnetic field region rapidly without sufficient time for magnetization. Consequently, M_x is relatively small.

6.2.7.4 Effects of Magnetization Relaxation Time

The magnetization relaxation time of ferrofluids also impacts the flow characteristics of Couette–Poiseuille flow, as demonstrated in Figures 6.17 and 6.18. Specifically, when the magnetization relaxation time is significantly shorter than the characteristic flow time (6.4×10^{-4} s in this case), the particles promptly align with the magnetic field, resulting in a weaker magnetization relaxation effect and a smaller magnetization M_x in the non-magnetic field direction as shown in Figure 6.17(b). At shorter magnetization relaxation times, particle rotation is primarily influenced by fluid drag. However, as the relaxation time increases, the magnetization relaxation increasingly dominates, potentially causing particles to rotate in the opposite direction as seen in Figure 6.17(a). Furthermore, an increase in magnetization relaxation time leads to a decrease in the dimensionless magnetic stress and total stress acting on the upper plate as shown in Figure 6.18(a). Additionally, both vorticity and fluid flow rate increase slightly with the magnetization relaxation time as illustrated in Figure 6.18(c) and (d).

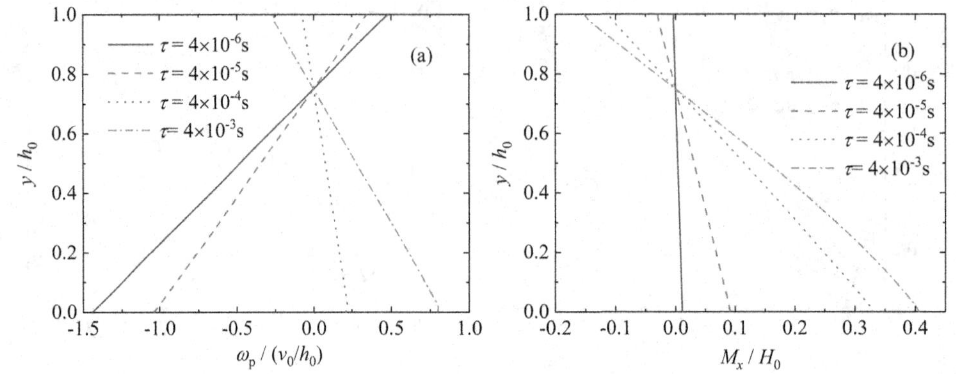

FIGURE 6.17 Distributions of velocity, particle spin, and M_x under the influence of different magnetization relaxation time ($P^* = 2$, $M_{nf} = 20$, $Re = 10$).

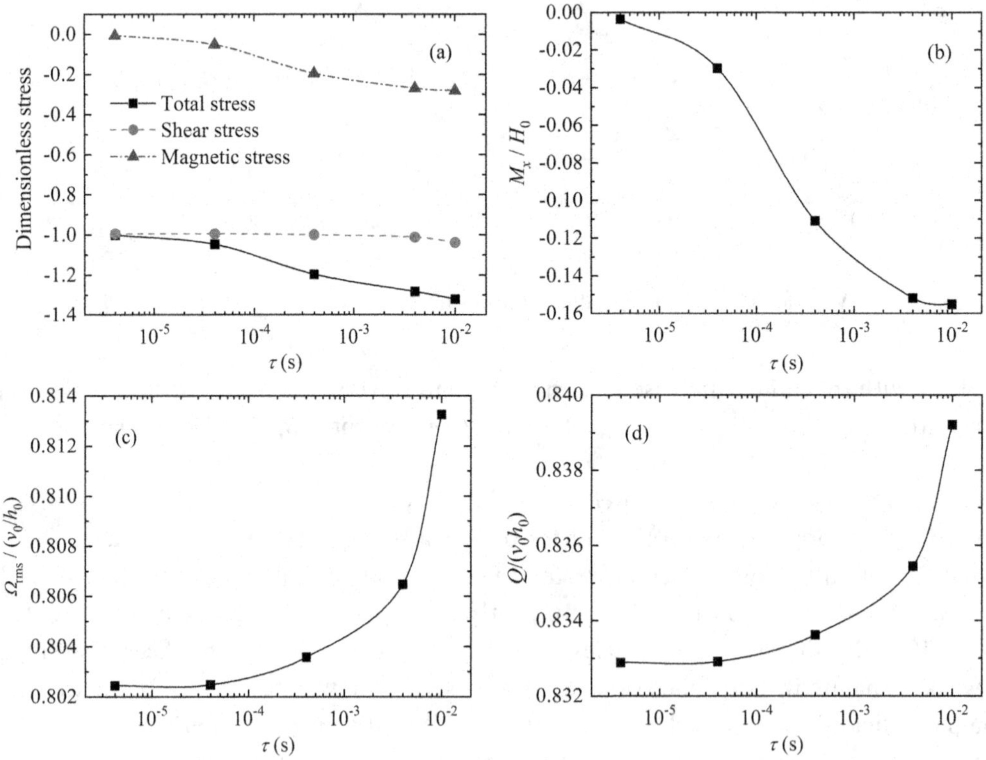

FIGURE 6.18 Variation of (a) dimensionless stress, (b) M_x, (c) mean vorticity, and (d) flow rate with magnetization relaxation time ($P^* = 2$, $M_{nf} = 20$, $Re = 10$).

6.3 FLOW CHARACTERISTICS IN MAGNETIC FIELDS WITH CONSTANT GRADIENT

6.3.1 Governing Equations

For the planar Couette–Poiseuille flow of ferrofluid under the influence of a gradient magnetic field as shown in Figure 6.19, the impact of the gradient magnetic field on flow characteristics is discussed in two scenarios: When the direction of the magnetic field gradient aligns with and opposes the direction of the wall driving velocity. Herein, the gradient of magnetic field is denoted as $\mathbf{G}_H$. Furthermore, considering a constant pressure difference maintained between the inlet and outlet, two additional cases are examined: When the pressure gradient direction is the same as and opposite to the direction of the magnetic field gradient.

In the steady-state ferrofluid flow illustrated in Figure 6.19, it is assumed that the fluid velocity solely exists in the x-direction, that is $v_y = 0$. Neglecting any alterations in H_x along the x-direction, as well as variations of M_y and H_y along the y-direction, that is $\frac{\partial H_x}{\partial x} = 0$, $\frac{\partial M_y}{\partial y} = 0$, and $\frac{\partial H_y}{\partial y} = 0$, allows for the simplification of the momentum Eq. (2.219) into the following component equations:

$$\rho v_x \frac{\partial v_x}{\partial x} = -\frac{\partial p}{\partial x} + \eta\left(\frac{\partial^2 v_x}{\partial x^2} + \frac{\partial^2 v_x}{\partial y^2}\right) + \frac{\mu_0}{2}\left(M_y \frac{\partial H_x}{\partial y} + H_y \frac{\partial M_x}{\partial y} - H_x \frac{\partial M_y}{\partial y}\right) \tag{6.56}$$

$$0 = -\frac{\partial p}{\partial y} + \frac{\mu_0}{2}\left(M_x \frac{\partial H_y}{\partial x} + H_x \frac{\partial M_y}{\partial x} - H_y \frac{\partial M_x}{\partial x}\right) \tag{6.57}$$

To describe the magnetization relaxation process of ferrofluids with the flow depicted in Figure 6.19, microscopic magnetization Eq. (4.47) is applied. Under the aforementioned assumptions, Eq. (4.47) can be reduced to the following component equations:

$$v_x \frac{\partial M_x}{\partial x} = \frac{1}{2} M_y \frac{\partial v_x}{\partial y} - \frac{1}{\tau_B} M_x - \frac{\mu_0}{6\eta\phi} \frac{1}{L(\varsigma)}\left(\frac{1}{L(\varsigma)} - \frac{3}{\varsigma}\right)\left(M_x M_y H_y - M_y^2 H_x\right) \tag{6.58}$$

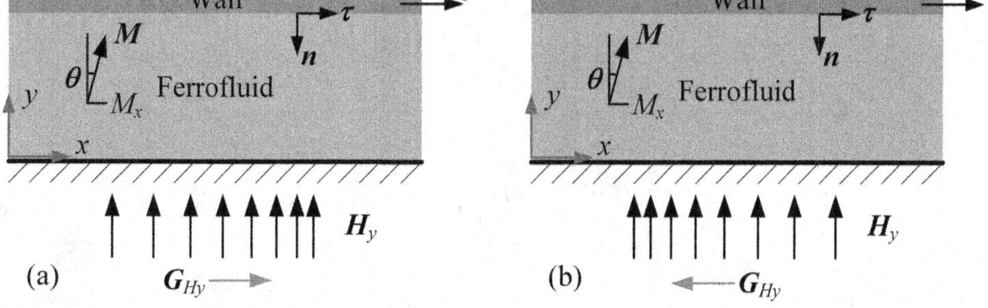

FIGURE 6.19 Ferrofluid Couette–Poiseuille flows under the application of field gradients. The direction of the magnetic field gradient is (a) the same as or (b) opposite to the velocity of the wall movement. (Reprinted from Yang et al., 2022 with permission from Elsevier.)

$$v_x \frac{\partial M_y}{\partial x} = -\frac{1}{2} M_x \frac{\partial v_x}{\partial y} - \frac{1}{\tau_B}\left(M_y - M_S \frac{L(\varsigma)}{\varsigma}\alpha\right) + \frac{\mu_0}{6\eta\phi} \frac{1}{L(\varsigma)}\left(\frac{1}{L(\varsigma)} - \frac{3}{\varsigma}\right)\left(M_x^2 H_y - M_x M_y H_x\right) \tag{6.59}$$

In addition, angular momentum Eq. (2.218) and magnetic field Eq. (6.9) are respectively simplified to

$$\omega_{\mathrm{p}} = -\frac{1}{2}\frac{\partial v_x}{\partial y} + \frac{\mu_0}{4\zeta}\left(M_x H_y - M_y H_x\right) \tag{6.60}$$

$$\frac{\partial^2 \varphi_m}{\partial x^2} + \frac{\partial^2 \varphi_m}{\partial y^2} - \frac{\partial M_x}{\partial x} - \frac{\partial M_y}{\partial y} = 0 \tag{6.61}$$

$$H_x = -\frac{\partial \varphi_m}{\partial x}, H_y = -\frac{\partial \varphi_m}{\partial y} \tag{6.62}$$

The Couette–Poiseuille flow of ferrofluids under the influence of a gradient magnetic field is thus described by a system of equations comprising Eqs. (6.56) to (6.62). To facilitate the identification of key parameters that govern flow characteristics, this system of equations is nondimensionalized. Referring to the definitions provided in Eq. (6.10), Eqs. (6.56) to (6.62) are transformed into the following nondimensional forms:

$$Re\, v_x^* \frac{\partial v_x^*}{\partial x^*} - \frac{\partial^2 v_x^*}{\partial x^{*2}} - \frac{\partial^2 v_x^*}{\partial y^{*2}} = P^* + \frac{3\phi Mn_f}{4}\left(M_y^* \frac{\partial H_y^*}{\partial x^*} + H_y^* \frac{\partial M_x^*}{\partial y^*} - H_x^* \frac{\partial M_y^*}{\partial y^*}\right) \tag{6.63}$$

$$0 = -\frac{\partial p^*}{\partial y^*} + \frac{3\phi Mn_f}{4}\left(M_x^* \frac{\partial H_y^*}{\partial x^*} + H_x^* \frac{\partial M_y^*}{\partial x^*} - H_y^* \frac{\partial M_x^*}{\partial x^*}\right) \tag{6.64}$$

$$v_x^* \frac{\partial M_x^*}{\partial x^*} = \frac{1}{2}\frac{\partial v_x^*}{\partial y^*} M_y^* - \frac{1}{\tau_B} M_x^* - \frac{Mn_f}{4}\frac{1}{L(\varsigma)}\left(\frac{1}{L(\varsigma)} - \frac{3}{\varsigma}\right)\left(M_x^* M_y^* H_y^* - M_y^{*2} H_x^*\right) \tag{6.65}$$

$$v_x^* \frac{\partial M_y^*}{\partial x^*} = -\frac{1}{2}\frac{\partial v_x^*}{\partial y^*} M_x^* - \frac{1}{\tau_B^*}\left(M_y^* - \frac{L(\varsigma)}{\varsigma}\alpha\right) + \frac{Mn_f}{4}\frac{1}{L(\varsigma)}\left(\frac{1}{L(\varsigma)} - \frac{3}{\varsigma}\right)\left(M_x^{*2} H_y^* - M_x^* M_y^* H_x^*\right) \tag{6.66}$$

$$\omega_{\mathrm{p},z}^* = -\frac{1}{2}\frac{\partial v_x^*}{\partial y^*} + \frac{Mn_f}{4}\left(M_x^* H_y^* - M_y^* H_x^*\right) \tag{6.67}$$

$$\frac{\partial^2 \varphi_m}{\partial x^{*2}} + \frac{\partial^2 \varphi_m^*}{\partial y^{*2}} - \frac{\partial M_x^*}{\partial x^*} - \frac{\partial M_y^*}{\partial y^*} = 0 \tag{6.68}$$

$$H_x^* = -\frac{\partial \varphi_m^*}{\partial x^*}, H_y^* = -\frac{\partial \varphi_m^*}{\partial y^*} \tag{6.69}$$

Assuming that the applied magnetic field is along the y-direction and possesses a constant gradient in the x-direction, as depicted in Figure 6.19, let H_0 represent the smallest magnitude of the magnetic field intensity, situated either at the left or right extremity of the flow region. Then, the magnetic field intensity at any arbitrary position within this region can be expressed as

$$H_y = H_0 + \frac{\partial H_y}{\partial x} x \tag{6.70}$$

Utilizing the relationship between the magnetic scalar potential and the magnetic field intensity, given by $H_y = -\partial \varphi_m / \partial y$, the relationship between the magnetic scalar potentials at the upper and lower boundaries can be derived from Eq. (6.70) that

$$\varphi_{m,h_0} = \varphi_{m,0} - \left(H_0 + \frac{\partial H_y}{\partial x} x \right) h_0 \tag{6.71}$$

Furthermore, the dimensionless form of Eq. (6.71) is

$$\varphi_{m,h_0}^* = \varphi_{m,0}^* - \left(1 + \frac{\partial H_y^*}{\partial x^*} x^* \right) \tag{6.72}$$

The magnitude of the non-dimensional magnetic field gradient can be represented as

$$G_H = \frac{\partial H_y^*}{\partial x^*} \tag{6.73}$$

With the aid of this expression and the definition given in Eqs. (6.11), (6.63) and (6.64) become

$$Re v_x^* \frac{\partial v_x^*}{\partial x^*} - \frac{\partial^2 v_x^*}{\partial x^{*2}} - \frac{\partial^2 v_x^*}{\partial y^{*2}} = P^* + \frac{3\phi Mn_f}{4} \left(M_y^* G_H + H_y^* \frac{\partial M_x^*}{\partial y^*} - H_x^* \frac{\partial M_y^*}{\partial y^*} \right) \tag{6.74}$$

$$0 = -\frac{\partial p^*}{\partial y^*} + \frac{3\phi Mn_f}{4} \left(M_x^* G_H + H_x^* \frac{\partial M_y^*}{\partial x^*} - H_y^* \frac{\partial M_x^*}{\partial x^*} \right) \tag{6.75}$$

By solving the system of equations consisting of Eqs. (6.74), (6.75), and (6.65)–(6.69), this section aims to investigate the flow characteristics of ferrofluid planar Couette–Poiseuille flow under constant values of H_0, Re, and Mn_f. Specifically, the variations in flow velocity, particle spin, flow vorticity, magnetization in the non-magnetic field direction, flow rate, and shear stress acting on the upper plate are examined when P^* and G_H are altered.

This analysis considers both alignment and opposition of the magnetic field gradient direction with respect to the plate's driving velocity. To demonstrate the effect of the gradient magnetic field, these results are compared with those obtained in uniform magnetic fields, varying in intensity from minimum ($H_{y,\min} = H_0$) to maximum ($H_{y,\max} = H_0 + L \cdot \partial H_y / \partial x$) and average ($H_{y,av} = \left(H_{y,\min} + H_{y,\max}\right)/2$).

For this problem, the boundary conditions pertaining to the magnetic scalar potential are

$$\varphi_{m,0} = 0;\ \varphi^*_{m,h_0} = \varphi^*_{m,0} - \left(1 + G_H x^*\right) \tag{6.76}$$

The remaining boundary conditions remain identical to those described in Section 6.2.2.

The system of governing equations is solved when $P^* = 0$, $H_0 = 1.4835 \times 10^3$ A/m, $L_0 = 1$ mm, $Re = 10$, and G_H is equal to 1, 3, 5, 7, and 9, respectively. The remaining parameter values are detailed in Table 6.1.

6.3.2 Magnetic Field and Flow Velocity Distribution Within Ferrofluids

Under the influence of a constant gradient magnetic field, a component perpendicular to the external magnetic field appears in the magnetic field intensity within the ferrofluid as illustrated in Figure 6.20. Furthermore, the magnetization is inhomogeneous in the x-direction. To compare the influence of different parameters, the magnetization along the x-direction near the upper plate at the midpoint of the geometric model is chosen as the comparison metrics.

Unlike the results observed in a uniform magnetic field, the velocity of ferrofluids under the influence of a gradient magnetic field does not remain constant along the x-direction as shown in Figure 6.21. Instead, there are slight variations in the distribution of velocity across the cross-section. For flow driven by a positive pressure gradient, the maximum velocity increases as the cross-section approaches the outlet, with a corresponding decrease in velocity at other y-positions to maintain a constant flow rate.

6.3.3 Flow Characteristics of Ferrofluids

6.3.3.1 Distribution of Velocity and Vorticity

As shown in Figure 6.22, the velocity distribution of ferrofluid within the microchannel is depicted for two scenarios: When the direction of the magnetic field gradient is the same

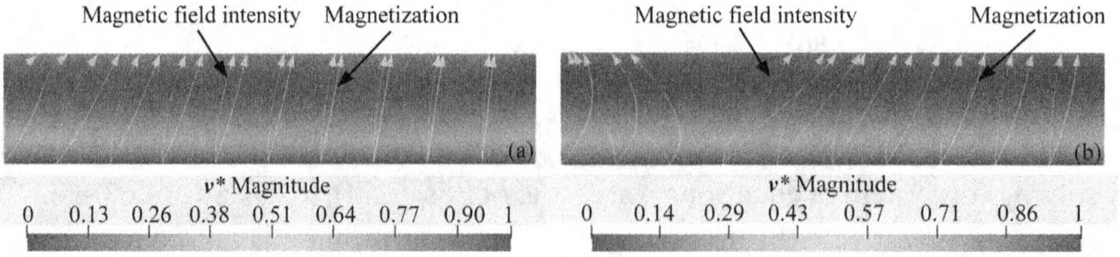

FIGURE 6.20 Streamlines of magnetic field intensity and magnetization under the conditions of $P^* = 2$ and (a) $G_H = 9$ or (b) $G_H = -9$. (Reprinted from Yang et al., 2022 with permission from Elsevier.)

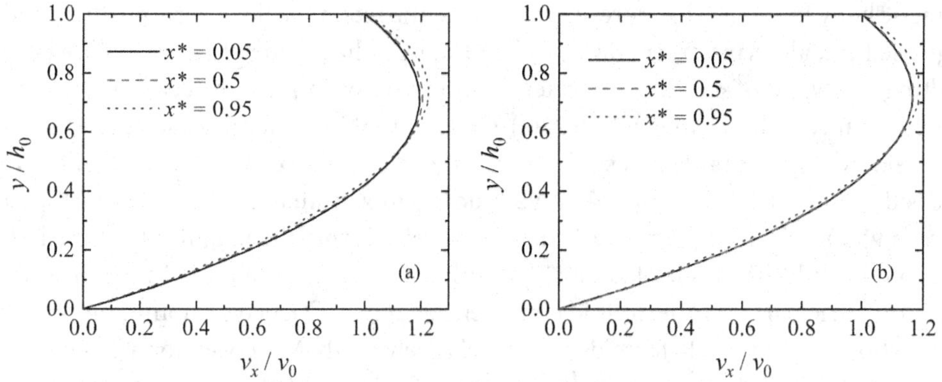

FIGURE 6.21 Comparison of flow velocity distributions at different longitudinal positions within the ferrofluid under a gradient magnetic field ($P^* = 2$). (a) $G_H = 9$ and (b) $G_H = -9$. (Reprinted from Yang et al., 2022 with permission from Elsevier.)

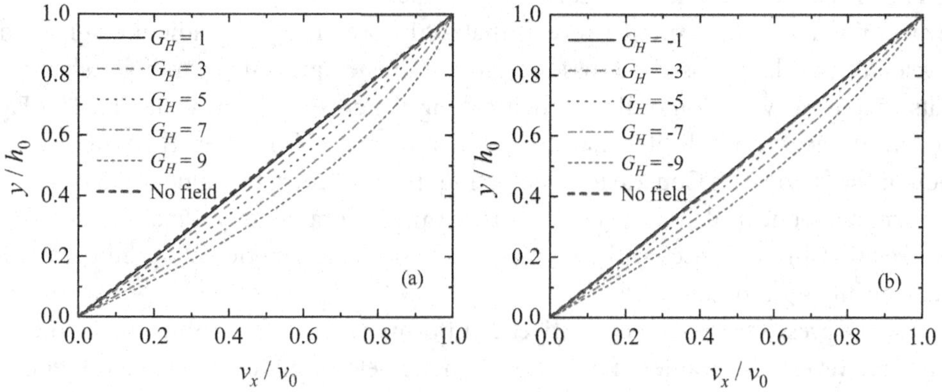

FIGURE 6.22 Velocity profiles of a ferrofluid Couette flow ($P^* = 0$) subject to (a) favourable and (b) unfavourable field gradients. (Reprinted from Yang et al., 2022 with permission from Elsevier.)

as and opposite to the plate driving velocity at $P^* = 0$. Notably, both gradient magnetic fields increase the velocity of the ferrofluid and promote its flow. However, in contrast to this observation, the conclusion drawn in Section 6.2.7 states that a uniform magnetic field does not alter the velocity distribution of a ferrofluid in a Couette flow. This distinction is the most pronounced when comparing the effects of gradient and uniform magnetic fields.

For Couette flows under the action of a gradient magnetic field, Eq. (6.74) becomes

$$\frac{\mathrm{d}^2 v_x^*}{\mathrm{d}y^{*2}} = -\frac{3\phi M n_f}{4}\left(M_y^* G_H + H_y^* \frac{\partial M_x^*}{\partial y^*} - H_x^* \frac{\partial M_y^*}{\partial y^*} \right) \tag{6.77}$$

Compared to the equation for a uniform magnetic field, the first term on the right-hand side of the equation incorporates a magnetic body force term represented by the magnetic field gradient, which is no longer zero. This term causes changes in the flow velocity distribution due to variations in the magnetic field, and the larger the gradient, the stronger the magnetic body force and the more significant its impact on the flow. For a positive gradient magnetic field, all terms except $\partial M_y^* / \partial y^*$ on the right-hand side of Eq. (6.77) are positive, resulting in $\mathrm{d}^2 v_x^* / \mathrm{d}y^{*2} < 0$. This indicates that the relationship curve between v_x^* and y^* has a convex shape, as shown in Figure 6.22(a). On the other hand, for a negative gradient magnetic field, when the gradient is sufficiently large, all terms except $\partial M_y^* / \partial y^*$ on the right-hand side of Eq. (6.77) become negative, leading to a similar convex relationship curve between v_x^* and y^*. A simple numerical analysis reveals that the magnitude of the last two terms on the right-hand side of Eq. (6.77) dominates over the first term, indicating that the anti-symmetric part of the ferrofluid stress tensor, $2(\omega_\mathrm{p} - \Omega)$, plays a dominant role, while the effect of magnetic body force density is relatively weak. Moreover, for a given gradient direction, the larger the gradient of the magnetic field, the more significant the increase in flow velocity. In Figure 6.22(a), the maximum increase in flow velocity is approximately 35%. When comparing the effects of magnetic field gradients in different directions, a positive gradient is found to have a more pronounced effect on increasing flow velocity compared to a negative gradient of the same magnitude.

Examining Figure 6.22 further reveals that in the presence of a gradient magnetic field, the velocity profile across the cross-section exhibits a quadratic parabolic shape. This results in a linear variation of the vorticity along the cross-section, as depicted in Figure 6.23. This observation indicates that the gradient magnetic field alters the vorticity distribution of the ferrofluid's Couette flow, deviating from the constant value of 0.5 observed in a uniform magnetic field. Moreover, within the range of gradients studied, the slope of this linear relationship increases with the magnitude of the magnetic field gradient, without exhibiting any signs of saturation.

When the pressure gradient is positive, a uniform magnetic field inhibits the flow of ferrofluid. However, in a gradient field, the magnetic field promotes the Couette–Poiseuille flow of ferrofluid once the magnitude of the magnetic field gradient exceeds a certain value as shown in Figure 6.24. The stronger the gradient value, the more pronounced this promotional effect becomes. These findings can also be corroborated by comparing the results of gradient magnetic field action with those of uniform magnetic field action in Figure 6.25.

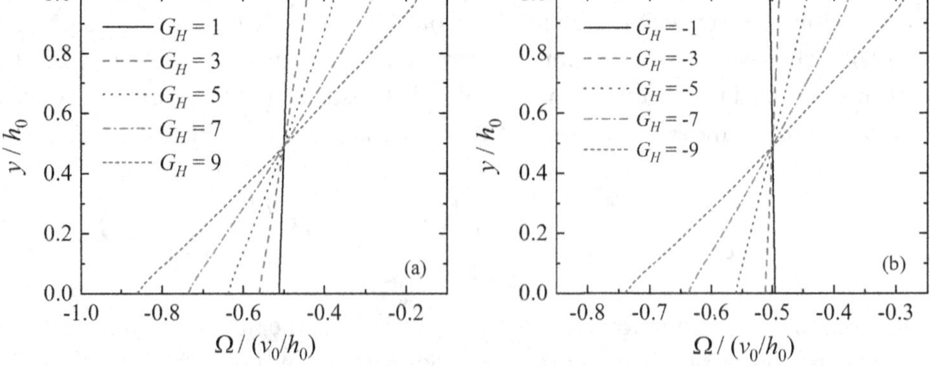

FIGURE 6.23 Vorticity profiles of ferrofluid Couette flow ($P^* = 0$) subject to (a) favourable and (b) unfavourable field gradients. (Reprinted from Yang et al., 2022 with permission from Elsevier.)

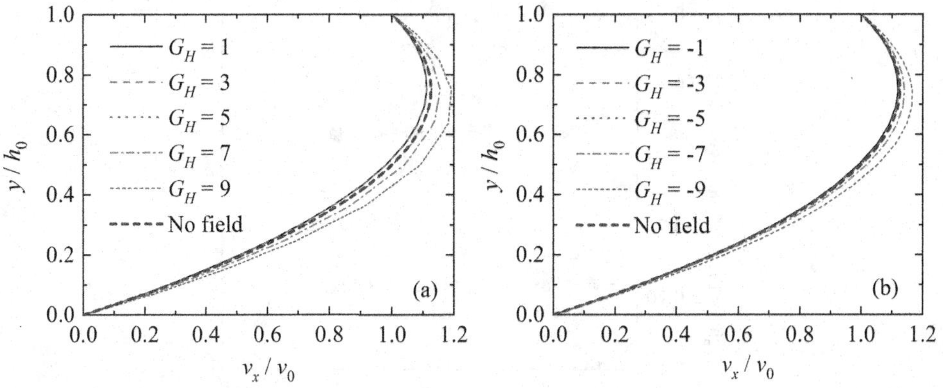

FIGURE 6.24 Velocity profiles of a ferrofluid Couette–Poiseuille flow ($P^* = 2$) subject to (a) positive and (b) negative gradient fields. (Reprinted from Yang et al., 2022 with permission from Elsevier.)

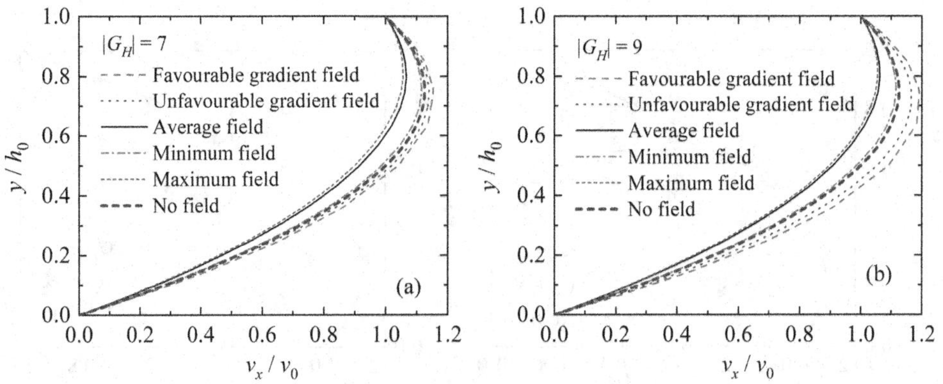

FIGURE 6.25 Flow velocity distribution of a ferrofluid in Couette–Poiseuille flows under the action of a gradient magnetic field at $P^* = 2$ compared to that under the action of a uniform magnetic field. (a) $|G_H| = 7$ and (b) $|G_H| = 9$. (Reprinted from Yang et al., 2022 with permission from Elsevier.)

When compared to a uniform average magnetic field, the maximum increase in flow velocity in a gradient magnetic field is approximately 19.4%. This effect is partly attributed to the magnetic body force and possibly partly due to the continuous rotation of particles as they advance with the fluid due to the inhomogeneity of the magnetic field. This rotation converts a portion of the magnetic field energy into mechanical energy, similar to the reduction in apparent viscosity of ferrofluid in an alternating magnetic field.

Upon examining Figure 6.24, it is evident that the position of the peak flow velocity within the velocity profile undergoes changes in response to variations in the magnetic field gradient. This is further illustrated in Figure 6.26, where a positive gradient magnetic field results in a decrease of approximately 7.7% in the position. This characteristic could potentially stem from the influence of magnetic body force, which is dependent on the magnetization. However, due to the actuation of the upper plate, the magnetization is no longer symmetrical with position.

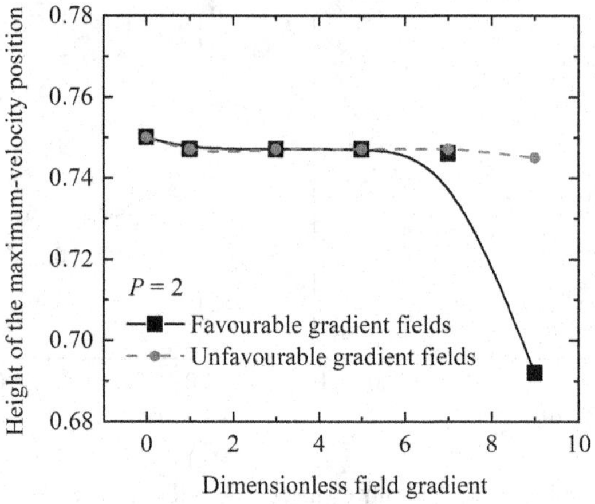

FIGURE 6.26 Effect of magnetic field gradient on the location of maximum flow velocity ($P^* = 2$). (Reprinted from Yang et al., 2022 with permission from Elsevier.)

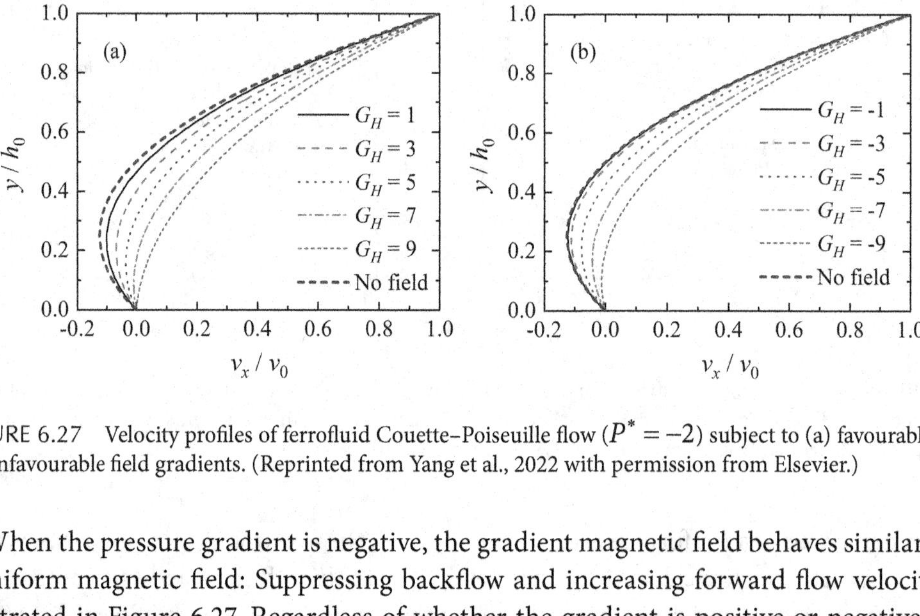

FIGURE 6.27 Velocity profiles of ferrofluid Couette–Poiseuille flow ($P^* = -2$) subject to (a) favourable and (b) unfavourable field gradients. (Reprinted from Yang et al., 2022 with permission from Elsevier.)

When the pressure gradient is negative, the gradient magnetic field behaves similarly to a uniform magnetic field: Suppressing backflow and increasing forward flow velocity as illustrated in Figure 6.27. Regardless of whether the gradient is positive or negative, the stronger the gradient, the more effective it is at suppressing backflow. Moreover, for the same gradient magnitude, the positive gradient exhibits a more pronounced impact. Compared to the uniform magnetic field, both positive and negative gradient magnetic fields outperform the uniform minimum, maximum, and average magnetic fields in terms of their effects as shown in Figure 6.28. Notably, when the gradient magnitude exceeds a certain threshold, backflow can be completely suppressed, such as in a positive gradient magnetic field with a gradient magnitude of 9. This eliminates the flow separation point that exists in negative pressure gradient flows.

As shown in Figure 6.29, the distribution of vorticity within the ferrofluid flow is presented for two cases: One with $P^* = 2$ and the other with $P^* = -2$, both under the influence

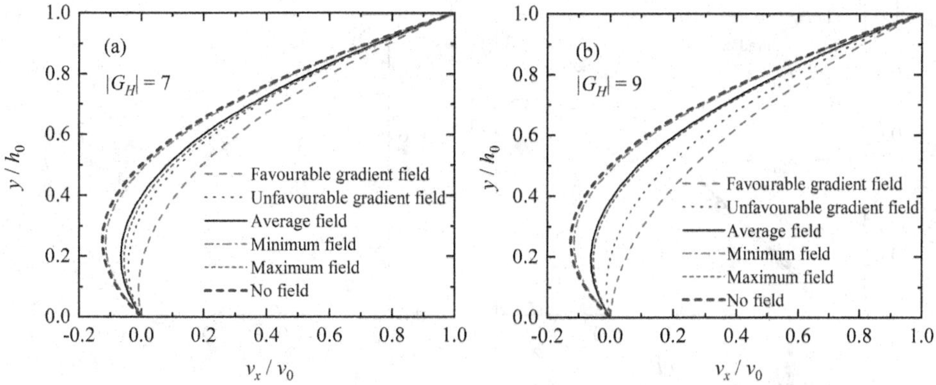

FIGURE 6.28 Comparison of flow velocity profiles between gradient and uniform magnetic fields ($P^* = -2$). (a) $|G_H| = 7$ and (b) $|G_H| = 9$. (Reprinted from Yang et al., 2022 with permission from Elsevier.)

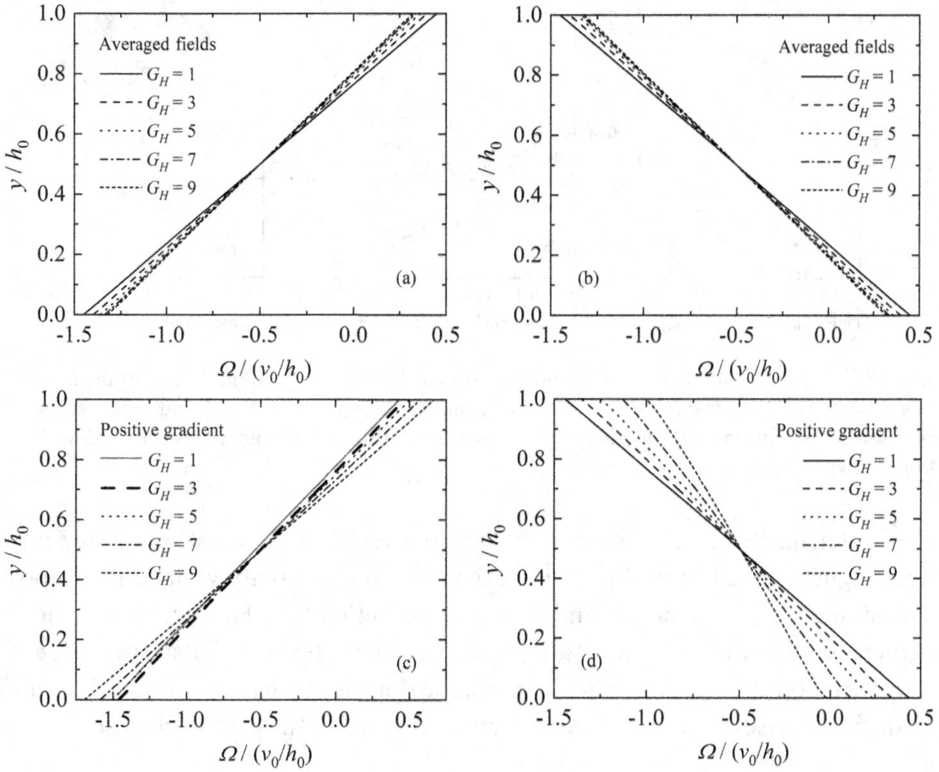

FIGURE 6.29 Distributions of vorticity in ferrofluid. (a) $P^* = 2$, Averaged fields; (b) $P^* = -2$, Averaged fields; (c) $P^* = 2$, Positive gradient fields; (d) $P^* = -2$, Positive gradient fields;

(Continued)

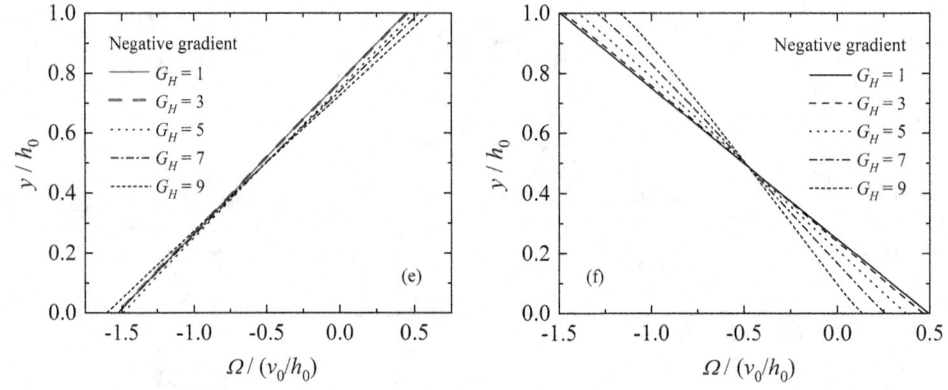

FIGURE 6.29 (Continued) (e) $P^* = 2$, Negative gradient fields; (f) $P^* = -2$, Negative gradient fields.

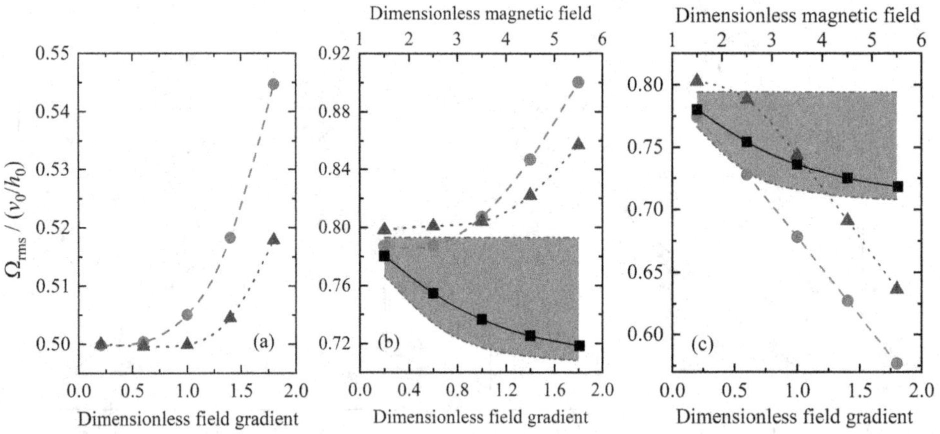

FIGURE 6.30 Variation of the root-mean-squared vorticity over the entire domain against the field gradient for different pressure gradients: (a) $P^* = 0$, (b) $P^* = 2$, and (c) $P^* = -2$. ─■─, ─●─, and ···▲··· represent averaged, favourable, and unfavourable gradient fields, respectively. (Reprinted from Yang et al., 2022 with permission from Elsevier.)

of a gradient magnetic field. A comparison is also made with the situation where a uniform average magnetic field is applied. Overall, the pattern of vorticity variation with position remains similar in both gradient and uniform magnetic fields. However, a notable difference emerges: When $P^* = 2$, the gradient magnetic field results in a steeper slope in the vorticity variation curve, whereas for $P^* = -2$, the slope is gentler compared to the uniform magnetic field. These observations are directly attributed to the previously discussed velocity distributions.

Figure 6.30 compares the variation of the root mean square vorticity, Ω_{rms}, with gradient magnitude under three different scenarios: $P^* = 0$, $P^* = 2$, and $P^* = -2$. The boundaries of the darkly shaded regions correspond to the cases of minimum and maximum uniform magnetic fields. It is evident that for non-negative pressure gradient, Ω_{rms} rises as the magnetic field gradient increases. However, for $P^* = -2$, Ω_{rms} decreases with the magnetic field gradient. These observations are direct consequences of the velocity distribution.

Specifically, for $P^* = -2$, larger magnetic field gradients result in reduced velocity gradients, leading to decreased flow vorticity. In contrast, positive magnetic field gradients produce larger vorticity variations, and both types of gradient magnetic fields cause Ω_{rms} to exceed the range defined by the minimum and maximum uniform magnetic field effects.

6.3.3.2 Distribution of $\mathbf{M}_x$

In pure Couette flows under a uniform magnetic field ($P^* = 0$), the value of M_x remains constant across a cross-section. However, when a gradient magnetic field is applied, M_x exhibits a linear variation with position, with higher values closer to the upper plate as depicted in Figure 6.31. A larger magnetic field gradient significantly alters the slope of these linear relationships, shifting the curves towards the direction of the higher M_x. This is a direct consequence of changes in magnetic field intensity induced by different gradients at the same location. The introduction of a pressure gradient in the Couette flow (resulting in Couette–Poiseuille flow) leads to a change in the sign of the slope of these linear relationships as illustrated in Figure 6.32.

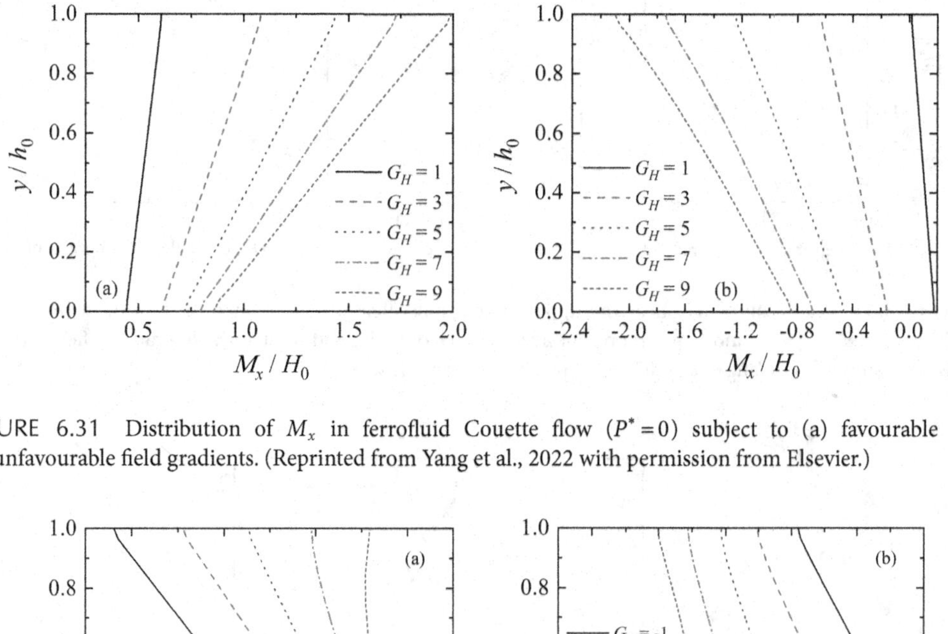

FIGURE 6.31 Distribution of M_x in ferrofluid Couette flow ($P^* = 0$) subject to (a) favourable and (b) unfavourable field gradients. (Reprinted from Yang et al., 2022 with permission from Elsevier.)

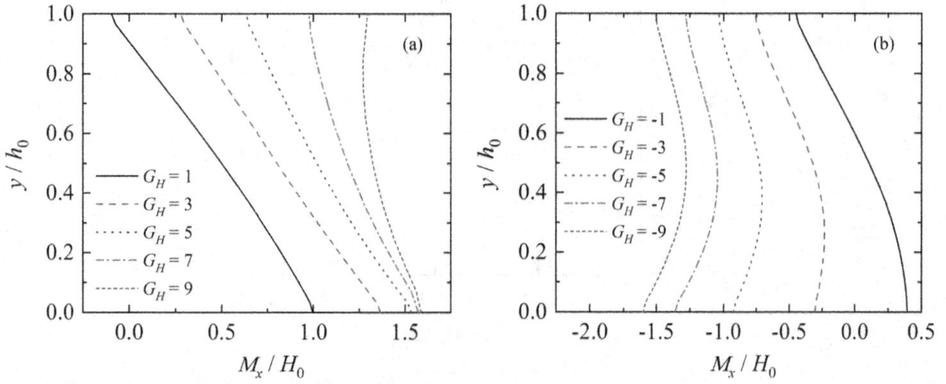

FIGURE 6.32 Distribution of M_x in ferrofluid Couette–Poiseuille flow ($P^* = 2$) subject to (a) favourable and (b) unfavourable field gradients. (Reprinted from Yang et al., 2022 with permission from Elsevier.)

When comparing the variation of M_x with magnetic field gradient near the upper plate in a ferrofluid under different pressure gradients, as shown in Figure 6.33, it is observed that both positive and negative gradient magnetic fields increase the magnitude of M_x. Moreover, all M_x results under gradient magnetic fields exceed the limits defined by the minimum and maximum magnetic fields in a uniform magnetic field. The larger the gradient magnitude, the more significant the exceedance becomes. A comparison of M_x distribution between gradient and uniform magnetic fields reveals that the M_x values induced by gradient magnetic fields far exceed the range observed in a uniform magnetic field as depicted in Figure 6.34. These observations suggest that magnetic field gradient can enhance the magnetization relaxation effect.

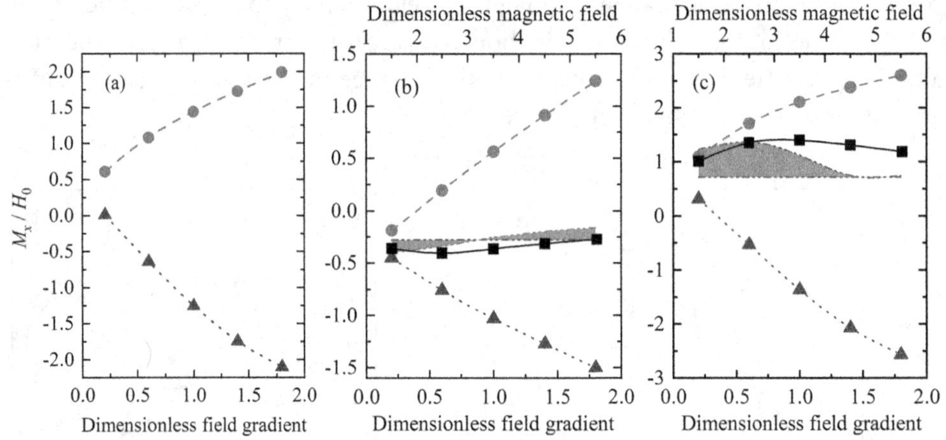

FIGURE 6.33 Variation of M_x near the upper plate against field gradient when (a) $P^* = 0$, (b) $P^* = 2$, and (c) $P^* = -2$. —■—, –●–, and ·▲· represent averaged, favourable, and unfavourable gradient fields, respectively. (Reprinted from Yang et al., 2022 with permission from Elsevier.)

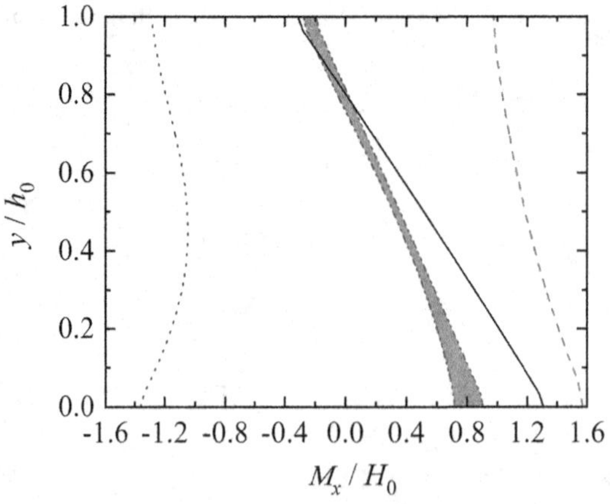

FIGURE 6.34 M_x profiles under the conditions of $P^* = 2$ and $G_H = 7$. - - - -, ······, and —— represent favourable gradient, unfavourable gradient, and averaged fields, respectively. (Reprinted from Yang et al., 2022 with permission from Elsevier.)

6.3.3.3 Distribution of Spin Velocity

In pure Couette flows under a uniform magnetic field ($P^* = 0$), the dimensionless particle spin velocity remains constant at −0.5. However, the application of a gradient magnetic field diminishes the local spin velocity of the ferrofluid in the Couette flow. As the gradient increases, the spin velocity gradually approaches zero. When the gradient becomes even stronger, the direction of spin reverses, as illustrated in Figure 6.35. In Eq. (6.67), the two terms on the right-hand side have opposite signs. When the magnetic field is weak, the particles rotate freely along with the local vorticity, with the same rotational direction as the vorticity. As the local magnetic field intensity increases due to the gradient, these two terms cancel each other out or the latter exceeds the former, ultimately causing the particle spin velocity to approach zero or increase in the opposite direction.

In Couette flows, the variation of spin velocity within the cross-section is relatively minimal for a given magnetic field gradient and only becomes significant when the gradient is relatively large. However, this pattern does not hold true for the Couette–Poiseuille flow of ferrofluids as shown in Figure 6.36. In this flow, the coupling of spin diffusion and vorticity leads to the emergence of a faint spin velocity boundary layer near the wall. Despite this, the spin velocity in Couette–Poiseuille flow maintains a linear dependence on position (excluding the boundary layer), with the slope of this linear relationship exhibiting significant variations with changes in the magnetic field gradient.

Upon further examination of Figure 6.36, it becomes evident that the distribution of spin velocity in a negative magnetic field gradient deviates significantly from that observed in a uniform average magnetic field. A notable contributor to this discrepancy is the significant deviation of M_x for the negative gradient from its counterpart in a uniform average magnetic field as depicted in Figure 6.34.

6.3.3.4 Variation of Flow Rate with Pressure Gradient and Magnetic Field Gradient

Since a uniform magnetic field does not alter the velocity profile of a ferrofluid in Couette flows, the flow rate remains constant in such a field. However, in a gradient magnetic field,

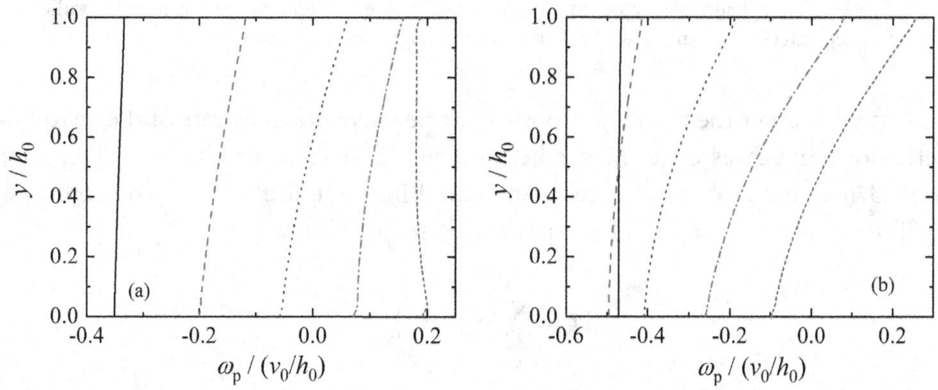

FIGURE 6.35 Distribution of spin velocity in ferrofluid Couette flow ($P^* = 0$) subject to (a) positive and (b) negative gradient fields. ——, - - - -, ·······, —·—·—, and ······ represent $|G_H| = 1$, $|G_H| = 3$, $|G_H| = 5$, $|G_H| = 7$, and $|G_H| = 9$, respectively. (Reprinted from Yang et al., 2022 with permission from Elsevier.)

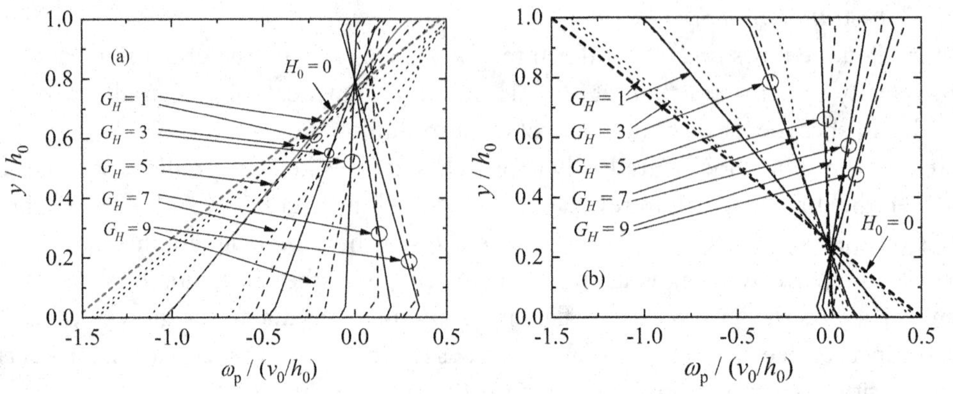

FIGURE 6.36 Distribution of spin velocity in ferrofluid Couette–Poiseuille flows subject to gradient magnetic fields for the cases of (a) $P^* = 2$ and (b) $P^* = -2$, Here, solid lines are for averaged fields, dashed lines for favourable field gradients, and dotted lines for unfavourable field gradients. (Reprinted from Yang et al., 2022 with permission from Elsevier.)

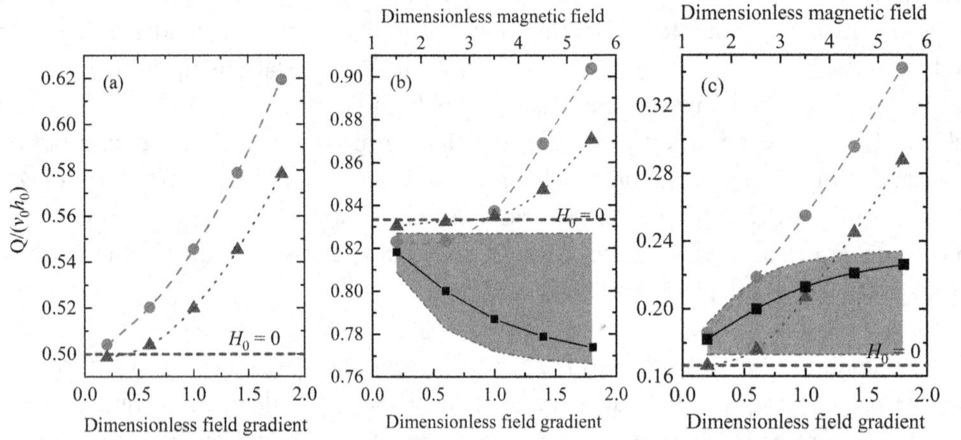

FIGURE 6.37 Effect of field gradient on the volumetric flow rate for the cases of (a) $P^* = 0$, (b) $P^* = 2$, and (c) $P^* = -2$. —■—, - ● -, and · ▲ · represent averaged, favourable gradient, and unfavourable gradient fields, respectively. (Reprinted from Yang et al., 2022 with permission from Elsevier.)

regardless of whether the gradient is positive or negative, the flow rate of the ferrofluid in Couette flows increases once the gradient exceeds a certain threshold, as illustrated in Figure 6.37(a). The maximum increases observed in the figure are 24% and 16%, respectively. The method used to calculate these flow rates is as follows:

$$Q = \sum_{k=1}^{CN} v_{x,k} \left(\Delta y\right)_k$$

Moreover, when the gradient magnitude of the magnetic field is the same, ferrofluids in a positive gradient magnetic field exhibit a greater flow rate compared to those in a negative gradient magnetic field.

For the Couette–Poiseuille flow of ferrofluids, a uniform magnetic field acts as a hindrance to the flow, reducing the flow rate when the pressure gradient is positive, as depicted in Figure 6.37(b). However, in a gradient magnetic field, the flow rate exhibits an increase once the gradient surpasses a certain threshold. The larger the gradient, the greater the flow rate becomes. In contrast, in negative pressure gradient-driven flow, both uniform and gradient magnetic fields promote the forward flow of ferrofluids as depicted in Figure 6.37(c). Nevertheless, when the gradient exceeds a certain value, the flow rate induced by the gradient magnetic field surpasses the limits set by the uniform magnetic fields. Positive gradient magnetic fields exhibit a more pronounced effect than negative ones. Within the computed range, the flow rate can increase up to twice its original value. In conclusion, gradient magnetic fields enhance the forward flow of ferrofluids within Couette–Poiseuille flows.

6.3.3.5 Stress Exerted on the Upper Plate

The shear stress acting on the upper plate has two components: Viscous stress from the ferrofluid and magnetic stress induced by abrupt changes in magnetization. Figure 6.38 illustrates the variations of magnetic stress, viscous stress, and total shear stress with respect to magnetic field gradient. In the case of Couette flow under a uniform magnetic field, only viscous drag is present as the uniform field does not alter the velocity distribution. However, in a gradient magnetic field, for all three pressure gradients, both magnetic stress and total stress increase with the increasing gradient magnitude, while viscous drag decreases. This is due to the fact that the rate of increase in magnetic stress outweighs the rate of decrease in viscous drag. Taking the example of a positive magnetic field gradient at $P^* = 0$, the viscous drag decreases by 60% compared to an ordinary fluid, while the magnetic stress increases by 8.6 times, ultimately leading to an increase in total stress by approximately 52%. Among all scenarios, the maximum increase in magnetic stress is observed to be 22 times, which corresponds to the case of $P^* = 0$ and $G_H = 7$.

In general, a positive magnetic field gradient tends to induce greater changes in stress compared to negative magnetic field gradient, except in the case of total stress when $P^* = -2$. In Couette–Poiseuille flows driven by pressure gradients, the shear stress and magnetic stress in gradient magnetic fields exceed the corresponding stress limits defined by the minimum and maximum uniform magnetic fields, respectively. However, the total stress remains bounded by these limits. Overall, gradient magnetic fields can significantly increase magnetic stress and decrease viscous stress compared to the effects of a uniform magnetic field. The balance between these two factors ultimately results in an increase in shear stress acting on the upper plate.

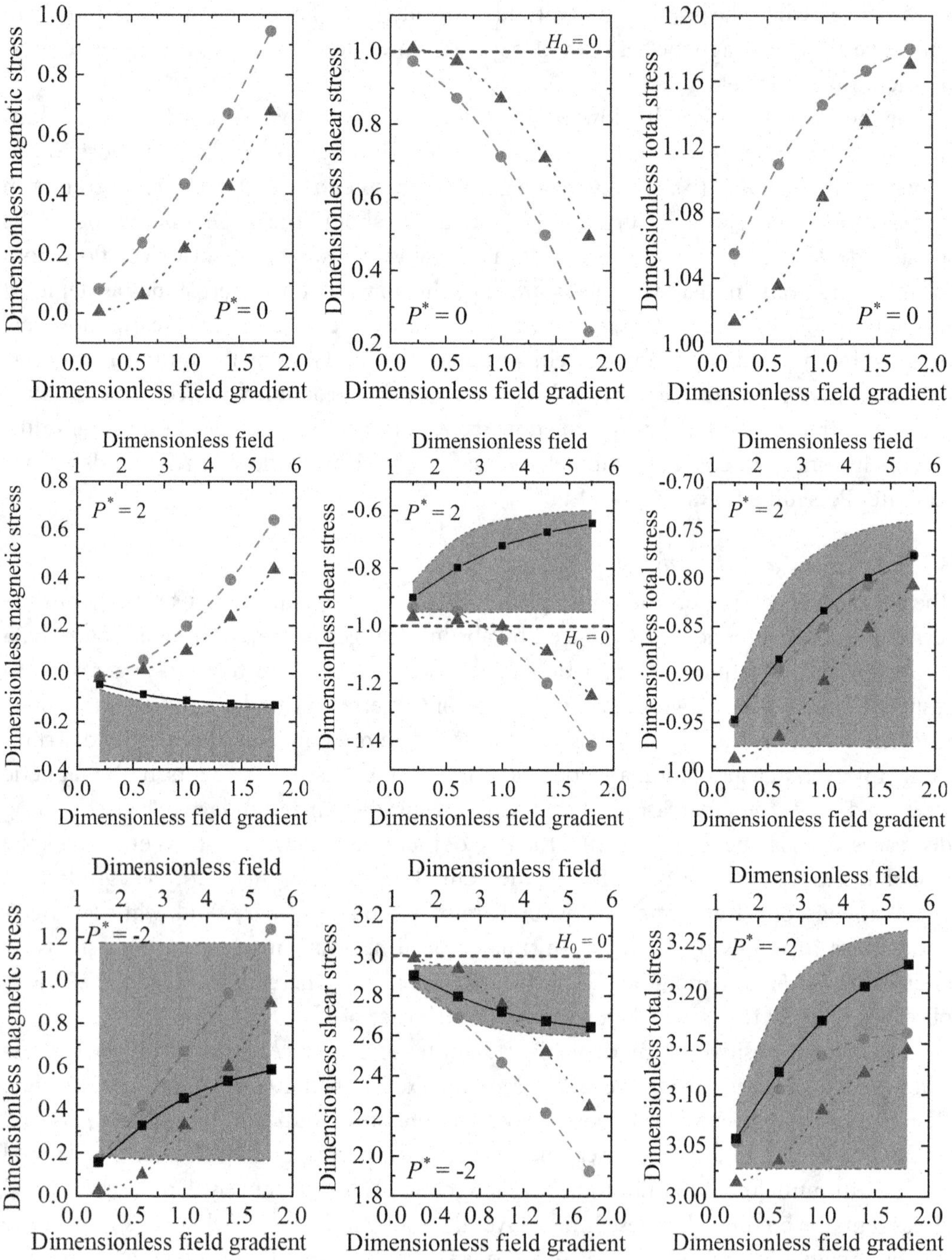

FIGURE 6.38 Effect of field gradient on the tangential stress acting on the moving plate (from top to bottom: $P^* = 0$, $P^* = 2$, and $P^* = -2$, from left to right: magnetic stress, shear stress, and total stress. —■—, - ● -, and - ▲ - represent the averaged field, favourable gradient, and unfavourable gradient fields, respectively). (Reprinted from Yang et al., 2022 with permission from Elsevier.)

6.4 FLOW CHARACTERISTICS IN TIME-VARYING MAGNETIC FIELDS

6.4.1 Ferrofluid Couette Flow in Weak Rotating Magnetic Fields

In the planar Couette flow depicted in Figure 6.1, when an external magnetic field rotates within the (x, y) plane, the magnetization of the ferrofluid rotates at the same rate within that plane. However, magnetization relaxation effect results in a phase lag ϑ between the magnetization and the magnetic field intensity. Under these conditions, the magnetic field intensity and magnetization can be represented as follows:

$$\boldsymbol{H} = \left(H_0 \cos wt, H_0 \sin wt, 0\right) \tag{6.78}$$

$$\boldsymbol{M} = \left(M \cos\left(wt - \vartheta\right), M \sin\left(wt - \vartheta\right), 0\right) \tag{6.79}$$

and

$$\boldsymbol{M}_0 = \left(M_0 \cos wt, M_0 \sin wt, 0\right) \tag{6.80}$$

where $M_0 = M_S L(\alpha) = \chi_0 H_0$.

Under the weak-field approximation, specifically when $\alpha \ll 1$, there exists a linear relationship between the equilibrium magnetization and the magnetic field intensity. Consequently, the magnetization equations can be linearized to Eq. (4.28), which can then be solved to determine M and ϑ. By substituting Eqs. (6.78) and (6.79) into the respective terms of Eq. (4.28), one obtains

$$\frac{\mathrm{D}\boldsymbol{M}}{\mathrm{D}t} = -Mw \sin\left(wt - \vartheta\right)\boldsymbol{e}_x + Mw \cos\left(wt - \vartheta\right)\boldsymbol{e}_y$$

$$\begin{aligned}
\boldsymbol{\Omega} \times \boldsymbol{M} &= \Omega \boldsymbol{e}_z \times \left[M \cos\left(wt - \vartheta\right)\boldsymbol{e}_x + M \sin\left(wt - \vartheta\right)\boldsymbol{e}_y\right] \\
&= -\Omega M\left[\boldsymbol{e}_x \sin\left(wt - \vartheta\right) - \boldsymbol{e}_y \cos\left(wt - \vartheta\right)\right]
\end{aligned}$$

$$\begin{aligned}
\boldsymbol{H} \cdot \left(\boldsymbol{M} - \boldsymbol{M}_0\right) &= H_0\left[\boldsymbol{e}_x \cos\left(wt\right) + \boldsymbol{e}_y \sin\left(wt\right)\right] \cdot \begin{Bmatrix} \left[M \cos\left(wt - \vartheta\right) - M_0 \cos\left(wt\right)\right]\boldsymbol{e}_x \\ -\left[M \sin\left(wt - \vartheta\right) - M_0 \sin\left(wt\right)\right]\boldsymbol{e}_y \end{Bmatrix} \\
&= H_0 \cos\left(wt\right)\left[M \cos\left(wt - \vartheta\right) - M_0 \cos\left(wt\right)\right] \\
&\quad + H_0 \sin\left(wt\right)\left[M \sin\left(wt - \vartheta\right) - M_0 \sin\left(wt\right)\right] \\
&= H_0 M\left[\cos wt \cos\left(wt - \vartheta\right) + \sin wt \sin\left(wt - \vartheta\right)\right] \\
&\quad - H_0 M_0\left[\cos^2\left(wt\right) + \sin^2\left(wt\right)\right] \\
&= H_0\left(M \cos\vartheta - M_0\right)
\end{aligned}$$

$$\boldsymbol{M} \times \boldsymbol{H} = \left[MH_0 \cos\left(wt - \vartheta\right)\sin\left(wt\right) - MH_0 \sin\left(wt - \vartheta\right)\cos\left(wt\right)\right]\boldsymbol{e}_z = MH_0 \sin\vartheta \boldsymbol{e}_z$$

$$\boldsymbol{H} \times \left(\boldsymbol{M} \times \boldsymbol{H}\right) = MH_0^2 \sin\vartheta\left[\boldsymbol{e}_x \sin\left(wt\right) - \boldsymbol{e}_y \cos\left(wt\right)\right]$$

Substituting the above expressions into Eq. (4.28) gives the component equation in the $\boldsymbol{e}_x$ direction

$$-Mw\sin\left(wt-\vartheta\right)=-\Omega M\sin\left(wt-\vartheta\right)-\frac{H_0^2\left(M\cos\vartheta-M_0\right)\cos\left(wt\right)}{\tau_{\parallel}H^2}-\frac{MH_0^2\sin\vartheta\sin\left(wt\right)}{\tau_{\perp}H^2} \tag{6.81}$$

Since $H^2=H_0^2$, Eq. (6.81) can be simplified to

$$M\left[-\left(w-\Omega\right)\sin\left(wt-\vartheta\right)+\frac{\cos\vartheta\cos\left(wt\right)}{\tau_{\parallel}}+\frac{\sin\vartheta\sin\left(wt\right)}{\tau_{\perp}}\right]=\frac{M_0\cos\left(wt\right)}{\tau_{\parallel}} \tag{6.82}$$

Similarly, the component equation in the $\boldsymbol{e}_j$ direction is

$$\begin{aligned}Mw\cos\left(wt-\vartheta\right)=\Omega M\cos\left(wt-\vartheta\right)-\frac{H_0^2\left(M\cos\vartheta-M_0\right)\sin\left(wt\right)}{\tau_{\parallel}H^2}\\+\frac{MH_0^2\sin\vartheta\cos\left(wt\right)}{\tau_{\perp}H^2}\end{aligned} \tag{6.83}$$

and its simplified form is

$$M\left[\left(w-\Omega\right)\cos\left(wt-\vartheta\right)+\frac{\cos\vartheta\sin\left(wt\right)}{\tau_{\parallel}}-\frac{\sin\vartheta\cos\left(wt\right)}{\tau_{\perp}}\right]=\frac{M_0\sin\left(wt\right)}{\tau_{\parallel}} \tag{6.84}$$

Combining Eq. (6.82) and (6.84) yields

$$\begin{aligned}&\frac{M_0\cos\left(wt\right)}{\tau_{\parallel}}\left[\left(w-\Omega\right)\cos\left(wt-\vartheta\right)+\frac{\cos\vartheta\sin\left(wt\right)}{\tau_{\parallel}}-\frac{\sin\vartheta\cos\left(wt\right)}{\tau_{\perp}}\right]\\&=\frac{M_0\sin\left(wt\right)}{\tau_{\parallel}}\left[-\left(w-\Omega\right)\sin\left(wt-\vartheta\right)+\frac{\cos\vartheta\cos\left(wt\right)}{\tau_{\parallel}}+\frac{\sin\vartheta\sin\left(wt\right)}{\tau_{\perp}}\right]\end{aligned} \tag{6.85}$$

which can be simplified to

$$\left(w-\Omega\right)\cos\vartheta=\frac{\sin\vartheta}{\tau_{\perp}} \tag{6.86}$$

Eventually, the equation satisfied by ϑ is obtained

$$\tan\vartheta=\tau_{\perp}\left(w-\Omega\right) \tag{6.87}$$

After substituting Eq. (6.87) into Eq. (6.84) and rearranging, one gets

$$M=\frac{\dfrac{M_0\cos(wt)}{\tau_{\|}}}{-\dfrac{\tan\vartheta}{\tau_{\perp}}\sin(wt-\vartheta)+\dfrac{\cos\vartheta\cos(wt)}{\tau_{\|}}+\dfrac{\sin\vartheta\sin(wt)}{\tau_{\perp}}}$$

$$=\frac{\dfrac{M_0\cos(wt)\cos\vartheta}{\tau_{\|}}}{-\dfrac{\sin\vartheta}{\tau_{\perp}}\left[\sin(wt-\vartheta)-\cos\vartheta\sin(wt)\right]+\dfrac{\cos^2\vartheta\cos(wt)}{\tau_{\|}}}$$

$$=\frac{\dfrac{M_0\cos(wt)\cos\vartheta}{\tau_{\|}}}{-\dfrac{\sin^2\vartheta}{\tau_{\perp}}\cos(wt)+\dfrac{\cos^2\vartheta\cos(wt)}{\tau_{\|}}}=\frac{\dfrac{M_0\cos\vartheta}{\tau_{\|}}}{\dfrac{\sin^2\vartheta}{\tau_{\perp}}+\dfrac{\cos^2\vartheta}{\tau_{\|}}}$$

Finally, one obtains

$$M=\frac{M_0\cos\vartheta}{\dfrac{\tau_{\|}}{\tau_{\perp}}\sin^2\vartheta+\cos^2\vartheta} \tag{6.88}$$

For the phenomenological magnetization equation I (4.28), when $\alpha \ll 1$, one has

$$\frac{\tau_{\|}}{\tau_{\perp}}=\frac{2}{2+\alpha L(\alpha)}\approx\frac{2}{2+\alpha^2/3}\approx 1$$

Regarding the microscopic magnetization equation, it can be derived from Eq. (4.55) that

$$\frac{\tau_{\|}}{\tau_{\perp}}=\frac{\dfrac{d\ln L(\alpha)}{d\ln\alpha}}{\dfrac{2L(\alpha)}{\alpha-L(\alpha)}}=\frac{\dfrac{\alpha L'(\alpha)}{L(\alpha)}}{\dfrac{2L(\alpha)}{\alpha-L(\alpha)}}=\frac{\alpha L'(\alpha)\left[\alpha-L(\alpha)\right]}{2L^2(\alpha)}\approx\frac{\alpha\dfrac{1}{3}\left[\alpha-\dfrac{\alpha}{3}\right]}{\dfrac{2\alpha^2}{9}}=1$$

Therefore, under weak field conditions, Eq. (6.88) is simplified to

$$M=M_0\cos\vartheta \tag{6.89}$$

Equation (6.89) represents the expression for magnetization of ferrofluid Couette–Poiseuille flows in a weak rotating magnetic field.

Substituting Eqs. (6.78) and (6.79) into the stress tensor expression (2.224) for ferrofluids, one obtains its non-diagonal components

$$\begin{aligned} T_{xy} &= \eta\left(\frac{\partial v_x}{\partial y}+\frac{\partial v_y}{\partial x}\right)+\frac{\mu_0}{2}MH_0\left[\cos(wt-\vartheta)\sin(wt)+\sin(wt-\vartheta)\cos(wt)\right] \\ &\quad +\mu_0 H_0^2\sin(wt)\cos(wt) \\ &= \eta\left(\frac{\partial v_x}{\partial y}+\frac{\partial v_y}{\partial x}\right)+\frac{\mu_0}{2}M_0H_0\cos\vartheta\sin(2wt-\vartheta)+\frac{\mu_0}{2}H_0^2\sin(2wt) \end{aligned} \tag{6.90}$$

When averaging the expression given in Eq. (6.90) over a magnetic field variation period of $2\pi/w$, the last two terms on the right-hand side of the equation vanish, thus indicating that a rotating magnetic field does not induce any flow within the ferrofluid.

For a stationary ferrofluid, substituting Eqs. (6.78), (6.79), (6.87), and (6.89) into Eq. (2.217) and setting $\Omega=0$, one obtains

$$\begin{aligned} \boldsymbol{\omega}_{\mathrm{p}} &= \frac{\mu_0}{6\eta\phi}\frac{1}{2}M_0H_0\sin(2\vartheta)\boldsymbol{e}_z \\ &= \frac{\mu_0}{6\eta\phi}\frac{1}{2}M_0H_0\frac{2\tan\vartheta}{\tan^2\vartheta+1}\boldsymbol{e}_z \\ &= \frac{\mu_0}{6\eta\phi}M_0H_0\frac{\tau_\perp(w-\Omega)}{\left[\tau_\perp(w-\Omega)\right]^2+1}\boldsymbol{e}_z \\ &= \frac{\mu_0}{6\eta\phi}M_0H_0\frac{\tau_\perp w}{(\tau_\perp w)^2+1}\boldsymbol{e}_z \end{aligned} \tag{6.91}$$

Equation (6.91) demonstrates that, even when a ferrofluid is stationary, the magnetic particles within it will rotate at this angular velocity under the influence of a rotating magnetic field. During this process, each rotating particle drags its surrounding carrier liquid into rotation, forming microscopic vortices centred around the rotating magnetic particles. The size of these vortices is limited to a scale of approximately $\phi^{-1/3}d_{\mathrm{p}}$. If a ferrofluid region with a volume of V is divided into cubes centred on each particle's centre, the side length of each cube would be $(V/N)^{1/3}$ and the width of the carrier liquid region between adjacent particles is equal to

$$(V/N)^{1/3}-d_{\mathrm{p}}=\left(\frac{\frac{NV_{\mathrm{d}}}{\phi}}{N}\right)^{\frac{1}{3}}-d_{\mathrm{p}}=d_{\mathrm{p}}\left[\left(\frac{\pi}{6\phi}\right)^{\frac{1}{3}}-1\right]=d_{\mathrm{p}}\phi^{-\frac{1}{3}}\left[\left(\frac{\pi}{6}\right)^{\frac{1}{3}}-\phi^{\frac{1}{3}}\right]\sim d_{\mathrm{p}}\phi^{-\frac{1}{3}} \tag{6.92}$$

Hence, the scale of micro-eddies will not exceed the dimensions represented by Eq. (6.92). However, these eddy movements do not induce macroscopic fluid flow, as the vortex motions induced by adjacent particles cancel each other out. Consequently, these physically minute eddies do not generate macroscopic vorticity, analogous to how adjacent molecular currents neutralize each other within a magnetic medium, as described in the molecular current model of magnetization in electromagnetics.

As shown in Eqs. (6.87) and (6.89), when the vorticity in the flow field is constant, the spin angular momentum $I\boldsymbol{\omega}_\mathrm{p}$ and magnetization $\boldsymbol{M}$ are uniform within the ferrofluid. However, this uniformity is disrupted at the interface between the ferrofluid and the non-magnetic solid wall, where a sudden change in magnetization causes a tangential magnetic surface force. This force also provides a coupling effect between the rotating magnetic field and the fluid flow. For example, in the Couette flow illustrated in Figure 6.1, the tangential force density acting on the upper plate is determined according to Eq. (2.229)

$$f_x = \eta \frac{\partial v_x}{\partial y} + \frac{\mu_0}{2}\left(M_x H_y - M_y H_x\right)$$

By substituting Eqs. (6.78), (6.79), (6.87), and (6.89) into it, one obtains

$$\begin{aligned} f_x &= 2\eta\Omega + \frac{\mu_0}{2} M H_0 \left[\cos(wt-\vartheta)\sin(wt) - \sin(wt-\vartheta)\cos(wt)\right] \\ &= 2\eta\Omega + \frac{\mu_0}{2} M_0 H_0 \cos\vartheta \sin\vartheta \\ &= 2\eta\Omega + \frac{\mu_0}{2} M_0 H_0 \frac{\tan\vartheta}{\tan^2\vartheta + 1} \\ &= 2\eta\Omega + \frac{\mu_0}{2} M_0 H_0 \frac{\tau_\perp (w-\Omega)}{\left[\tau_\perp (w-\Omega)\right]^2 + 1} \end{aligned} \tag{6.93}$$

If the ferrofluid has a free surface at the location of the upper plate, where $f_x = 0$ (when $x = h_0$), the rotating magnetic field will induce a flow with a constant vorticity, whose magnitude is given by

$$\Omega = -\frac{\mu_0}{4\eta} M_0 H_0 \frac{\tau_\perp (w-\Omega)}{\left[\tau_\perp (w-\Omega)\right]^2 + 1} \approx -\frac{\mu_0}{4\eta} M_0 H_0 \frac{w\tau_\perp}{(\tau_\perp w)^2 + 1} \tag{6.94}$$

Due to the condition $\alpha \ll 1$, it follows that $\frac{\Omega}{w} \ll \frac{1}{4}\phi\alpha^2$. Consequently, in obtaining the first equality of Eq. (6.94), the Ω term on the right-hand side can be neglected. Using the relationship $\Omega = \frac{1}{2}\frac{\partial v_x}{\partial y}$, one obtains a linear flow pattern in a planar Couette flow induced by a rotating magnetic field, given by $v_x = 2\Omega y$. If the upper plate is fixed, resulting in $\Omega = 0$ (when $y = h_0$), then the upper plate will experience a tangential force

$$f_x = \frac{\mu_0}{2} M_0 H_0 \frac{w\tau_\perp}{(\tau_\perp w)^2 + 1} \tag{6.95}$$

To maintain the stability of the upper plate, it is necessary to apply tangential forces that are equal to Eq. (6.95) in magnitude but opposite in direction.

6.4.2 Flow Characteristics under the Influence of Alternating and Rotating Magnetic Fields When Disregarding the Effects of Spin Viscosity

This section delves into the planar Couette–Poiseuille flow of ferrofluids under the influence of a uniformly sinusoidal time-varying magnetic field. For the flow depicted in Figure 6.1, an assumption is made that the velocity of the ferrofluid is along the x-direction, while the particles' spin velocity is aligned with the z-direction. They can be represented as

$$\boldsymbol{v} = v_x\left(y\right)\boldsymbol{e}_x \tag{6.96}$$

$$\boldsymbol{\omega}_{\mathrm{p}} = \omega_{\mathrm{p},z}\left(y\right)\boldsymbol{e}_z \tag{6.97}$$

Three scenarios are considered: Uniform time-varying magnetic fields in the x-direction, uniform time-varying magnetic fields in the y-direction, and rotating magnetic fields. It is assumed that the frequency of the magnetic fields is significantly faster than the macroscopic flow of the ferrofluid. Due to the viscous nature of the fluid, it exhibits negligible response to second-order or higher harmonic magnetic field forces and torques. Consequently, Eqs. (6.96) and (6.97), respectively, represent the time-averaged flow velocity and time-averaged spin velocity.

6.4.2.1 Representation of Sinusoidal Time-varying Magnetic Fields

A sinusoidal time-varying magnetic field is described using a complex representation; for example, the x-component of the magnetic field intensity, H_x, can be expressed as

$$H_x = H_{0x}\cos\left(wt+\phi_x\right) = \mathrm{Re}\left(H_{0x}e^{\mathrm{i}\phi_x}e^{\mathrm{i}wt}\right)$$

Letting

$$\hat{H}_x = H_{0x}e^{\mathrm{i}\phi_x}$$

where i is the imaginary unit, then H_x can be expressed in its complex form

$$H_x = \mathrm{Re}\left(\hat{H}_x e^{\mathrm{i}wt}\right)$$

After expressing the y-component of the magnetic field intensity in a similar manner, the magnetic field intensity vector $\boldsymbol{H}$ can be represented as

$$\boldsymbol{H} = \mathrm{Re}\left[\left(\hat{H}_x\boldsymbol{e}_x + \hat{H}_y\boldsymbol{e}_y\right)e^{\mathrm{i}wt}\right] \tag{6.98}$$

Assuming a uniform distribution of the x-component of magnetic field intensity and the y-component of magnetic induction in space, then $\boldsymbol{H}$ and $\boldsymbol{B}$ can be represented accordingly (Zahn and Greer, 1995)

$$\boldsymbol{H} = \mathrm{Re}\left\{\left[\hat{H}_x \boldsymbol{e}_x + \hat{H}_y(y)\boldsymbol{e}_y\right]e^{iwt}\right\} \tag{6.99}$$

$$\boldsymbol{B} = \mathrm{Re}\left\{\left[\hat{B}_x(y)\boldsymbol{e}_x + \hat{B}_y \boldsymbol{e}_y\right]e^{iwt}\right\} \tag{6.100}$$

Based on the representation given by Eq. (6.98), the spatially uniform time-varying magnetic field in the x-direction can be expressed as

$$\hat{H}_x = H_0, \hat{B}_y = 0 \tag{6.101}$$

and that in the y-direction can be expressed as

$$\hat{H}_x = 0, \hat{B}_y = \mu_0 H_0 \tag{6.102}$$

the rotating magnetic field can be represented as

$$\hat{H}_x = \mathrm{i}H_0, \hat{B}_y = \mu_0 H_0 \tag{6.103}$$

6.4.2.2 Solving the Magnetization Equation

An assumption is made that the magnetization of ferrofluids in a time-varying magnetic field can be represented as

$$\boldsymbol{M} = \mathrm{Re}\left\{\left[\hat{M}_x(y)\boldsymbol{e}_x + \hat{M}_y(y)\boldsymbol{e}_y\right]e^{iwt}\right\} \tag{6.104}$$

For the phenomenological magnetization equation I (4.17), assuming that the magnetic field intensity is much smaller than the saturation magnetic field intensity of the ferrofluid, and considering the magnetization susceptibility χ_0 to be constant, the equilibrium magnetization can be represented linearly as

$$\boldsymbol{M}_0 = \chi_0 \boldsymbol{H}$$

Substituting it into Eq. (4.17) and assuming the ferrofluid is incompressible, one gets

$$\frac{\mathrm{D}\boldsymbol{M}}{\mathrm{D}t} = \boldsymbol{\omega}_\mathrm{p} \times \boldsymbol{M} + \frac{1}{\tau}\left(\chi_0 \boldsymbol{H} - \boldsymbol{M}\right) \tag{6.105}$$

Here

$$\frac{\mathrm{D}\boldsymbol{M}}{\mathrm{D}t} = \frac{\partial \boldsymbol{M}}{\partial t} + (\boldsymbol{v}\cdot\nabla)\boldsymbol{M}$$

and

$$(\boldsymbol{v}\cdot\nabla)\boldsymbol{M}=v_x\frac{\partial\boldsymbol{M}}{\partial x}+v_y\frac{\partial\boldsymbol{M}}{\partial y}$$

Assuming that $\boldsymbol{M}$ varies solely in the y-direction, it follows from Eq. (6.96) that $(\boldsymbol{v}\cdot\nabla)\boldsymbol{M}=\boldsymbol{0}$, thereby Eq. (6.105) can be simplified to

$$\frac{\partial\boldsymbol{M}}{\partial t}=\boldsymbol{\omega}_{\mathrm{p}}\times\boldsymbol{M}+\frac{1}{\tau}(\chi_0\boldsymbol{H}-\boldsymbol{M})\tag{6.106}$$

Rewriting Eq. (6.106) into its component form,

$$\frac{\partial M_x}{\partial t}=-\omega_{\mathrm{p}}M_z+\frac{1}{\tau}(\chi_0 H_x-M_x)\tag{6.107}$$

$$\frac{\partial M_y}{\partial t}=\omega_{\mathrm{p}}M_x+\frac{1}{\tau}(\chi_0 H_y-M_y)\tag{6.108}$$

Substituting the components of each quantity from Eqs. (6.97), (6.99), (6.100), and (6.104) into Eqs. (6.107) and (6.108) gives

$$\left(\mathrm{i}w+\frac{1}{\tau}+\frac{\chi_0}{\tau}\right)\hat{M}_y-\omega_{\mathrm{p},z}\hat{M}_x=\frac{\chi_0}{\tau}\frac{\hat{B}_y}{\mu_0}\tag{6.109}$$

$$\omega_{\mathrm{p},z}\hat{M}_y+\left(\mathrm{i}w+\frac{1}{\tau}\right)\hat{M}_x=\frac{\chi_0}{\tau}\hat{H}_x\tag{6.110}$$

where use has been made of

$$\hat{H}_y=\frac{\hat{B}_y}{\mu_0}-\hat{M}_y\tag{6.111}$$

Solving the system of equations consisting of Eqs. (6.109) and (6.110) yields

$$\hat{M}_x=\frac{\chi_0\left[(\mathrm{i}w\tau+1+\chi_0)\hat{H}_x-\omega_{\mathrm{p},z}\tau\frac{\hat{B}_y}{\mu_0}\right]}{(\mathrm{i}w\tau+1+\chi_0)(\mathrm{i}w\tau+1)+(\omega_{\mathrm{p},z}\tau)^2}\tag{6.112}$$

$$\hat{M}_y=\frac{\chi_0\left[\omega_{\mathrm{p},z}\tau\hat{H}_x+(\mathrm{i}w\tau+1)\frac{\hat{B}_y}{\mu_0}\right]}{(\mathrm{i}w\tau+1+\chi_0)(\mathrm{i}w\tau+1)+(\omega_{\mathrm{p},z}\tau)^2}\tag{6.113}$$

From Eqs. (6.112) and (6.113), it is evident that magnetization is a function of the applied magnetic fields $\hat{H}_x$ and $\hat{B}_y$, as well as the unknown variable $\omega_{p,z}$. Furthermore, there exists a mutual coupling between the magnetization and $\omega_{p,z}$: Changes in magnetization generate magnetic torques within the ferrofluid, which in turn induce fluid flow, resulting in non-zero values of $\omega_{p,z}$. Conversely, $\omega_{p,z}$ also modifies the magnetization.

6.4.2.3 Time-averaged Magnetic Force and Time-averaged Magnetic Moment

To solve the momentum equation, it is necessary to first represent the time-averaged magnetic force and time-averaged magnetic moment of the ferrofluid. The magnetic force term on the right-hand side of Eq. (2.208) requires expansion as

$$f_m = \mu_0 \boldsymbol{M} \cdot \nabla \boldsymbol{H} = \mu_0 \left(M_x \frac{\partial H_x}{\partial x} + M_y \frac{\partial H_x}{\partial y} \right) \boldsymbol{e}_x + \mu_0 \left(M_x \frac{\partial H_y}{\partial x} + M_y \frac{\partial H_y}{\partial y} \right) \boldsymbol{e}_y$$

As the x-component of the magnetic field intensity is assumed to be uniformly distributed, the x-component of the magnetic force is zero, that is

$$f_{m,x} = 0 \tag{6.114}$$

Since H_y does not vary with x, the y-component of the magnetic force can be expressed as

$$f_{m,y} = \mu_0 M_y \frac{\mathrm{d}H_y}{\mathrm{d}y} = \mu_0 M_y \frac{\mathrm{d}}{\mathrm{d}y}\left(\frac{B_y}{\mu_0} - M_y \right) \tag{6.115}$$

Due to the uniform distribution of the y-component of the magnetic induction, Eq. (6.115) becomes

$$f_{m,y} = -\mu_0 M_y \frac{\mathrm{d}M_y}{\mathrm{d}y} = -\frac{\mathrm{d}}{\mathrm{d}y}\left(\frac{1}{2} \mu_0 M_y^2 \right) \tag{6.116}$$

The components of the time-averaged magnetic force can be obtained according to Eqs. (6.114) and (6.116), respectively,

$$\langle f_{m,x} \rangle = 0 \tag{6.117}$$

$$\langle f_{m,y} \rangle = \frac{1}{t_m} \int_0^{t_m} \left[-\frac{\mathrm{d}}{\mathrm{d}y}\left(\frac{1}{2} \mu_0 M_y^2 \right) \right] \mathrm{d}t = -\frac{1}{2} \mu_0 \frac{1}{t_m} \frac{\mathrm{d}}{\mathrm{d}y} \int_0^{t_m} M_y^2 \mathrm{d}t \tag{6.118}$$

where t_m is the period of the magnetic field, $t_m = 2\pi / w$. Substituting the magnetic field intensity component $M_y = \hat{M}_y \cos wt$ expressed in Eq. (6.104) into Eq. (6.118) yields

$$\langle f_{m,y} \rangle = -\frac{1}{2} \mu_0 \frac{1}{t_m} \frac{\mathrm{d}}{\mathrm{d}y} \int_0^{t_m} \hat{M}_y^2 \cos^2(wt) \mathrm{d}t$$

Calculating the integral therein gives the time-averaged quantity of $f_{m,y}$ as

$$\langle f_{m,y}\rangle = -\frac{\mathrm{d}}{\mathrm{d}y}\left(\frac{1}{4}\mu_0 \hat{M}_y^2\right) \tag{6.119}$$

For the magnetic moment of the last term at the right-hand side of Eq. (2.210), it can be expanded as

$$\boldsymbol{L}_m = \mu_0 \boldsymbol{M} \times \boldsymbol{H} = \mu_0\left(M_x H_y - M_y H_x\right)\boldsymbol{e}_z \tag{6.120}$$

Substituting $H_y = \frac{B_y}{\mu_0} - M_y$ into Eq. (6.120) and writing it in component form gives

$$L_{m,z} = M_x B_y - \mu_0 M_y\left(H_x + M_x\right) \tag{6.121}$$

Then the time-averaged form of Eq. (6.120) is

$$\langle L_{m,z}\rangle = \frac{1}{t_m}\int_0^{t_m}\left[M_x B_y - \mu_0 M_y\left(H_x + M_x\right)\right]\mathrm{d}t \tag{6.122}$$

Taking the first integral in Eq. (6.122) as an example to demonstrate the integration method, one can substitute the components represented by Eqs. (6.99) and (6.104) into this integral, resulting in the following expression:

$$M_x B_y = \mathrm{Re}\left(\hat{M}_x e^{\mathrm{i}wt}\right)\cdot \mathrm{Re}\left(\hat{B}_y e^{\mathrm{i}wt}\right) \tag{6.123}$$

According to the identity

$$\mathrm{Re}\left(\hat{M}_x e^{\mathrm{i}wt}\right) = \frac{1}{2}\left(\hat{M}_x e^{\mathrm{i}wt} + \overline{\hat{M}_x} e^{-\mathrm{i}wt}\right)$$

where $\overline{\hat{M}_x}$ denotes the conjugate complex of $\hat{M}_x$, Eq. (6.123) becomes

$$\begin{aligned} M_x B_y &= \frac{1}{4}\left(\hat{M}_x e^{\mathrm{i}wt} + \overline{\hat{M}_x} e^{-\mathrm{i}wt}\right)\cdot\left(\hat{B}_y e^{\mathrm{i}wt} + \overline{\hat{B}_y} e^{-\mathrm{i}wt}\right) \\ &= \frac{1}{4}\left(\hat{M}_x \overline{\hat{B}_y} + \overline{\hat{M}_x}\hat{B}_y + \hat{M}_x \hat{B}_y e^{\mathrm{i}2wt} + \overline{\hat{M}_x}\cdot\overline{\hat{B}_y} e^{-\mathrm{i}2wt}\right) \end{aligned}$$

Since

$$\int_0^{t_m}\left(\hat{M}_x \hat{B}_y e^{\mathrm{i}2wt} + \overline{\hat{M}_x}\cdot\overline{\hat{B}_y} e^{-\mathrm{i}2wt}\right)\mathrm{d}t = 0$$

and

$$\hat{M}_x\overline{\hat{B}_y} + \overline{\hat{M}_x}\hat{B}_y = 2\,\mathrm{Re}\left(\hat{M}_x\overline{\hat{B}_y}\right) = 2\,\mathrm{Re}\left(\overline{\hat{M}_x}\hat{B}_y\right)$$

then one derives

$$M_xB_y = \frac{1}{2}\mathrm{Re}\left(\hat{M}_x\overline{\hat{B}_y}\right)$$

Based on this equation, the time-averaged magnetic torque expressed in Eq. (6.122) can be calculated as

$$\langle L_{m,z}\rangle = \frac{1}{2}\mathrm{Re}\left[\hat{M}_x\overline{\hat{B}_y} - \mu_0\hat{M}_y\left(\hat{H}_x + \hat{M}_x\right)\right] \tag{6.124}$$

Figure 6.39 illustrates the curves of the time-averaged magnetic torque density plotted against the dimensionless spin velocity for the three time-varying magnetic fields. It is evident that when the spin velocity is zero ($\omega_{p,z}\tau = 0$), the slope of the magnetic torque variation is negative in the low-frequency magnetic field and positive in the high-frequency magnetic field. Additionally, only in the rotating magnetic field, the ferrofluid exhibits a non-zero time-averaged magnetic torque when the spin velocity is zero.

6.4.2.4 Governing Equations and Their Normalized Forms

Given the momentum equation (2.208) and the angular momentum equation (2.210), it is assumed that the ferrofluid is incompressible and exhibits steady-state flow, with viscous forces dominating over inertial forces. Under these assumptions, the terms on the left-hand side of Eq. (2.208) vanish. Furthermore, since $\omega_{p,z}$ varies primarily in the y-direction, the terms on the left-hand side of Eq. (2.210) also approximately zero out. Additionally, due to the assumption in Eq. (6.96) that the velocity varies only in the x-direction, the presence

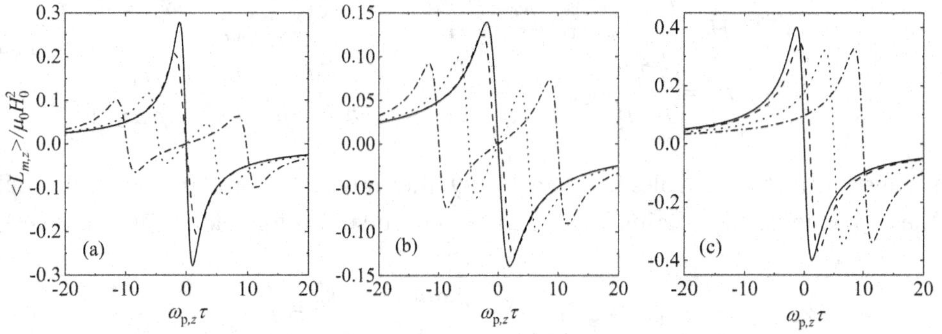

FIGURE 6.39 Time-averaged magnetic moment density versus dimensionless spin velocity (where $\chi_0 = 1$, solid lines: $w\tau = 0$, dashed lines: $w\tau = 1$, dotted line: $w\tau = 5$, dotted dash line: $w\tau = 10$). (a) Uniform magnetic field in x-direction. (b) Uniform magnetic field in y-direction. (c) Rotating magnetic field. (Reprinted from Zahn and Greer, 1995 with permission from Elsevier.)

of a non-zero $\omega_{p,y}$ component would lead to non-zero v_y and v_z values, contradicting the assumption of Eq. (6.96). Therefore, it must hold that $\nabla\cdot\boldsymbol{\omega}_p = 0$. Under these approximations and neglecting the effect of gravity, one can apply Eqs. (6.96) and (6.97) to derive the x-component of the momentum equation and the z-component of the angular momentum equation, respectively,

$$(\eta+\zeta)\frac{d^2v_x}{dy^2}+2\zeta\frac{d\omega_{p,z}}{dy}-\frac{\partial p}{\partial x}+f_{m,x}=0 \tag{6.125}$$

$$\eta'\frac{d^2\omega_{p,z}}{dy^2}-2\zeta\left(2\omega_{p,z}+\frac{dv_x}{dy}\right)+L_{m,z}=0 \tag{6.126}$$

Assuming that the flow velocity and spin velocity of the macroscopically steady flow only respond to the time-averaged magnetic force and magnetic moment, the results of Eqs. (6.117) and (6.124) for the magnetic force and magnetic moment can be substituted in Eqs. (6.125) and (6.126), resulting in the following component equations in their final forms:

$$(\eta+\zeta)\frac{d^2v_x}{dy^2}+2\zeta\frac{d\omega_{p,z}}{dy}-\frac{\partial p}{\partial x}=0 \tag{6.127}$$

$$\eta'\frac{d^2\omega_{p,z}}{dy^2}-2\zeta\left(2\omega_{p,z}+\frac{dv_x}{dy}\right)+\langle L_{m,z}\rangle=0 \tag{6.128}$$

To facilitate the solution, the dimensionless forms of the variables in the equations are derived by scaling them with respect to the characteristic length h_0 (the distance between the two plates), characteristic time τ (the magnetization relaxation time), and characteristic magnetic field strength H_0

$$\begin{gathered} w^*=w\tau,\ \hat{H}^*=\frac{\hat{H}}{H_0},\ \hat{B}^*=\frac{\hat{B}}{\mu_0H_0},\ \hat{M}^*=\frac{\hat{M}}{H_0},\ y^*=\frac{y}{h_0},\ v_x^*=\frac{v_x}{h_0/\tau},\ \omega_{p,z}^*=\omega_{p,z}\tau \\ L_{m,z}^*=\frac{L_{m,z}}{\mu_0H_0^2},\ \eta^*=\frac{2\eta}{\mu_0H_0^2\tau},\ \zeta^*=\frac{2\zeta}{\mu_0H_0^2\tau},\ \eta'^*=\frac{\eta'}{\mu_0H_0^2\tau h_0^2},\ \frac{\partial p^*}{\partial x^*}=\frac{h_0}{\mu_0H_0^2}\frac{\partial p}{\partial x} \end{gathered} \tag{6.129}$$

By substituting the variables from Eq. (6.129) into Eqs. (6.127) and (6.128), and simplifying the expressions, one obtains the dimensionless equations for velocity and spin velocity

$$\frac{1}{2}(\eta^*+\zeta^*)\frac{d^2v_x^*}{dy^{*2}}+\zeta^*\frac{d\omega_{p,z}^*}{dy^*}-\frac{\partial p^*}{\partial x^*}=0 \tag{6.130}$$

$$\eta'^*\frac{d^2\omega_{p,z}^*}{dy^{*2}}-\zeta^*\left(2\omega_{p,z}^*+\frac{dv_x^*}{dy^*}\right)+\langle L_{m,z}^*\rangle=0 \tag{6.131}$$

Putting the quantities from Eq. (6.129) into Eqs. (6.124), (6.112), and (6.113) gives

$$\langle L^*_{m,z} \rangle = \frac{1}{2}\mathrm{Re}\left[\hat{M}^*_x \overline{\hat{B}^*_y} - \overline{\hat{M}^*_y}\left(\hat{H}^*_x + \hat{M}^*_x \right) \right] \tag{6.132}$$

$$\hat{M}^*_x = \frac{\chi_0\left[\left(\mathrm{i}w^* + 1 + \chi_0 \right)\hat{H}^*_x - \omega^*_{\mathrm{p},z}\hat{B}^*_y \right]}{\left(\mathrm{i}w^* + 1 + \chi_0 \right)\left(\mathrm{i}w^* + 1 \right) + \left(\omega^*_{\mathrm{p},z} \right)^2} \tag{6.133}$$

$$\hat{M}^*_y = \frac{\chi_0\left[\omega^*_{\mathrm{p},z}\hat{H}^*_x + \left(\mathrm{i}w^* + 1 \right)\hat{B}^*_y \right]}{\left(\mathrm{i}w^* + 1 + \chi_0 \right)\left(\mathrm{i}w^* + 1 \right) + \left(\omega^*_{\mathrm{p},z} \right)^2} \tag{6.134}$$

The challenge in solving the system of equations ranging from Eq. (6.130) to Eq. (6.134) lies in the presence of the time-averaged magnetic torque $\langle L^*_{m,z} \rangle$ in the spin velocity equation. This torque varies with the spin velocity $\omega^*_{\mathrm{p},z}$, which itself is a function of y^* such that further simplification of the $\langle L^*_{m,z} \rangle$ term is required. To initiate this refinement, one has to rewrite Eq. (6.132) according to complex arithmetic rules as

$$\begin{aligned} L^*_{m,z} &= \frac{1}{4}\left\{ \left[\hat{M}^*_x \overline{\hat{B}^*_y} - \overline{\hat{M}^*_y}\left(\hat{H}^*_x + \hat{M}^*_x \right) \right] + \left[\overline{\hat{M}^*_x \overline{\hat{B}^*_y} - \overline{\hat{M}^*_y}\left(\hat{H}^*_x + \hat{M}^*_x \right)} \right] \right\} \\ &= \frac{1}{4}\left\{ \left[\hat{M}^*_x \overline{\hat{B}^*_y} - \overline{\hat{M}^*_y}\left(\hat{H}^*_x + \hat{M}^*_x \right) \right] + \left[\overline{\hat{M}^*_x}\hat{B}^*_y - \hat{M}^*_y\left(\overline{\hat{H}^*_x} + \overline{\hat{M}^*_x} \right) \right] \right\} \\ &= \frac{1}{4}\left[\left(\hat{M}^*_x \overline{\hat{B}^*_y} + \overline{\hat{M}^*_x}\hat{B}^*_y \right) - \left(\overline{\hat{M}^*_y}\hat{H}^*_x + \hat{M}^*_y \overline{\hat{H}^*_x} \right) - \left(\overline{\hat{M}^*_y}\hat{M}^*_x + \hat{M}^*_y \overline{\hat{M}^*_x} \right) \right] \end{aligned} \tag{6.135}$$

Substituting Eqs. (6.133) and (6.134) into Eq. (6.135), where for the first term in the parentheses at the right-hand side of Eq. (6.135), one has

$$\begin{aligned} \hat{M}^*_x \overline{\hat{B}^*_y} + \overline{\hat{M}^*_x}\hat{B}^*_y &= \frac{\chi_0\left[\left(\mathrm{i}w^* + 1 + \chi_0 \right)\hat{H}^*_x - \omega^*_{\mathrm{p},z}\hat{B}^*_y \right]\overline{\hat{B}^*_y}}{\left(\mathrm{i}w^* + 1 + \chi_0 \right)\left(\mathrm{i}w^* + 1 \right) + \left(\omega^*_{\mathrm{p},z} \right)^2} + \frac{\chi_0\left[\left(1 + \chi_0 - \mathrm{i}w^* \right)\overline{\hat{H}^*_x} - \omega^*_{\mathrm{p},z}\overline{\hat{B}^*_y} \right]\hat{B}^*_y}{\left(1 + \chi_0 - \mathrm{i}w^* \right)\left(1 - \mathrm{i}w^* \right) + \left(\omega^*_{\mathrm{p},z} \right)^2} \\ &= \frac{\chi_0\left[\left(\mathrm{i}w^* + 1 + \chi_0 \right)\hat{H}^*_x \overline{\hat{B}^*_y} - \omega^*_{\mathrm{p},z}\hat{B}^*_y \overline{\hat{B}^*_y} \right]}{\left[1 + \chi_0 - w^{*2} + \left(\omega^*_{\mathrm{p},z} \right)^2 \right] + \left(2 + \chi_0 \right)w^*\mathrm{i}} + \frac{\chi_0\left[\left(1 + \chi_0 - \mathrm{i}w^* \right)\overline{\hat{H}^*_x}\hat{B}^*_y - \omega^*_{\mathrm{p},z}\overline{\hat{B}^*_y}\hat{B}^*_y \right]}{\left[1 + \chi_0 - w^{*2} + \left(\omega^*_{\mathrm{p},z} \right)^2 \right] - \left(2 + \chi_0 \right)w^*\mathrm{i}} \end{aligned} \tag{6.136}$$

Letting

$$A_1 = 1 + \chi_0 - w^{*2} + \left(\omega^*_{\mathrm{p},z} \right)^2 \tag{6.137}$$

$$A_2 = \left(2 + \chi_0 \right)w^* \tag{6.138}$$

then Eq. (6.136) becomes

$$\begin{aligned}
\hat{M}_x^*\overline{\hat{B}_y^*}+\overline{\hat{M}_x^*}\hat{B}_y^* &= \chi_0\frac{\left\{\begin{aligned}&\left\{\left[(1+\chi_0)\hat{H}_x^*\overline{\hat{B}_y^*}-\omega_{\mathrm{p},z}^*\hat{B}_y^*\overline{\hat{B}_y^*}\right]+\mathrm{i}w^*\hat{H}_x^*\overline{\hat{B}_y^*}\right\}(A_1-A_2\mathrm{i})+\\&\left\{\left[(1+\chi_0)\overline{\hat{H}_x^*}\hat{B}_y^*-\omega_{\mathrm{p},z}^*\overline{\hat{B}_y^*}\hat{B}_y^*\right]-\mathrm{i}w^*\overline{\hat{H}_x^*}\hat{B}_y^*\right\}(A_1+A_2\mathrm{i})\end{aligned}\right\}}{A_1^2+A_2^2}\\
&= \chi_0\frac{\left\{\begin{aligned}&\left\{\left[(1+\chi_0)A_1+A_2w^*\right]\hat{H}_x^*\overline{\hat{B}_y^*}-A_1\omega_{\mathrm{p},z}^*\hat{B}_y^*\overline{\hat{B}_y^*}\right\}+\mathrm{i}\left\{\left[A_1w^*-(1+\chi_0)A_2\right]\hat{H}_x^*\overline{\hat{B}_y^*}+A_2\omega_{\mathrm{p},z}^*\hat{B}_y^*\overline{\hat{B}_y^*}\right\}+\\&\left\{\left[(1+\chi_0)A_1+A_2w^*\right]\overline{\hat{H}_x^*}\hat{B}_y^*-A_1\omega_{\mathrm{p},z}^*\overline{\hat{B}_y^*}\hat{B}_y^*\right\}+\mathrm{i}\left\{\left[(1+\chi_0)A_2-A_1w^*\right]\overline{\hat{H}_x^*}\hat{B}_y^*-A_2\omega_{\mathrm{p},z}^*\overline{\hat{B}_y^*}\hat{B}_y^*\right\}\end{aligned}\right\}}{A_1^2+A_2^2}\\
&= \chi_0\frac{\left\{\begin{aligned}&\left\{\left[(1+\chi_0)A_1+A_2w^*\right]\left(\hat{H}_x^*\overline{\hat{B}_y^*}+\overline{\hat{H}_x^*}\hat{B}_y^*\right)-2A_1\omega_{\mathrm{p},z}^*\hat{B}_y^*\overline{\hat{B}_y^*}\right\}\\&+\mathrm{i}\left\{\left[A_1w^*-A_2(1+\chi_0)\right]\left(\hat{H}_x^*\overline{\hat{B}_y^*}-\overline{\hat{H}_x^*}\hat{B}_y^*\right)\right\}\end{aligned}\right\}}{A_1^2+A_2^2}
\end{aligned} \tag{6.139}$$

Substituting Eqs. (6.137) and (6.138) into related coefficients in Eq. (6.139) yields

$$(1+\chi_0)A_1+A_2w^* = (1+\chi_0)\left[1+\chi_0+\left(\omega_{\mathrm{p},z}^*\right)^2\right]+w^{*2} \tag{6.140}$$

$$A_1w^*-A_2(1+\chi_0) = w^*\left[\left(\omega_{\mathrm{p},z}^*\right)^2-w^{*2}-(1+\chi_0)^2\right] \tag{6.141}$$

Similarly, regarding the second term enclosed within the parentheses on the right-hand side of Eq. (6.135), it holds that

$$\begin{aligned}
\overline{\hat{M}_y^*}\hat{H}_x^*+\hat{M}_y^*\overline{\hat{H}_x^*} &= \frac{\chi_0\left[\omega_{\mathrm{p},z}^*\overline{\hat{H}_x^*}+(1-\mathrm{i}w^*)\overline{\hat{B}_y^*}\right]\hat{H}_x^*}{(1+\chi_0-\mathrm{i}w^*)(1-\mathrm{i}w^*)+\left(\omega_{\mathrm{p},z}^*\right)^2}+\frac{\chi_0\left[\omega_{\mathrm{p},z}^*\hat{H}_x^*+(\mathrm{i}w^*+1)\hat{B}_y^*\right]\overline{\hat{H}_x^*}}{(\mathrm{i}w^*+1+\chi_0)(\mathrm{i}w^*+1)+\left(\omega_{\mathrm{p},z}^*\right)^2}\\
&= \chi_0\frac{\left\{\begin{aligned}&\left\{\left[\overline{\hat{B}_y^*}\hat{H}_x^*+\omega_{\mathrm{p},z}^*\overline{\hat{H}_x^*}\hat{H}_x^*\right]-\mathrm{i}w^*\overline{\hat{B}_y^*}\hat{H}_x^*\right\}(A_1+A_2\mathrm{i})+\\&\left\{\left[\hat{B}_y^*\overline{\hat{H}_x^*}+\omega_{\mathrm{p},z}^*\hat{H}_x^*\overline{\hat{H}_x^*}\right]+\mathrm{i}w^*\hat{B}_y^*\overline{\hat{H}_x^*}\right\}(A_1-A_2\mathrm{i})\end{aligned}\right\}}{A_1^2+A_2^2}\\
&= \chi_0\frac{\left\{\begin{aligned}&\left[(A_1+A_2w^*)\overline{\hat{B}_y^*}\hat{H}_x^*+A_1\omega_{\mathrm{p},z}^*\overline{\hat{H}_x^*}\hat{H}_x^*\right]+\mathrm{i}\left[(A_2-A_1w^*)\overline{\hat{B}_y^*}\hat{H}_x^*+A_2\omega_{\mathrm{p},z}^*\overline{\hat{H}_x^*}\hat{H}_x^*\right]+\\&\left[(A_1+A_2w^*)\hat{B}_y^*\overline{\hat{H}_x^*}+A_1\omega_{\mathrm{p},z}^*\hat{H}_x^*\overline{\hat{H}_x^*}\right]+\mathrm{i}\left[(A_1w^*-A_2)\hat{B}_y^*\overline{\hat{H}_x^*}-A_2\omega_{\mathrm{p},z}^*\hat{H}_x^*\overline{\hat{H}_x^*}\right]\end{aligned}\right\}}{A_1^2+A_2^2}\\
&= \chi_0\frac{\left[(A_1+A_2w^*)\left(\overline{\hat{B}_y^*}\hat{H}_x^*+\hat{B}_y^*\overline{\hat{H}_x^*}\right)+2A_1\omega_{\mathrm{p},z}^*\overline{\hat{H}_x^*}\hat{H}_x^*\right]+\mathrm{i}\left[(A_2-A_1w^*)\left(\overline{\hat{B}_y^*}\hat{H}_x^*-\hat{B}_y^*\overline{\hat{H}_x^*}\right)\right]}{A_1^2+A_2^2}
\end{aligned} \tag{6.142}$$

where

$$A_1 + A_2 w^* = (1+\chi_0)(1+w^*)^2 + (\omega_{\mathrm{p},z}^*)^2 \tag{6.143}$$

$$A_1 w^* - A_2 = w^*\left[(\omega_{\mathrm{p},z}^*)^2 - w^{*2} - 1\right] \tag{6.144}$$

Regarding the third term enclosed in parentheses on the right-hand side of Eq. (6.135), one has

$$\begin{aligned}
\overline{\hat{M}_y^*}\hat{M}_x^* + \hat{M}_y^*\overline{\hat{M}_x^*} &= \frac{\chi_0\left[\omega_{\mathrm{p},z}^*\overline{\hat{H}_x^*} + (1-\mathrm{i}w^*)\overline{\hat{B}_y^*}\right]}{(1+\chi_0-\mathrm{i}w^*)(1-\mathrm{i}w^*)+(\omega_{\mathrm{p},z}^*)^2} \times \frac{\chi_0\left[(\mathrm{i}w^*+1+\chi_0)\hat{H}_x^* - \omega_{\mathrm{p},z}^*\hat{B}_y^*\right]}{(\mathrm{i}w^*+1+\chi_0)(\mathrm{i}w^*+1)+(\omega_{\mathrm{p},z}^*)^2} \\
&+ \frac{\chi_0\left[\omega_{\mathrm{p},z}^*\hat{H}_x^* + (\mathrm{i}w^*+1)\hat{B}_y^*\right]}{(\mathrm{i}w^*+1+\chi_0)(\mathrm{i}w^*+1)+(\omega_{\mathrm{p},z}^*)^2} \times \frac{\chi_0\left[(1+\chi_0-\mathrm{i}w^*)\overline{\hat{H}_x^*} - \omega_{\mathrm{p},z}^*\overline{\hat{B}_y^*}\right]}{(1+\chi_0-\mathrm{i}w^*)(1-\mathrm{i}w^*)+(\omega_{\mathrm{p},z}^*)^2} \\
&= \chi_0^2 \frac{\left\{\begin{array}{l}\left[\left(\omega_{\mathrm{p},z}^*\overline{\hat{H}_x^*} + \overline{\hat{B}_y^*}\right) - \mathrm{i}w^*\overline{\hat{B}_y^*}\right]\left\{\left[(1+\chi_0)\hat{H}_x^* - \omega_{\mathrm{p},z}^*\hat{B}_y^*\right] + \mathrm{i}w^*\hat{H}_x^*\right\} + \\ \left[\left(\omega_{\mathrm{p},z}^*\hat{H}_x^* + \hat{B}_y^*\right) + \mathrm{i}w^*\hat{B}_y^*\right]\left\{\left[(1+\chi_0)\overline{\hat{H}_x^*} - \omega_{\mathrm{p},z}^*\overline{\hat{B}_y^*}\right] - \mathrm{i}w^*\overline{\hat{H}_x^*}\right\}\end{array}\right\}}{A_1^2 + A_2^2} \\
&= \chi_0 \frac{\left\{\begin{array}{l}\left\{\left(\omega_{\mathrm{p},z}^*\overline{\hat{H}_x^*} + \overline{\hat{B}_y^*}\right)\left[(1+\chi_0)\hat{H}_x^* - \omega_{\mathrm{p},z}^*\hat{B}_y^*\right] + w^{*2}\overline{\hat{B}_y^*}\hat{H}_x^*\right\} \\ +\mathrm{i}\left\{w^*\hat{H}_x^*\left(\omega_{\mathrm{p},z}^*\overline{\hat{H}_x^*} + \overline{\hat{B}_y^*}\right) - w^*\overline{\hat{B}_y^*}\left[(1+\chi_0)\hat{H}_x^* - \omega_{\mathrm{p},z}^*\hat{B}_y^*\right]\right\} \\ +\left\{\left(\omega_{\mathrm{p},z}^*\hat{H}_x^* + \hat{B}_y^*\right)\left[(1+\chi_0)\overline{\hat{H}_x^*} - \omega_{\mathrm{p},z}^*\overline{\hat{B}_y^*}\right] + w^{*2}\hat{B}_y^*\overline{\hat{H}_x^*}\right\} \\ +\mathrm{i}\left\{-w^*\overline{\hat{H}_x^*}\left(\omega_{\mathrm{p},z}^*\hat{H}_x^* + \hat{B}_y^*\right) + w^*\hat{B}_y^*\left[(1+\chi_0)\overline{\hat{H}_x^*} - \omega_{\mathrm{p},z}^*\overline{\hat{B}_y^*}\right]\right\}\end{array}\right\}}{A_1^2 + A_2^2} \\
&= \chi_0^2 \frac{\left\{\begin{array}{l}\left\{2(1+\chi_0)\omega_{\mathrm{p},z}^*\overline{\hat{H}_x^*}\hat{H}_x^* + \left[1+\chi_0+w^{*2}-(\omega_{\mathrm{p},z}^*)^2\right]\left(\hat{H}_x^*\overline{\hat{B}_y^*} + \hat{B}_y^*\overline{\hat{H}_x^*}\right) - 2\omega_{\mathrm{p},z}^*\overline{\hat{B}_y^*}\hat{B}_y^*\right\} \\ +\mathrm{i}\left[\chi_0 w^*\left(\hat{B}_y^*\overline{\hat{H}_x^*} - \hat{H}_x^*\overline{\hat{B}_y^*}\right)\right]\end{array}\right\}}{A_1^2 + A_2^2}
\end{aligned} \tag{6.145}$$

Substituting Eqs. (6.139) to (6.145) into Eq. (6.135) and replacing A_1 and A_2 with Eqs. (6.137) and (6.138), respectively, obtains

$$\langle L_{m,z}^* \rangle = \frac{1}{2}\frac{\left\{\begin{array}{c}-\chi_0\omega_{\mathrm{p},z}^*\left[(\omega_{\mathrm{p},z}^*)^2 - w^{*2} + 1\right]\hat{B}_y^*\overline{\hat{B}_y^*} - \chi_0\omega_{\mathrm{p},z}^*\left[(\omega_{\mathrm{p},z}^*)^2 - w^{*2} + (1+\chi_0)^2\right]\hat{H}_x^*\overline{\hat{H}_x^*} \\ +\chi_0^2\left[(\omega_{\mathrm{p},z}^*)^2 - w^{*2}\right]\left(\hat{H}_x^*\overline{\hat{B}_y^*} + \overline{\hat{H}_x^*}\hat{B}_y^*\right) \\ +\mathrm{i}\chi_0 w^*\left[(\omega_{\mathrm{p},z}^*)^2 - w^{*2} - 1 - \chi_0\right]\left(\hat{H}_x^*\overline{\hat{B}_y^*} - \overline{\hat{H}_x^*}\hat{B}_y^*\right)\end{array}\right\}}{\left[1+\chi_0 - w^{*2} + (\omega_{\mathrm{p},z}^*)^2\right]^2 + (2+\chi_0)^2 w^{*2}} \tag{6.146}$$

According to the identities

$$\hat{B}_y^*\overline{\hat{B}_y^*}=\left|\hat{B}_y^*\right|^2,\ \hat{H}_x^*\overline{\hat{H}_x^*}=\left|\hat{H}_x^*\right|^2,\ \hat{H}_x^*\overline{\hat{B}_y^*}+\overline{\hat{H}_x^*}\hat{B}_y^*=2\mathrm{Re}\left(\hat{H}_x^*\overline{\hat{B}_y^*}\right),\ \hat{H}_x^*\overline{\hat{B}_y^*}-\overline{\hat{H}_x^*}\hat{B}_y^*=2\mathrm{iIm}\left(\hat{H}_x^*\overline{\hat{B}_y^*}\right)$$

Eq. (6.146) becomes

$$\langle L_{m,z}^*\rangle=\frac{\chi_0}{2}\frac{\left\{\begin{array}{l}-\omega_{\mathrm{p},z}^*\left[\left(\omega_{\mathrm{p},z}^*\right)^2-w^{*2}+1\right]\left|\hat{B}_y^*\right|^2-\omega_{\mathrm{p},z}^*\left[\left(\omega_{\mathrm{p},z}^*\right)^2-w^{*2}+\left(1+\chi_0\right)^2\right]\left|\hat{H}_x^*\right|^2+\\2\chi_0\left[\left(\omega_{\mathrm{p},z}^*\right)^2-w^{*2}\right]\mathrm{Re}\left(\hat{H}_x^*\overline{\hat{B}_y^*}\right)-2w^*\left[\left(\omega_{\mathrm{p},z}^*\right)^2-w^{*2}-1-\chi_0\right]\mathrm{Im}\left(\hat{H}_x^*\overline{\hat{B}_y^*}\right)\end{array}\right\}}{\left[1+\chi_0-w^{*2}+\left(\omega_{\mathrm{p},z}^*\right)^2\right]^2+\left(2+\chi_0\right)^2w^{*2}}\tag{6.147}$$

The phase relationship between $\hat{H}_x^*$ and $\hat{B}_y^*$ determines the values of the third and fourth terms in the numerator of Eq. (6.147).

6.4.2.5 Solving the Equations of Motion

When $\eta'=0$, taking the derivative of both sides of Eq. (6.131) with respect to y^* yields

$$\frac{\mathrm{d}\omega_{\mathrm{p},z}^*}{\mathrm{d}y^*}=-\frac{1}{2}\frac{\mathrm{d}^2v_x^*}{\mathrm{d}y^{*2}}+\frac{1}{2\zeta^*}\frac{\mathrm{d}\langle L_{m,z}^*\rangle}{\mathrm{d}y^*}\tag{6.148}$$

Substituting Eq. (6.148) into Eq. (6.130) gives

$$\frac{\mathrm{d}^2v_x^*}{\mathrm{d}y^{*2}}=\frac{2}{\eta^*}\frac{\partial p^*}{\partial x^*}-\frac{1}{\eta^*}\frac{\mathrm{d}\langle L_{m,z}^*\rangle}{\mathrm{d}y^*}\tag{6.149}$$

After integrating Eq. (6.149) twice with respect to y^*, the general solution for v_x^* can be obtained

$$v_x^*=\frac{1}{\eta^*}\frac{\partial p^*}{\partial x^*}y^{*2}-\frac{1}{\eta^*}\int_0^{y^*}\langle L_{m,z}^*\rangle\mathrm{d}y^*+K_1y^*+K_2\tag{6.150}$$

where K_1 and K_2 are coefficients to be determined, respectively. Based on the boundary conditions of the Couette–Poiseuille flow of ferrofluids,

$$v_x^*=0,\text{when }y^*=0$$

$$v_x^*=\frac{v_0}{h_0/\tau},\text{when }y^*=1$$

the coefficients K_1 and K_2 are obtained as

$$K_1 = -\frac{1}{\eta^*}\frac{\partial p^*}{\partial x^*} + \frac{1}{\eta^*}\int_0^1 \langle L^*_{m,z}\rangle \mathrm{d}y^* + \frac{v_0}{h_0/\tau}, K_2 = 0$$

Substituting them into Eq. (6.150) gives the general solution of v_x^* as

$$v_x^* = \frac{1}{\eta^*}\frac{\partial p^*}{\partial x^*}\left(y^{*2} - y^*\right) + \frac{1}{\eta^*}\left(y^*\int_0^1 \langle L^*_{m,z}\rangle \mathrm{d}y^* - \int_0^{y^*} \langle L^*_{m,z}\rangle \mathrm{d}y^*\right) + v_0^* y^* \qquad (6.151)$$

where $v_0^* = \frac{v_0}{h_0/\tau}$.

To find the solution for $\omega^*_{p,z}$, integrating Eq. (6.148) once with respect to y^* gives

$$\omega^*_{p,z} = -\frac{1}{2}\frac{\mathrm{d}v_x^*}{\mathrm{d}y^*} + \frac{1}{2\zeta^*}\langle L^*_{m,z}\rangle \qquad (6.152)$$

After substituting Eq. (6.151) into Eq. (6.152), the general solution for the spin velocity is obtained,

$$\omega^*_{p,z} = -\frac{1}{2\eta^*}\left[\frac{\partial p^*}{\partial x^*}\left(2y^* - 1\right) - \frac{\eta^* + \zeta^*}{\zeta^*}\langle L^*_{m,z}\rangle + \int_0^1 \langle L^*_{m,z}\rangle \mathrm{d}y^* + \eta^* v_0^*\right] \qquad (6.153)$$

In Eqs. (6.151) and (6.153), $\langle L^*_{m,z}\rangle$ is not an analytical result as it is itself a function of $\omega^*_{p,z}$. If the pressure gradient $\frac{\partial p^*}{\partial x^*}$ is significantly greater than the magnetic torque $\langle L^*_{m,z}\rangle$, the dimensional form of Eq. (6.151) can be approximately simplified to Eq. (6.1). This indicates that, under these conditions, the velocity distribution of the ferrofluid coincides with the Couette–Poiseuille pattern of the ordinary fluid, exhibiting a parabolic flow pattern. Furthermore, the spin velocity $\omega^*_{p,z}$ varies linearly with the position y^* under these circumstances.

If the time-averaged magnetic torque $\langle L^*_{m,z}\rangle$ remains constant and independent of position, the second term on the right-hand side of Eq. (6.151) vanishes, rendering the magnetic torque ineffective on the flow velocity of the ferrofluid. Instead, it exerts a constant magnetic torque of $\frac{1}{2\zeta^*}\langle L^*_{m,z}\rangle$ on the spin velocity. Under the condition of a constant time-averaged magnetic torque $\langle L^*_{m,z}\rangle$, the absence of a pressure gradient within the ferrofluid leads to a Couette flow scenario, where the fluid velocity varies linearly with position ($v_x^* = v_0^* y^*$). This leads to a spatially uniform spin velocity $\frac{1}{2\zeta^*}\langle L^*_{m,z}\rangle - \frac{1}{2}v_0^*$. Conversely, in the absence of both a pressure gradient and the driving force from the upper plate, under a constant time-averaged magnetic torque $\langle L^*_{m,z}\rangle$, the fluid velocity becomes zero ($v_x^* = 0$), stabilizing the spin velocity at a constant value of $\frac{1}{2\zeta^*}\langle L^*_{m,z}\rangle$. These observations suggest that in a stationary ferrofluid, a time-varying magnetic field can induce particle spin, consistent with the findings in Section 6.4.1.

6.4.2.6 Numerical Solutions of the Equations of Motion

Due to the presence of integral terms associated with magnetic torque in the Eqs. (6.151) and (6.153) obtained by directly solving the governing equations, further simplification to obtain analytical expressions is not feasible. Therefore, numerical methods are employed to solve the governing equations. To accomplish this, Eq. (6.131) at $\eta' = 0$ is first transformed into

$$\frac{dv_x^*}{dy^*} = -2\omega_{p,z}^* + \frac{\langle L_{m,z}^* \rangle}{\zeta^*} \tag{6.154}$$

Substituting Eq. (6.154) into Eq. (6.130) yields

$$\frac{d\omega_{p,z}^*}{dy^*} = \frac{\eta^* + \zeta^*}{2\zeta^* \eta^*} \frac{d\langle L_{m,z}^* \rangle}{dy^*} - \frac{1}{\eta^*} \frac{\partial p^*}{\partial x^*} \tag{6.155}$$

Since

$$\frac{d\langle L_{m,z}^* \rangle}{dy^*} = \frac{d\langle L_{m,z}^* \rangle}{d\omega_{p,z}^*} \frac{d\omega_{p,z}^*}{dy^*}$$

Eq. (6.155) can be further expressed as

$$\frac{d\omega_{p,z}^*}{dy^*} = \frac{-\dfrac{1}{\eta^*} \dfrac{\partial p^*}{\partial x^*}}{1 - \dfrac{\eta^* + \zeta^*}{2\zeta^* \eta^*} \cdot \dfrac{d\langle L_{m,z}^* \rangle}{d\omega_{p,z}^*}} \tag{6.156}$$

According to

$$\frac{dv_x^*}{d\omega_{p,z}^*} = \frac{dv_x^*}{dy^*} \frac{dy^*}{d\omega_{p,z}^*}$$

Eq. (6.154) can be converted into

$$\frac{dv_x^*}{d\omega_{p,z}^*} = \frac{-2\omega_{p,z}^* + \dfrac{\langle L_{m,z}^* \rangle}{\zeta^*}}{\dfrac{d\omega_{p,z}^*}{dy^*}} \tag{6.157}$$

Substituting Eq. (6.156) into Eq. (6.157) yields

$$\frac{dv_x^*}{d\omega_{p,z}^*}=\frac{\left(-2\omega_{p,z}^*+\frac{\langle L_{m,z}^*\rangle}{\varsigma^*}\right)\left(1-\frac{\eta^*+\varsigma^*}{2\varsigma^*\eta^*}\cdot\frac{d\langle L_{m,z}^*\rangle}{d\omega_{p,z}^*}\right)}{-\frac{1}{\eta^*}\frac{\partial p^*}{\partial x^*}} \tag{6.158}$$

To numerically solve the system of equations consisting of Eqs. (6.156) and (6.158), it is necessary to individually expand the derivative term $\frac{d\langle L_{m,z}^*\rangle}{d\omega_{p,z}^*}$ across three distinct time-varying magnetic field configurations. For a spatially uniform time-varying magnetic field in the x-direction, Eq. (6.101) and the dimensionless variable representation Eq. (6.129) can be invoked to yield

$$\hat{H}_y^*=1,\ \hat{B}_y^*=0 \tag{6.159}$$

Introducing Eq. (6.159) into Eq. (6.147) yields

$$\langle L_{m,z}^*\rangle_x=\frac{\chi_0}{2}\frac{-\omega_{p,z}^*\left[\left(\omega_{p,z}^*\right)^2-w^{*2}+\left(1+\chi_0\right)^2\right]}{\left[1+\chi_0-w^{*2}+\left(\omega_{p,z}^*\right)^2\right]^2+\left(2+\chi_0\right)^2w^{*2}} \tag{6.160}$$

For a spatially uniform time-varying magnetic field in the y-direction, according to Eqs. (6.102) and (6.129), it follows that $\hat{H}_x^*=0$, $\hat{B}_y^*=1$. Substituting them into Eq. (6.147), one obtains

$$\langle L_{m,z}^*\rangle_y=\frac{\chi_0}{2}\frac{-\omega_{p,z}^*\left[\left(\omega_{p,z}^*\right)^2-w^{*2}+1\right]}{\left[1+\chi_0-w^{*2}+\left(\omega_{p,z}^*\right)^2\right]^2+\left(2+\chi_0\right)^2w^{*2}} \tag{6.161}$$

When a rotating magnetic field is applied, according to Eqs. (6.103) and (6.129), it follows that

$$\hat{H}_x^*=\mathrm{i},\hat{B}_y^*=1 \tag{6.162}$$

Putting Eq. (6.162) into Eq. (6.147) gives

$$\langle L_{m,z}^*\rangle_{\text{rot}}=\frac{\chi_0}{2}\frac{\left\{\begin{array}{l}-\omega_{p,z}^*\left[\left(\omega_{p,z}^*\right)^2-w^{*2}+1\right]+\omega_{p,z}^*\left[\left(\omega_{p,z}^*\right)^2-w^{*2}+\left(1+\chi_0\right)^2\right]\\-2w^*\left[\left(\omega_{p,z}^*\right)^2-w^{*2}-1-\chi_0\right]\end{array}\right\}}{\left[1+\chi_0-w^{*2}+\left(\omega_{p,z}^*\right)^2\right]^2+\left(2+\chi_0\right)^2w^{*2}} \tag{6.163}$$

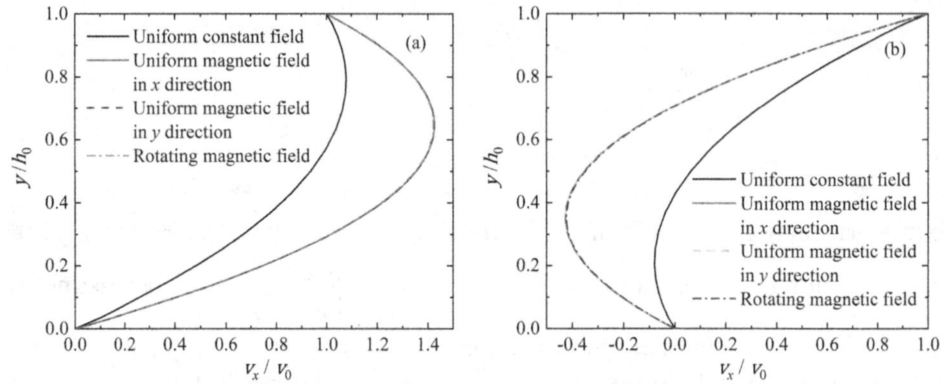

FIGURE 6.40 Comparison of the flow velocity distribution within the cross-section of a ferrofluid Couette–Poiseuille flow in a time-varying magnetic field with the flow velocity in a uniform constant magnetic field (the magnitude of the magnetic field is the same as the amplitude of the time-averaged magnetic field intensity and the direction is along the y-direction) ($\chi_0 = 1.01, w\tau = 10, \eta^* / \zeta^* = 5.65$). (a) $P^* = 2$, (b) $P^* = -2$.

By substituting Eqs. (6.160) to (6.163) into Eq. (6.156), one can obtain the spin velocity equations under three different magnetic field conditions. These equations are then solved using the fourth-order Runge–Kutta method, with the boundary conditions

$$\omega_{p,z}^* = 0, \text{ when } y^* = 0$$

Then the distribution of $\omega_{p,z}^*$ can be obtained. Next, given the distribution of $\omega_{p,z}^*$, one can solve Eq. (6.158) to obtain the distribution of v_x^*.

Figure 6.40 compares the velocity profiles of the Couette–Poiseuille flow of ferrofluids influenced by three different time-varying magnetic fields with those observed under a uniform constant magnetic field of the same intensity as described in Section 6.2.6. It is evident that a high-frequency time-varying magnetic field significantly enhances the Couette–Poiseuille flow of ferrofluids. This enhancement is attributed to the particle rotation induced by the time-varying magnetic field, which in turn leads to a negative rotational viscosity within the ferrofluid, effectively reducing its apparent viscosity. This observation is supported by experimental evidence.

Figure 6.41 illustrates the distribution of velocity and spin velocity within the cross-section of a ferrofluid Couette–Poiseuille flow under the influence of three types of time-varying magnetic fields varying in frequency. For both uniform time-varying magnetic fields in the x- and y-directions, a higher frequency results in greater spin velocities and fluid velocities at the same y-position within the cross-section. Moreover, in terms of promoting flow, a uniform time-varying magnetic field in the y-direction is more effective than one in the x-direction within the low-frequency range.

Regarding the rotating magnetic field, the velocity and spin velocity exhibit non-monotonic behaviour with respect to the frequency of the magnetic field. At low frequencies, the magnetic field hinders flow and produces negative spin velocities. However, at higher frequencies, the magnetic field significantly enhances flow, and the spin velocities are comparable to those induced by the other two types of magnetic fields.

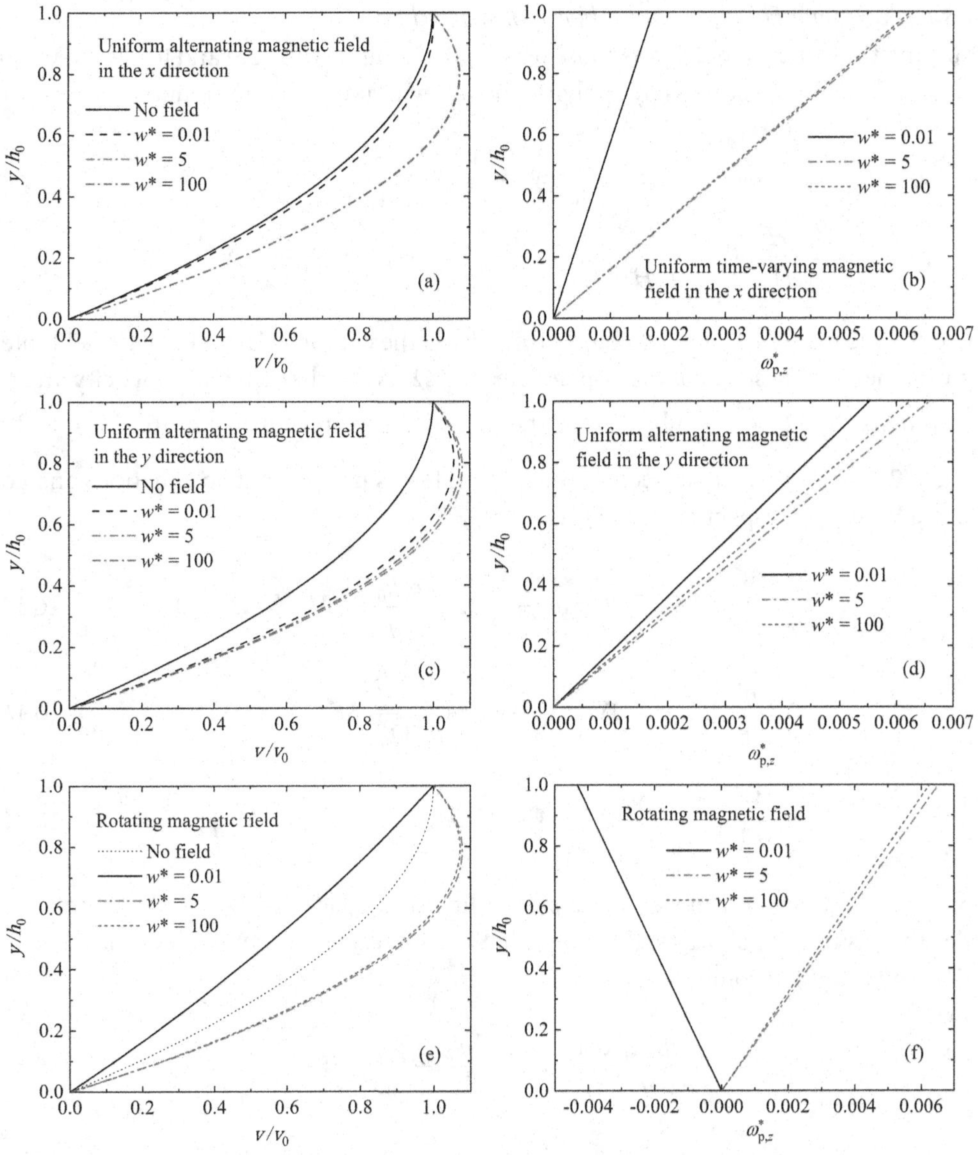

FIGURE 6.41 Flow velocity and spin velocity distributions within the cross-section of ferrofluid Couette–Poiseuille flows under time-varying magnetic fields of different frequencies ($\chi_0 = 3.5$, $\eta^* / \zeta^* = 5.65$, $P^* = 1$, $Re = 0.1$, $Mn_f = 800$, $Pe = 0.00625$). (a) and (b) Uniform time-varying magnetic fields in the x-direction. (c) and (d) Uniform time-varying magnetic fields in the y-direction. (e) and (f) Rotating magnetic fields. (Reprinted from Yang et al., 2024 with permission from Elsevier.)

6.4.3 Flow Characteristics under the Influence of Alternating and Rotating Magnetic Fields When Considering the Effects of Spin Viscosity

This section examines the Couette–Poiseuille flow characteristics of ferrofluids under the same magnetic field conditions as Section 6.4.2, taking into account the spin viscosity. The complex representations of the time-varying magnetic fields and magnetization maintain their format as specified in Eqs. (6.99) to (6.104), but the regularization perturbation method is required to solve the magnetization equation when considering spin viscosity.

6.4.3.1 Governing Equations and Their Normalized Forms

Assuming the same conditions as those used in deriving Eqs. (6.125) and (6.126), the equations are made dimensionless by applying the following dimensionless quantities:

$$w^* = w\tau,\ x^* = \frac{x}{h_0},\ y^* = \frac{y}{h_0},\ \boldsymbol{v}^* = \frac{\boldsymbol{v}}{v_0},\ \boldsymbol{\omega}_\mathrm{p}^* = \frac{\boldsymbol{\omega}_\mathrm{p}}{\Omega_0}$$
$$\boldsymbol{H}^* = \frac{\boldsymbol{H}}{H_0},\ \boldsymbol{B}^* = \frac{\boldsymbol{B}}{\mu_0 H_0},\ \boldsymbol{M}^* = \frac{\boldsymbol{M}}{H_0},\ t^* = \frac{t}{\tau},\ p^* = \frac{p}{\eta v_0 / h_0},\ \nabla^* = h_0 \nabla \tag{6.164}$$

where τ is the magnetization relaxation time, H_0 is the characteristic magnetic field intensity, v_0 is the moving speed of the upper plate, and Ω_0 is the characteristic vorticity and can be expressed by $\Omega_0 = \frac{v_0}{h_0}$. Substituting the quantities expressed in Eq. (6.164) into Eqs. (2.208), (2.210) and (4.17), neglecting the inertial terms in the first two equations and considering $\nabla \cdot \boldsymbol{\omega}_\mathrm{p} = 0$, respectively, one gets

$$\frac{\eta+\zeta}{\eta}\nabla^{*2}\boldsymbol{v}^* + 2\frac{\zeta}{\eta}\nabla^* \times \boldsymbol{\omega}_\mathrm{p}^* - \nabla^* p^* + \frac{\mu_0 h_0 H_0^2}{\eta v_0}\boldsymbol{M}^* \cdot \nabla^* \boldsymbol{H}^* = \boldsymbol{0} \tag{6.165}$$

$$\frac{\eta'}{\zeta h_0^2}\nabla^{*2}\boldsymbol{\omega}_\mathrm{p}^* + 2\nabla^* \times \boldsymbol{v}^* - 4\boldsymbol{\omega}_\mathrm{p}^* + \frac{\mu_0 H_0^2}{\zeta \Omega_0}\boldsymbol{M}^* \times \boldsymbol{H}^* = \boldsymbol{0} \tag{6.166}$$

$$\frac{\partial \boldsymbol{M}^*}{\partial t^*} + \frac{v_0 \tau}{h_0}\left(\boldsymbol{v}^* \cdot \nabla^*\right)\boldsymbol{M}^* = \Omega_0 \tau \boldsymbol{\omega}_\mathrm{p}^* \times \boldsymbol{M}^* + \left(\boldsymbol{M}_0^* - \boldsymbol{M}^*\right) \tag{6.167}$$

Since the last term on the left-hand side of Eq. (6.165) lacks an x-direction component, under the assumptions of Eqs. (6.96) and (6.97), the respective component forms of Eqs. (6.165) and (6.166) become

$$\frac{\eta+\zeta}{\eta}\frac{\mathrm{d}^2 v_x^*}{\mathrm{d}y^{*2}} + 2\frac{\zeta}{\eta}\frac{\mathrm{d}\omega_{\mathrm{p},z}^*}{\mathrm{d}y^*} - \frac{\partial p^*}{\partial x^*} = 0 \tag{6.168}$$

$$\frac{\eta'}{\zeta h_0^2}\frac{\mathrm{d}^2 \omega_{\mathrm{p},z}^*}{\mathrm{d}y^{*2}} - 2\frac{\mathrm{d}v_x^*}{\mathrm{d}y^*} - 4\omega_{\mathrm{p},z}^* + \frac{\mu_0 H_0^2}{\zeta \Omega_0}\langle L_{m,z}^* \rangle = 0 \tag{6.169}$$

where $\langle T_{m,z}^* \rangle$ denotes the time-averaged value of $\boldsymbol{M}^* \times \boldsymbol{H}^*$.

Assuming a linear relationship exists between magnetization and magnetic field intensity, one has

$$\boldsymbol{M}_0^* = \chi_0 \boldsymbol{H}^*$$

Given that $\boldsymbol{M}$ varies solely in the y-direction and the velocity has only an x-component, it follows that the convective term $\left(\boldsymbol{v}^* \cdot \nabla^*\right)\boldsymbol{M}^*$ in Eq. (6.167) vanishes. Consequently, Eq. (6.167) can be simplified to

$$\frac{\partial \boldsymbol{M}^*}{\partial t^*} = \epsilon \boldsymbol{\omega}_{\mathrm{p}}^* \times \boldsymbol{M}^* + \left(\chi_0 \boldsymbol{H}^* - \boldsymbol{M}^*\right) \tag{6.170}$$

where the parameter $\epsilon = \Omega_0 \tau$.

6.4.3.2 Solving the Magnetization Equation Using the Regular Perturbation Method

The regular perturbation method is utilized to solve the magnetization equation (6.170). Assuming that the magnetization can be represented as a power series in the small parameter ϵ,

$$\boldsymbol{M}^* = \sum_{n=0}^{\infty} \epsilon^n \boldsymbol{M}_n^* \tag{6.171}$$

It is supposed that the nth-order term of the perturbative expansion exhibits the identical temporal dependence as the magnetic field intensity, which can be represented as

$$\boldsymbol{M}_n^* = \mathrm{Re}\left\{\left[\hat{M}_{x,n}(y)\boldsymbol{e}_x + \hat{M}_{y,n}(y)\boldsymbol{e}_y\right]e^{\mathrm{i}w^* t^*}\right\} \tag{6.172}$$

Substituting Eq. (6.171) into Eq. (6.170), one has

$$\begin{aligned} \frac{\partial \boldsymbol{M}_0^*}{\partial t^*} + \epsilon \frac{\partial \boldsymbol{M}_1^*}{\partial t^*} + \epsilon^2 \frac{\partial \boldsymbol{M}_2^*}{\partial t^*} + \cdots + \epsilon^n \frac{\partial \boldsymbol{M}_n^*}{\partial t^*} &= \epsilon \boldsymbol{\omega}_{\mathrm{p}}^* \times \left(\boldsymbol{M}_0^* + \epsilon \boldsymbol{M}_1^* + \epsilon^2 \boldsymbol{M}_2^* + \cdots + \epsilon^n \boldsymbol{M}_n^*\right) \\ &\quad - \left(\boldsymbol{M}_0^* + \epsilon \boldsymbol{M}_1^* + \epsilon^2 \boldsymbol{M}_2^* + \cdots + \epsilon^n \boldsymbol{M}_n^*\right) + \chi_0 \boldsymbol{H}^* \end{aligned} \tag{6.173}$$

Since Eq. (6.173) is valid for arbitrary values of ϵ, the corresponding terms of the same order on both sides must be equal, so that equations of different orders can be derived. For example, the equations of the zeroth and the nth orders are respectively

$$\frac{\partial \boldsymbol{M}_0^*}{\partial t^*} = -\boldsymbol{M}_0^* + \chi_0 \boldsymbol{H}^* \tag{6.174}$$

$$\frac{\partial \boldsymbol{M}_n^*}{\partial t^*} = \boldsymbol{\omega}_{\mathrm{p}}^* \times \boldsymbol{M}_{n-1}^* - \boldsymbol{M}_n^* + \chi_0 \boldsymbol{H}^*, \ (n \geq 1) \tag{6.175}$$

Furthermore, given that $\boldsymbol{B}^* = \boldsymbol{H}^* + \boldsymbol{M}^*$, where $\boldsymbol{M}^*$ is defined in Eq. (6.171), the zeroth-order and nth-order equations can be respectively written as

$$\frac{\partial \boldsymbol{M}_0^*}{\partial t^*} = -(1+\chi_0)\boldsymbol{M}_0^* + \chi_0 \boldsymbol{B}^* \tag{6.176}$$

$$\frac{\partial \boldsymbol{M}_n^*}{\partial t^*} = \boldsymbol{\omega}_{\mathrm{p}}^* \times \boldsymbol{M}_{n-1}^* - (1+\chi_0)\boldsymbol{M}_n^* + \chi_0 \boldsymbol{B}^*, (n \geq 1) \tag{6.177}$$

The solution for the zeroth-order problem defined by Eq. (6.176) is independent of spin velocity. Considering the constant values of $\hat{H}_x$ and $\hat{B}_y$ in Eqs. (6.99) and (6.100), respectively, the zeroth-order components of the magnetization in the x and y directions can be written by applying Eqs. (6.174) and (6.176)

$$\frac{\partial M^*_{x,0}}{\partial t^*} = -M^*_{x,0} + \chi_0 H^*_x \tag{6.178}$$

$$\frac{\partial M^*_{y,0}}{\partial t^*} = -(1+\chi_0) M^*_{y,0} + \chi_0 B^*_y \tag{6.179}$$

Substituting the zero-order magnetization represented by Eq. (6.172), as well as Eqs. (6.99) and (6.100), into Eqs. (6.178) and (6.179), respectively, one obtains

$$\mathrm{Re}\left\{ \mathrm{i}w^* \hat{M}^*_{x,0} e^{\mathrm{i}w^*t^*} \right\} = \mathrm{Re}\left\{ -\hat{M}^*_{x,0} e^{\mathrm{i}w^*t^*} + \chi_0 \hat{H}^*_x e^{\mathrm{i}w^*t^*} \right\} \tag{6.180}$$

$$\mathrm{Re}\left\{ \mathrm{i}w^* \hat{M}^*_{y,0} e^{\mathrm{i}w^*t^*} \right\} = \mathrm{Re}\left\{ -(1+\chi_0) \hat{M}^*_{y,0} e^{\mathrm{i}w^*t^*} + \chi_0 \hat{B}^*_y e^{\mathrm{i}w^*t^*} \right\} \tag{6.181}$$

Since Eqs. (6.180) and (6.181) are valid for any arbitrary value of w^*, it follows from complex number arithmetic that the expressions enclosed in parentheses on both sides of the equations are equal. Consequently, the zeroth-order solution for magnetization can be obtained

$$\hat{M}^*_{x,0} = \frac{\chi_0}{1+\mathrm{i}w^*} \hat{H}^*_x \tag{6.182}$$

$$\hat{M}^*_{y,0} = \frac{\chi_0}{1+\chi_0+\mathrm{i}w^*} \hat{B}^*_y \tag{6.183}$$

For the nth-order problem described by Eq. (6.175), the component equations for the nth-order term of magnetization in the x and y directions are derived by applying Eqs. (6.99) and (6.100)

$$\frac{\partial M^*_{x,n}}{\partial t^*} = -\omega^*_{\mathrm{p,z}} M^*_{y,(n-1)} - M^*_{x,n} \tag{6.184}$$

$$\frac{\partial M^*_{y,n}}{\partial t^*} = \omega^*_{\mathrm{p,z}} M^*_{x,(n-1)} - (1+\chi_0) M^*_{y,n} \tag{6.185}$$

where the nth-order terms of both H^*_x and B^*_y are zero. Utilizing a similar approach to Eqs. (6.180) and (6.181), one can obtain the recursive solution for the nth-order equation

$$\hat{M}^*_{x,n} = -\frac{\omega^*_{\mathrm{p,z}}}{1+\mathrm{i}w^*} \hat{M}^*_{y,(n-1)} \tag{6.186}$$

$$\hat{M}_{y,n}^{*} = \frac{\omega_{\mathrm{p,z}}^{*}}{1+\chi_0+\mathrm{i}w^{*}}\hat{M}_{x,(n-1)}^{*} \tag{6.187}$$

Based on the recursive relations given by Eqs. (6.186) and (6.187), the nth-order approximate solutions for magnetization can be derived

$$\hat{M}_{x,n}^{*} = \begin{cases} (-1)^{\frac{n+1}{2}}\chi_0\left(\omega_{\mathrm{p,z}}^{*}\right)^{n}\left(\frac{1}{1+\mathrm{i}w^{*}}\right)^{\frac{n+1}{2}}\left(\frac{1}{1+\chi_0+\mathrm{i}w^{*}}\right)^{\frac{n+1}{2}}\hat{B}_{y}^{*}, & n \text{ is odd} \\ (-1)^{\frac{n}{2}}\chi_0\left(\omega_{\mathrm{p,z}}^{*}\right)^{n}\left(\frac{1}{1+\mathrm{i}w^{*}}\right)^{\frac{n}{2}+1}\left(\frac{1}{1+\chi_0+\mathrm{i}w^{*}}\right)^{\frac{n}{2}}\hat{H}_{x}^{*}, & n \text{ is even} \end{cases} \tag{6.188}$$

$$\hat{M}_{y,n}^{*} = \begin{cases} (-1)^{\frac{n-1}{2}}\chi_0\left(\omega_{\mathrm{p,z}}^{*}\right)^{n}\left(\frac{1}{1+\mathrm{i}w^{*}}\right)^{\frac{n+1}{2}}\left(\frac{1}{1+\chi_0+\mathrm{i}w^{*}}\right)^{\frac{n+1}{2}}\hat{H}_{x}^{*}, & n \text{ is odd} \\ (-1)^{\frac{n}{2}}\chi_0\left(\omega_{\mathrm{p,z}}^{*}\right)^{n}\left(\frac{1}{1+\mathrm{i}w^{*}}\right)^{\frac{n}{2}}\left(\frac{1}{1+\chi_0+\mathrm{i}w^{*}}\right)^{\frac{n}{2}+1}\hat{B}_{y}^{*}, & n \text{ is even} \end{cases} \tag{6.189}$$

When $n=1$, the first-order term of the regular perturbation method can be obtained from Eqs. (6.188) and (6.189)

$$\hat{M}_{x,1}^{*} = -\frac{\chi_0\omega_{\mathrm{p,z}}^{*}\hat{B}_{y}^{*}}{\left(1+\mathrm{i}w^{*}\right)\left(1+\chi_0+\mathrm{i}w^{*}\right)} \tag{6.190}$$

$$\hat{M}_{y,1}^{*} = \frac{\chi_0\omega_{\mathrm{p,z}}^{*}\hat{H}_{x}^{*}}{\left(1+\mathrm{i}w^{*}\right)\left(1+\chi_0+\mathrm{i}w^{*}\right)} \tag{6.191}$$

From Eqs. (6.182), (6.186), (6.190), and (6.191), one can obtain the first-order approximate solution for the magnetization when the small parameter $\epsilon=\Omega_0\tau$,

$$\boldsymbol{M}^{*} = \mathrm{Re}\left\{\left[\hat{M}_{x}^{*}\boldsymbol{e}_x+\hat{M}_{y}^{*}\boldsymbol{e}_y\right]e^{\mathrm{i}w^{*}t^{*}}\right\} \tag{6.192}$$

where

$$\hat{M}_{x}^{*} = \hat{M}_{x,0}^{*}+\epsilon\hat{M}_{x,1}^{*}+O\left(\epsilon^{2}\right) = \chi_0\frac{\left(1+\chi_0+\mathrm{i}w^{*}\right)\hat{H}_{x}^{*}-\epsilon\omega_{\mathrm{p,z}}^{*}\hat{B}_{y}^{*}}{\left(1+\mathrm{i}w^{*}\right)\left(1+\chi_0+\mathrm{i}w^{*}\right)}+O\left(\epsilon^{2}\right) \tag{6.193}$$

$$\hat{M}_{y}^{*} = \hat{M}_{y,0}^{*}+\epsilon\hat{M}_{y,1}^{*}+O\left(\epsilon^{2}\right) = \chi_0\frac{\left(1+\mathrm{i}w^{*}\right)\hat{B}_{y}^{*}+\epsilon\omega_{\mathrm{p,z}}^{*}\hat{H}_{x}^{*}}{\left(1+\mathrm{i}w^{*}\right)\left(1+\chi_0+\mathrm{i}w^{*}\right)}+O\left(\epsilon^{2}\right) \tag{6.194}$$

6.4.3.3 Time-averaged Magnetic Moment

The time-averaged magnetic torque density of the ferrofluid can still be represented by Eq. (6.124), as the characteristic time related to magnetic field quantities, $1/w^*$, is generally much shorter than the characteristic time of the fluid's macroscopic motion, h_0/v_0. Therefore, it is reasonable to assume that the spin velocity $\omega^*_{p,z}$ in Eqs. (6.188) and (6.189) is independent of time. Under this assumption, given the representation in Eq. (6.164), the magnetic torque density term in Eq. (6.169) becomes

$$\langle L^*_{m,z}\rangle = \frac{1}{2}\mathrm{Re}\left[\hat{M}^*_x\overline{\hat{B}^*_y} - \overline{\hat{M}^*_y}\left(\hat{H}^*_x + \hat{M}^*_x\right)\right] \tag{6.195}$$

Since the magnetic torque is directly proportional to the square of the magnetization, it can also be expressed as a function of the small parameter ϵ,

$$\langle L^*_{m,z}\rangle = \sum_{n=0}^{\infty}\epsilon^n\langle L^*_{m,z}\rangle_n \tag{6.196}$$

where the nth-order term is

$$\langle L^*_{m,z}\rangle_n = \frac{1}{2}\mathrm{Re}\left[\hat{M}^*_{x,n}\overline{\hat{B}^*_y} - \overline{\hat{M}^*_{y,n}}\hat{H}^*_x - \sum_{k=0}^{n}\overline{\hat{M}^*_{y,n-k}}\hat{M}^*_{x,k}\right] \tag{6.197}$$

Substituting Eqs. (6.182) and (6.183) into Eq. (6.197) yields the zero-order term of the magnetic torque

$$\begin{aligned}\left\langle L^*_{m,z}\right\rangle_0 &= \frac{1}{2}\mathrm{Re}\left(\hat{M}^*_{x,0}\overline{\hat{B}^*_y} - \overline{\hat{M}^*_{y,0}}\hat{H}^*_x - \overline{\hat{M}^*_{y,0}}\hat{M}^*_{x,0}\right)\\ &= -\chi_0\,\mathrm{Re}\left[\frac{\mathrm{i}w^*}{\left(1+\chi_0+w^{*2}\right)+\mathrm{i}\chi_0 w^*}\overline{\hat{B}^*_y}\hat{H}^*_x\right]\\ &= -\chi_0\frac{\mathrm{Re}\left\{\left[\chi_0 w^{*2} + \mathrm{i}w^*\left(1+\chi_0+w^{*2}\right)\right]\overline{\hat{B}^*_y}\hat{H}^*_x\right\}}{\left(1+\chi_0+w^{*2}\right)^2+\chi_0^2 w^{*2}}\end{aligned} \tag{6.198}$$

Figure 6.42 presents two curves depicting the relationship between the magnitude of the zero-order magnetic torque term and the frequency of magnetic field, w^*, under the influence of two time-varying magnetic fields. Upon examination, it is observed that these curves exhibit extreme points. To pinpoint these extreme points, one lets

$$\frac{\partial\langle L^*_{m,z}\rangle_0}{\partial w^*} = 0 \tag{6.199}$$

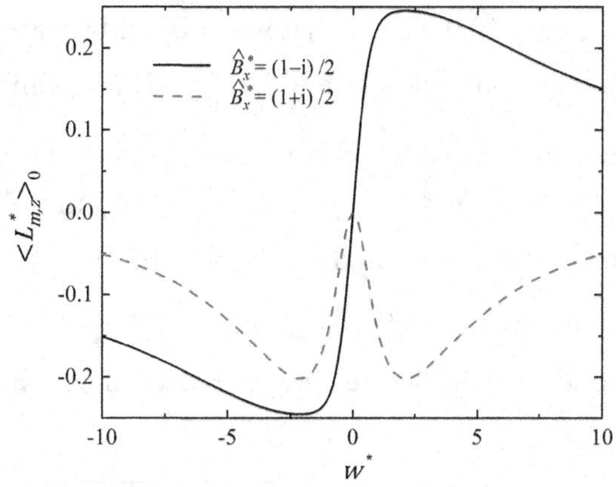

FIGURE 6.42 Variation of the zero-order magnetic moment term with w^*. ($\chi_0 = 3.5$, $\hat{H}_x^* = (1+\mathrm{i})/2$). (Reprinted from Rinaldi and Zahn, 2002 with permission from AIP.)

Substituting Eq. (6.198) into Eq. (6.199) yields

$$\begin{aligned}&\left[\left(1+\chi_0+w^{*2}\right)^2+\chi_0^2 w^{*2}\right]\mathrm{Re}\left\{\left[2\chi_0 w^*+\mathrm{i}\left(1+\chi_0+3w^{*2}\right)\right]\overline{\hat{B}_y^*\hat{H}_x^*}\right\}\\&-2\left[2w^*\left(1+\chi_0+w^{*2}\right)+\chi_0^2 w^*\right]\mathrm{Re}\left\{\left[\chi_0 w^{*2}+\mathrm{i}w^*\left(1+\chi_0+w^{*2}\right)\right]\overline{\hat{B}_y^*\hat{H}_x^*}\right\}=0\end{aligned} \tag{6.200}$$

Based on the rules of complex number operation

$$\mathrm{Re}\left(C_1C_2\right)=\mathrm{Re}\left(C_1\right)\mathrm{Re}\left(C_2\right)-\mathrm{Im}\left(C_1\right)\mathrm{Im}\left(C_2\right) \tag{6.201}$$

Eq. (6.200) is further written as

$$\begin{aligned}&\left\{\begin{matrix}\left[\left(1+\chi_0+w^{*2}\right)^2+\chi_0^2 w^{*2}\right]2\chi_0 w^*-\\ 2\left[2w^*\left(1+\chi_0+w^{*2}\right)+\chi_0^2 w^*\right]\chi_0 w^{*2}\end{matrix}\right\}\mathrm{Re}\left(\overline{\hat{B}_y^*\hat{H}_x^*}\right)\\&-\left\{\begin{matrix}\left[\left(1+\chi_0+w^{*2}\right)^2+\chi_0^2 w^{*2}\right]\left(1+\chi_0+3w^{*2}\right)-\\ 2\left[2w^*\left(1+\chi_0+w^{*2}\right)+\chi_0^2 w^*\right]\left(1+\chi_0+w^{*2}\right)w^*\end{matrix}\right\}\mathrm{Im}\left(\overline{\hat{B}_y^*\hat{H}_x^*}\right)=0\end{aligned} \tag{6.202}$$

It can be simplified as

$$\left(1+\chi_0-w^{*2}\right)\left[\begin{matrix}2\chi_0 w^*\left(1+\chi_0+w^{*2}\right)\mathrm{Re}\left(\overline{\hat{B}_y^*\hat{H}_x^*}\right)\\ -\left(1+\chi_0+w^{*2}+\chi_0 w^*\right)\left(1+\chi_0+w^{*2}-\chi_0 w^*\right)\mathrm{Im}\left(\overline{\hat{B}_y^*\hat{H}_x^*}\right)\end{matrix}\right]=0 \tag{6.203}$$

Solving the algebraic Eq. (6.203) reveals that when $\hat{B}_y^*$ and $\hat{H}_x^*$ are in phase, the product $\overline{\hat{B}_y^*}\hat{H}_x^*$ is a pure real number. Consequently, $\mathrm{Re}\left(\overline{\hat{B}_y^*}\hat{H}_x^*\right) = \left|\overline{\hat{B}_y^*}\hat{H}_x^*\right|$, while $\mathrm{Im}\left(\overline{\hat{B}_y^*}\hat{H}_x^*\right) = 0$. This indicates that the extrema of $\langle L_{m,z}^* \rangle_0$ occurs at $w^* = \pm\sqrt{1+\chi_0}$ and $w^* = 0$, as exemplified by the dashed lines in Figure 6.42. On the other hand, when $\hat{B}_y^*$ and $\hat{H}_x^*$ are out of phase, the product $\overline{\hat{B}_y^*}\hat{H}_x^*$ becomes a pure imaginary number. In this case, $\mathrm{Re}\left(\overline{\hat{B}_y^*}\hat{H}_x^*\right) = 0$ and $\mathrm{Im}\left(\overline{\hat{B}_y^*}\hat{H}_x^*\right) = -\left|\overline{\hat{B}_y^*}\hat{H}_x^*\right|$. If $0 < \chi_0 < 2+2\sqrt{2}$, the extrema of $\langle L_{m,z}^* \rangle_0$ occurs at $w^* = \pm\sqrt{1+\chi_0}$, as demonstrated by the solid lines in Figure 6.42. However, if $\chi_0 > 2+2\sqrt{2}$, the extrema of $\langle L_{m,z}^* \rangle_0$ are located at

$$w^* = \frac{\pm\chi_0 \pm \sqrt{\left(\chi_0 - 2 - 2\sqrt{2}\right)\left(\chi_0 - 2 + 2\sqrt{2}\right)}}{2}$$

By substituting Eqs. (6.182), (6.183), (6.190), and (6.191) into Eq. (6.197), the first-order term of the magnetic torque can be obtained

$$\begin{aligned}
\left\langle L_{m,z}^* \right\rangle_1 &= \frac{1}{2}\mathrm{Re}\left(\hat{M}_{x,1}^*\overline{\hat{B}_y^*} - \hat{M}_{y,1}^*\overline{\hat{H}_x^*} - \hat{M}_{y,1}^*\overline{\hat{M}_{x,0}^*} - \hat{M}_{y,0}^*\overline{\hat{M}_{x,1}^*}\right) \\
&= -\frac{1}{2}\chi_0\omega_{\mathrm{p},z}^*\,\mathrm{Re}\left[\frac{\left(1-\mathrm{i}w^*\right)^2 \hat{B}_y^*\overline{\hat{B}_y^*} + \left(1+\chi_0+\mathrm{i}w^*\right)^2 \hat{H}_x^*\overline{\hat{H}_x^*}}{\left(1-\mathrm{i}w^*\right)\left(1+\mathrm{i}w^*\right)\left(1+\chi_0+\mathrm{i}w^*\right)\left(1+\chi_0-\mathrm{i}w^*\right)}\right] \\
&= \frac{1}{2}\chi_0\omega_{\mathrm{p},z}^*\frac{\left(w^{*2}-1\right)\left|\hat{B}_y^*\right|^2 + \left[w^{*2} - \left(1+\chi_0\right)^2\right]\left|\hat{H}_x^*\right|^2}{\left(1+\chi_0+w^{*2}\right)^2 + \chi_0^2 w^{*2}}
\end{aligned} \tag{6.204}$$

Letting

$$\xi = \frac{1}{2}\chi_0\frac{\left(w^{*2}-1\right)\left|\hat{B}_y^*\right|^2 + \left[w^{*2} - \left(1+\chi_0\right)^2\right]\left|\hat{H}_x^*\right|^2}{\left(1+\chi_0+w^{*2}\right)^2 + \chi_0^2 w^{*2}} \tag{6.205}$$

then $\langle L_{m,z}^* \rangle_1 = \xi\omega_{\mathrm{p},z}^*$. Figure 6.43 presents the relationship between the first-order magnetic torque term $\langle L_{m,z}^* \rangle_1$ and the frequency of the magnetic field, w^*, under the influence of two types of time-varying magnetic fields. Similarly, these curves exhibit points of extrema.

To determine the positions of extreme points on the curve of the first-order magnetic moment term $\langle L_{m,z}^* \rangle_1$, assuming a constant $\omega_{\mathrm{p},z}^*$, one can take the derivative of Eq. (6.204) with respect to w^* and equate the resulting derivative to zero

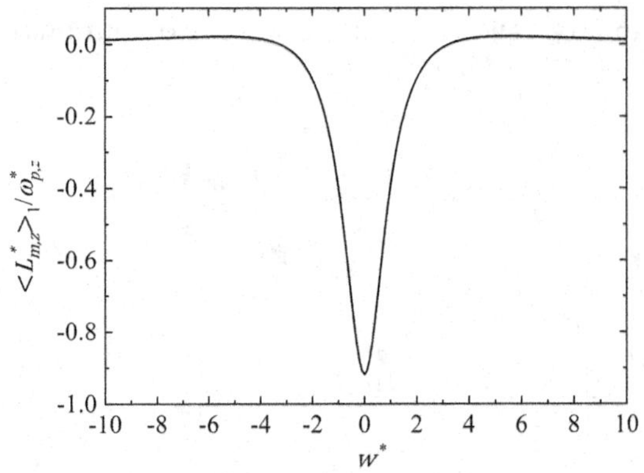

FIGURE 6.43 Variation of the first-order magnetic moment term with w^*. ($\chi_0 = 3.5$, $\hat{H}_x^* = (1+\mathrm{i})/2$, $\hat{B}_y^* = (1\pm\mathrm{i})/2$). (Reprinted from Rinaldi and Zahn, 2002 with permission from AIP.)

$$\begin{aligned} & w^*\left(\left|\hat{B}_y^*\right|^2+\left|\hat{H}_x^*\right|^2\right)\left[\left(1+\chi_0+w^{*2}\right)^2+\chi_0^2 w^{*2}\right] \\ & -\left[2w^*\left(1+\chi_0+w^{*2}\right)+\chi_0^2 w^*\right]\left\{\left(w^{*2}-1\right)\left|\hat{B}_y^*\right|^2+\left[w^{*2}-\left(1+\chi_0\right)^2\right]\left|\hat{H}_x^*\right|^2\right\}=0 \end{aligned} \tag{6.206}$$

After a rearrangement of Eq. (6.206), it can be found that

$$w^*\left\{\begin{array}{c} \left\{\left[\left(1+\chi_0+w^{*2}\right)^2+\chi_0^2 w^{*2}\right]-\left(w^{*2}-1\right)\left[2w^*\left(1+\chi_0+w^{*2}\right)+\chi_0^2 w^*\right]\right\}\left|\hat{B}_y^*\right|^2+ \\ \left\{\left[\left(1+\chi_0+w^{*2}\right)^2+\chi_0^2 w^{*2}\right]-\left[w^{*2}-\left(1+\chi_0\right)^2\right]\left[2w^*\left(1+\chi_0+w^{*2}\right)+\chi_0^2 w^*\right]\right\}\left|\hat{H}_x^*\right|^2 \end{array}\right\}=0$$

which can be further reduced to

$$w^*\left\{\begin{array}{c} \left[-w^{*4}+2w^{*2}+2\left(1+\chi_0\right)^2+1\right]\left|\hat{B}_y^*\right|^2+ \\ \left[-w^{*4}+2\left(1+\chi_0\right)^2 w^{*2}+\left(1+\chi_0\right)^4+2\left(1+\chi_0\right)^2\right]\left|\hat{H}_x^*\right|^2 \end{array}\right\}=0$$

The final polynomial equation satisfied by w^* is

$$w^*\left\{\begin{array}{c} \left(\left|\hat{B}_y^*\right|^2+\left|\hat{H}_x^*\right|^2\right)w^{*4}-2\left[\left|\hat{B}_y^*\right|^2+\left(1+\chi_0\right)^2\left|\hat{H}_x^*\right|^2\right]w^{*2}- \\ \left\{\left[2\left(1+\chi_0\right)^2+1\right]\left|\hat{B}_y^*\right|^2+\left(1+\chi_0\right)^2\left[\left(1+\chi_0\right)^2+2\right]\left|\hat{H}_x^*\right|^2\right\} \end{array}\right\}=0 \tag{6.207}$$

One solution of Eq. (6.207) is $w^* = 0$, and the other two solutions can be calculated using the quadratic formula

$$
\begin{aligned}
w^* &= \pm\left\{\frac{|\hat{B}_y^*|^2 + (1+\chi_0)^2|\hat{H}_x^*|^2}{|\hat{B}_y^*|^2 + |\hat{H}_x^*|^2} + \frac{\left\{\left[|\hat{B}_y^*|^2 + (1+\chi_0)^2|\hat{H}_x^*|^2\right]^2 + \left(|\hat{B}_y^*|^2 + |\hat{H}_x^*|^2\right)\left\{\left[2(1+\chi_0)^2+1\right]|\hat{B}_y^*|^2 + (1+\chi_0)^2\left[(1+\chi_0)^2+2\right]|\hat{H}_x^*|^2\right\}\right\}^{\frac{1}{2}}}{|\hat{B}_y^*|^2 + |\hat{H}_x^*|^2}\right\}^{\frac{1}{2}} \\
&= \pm\left\{\frac{|\hat{B}_y^*|^2 + (1+\chi_0)^2|\hat{H}_x^*|^2}{|\hat{B}_y^*|^2 + |\hat{H}_x^*|^2} + \frac{\left\{2\left[(1+\chi_0)^2+1\right]|\hat{B}_y^*|^4 + 2(1+\chi_0)^2\left[(1+\chi_0)^2+1\right]|\hat{H}_x^*|^4 + \left\{4(1+\chi_0)^2 + \left[(1+\chi_0)^2+1\right]^2\right\}|\hat{B}_y^*|^2|\hat{H}_x^*|^2\right\}^{\frac{1}{2}}}{|\hat{B}_y^*|^2 + |\hat{H}_x^*|^2}\right\}^{\frac{1}{2}} \\
&= \pm\left\{\frac{|\hat{B}_y^*|^2 + (1+\chi_0)^2|\hat{H}_x^*|^2}{|\hat{B}_y^*|^2 + |\hat{H}_x^*|^2} + \frac{\left\{2|\hat{B}_y^*|^2 + \left[(1+\chi_0)^2+1\right]|\hat{H}_x^*|^2\right\}^{\frac{1}{2}}\left\{\left[(1+\chi_0)^2+1\right]|\hat{B}_y^*|^2 + 2(1+\chi_0)^2|\hat{H}_x^*|^2\right\}^{\frac{1}{2}}}{|\hat{B}_y^*|^2 + |\hat{H}_x^*|^2}\right\}^{\frac{1}{2}}
\end{aligned}
\tag{6.208}
$$

These solutions correspond to the positions of the three extrema on the curve of the first-order magnetic moment term $\langle L_{m,z}^* \rangle_1$ as illustrated in Figure 6.43 as an example.

Retaining only the zeroth-order and first-order terms in Eq. (6.196), one can obtain an approximation for the time-averaged magnetic torque as follows:

$$
\langle L_{m,z}^* \rangle = \langle L_{m,z}^* \rangle_0 + \epsilon \langle L_{m,z}^* \rangle_1 = \langle L_{m,z}^* \rangle_0 + \epsilon \xi \omega_{\mathrm{p},z}^* \tag{6.209}
$$

where $\langle T_{m,z}^* \rangle_0$ and ξ are given by Eqs. (6.198) and (6.205), respectively.

6.4.3.4 Velocity and Spin Velocity When No-slip Spin Velocity Boundary Condition Is Used

Substituting the approximate value of the time-averaged magnetic torque represented by Eq. (6.209) into the angular momentum equation (6.169) yields

$$
\frac{\eta'}{\zeta h_0^2}\frac{\mathrm{d}^2\omega_{\mathrm{p},z}^*}{\mathrm{d}y^{*2}} - 2\frac{\mathrm{d}v_x^*}{\mathrm{d}y^*} + \left(\frac{\mu_0 H_0^2}{\zeta\Omega_0}\epsilon\xi - 4\right)\omega_{\mathrm{p},z}^* + \frac{\mu_0 H_0^2}{\zeta\Omega_0}\langle L_{m,z}^* \rangle_0 = 0 \tag{6.210}
$$

In the system of governing equations consisting of Eqs. (6.168) and (6.210), there exist three dimensionless parameters that influence the solutions for flow velocity and spin velocity. The first parameter is the ratio of vortex viscosity to shear viscosity, ζ / η, which is proportional to the volume fraction of magnetic particles in ferrofluids. According to Eq. (2.213), it can be approximately calculated as

$$\frac{\zeta}{\eta} = \frac{3}{2}\phi \tag{6.211}$$

The second dimensionless parameter is the ratio of spin viscosity to vortex viscosity multiplied by the square of the characteristic length, $\eta' / \left(\zeta h_0^2\right)$, which is directly proportional to the square of the ratio between the intrinsic characteristic length of the ferrofluid and the flow characteristic length, namely,

$$\frac{\eta'}{\zeta h_0^2} \propto \frac{L_\mathrm{p}^2}{h_0^2} \tag{6.212}$$

The intrinsic characteristic length L_p is comparable to the diameter of particles in the ferrofluid, which is the reason why many studies overlook the first term on the left-hand side of Eq. (6.210). However, some research has shown that even when this parameter is small, spin diffusion and spin-vorticity coupling can lead to the formation of boundary layers on walls and interfaces. These boundary layers can affect the shear stress exerted on walls and interfaces, potentially influencing flow experiment results aimed at assessing ferrofluid properties (Rinaldi and Zahn, 2002). The third parameter is $\mu_0 H_0^2 / \left(\zeta \Omega_0\right)$, representing the ratio between the intrinsic angular momentum "pumped" by the magnetic moment and the interconversion of external/internal angular momentum through the vorticity-spin coupling effect. To ensure a relatively strong interaction between magnetization and spin velocity, let

$$\frac{\mu_0 H_0^2}{\zeta \Omega_0}\epsilon = \frac{\mu_0 H_0^2 \tau}{\zeta} \sim 1 \tag{6.213}$$

However, this assumption necessitates a significant magnetic field intensity, potentially violating the linear assumption between equilibrium magnetization and magnetic field intensity, as they exhibit a nonlinear relationship in the presence of strong magnetic fields.

As shown in Figures 6.44 and 6.45, the flow velocity distribution and spin velocity distribution within the cross-section of a ferrofluid Couette–Poiseuille flow are presented for various magnetic field intensities and frequencies of $w^* = 0.1$ and $w^* = 5$, respectively. From the flow velocity distribution, it is evident that the low-frequency time-varying magnetic field exhibits a slight hindrance to flow, which becomes more pronounced as the magnetic field intensity increases. Notably, in a rotating magnetic field, the flow rate decreases by approximately 2.9% compared to the case without a magnetic field. However, when the frequency is $w^* = 5$, there is no significant difference in the flow velocity distribution across different magnetic field intensities. The reason for the insignificant effect of magnetic field intensity in Figure 6.44 lies in the fact that, for typical ferrofluids with magnetization relaxation times ranging from 10^{-6} to 10^{-4} s and a dynamic viscosity of 0.01 Pa·s, the

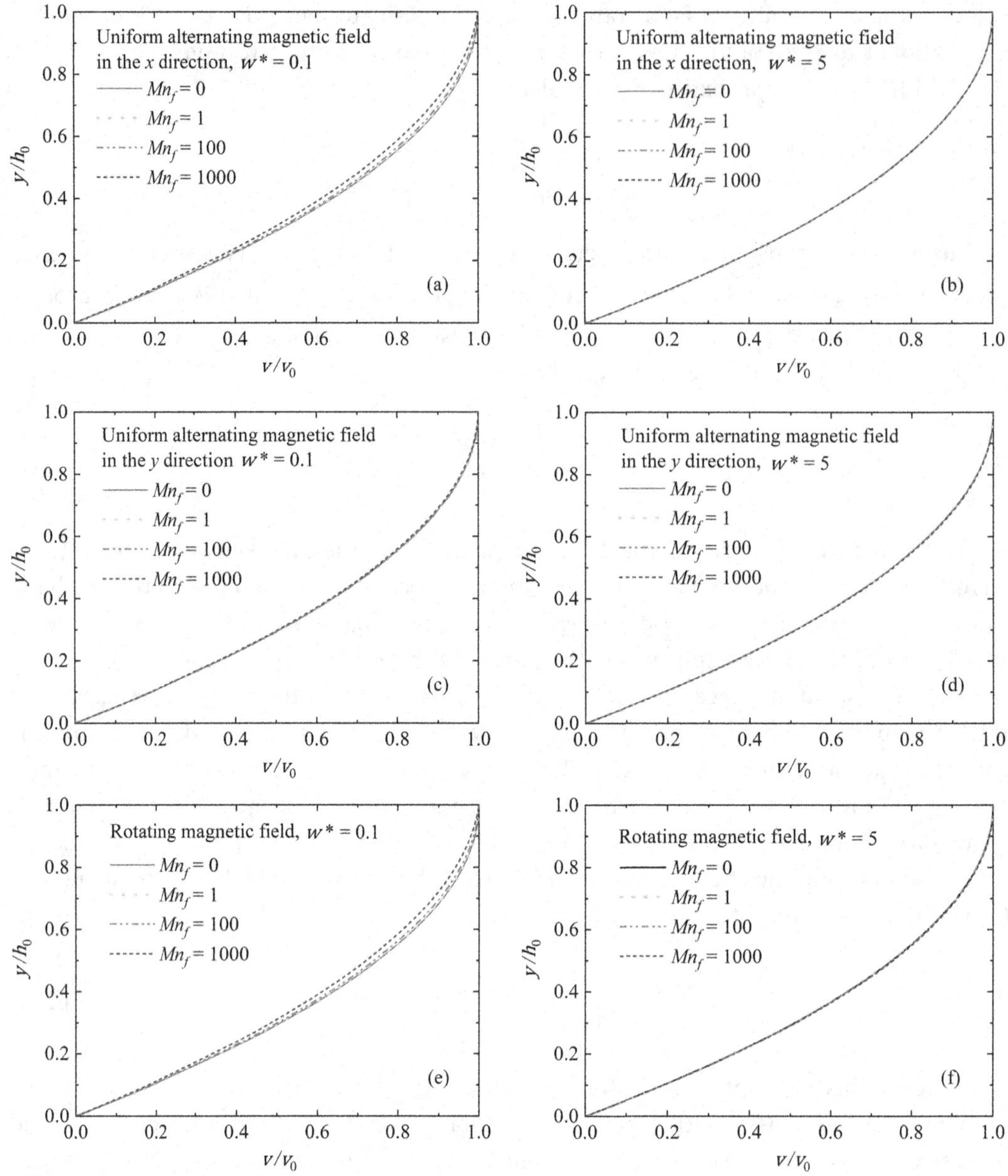

FIGURE 6.44 Flow velocity distributions within the flow cross-section of a ferrofluid Couette–Poiseuille flow under the action of time-varying magnetic fields of two frequencies and different magnitudes ($\chi_0 = 3.5$, $\eta^* / \zeta^* = 5.65$, $P^* = 1$, $Re = 0.1$, $Pe = 0.00625$, $\eta' = 10^{-20}$ N·s). (Reprinted from Yang et al., 2024 with permission from Elsevier.)

interaction between magnetization and spin velocity is minimal when $Mn_f \leq 100$ and $Mn_f \epsilon \ll 1$. Under these conditions, the magnetization and spin velocity exhibit weak coupling. However, it is important to note that extremely strong magnetic fields cannot be applied as they may violate the assumption of a linear relationship between the equilibrium magnetization and magnetic field intensity.

Upon examining the spin velocity distribution depicted in Figure 6.45, it becomes evident that throughout almost the entire flow cross-section, the curves representing the

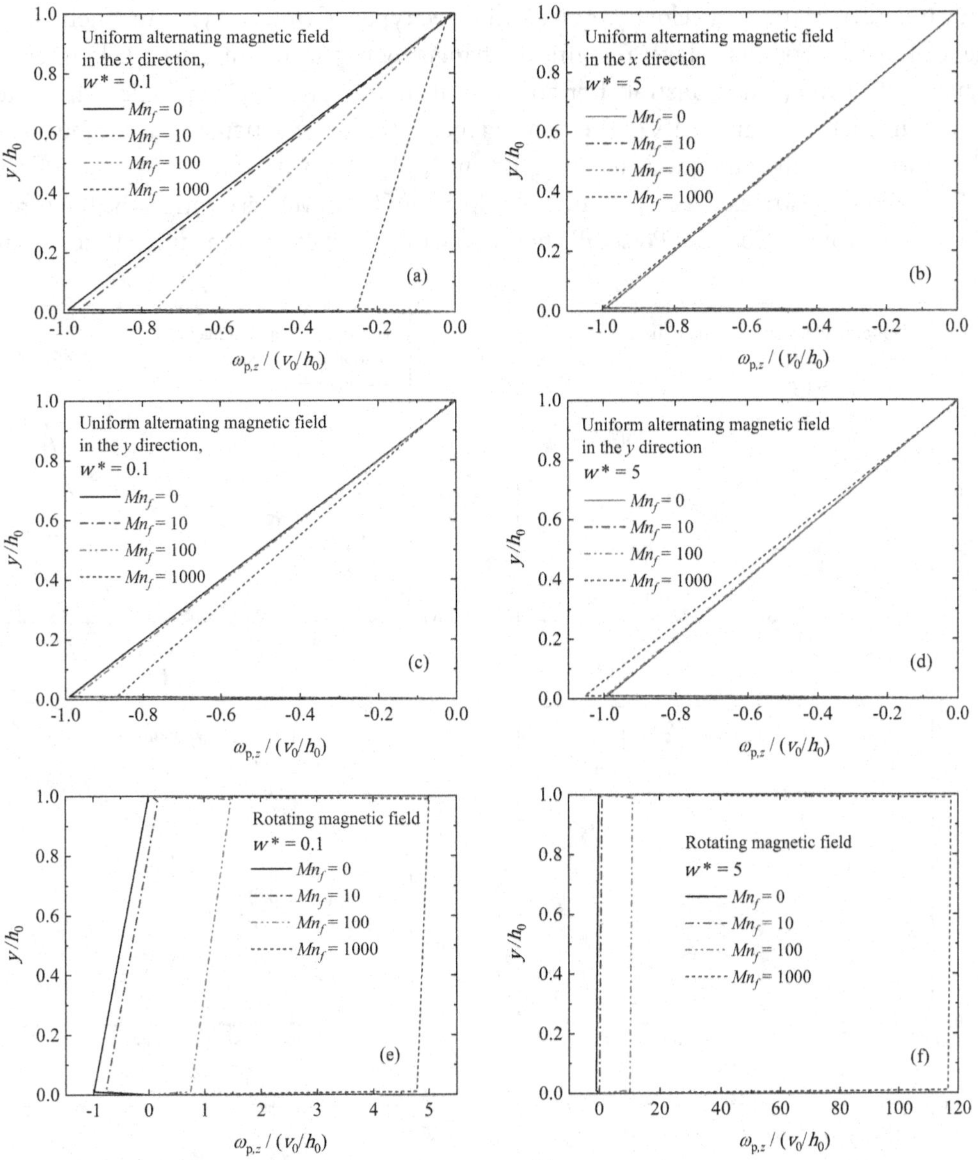

FIGURE 6.45 Spin velocity distributions within the flow cross-section of a ferrofluid Couette–Poiseuille flow under the action of time-varying magnetic fields of two frequencies and different magnitudes ($\chi_0 = 3.5$, $\eta^* / \zeta^* = 5.65$, $P^* = 1$, $Re = 0.1$, $Pe = 0.00625$, $\eta' = 10^{-20}$ N·s). (Reprinted from Yang et al., 2024 with permission from Elsevier.)

variation of spin velocity with position maintain a linear relationship with a constant slope. However, due to the boundary condition of zero spin velocity at the stationary wall, the adjacent boundary layer is extremely thin. In the homogeneously alternating magnetic field along the x and y directions, higher amplitudes of magnetic field intensity at lower frequencies result in a smaller slope of spin velocity variation within the cross-section. In contrast, at higher frequencies, although the change is minor, the trend of slope variation reverses. In the rotating magnetic field, a significant increase in magnetic field intensity leads to a reversal of spin velocity. Across all three types of time-varying magnetic fields, the width of the boundary layer remains unchanged with varying magnetic field intensity. Figure 6.45(e) and (f) demonstrate horizontal shifts in the spin velocity profiles when altering the magnetic field intensity of the rotating magnetic field, particularly at a frequency of $w^* = 5$, where stronger magnetic fields significantly alter the value of $\omega^*_{p,z}$.

Figure 6.46 illustrates the distribution of velocity and spin velocity within the flow cross-section of ferrofluid Couette–Poiseuille flows under the influence of the three time-varying

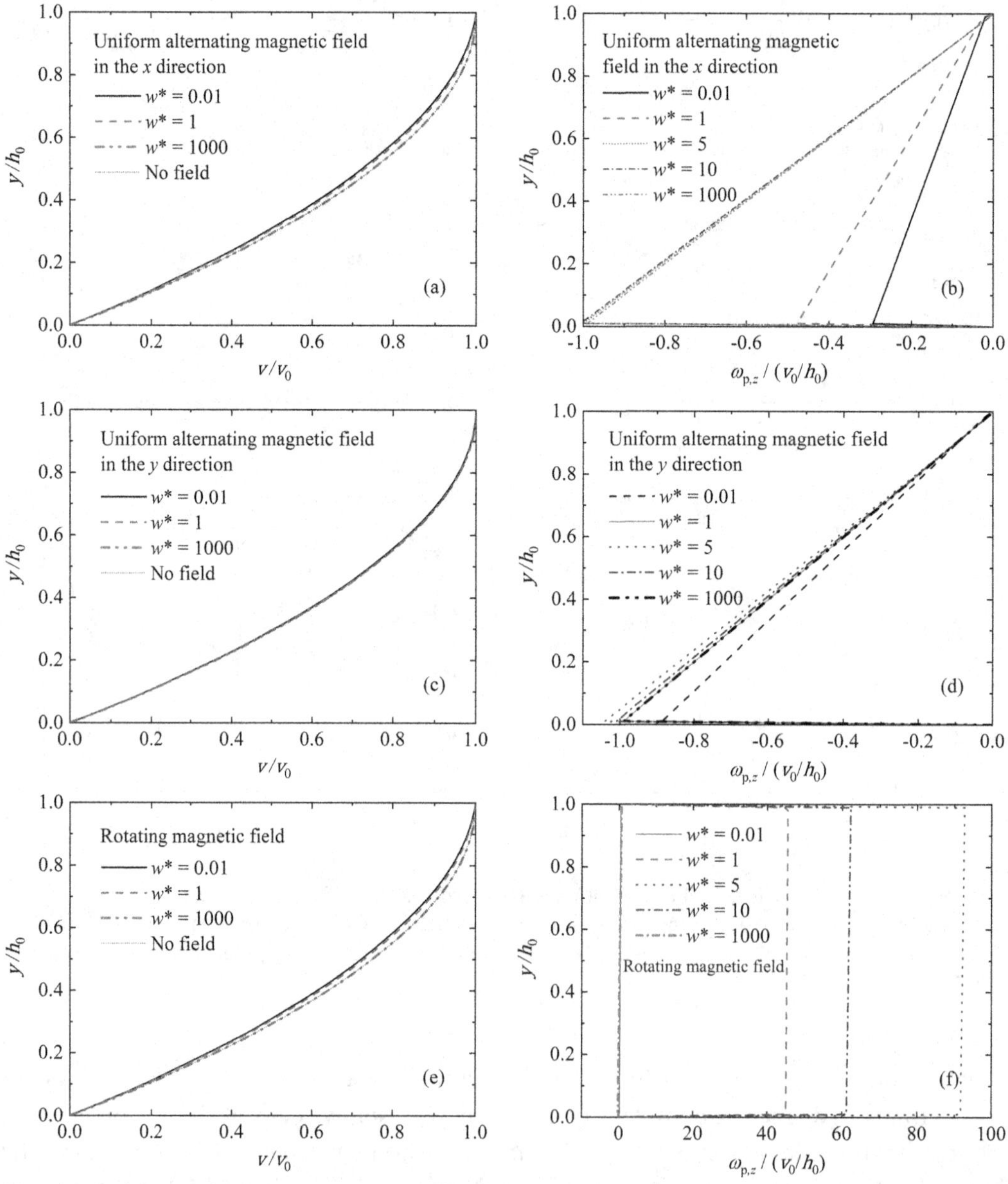

FIGURE 6.46 Velocity and spin velocity distributions within the flow cross-section of a ferrofluid Couette–Poiseuille flow under time-varying magnetic fields of different frequencies ($\chi_0 = 3.5$, $\eta^* / \zeta^* = 5.65$, $P^* = 1$, $Re = 0.1$, $Mn_f = 800$, $Pe = 0.00625$, $\eta' = 10^{-20}$ N·s). (Reprinted from Yang et al., 2024 with permission from Elsevier.)

magnetic fields at different frequencies. It is evident that the impact of frequency on the velocity distribution is minimal, with only a slight hindrance to flow at lower frequencies. This suggests that the assumption of zero spin viscosity overestimates the promotional effect of the time-varying magnetic field on the Couette–Poiseuille flow of ferrofluids. In scenarios where spin viscosity is non-zero, the spin velocity exhibits a linear distribution within the flow cross-section, with its slope peaking at $w^* = 5$ in a homogeneously time-varying magnetic field in the x or y-direction. However, a comparison reveals that the value of $\omega^*_{\mathrm{p},z}$ is significantly higher compared to the case of zero spin viscosity, and there exists a thin boundary layer near the stationary wall. Notably, in rotating magnetic fields, the assumption of zero spin viscosity causes a significant deviation in the spin velocity distribution from the nearly constant value observed in non-zero spin viscosity scenarios. It is important to note that although spin viscosity values are typically very small, their non-negligible impact on both velocity and spin velocity distributions arises because the governing equation for $\omega^*_{\mathrm{p},z}$ has reduced order by two, necessitating the use of distinct boundary conditions.

Spin velocity/vorticity matching boundary conditions have been used in some studies when given the boundary conditions for the spin velocity, that is

$$\omega^*_{\mathrm{p},z}(0) = \Omega^*(0), \text{when } y^* = 0$$
$$\omega^*_{\mathrm{p},z}(1) = \Omega^*(1), \text{when } y^* = 1$$

where $\Omega^* = \Omega / \Omega_0$. This boundary condition is derived based on the assumption of a constant anti-symmetric component in the stress tensor $(2\boldsymbol{\omega}_\mathrm{p} - \nabla \times \boldsymbol{v})$. For the flow configuration in Figure 6.1, they correspond to

$$\omega_{\mathrm{p},z} = -\frac{1}{2} v_x, \text{when } y^* = 0 \text{ or } y^* = 1$$

The results of both the velocity and the spin velocity distributions are similar to those described above for the no-slip spin velocity boundary conditions. The only difference is that the value of the spin velocity is non-zero at the stationary plate boundary as shown in Figure 6.47.

6.4.3.5 *Rotational Viscosity*

To determine the rotational viscosity of a ferrofluid in a Couette–Poiseuille flow under the influence of a time-varying magnetic field, it is assumed that the velocity distribution remains parabolic and then it is expressed in the form given by Eq. (6.1) as

$$\frac{v_x}{v_0} = \frac{y}{h_0} - \frac{h_0^2}{2(\eta + \eta_r) v_0} \frac{\mathrm{d}p}{\mathrm{d}x} \frac{y}{h_0}\left(1 - \frac{y}{h_0}\right) \tag{6.214}$$

Letting

$$P^{*\prime} = -\frac{h_0^2}{2(\eta + \eta_r) v_0} \frac{\mathrm{d}p}{\mathrm{d}x} \tag{6.215}$$

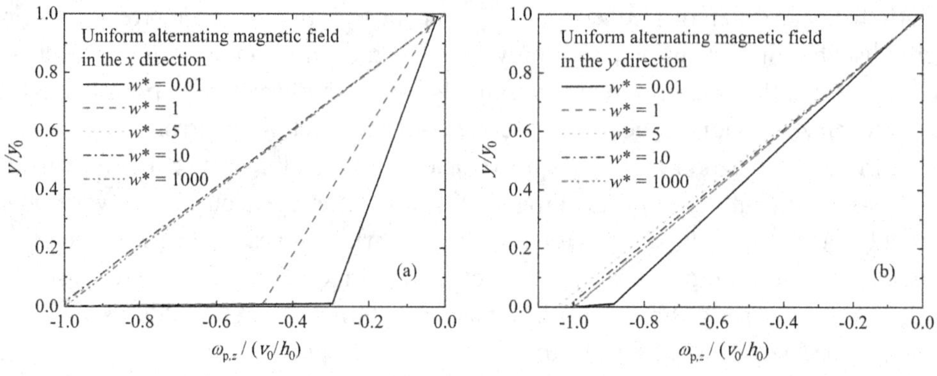

FIGURE 6.47 Spin velocity distributions within the flow cross-section of a ferrofluid Couette–Poiseuille flow in the uniform time-varying magnetic fields in the x and y directions under the velocity/vorticity matching boundary conditions ($\chi_0 = 3.5$, $\eta / \zeta = 5.65$, $\partial p^* / \partial x^* = -2$, $Re = 0.1$, $Mn_f = 800$, $Pe = 0.00625$, $\eta' = 10^{-20}$ N·s).

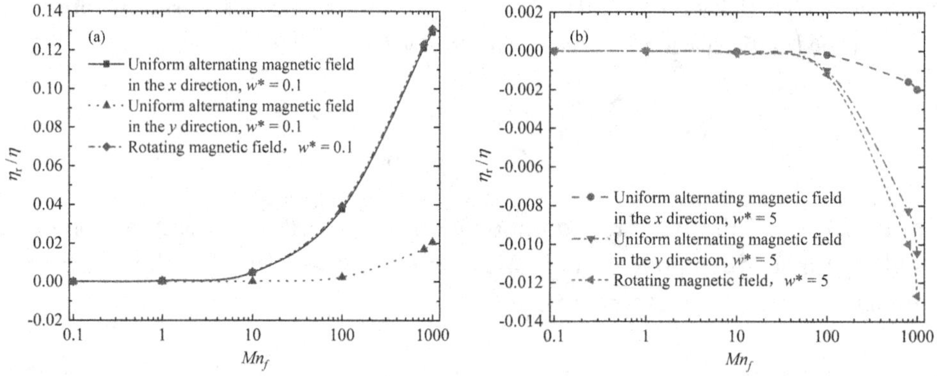

FIGURE 6.48 Variation of rotational viscosity with magnitude of magnetic field intensity in a ferrofluid Couette–Poiseuille flow under the action of time-varying magnetic fields at two frequencies ($\chi_0 = 3.5$, $\eta^* / \zeta^* = 5.65$, $P^* = 1$, $Re = 0.1$, $Pe = 0.00625$, $\eta' = 10^{-20}$ N·s). (a) $w^* = 0.1$; (b) $w^* = 5$. (Reprinted from Yang et al., 2024 with permission from Elsevier.)

then the comparison of Eqs. (6.1) and (6.218) gives the ratio of rotational viscosity to the intrinsic viscosity as

$$\frac{\eta_r}{\eta} = \frac{P^*}{P^{*\prime}} - 1 \tag{6.216}$$

Based on Eq. (6.216), the value of $P^{*\prime}$ is obtained by fitting the flow velocity distribution curve with a quadratic polynomial, which can then be used to calculate η_r / η.

Figure 6.48 illustrates the relationship between the variation of η_r / η within a ferroluid Couette–Poiseuille flow in time-varying magnetic fields, considering two different frequencies of magnetic field: $w^* = 0.1$ and $w^* = 5$. It is evident that when $Mn_f < 10$, the rotational viscosity remains nearly zero. However, for $Mn_f > 10$, the rotational viscosity increases with the increase in magnetic field intensity in the low-frequency magnetic field, while it slightly decreases in the high-frequency magnetic field. This explains why the low-frequency magnetic field hinders the Couette–Poiseuille flow while the high-frequency magnetic field promotes it. At a Mn_f value of 1000, the rotational viscosity increases by 13% in the rotating magnetic field with a frequency of $w^* = 0.1$, whereas it decreases by 1.4% in the rotating magnetic field with a frequency of $w^* = 5$.

To clarify the relationship between rotational viscosity and magnetic field frequency, Figures 6.49 and 6.50 depict the variation curves of rotational viscosity under three types of time-varying magnetic fields when the magnetization relaxation times are $\tau = 4\times10^{-4}$ s and $\tau = 4\times10^{-3}$ s, respectively. Both curves exhibit a positive value at low frequencies, a negative value with a minimum within the range of $w^* = 1 \sim 10$ and a gradual approach to zero beyond this range. This implies the existence of an optimal magnetic field frequency that maximizes the flow rate of Couette–Poiseuille flow. A comparison of Figures 6.49 and 6.50 suggests that this optimal value seems independent of both the magnetization relaxation time and the type of time-varying magnetic field. Among the three types of magnetic fields, the rotating magnetic field achieves the smallest "negative viscosity", reaching a minimum of 1.9% and 20% for magnetization relaxation times of $\tau = 4\times10^{-4}$s and $\tau = 4\times10^{-3}$s, respectively. Furthermore, when the magnetic field frequency is held constant, longer magnetization relaxation times lead to larger absolute values of negative rotational viscosity, indicating a more significant effect of the magnetic field on flow promotion.

6.4.3.6 *Tangential Stress on the Upper Plate*

Based on Eq. (2.229), the shear stress acting on a moving wall by a planar Couette–Poiseuille flow in a time-varying magnetic field can be expressed as

$$f_x = \eta \frac{\partial v_x}{\partial y} + \frac{\mu_0}{2}\left(M_x H_y - M_y H_x\right) \tag{6.217}$$

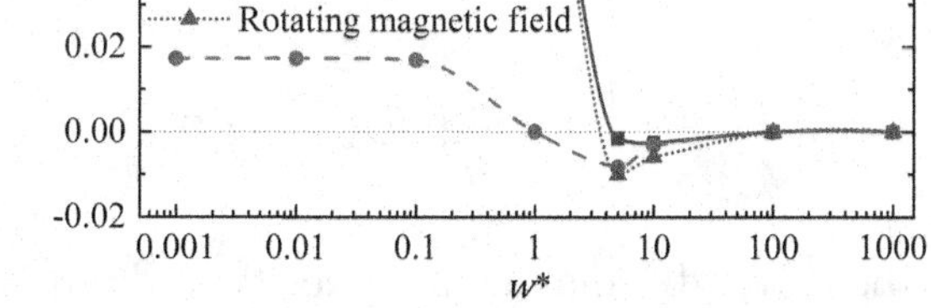

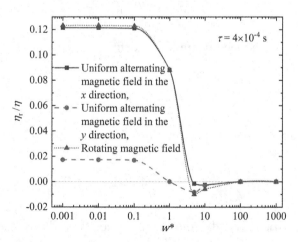

FIGURE 6.49 Variation of rotational viscosity with magnetic field frequency in a ferrofluid Couette–Poiseuille flow for $\tau = 4\times10^{-4}$ s ($\chi_0 = 3.5, \eta^*/\zeta^* = 5.65, P^* = 1, Re = 0.1, Pe = 0.00625, Mn_f = 800, \eta' = 10^{-20}$ N·s). (Reprinted from Yang et al., 2024 with permission from Elsevier.)

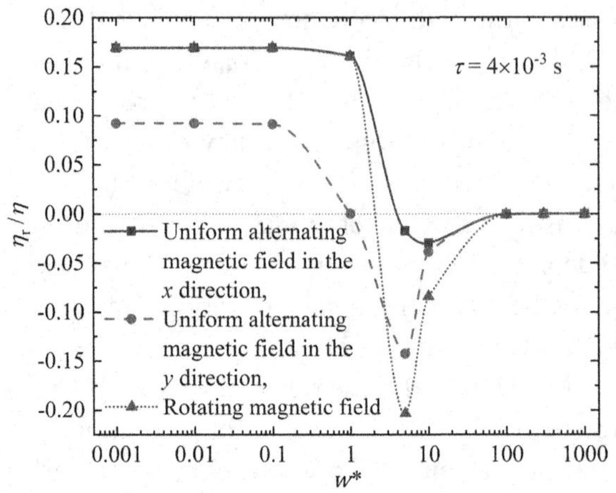

FIGURE 6.50 Variation of rotational viscosity with magnetic field frequency in a ferrofluid Couette–Poiseuille flow for $\tau = 4\times10^{-3}$s ($\chi_0 = 3.5, \eta^*/\zeta^* = 5.65, P^* = 1$, $Re = 0.1$, $Pe = 0.0625$, $Mn_f = 800, \eta' = 10^{-20}$ N·s). (Reprinted from Yang et al., 2024 with permission from Elsevier.)

Upon comparing Eqs. (6.120) and (6.217), it is observed that the second term on the right-hand side of Eq. (6.217) can be expressed as the magnetic torque,

$$f_x = \eta\frac{\partial v_x}{\partial y} + \frac{1}{2}L_{m,z} \tag{6.218}$$

Then the results of the time-averaged magnetic torque can be used to calculate the time-averaged shear stress exerted on the upper plate. Using the dimensionless representation in Eq. (6.164), the shear stress can be expressed as

$$f_x^* = \frac{dv_x^*}{dy^*} + \frac{1}{2}Mn_f\frac{\zeta}{\eta}\langle L_{m,z}^*\rangle \tag{6.219}$$

where the dimensionless tangential stress f_x^* is defined as

$$f_x^* = \frac{f_x}{\eta v_0 / h_0}$$

Substituting the approximate representation of Eq. (6.209) into Eq. (6.219) yields

$$f_x^* = \frac{dv_x^*}{dy^*} + \frac{1}{2}Mn_f\frac{\zeta}{\eta}\left(\langle L_{m,z}^*\rangle_0 + \epsilon\xi\omega_{p,z}^*\right) \tag{6.220}$$

where the velocity gradient dv_x^*/dy^* and the spin velocity $\omega_{p,z}^*$ both take values at $y^* = 1$.

Figure 6.51 depicts the effect of the magnitude of time-varying magnetic fields on the tangential stress exerted by the upper plate in ferrofluid Couette–Poiseuille flows when the field frequency is $w^* = 0.1$ and $w^* = 5$, respectively. In a uniform alternating magnetic field,

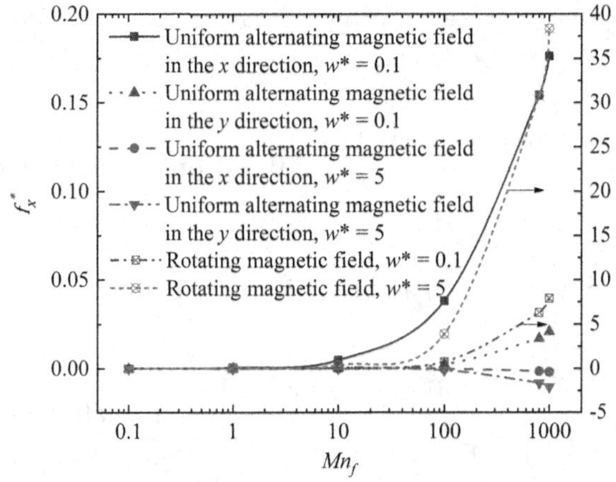

FIGURE 6.51 Effect of the magnitude of the time-varying magnetic field on the tangential stress on the upper plate in ferrofluid Couette–Poiseuille flows ($\chi_0 = 3.5$, $\eta/\zeta = 5.65$, $\partial p^*/\partial x^* = -2$, $Re = 0.1$, $Pe = 0.00625$, $Mn_f = 800$, $\eta' = 10^{-20}$ N·s). (Reprinted from Yang et al., 2024 with permission from Elsevier.)

the shear stress monotonically increases at low frequencies with increasing magnetic field amplitude, while it decreases slowly at high frequencies. However, in a rotating magnetic field, the shear stress increases with magnetic field intensity at both frequencies. Observing Figure 6.43(b), (d), and (f), it becomes evident that the velocity gradient near the upper plate is nearly independent of Mn_f. This implies that the changes in f_x^* shown in Figure 6.51 are primarily attributed to the second term on the right-hand side of Eq. (6.219), which represents the variation in magnetic stress.

In uniformly alternating magnetic fields, according to the representations given by Eqs. (6.101), (6.102), and (6.198), the term $\langle L_{m,z}^* \rangle_0$ is zero, and consequently, the coefficient ξ becomes

$$\xi = \frac{1}{2}\chi_0 \frac{w^{*2} - (1+\chi_0)^2}{\left(1+\chi_0+w^{*2}\right)^2 + \chi_0^2 w^{*2}} \begin{cases} <0, w^* = 0.1 \\ >0, w^* = 5 \end{cases}, \text{for the magnetic field in the } x \text{ direction} \tag{6.221}$$

$$\xi = \frac{1}{2}\chi_0 \frac{w^{*2} - 1}{\left(1+\chi_0+w^{*2}\right)^2 + \chi_0^2 w^{*2}} \begin{cases} <0, w^* = 0.1 \\ >0, w^* = 5 \end{cases}, \text{for the magnetic field in the } y \text{ direction} \tag{6.222}$$

Furthermore, Figure 6.47 reveals that the spin velocity adjacent to the upper plate in both magnetic fields is negative. Consequently, the coefficient of Mn_f in the second term of Eq. (6.220) is positive when $w^* = 0.1$ and negative when $w^* = 5$. This ultimately leads to a positive correlation between f_x^* and Mn_f at low frequencies and a negative correlation at high frequencies.

In the rotating magnetic field, according to Eqs. (6.102) and (6.198), the corresponding $\langle L^*_{m,z}\rangle_0$ is non-zero and can be expressed as

$$\langle L^*_{m,z}\rangle_0 = \frac{\chi_0 w^*\left(1+\chi_0+w^{*2}\right)}{\left(1+\chi_0+w^{*2}\right)^2+\chi_0^2 w^{*2}} > 0, \text{for rotating magnetic field} \tag{6.223}$$

In this case, based on Eqs. (6.103) and (6.205), the coefficient ξ becomes

$$\xi = \frac{1}{2}\chi_0 \frac{\left(w^{*2}-1\right)+\left[w^{*2}-\left(1+\chi_0\right)^2\right]}{\left(1+\chi_0+w^{*2}\right)^2+\chi_0^2 w^{*2}} \begin{cases} <0, w^*=0.1 \\ >0, w^*=5 \end{cases}, \text{for rotating magnetic field} \tag{6.224}$$

Since $\langle L^*_{m,z}\rangle_0$ is much greater than $\epsilon\xi\omega^*_{p,z}$, its influence on Mn_f in Eq. (6.220) is significantly stronger than that of $\epsilon\xi\omega^*_{p,z}$. Consequently, this leads to a positive correlation between f^*_x and Mn_f, and their relationship is independent of the magnetic field frequency.

When $w^*=0.1$, the absolute value of the numerator in Eq. (6.221) is significantly larger than that in Eq. (6.222), that is $\left|w^{*2}-\left(1+\chi_0\right)^2\right| > \left|w^{*2}-1\right|$. Consequently, in a uniform alternating magnetic field along the x-direction, f^*_x should be much larger than that in the y-direction field. This can also be observed from Figure 6.51. On the other hand, when $w^*=5$, the condition $\left|w^{*2}-\left(1+\chi_0\right)^2\right| < \left|w^{*2}-1\right|$ holds, leading to a larger f^*_x in the y-direction uniform alternating magnetic field at higher frequencies. In the rotating magnetic field, at $w^*=5$, the absolute value of the numerator in Eq. (6.224) is significantly larger than that at $w^*=0.1$, which explains why the curve corresponding to $w^*=5$ lies above the curve for $w^*=0.1$ in Figure 6.51.

Another qualitative conclusion is that the tangential stress can be reversed in strong uniform alternating magnetic fields at high frequencies. Overall, the rotating magnetic field can induce larger tangential stresses compared to alternating magnetic fields for the same magnetic field magnitude.

Figure 6.52 depicts the variation of shear stress acting on the upper plate under three types of time-varying magnetic fields as a function of magnetic field frequency. In both uniform alternating magnetic fields along the x and y directions, the shear stress exhibits similar trends. It exhibits positive values at low frequencies, experiences a rapid decline until reaching a minimum (within the range of $w^*=2\sim 8$), and then gradually rises to settle around zero. However, in the rotating magnetic field, the tangential stress initially rises with the frequency, followed by a significant drop back to its initial value. In this case, there is an extreme value around $w^*=1$.

The location of the extrema points depicted in Figure 6.52 can be determined from Eq. (6.220), which indicates that the position of the extrema depends on the term $\langle L^*_{m,z}\rangle_0+\epsilon\xi\omega^*_{p,z}$. In a homogeneously alternating magnetic field, this term is equivalent to ξ, as $\langle L^*_{m,z}\rangle_0$ is zero and $\epsilon\omega^*_{p,z}$ remains constant. Under these conditions, it naturally requires

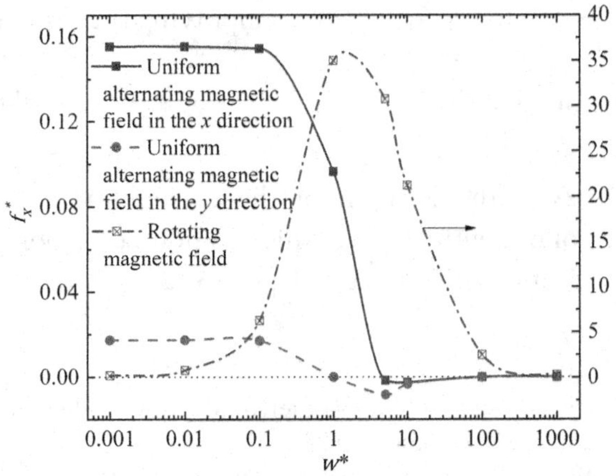

FIGURE 6.52 Effect of the field frequency on the tangential stress on the upper plate in ferrofluid Couette–Poiseuille flows ($\chi_0 = 3.5$, $\eta/\zeta = 5.65$, $\partial p^*/\partial x^* = -2$, $Re = 0.1$, $Pe = 0.00625$, $Mn_f = 800$, $\eta' = 10^{-20}$ N·s). (Reprinted from Yang et al., 2024 with permission from Elsevier.)

$$\frac{\partial \xi}{\partial w^*} = 0 \tag{6.225}$$

Substituting Eqs. (6.221) and (6.222) into Eq. (6.225), respectively, yields

$$2w^*\left\{\begin{matrix} w^{*4} - 2(1+\chi_0)^2 w^{*2} \\ -(1+\chi_0)^2\left[(1+\chi_0)^2 + 2\right]\end{matrix}\right\} = 0, \text{for the magnetic field in the } x \text{ direction} \tag{6.226}$$

$$2w^*\left\{w^{*4} - 2w^{*2} - \left[2(1+\chi_0)^2 + 1\right]\right\} = 0, \text{for the magnetic field in the } y \text{ direction} \tag{6.227}$$

Thus, the locations of the extreme points for these two magnetic fields can be found at

$$w^* = (1+\chi_0)^2 + \sqrt{2}(1+\chi_0)\sqrt{(1+\chi_0)^2 + 1}, \text{for the magnetic field in the } x \text{ direction}$$

$$w^* = 1 + \sqrt{2}\sqrt{(1+\chi_0)^2 + 1}, \text{for the magnetic field in the } y \text{ direction}$$

For a typical ferrofluid with $\chi_0 = 3.5$, the corresponding positions are located at $w^* = 7.0$ and $w^* = 2.7$, respectively, which are consistent with the results shown in Figure 6.52.

In a rotating magnetic field, the location of the extrema points on the curve is determined by $\langle L^*_{m,z}\rangle_0$, and it is required that

$$\frac{\partial \langle L^*_{m,z}\rangle_0}{\partial w^*} = 0 \tag{6.228}$$

Substituting Eq. (6.198) into Eq. (6.228) and simplifying the resultant equation yields

$$\left(1+\chi_0-w^{*2}\right)\left(1+\chi_0+w^{*2}+\chi_0 w^*\right)\left(1+\chi_0+w^{*2}-\chi_0 w^*\right)=0 \tag{6.229}$$

Because $0<\chi_0<2+2\sqrt{2}$, the location of the extremum point of $\langle L^*_{m,z}\rangle_0$ lies at $w^*=\sqrt{1+\chi_0}$. For a ferrofluid with $\chi_0=3.5$, this position corresponds to $w^*=2.1$, which coincides with the observations illustrated in Figure 6.52.

REFERENCES

Amirat Y., Hamdache K., Steady state solutions of ferrofluid flow models, *Communications on Pure and Applied Analysis*, 15(6): 2329–2355, 2016.

Carvalho D. D. de, Gontijo R. G., Magnetization diffusion in duct flow: The magnetic entrance length and the interplay between hydrodynamic and magnetic timescales, *Physics of Fluids*, 32: 072007, 2020.

Cunha F. R., Sobral Y. D., Characterization of the physical parameters in a process of magnetic separation and pressure-driven flow of a magnetic fluid, *Physica A*, 343: 36–64, 2004.

Ghosh D., Das P. K., Control of flow and suppression of separation for Couette-Poiseuille hydrodynamics of ferrofluids using tunable magnetic fields, *Physics of Fluids*, 31: 083609, 2019.

Ghosh D., Meena P. R., Das P. K., A fully analytical solution of convection in ferrofluids during Couette-Poiseuille flow subjected to an orthogonal magnetic field, *International Communications in Heat and Mass Transfer*, 130, 105793, 2022.

Korlie M. S., Mukherjee A., Nita B. G., Stevens J. G., Trubatch A. D., Yecko P., Analysis of flows of ferrofluids under simple shear, *Magnetohydrodynamics*, 44(1): 51–59, 2008.

Landau L. D., Lifshitz E. M., *Fluid mechanics*. Sebastopol: Butterworth-Heinemann. 1987.

Matsuno Y., Araki K., Yamamoto H., Magnetic fluid flows in a two-dimensional channel, *Journal of Magnetism and Magnetic Materials*, 122(1–3): 204–206, 1993.

Paz P. Z. S., Cunha F. R., Sobral Y. D., Stability of plane-parallel flow of magnetic fluids under external magnetic fields, *Applied Mathematics and Mechanics*, 43(2): 295–310, 2022.

Rinaldi C., Zahn M., Effects of spin viscosity on ferrofluid flow profiles in alternating and rotating magnetic fields, *Physics of Fluids*, 14: 2847, 2002.

Weng H. C., Chen C. L., Chen C. K., Non-Newtonian flow of dilute ferrofluids in a uniform magnetic field, *Physical Review E*, 78: 056305, 2008.

Yang W. M., A finite volume solver for ferrohydrodynamics coupled with microscopic magnetization dynamics, *Applied Mathematics and Computation*, 441: 127704, 2023.

Yang W. M., Fang B. S., Liu B. Y., Yang Z., Promotion of ferrofluid microchannel flow by magnetic fields with constant gradient, *Journal of Non-Newtonian Fluid Mechanics*, 300: 104730, 2022.

Yang W. M., Li Y. F., Ren J. T., Yang X. L., Non-Newtonian behaviors of ferrofluid Couette–Poiseuille flows in time-varying magnetic fields, *Journal of Non-Newtonian Fluid Mechanics*, 332: 105306, 2024.

Yang W. M., Liu B. Y., Effects of magnetization relaxation in ferrofluid film flows under a uniform magnetic field, *Physics of Fluids*, 32: 062003, 2020.

Zahn M., Greer D. R., Ferrohydrodynamic pumping in spatially uniform sinusoidally time-varying magnetic fields, *Journal of Magnetism and Magnetic Materials*, 149: 165–173, 1995.

CHAPTER 7

Ferrofluid Flow in Pipes

THE FLOW OF FERROFLUIDS in pipes under the influence of a magnetic field is one of the fundamental flows of ferrofluids. This particular flow pattern serves as a theoretical model for many engineering applications such as energy conversion systems, viscous dampers, and actuators (Kamiyama, 2001; Bianco et al., 2021; Dalvi et al., 2022; Wang et al., 2021). In this chapter, the flow characteristics of ferrofluids in pipes in a cylindrical coordinate system (r, θ, z), as depicted in Figure 7.1, are investigated, where the z-axis coincides with the pipe's axis.

7.1 STEADY FLOW IN CONSTANT UNIFORM MAGNETIC FIELDS

7.1.1 Flow Characteristics under the Influence of Axial Magnetic Fields

In the tubular flow illustrated in Figure 7.1, when a constant uniform magnetic field is applied along the axial direction, the magnetic field intensity can be represented as

$$\boldsymbol{H} = \left(0, 0, H_z\right) \tag{7.1}$$

Assuming the laminar flow of the ferrofluid, it can be described by the motion equation (2.219), the continuity equation (2.205), and the magnetization equation (4.17). However, if the ferrofluid is assumed to be incompressible, the second term on the right-hand side of Eq. (4.17) becomes zero, thereby simplifying this equation into

$$\frac{\mathrm{D}\boldsymbol{M}}{\mathrm{D}t} = \boldsymbol{\omega}_{\mathrm{p}} \times \boldsymbol{M} + \frac{1}{\tau}\left(\boldsymbol{M}_0 - \boldsymbol{M}\right) \tag{7.2}$$

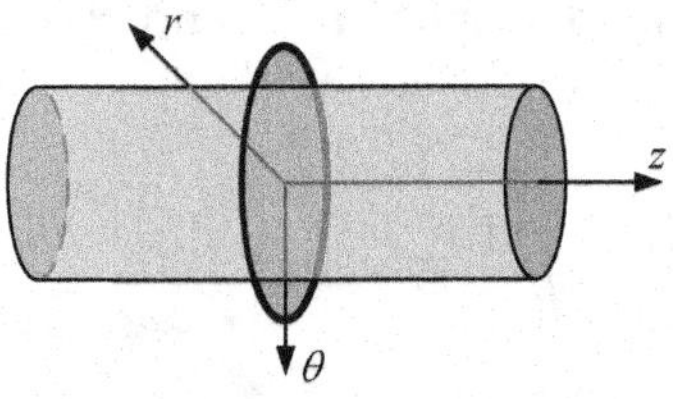

FIGURE 7.1 A model for the flow of ferrofluid in a pipe.

DOI: 10.1201/9781003540342-7

where the equilibrium magnetization, $\boldsymbol{M}_0$, is aligned with the direction of the external magnetic field, and its magnitude is determined by Eq. (4.1).

In the cylindrical coordinate, the component equation of Eq. (2.219) in the z-direction is given by

$$\rho\left(\frac{\partial v_z}{\partial t}+v_r\frac{\partial v_z}{\partial r}+\frac{v_\theta}{r}\frac{\partial v_z}{\partial \theta}+v_z\frac{\partial v_z}{\partial z}\right)=-\frac{\partial p}{\partial z}+\eta\left(\frac{1}{r}\frac{\partial v_z}{\partial r}+\frac{\partial^2 v_z}{\partial r^2}+\frac{1}{r^2}\frac{\partial^2 v_z}{\partial \theta^2}+\frac{\partial^2 v_z}{\partial z^2}\right)$$
$$+\mu_0\left(M_r\frac{\partial H_z}{\partial r}+\frac{M_\theta}{r}\frac{\partial H_z}{\partial \theta}+M_z\frac{\partial H_z}{\partial z}\right) \tag{7.3}$$
$$-\frac{\mu_0}{2r}\frac{\partial}{\partial r}\left[r\left(M_rH_z-M_zH_r\right)\right]-\frac{\mu_0}{2r}\frac{\partial}{\partial \theta}\left(M_\theta H_z-M_zH_\theta\right)$$

Assuming that the flow velocity of the ferrofluid is aligned with the z-direction, the velocity vector can be represented as

$$\boldsymbol{v}=\left(0,0,v_z(r)\right) \tag{7.4}$$

At this moment, the flow vorticity $\boldsymbol{\Omega}$ is aligned in the θ direction, with a magnitude of

$$\Omega=-\frac{1}{2}\frac{\mathrm{d}v_z}{\mathrm{d}r} \tag{7.5}$$

Based on the continuity equation for steady, incompressible fluid flow in the cylindrical coordinate

$$\frac{1}{r}\frac{\partial(rv_r)}{\partial r}+\frac{1}{r}\frac{\partial v_\theta}{\partial \theta}+\frac{\partial v_z}{\partial z}=0$$

one gets

$$\frac{\partial v_z}{\partial z}=0$$

Due to the axial symmetry of pipe flow, the velocity v_z does not vary along the θ direction. Likewise, the magnitude of magnetic field intensity H_z and the magnetization component M_r remain constant with respect to r and θ. Consequently, Eq. (7.3) can be simplified as

$$\frac{\partial p}{\partial z}=\eta\frac{1}{r}\frac{\mathrm{d}v_z}{\mathrm{d}r}+\eta\frac{\mathrm{d}^2v_z}{\mathrm{d}r^2}+\mu_0M_z\frac{\mathrm{d}H_z}{\mathrm{d}z}-\frac{\mu_0}{2}\frac{1}{r}M_rH_z \tag{7.6}$$

Similarly, the component equation in the r-direction of the motion equation (2.219) can be simplified to

$$\frac{\partial p}{\partial r}=0$$

It follows that the pressure p does not vary with position r, thereby allowing the substitution of $\frac{\mathrm{d}p}{\mathrm{d}z}$ for $\frac{\partial p}{\partial z}$ in Eq. (7.6).

The component equations of Eq. (7.2) in the r and z directions are given as follows:

$$\frac{\partial M_r}{\partial t}+v_r\frac{\partial M_r}{\partial r}+\frac{v_\theta}{r}\frac{\partial M_r}{\partial \theta}+v_z\frac{\partial M_r}{\partial z}=\omega_{\mathrm{p},\theta}M_z-\omega_{\mathrm{p},z}M_\theta+\frac{1}{\tau}\left(M_{0,r}-M_r\right) \tag{7.7}$$

$$\frac{\partial M_z}{\partial t}+v_r\frac{\partial M_z}{\partial r}+\frac{v_\theta}{r}\frac{\partial M_z}{\partial \theta}+v_z\frac{\partial M_z}{\partial z}=\omega_{\mathrm{p},r}M_\theta-\omega_{\mathrm{p},\theta}M_r+\frac{1}{\tau}\left(M_{0,z}-M_z\right) \tag{7.8}$$

To simplify Eqs. (7.7) and (7.8), the influence of spin viscosity is neglected (Schumacher et al., 2003). Subsequently, the component equation in the θ direction of the angular momentum equation (2.217) is

$$\omega_{\mathrm{p},\theta}=\Omega-\frac{\mu_0}{4\zeta}\left(M_rH_z-M_zH_r\right) \tag{7.9}$$

Under the assumptions of steady-state conditions and particle rotation velocity along the θ-direction, with the magnetic field along the z-direction and the magnetization component M_r independent on z, Eqs. (7.7)–(7.9) can be simplified to

$$0=\omega_{\mathrm{p},\theta}M_z-\frac{1}{\tau}M_r \tag{7.10}$$

$$v_z\frac{\partial M_z}{\partial z}=-\omega_{\mathrm{p},\theta}M_r-\frac{1}{\tau}\left(M_z-M_0\right) \tag{7.11}$$

$$\omega_{\mathrm{p},\theta}=\Omega-\frac{\mu_0}{4\zeta}M_rH_z \tag{7.12}$$

Substituting Eq. (7.12) into Eq. (7.10) yields

$$M_r=\frac{\tau\Omega M_z}{1+\frac{\mu_0\tau}{4\zeta}M_zH_z} \tag{7.13}$$

Introducing Eqs. (7.12) and (7.13) into Eq. (7.11) gives

$$v_z \frac{\partial M_z}{\partial z} = \frac{1}{\tau}\left[M_0 - M_z - \frac{(\tau\Omega)^2 M_z}{\left(1+\frac{\mu_0 \tau}{4\zeta} M_z H_z\right)^2} \right] \tag{7.14}$$

By substituting Eq. (7.13) into Eq. (7.6) and applying Eq. (7.5), the motion equation can be further expressed as

$$\frac{dp}{dz} = \eta \frac{1}{r}\frac{dv_z}{dr} + \eta \frac{d^2 v_z}{dr^2} + \mu_0 M_z \frac{dH_z}{dz} + \frac{\mu_0}{4r} \frac{\tau M_z H_z}{1+\frac{\mu_0 \tau}{4\zeta} M_z H_z} \frac{dv_z}{dr} \tag{7.15}$$

To facilitate the solution of Eq. (7.15), the following dimensionless quantities are introduced:

$$r^* = \frac{r}{R_0}, z^* = \frac{z}{R_0}, v_z^* = \frac{v_z}{v_m}, H_z^* = \frac{H_z}{H_{\max}}, M_z^* = \frac{M_z}{M_{0\max}}, p^* = \frac{p}{\eta v_m / R_0} \tag{7.16}$$

where R_0 is the radius of the circular tube, v_m is the average velocity of the ferrofluid within the tube, and $H_{\max}$ and $M_{0\max}$ are the maximum magnetic field intensity and the maximum equilibrium magnetization, respectively. From the definition in Eqs. (4.1) and (4.2), one has

$$M_{0\max} = N m_p L(\alpha_{\max}), \alpha_{\max} = \frac{\mu_0 m_p H_{\max}}{k_B T} \tag{7.17}$$

Also, for ease of representation, the following definitions are given

$$L^* = \frac{L(\alpha)}{L(\alpha_{\max})} \tag{7.18}$$

$$\hbar = \frac{\mu_0 \tau}{4\zeta} M_{0\max} H_{\max} \tag{7.19}$$

7.1.1.1 Flow Velocity and Pressure in a Uniform Axial Magnetic Field When $\tau\Omega \ll 1$

Under the assumption that $\tau\Omega \ll 1$, the final term within the parentheses on the right-hand side of Eq. (7.14) can be neglected. Furthermore, given a uniform magnetic field, the term on the left-hand side of Eq. (7.14) can also be disregarded. Consequently, one obtains

$$M_z = M_0 \tag{7.20}$$

Now simplify the last two terms on the right-hand side of Eq. (7.15) using Eq. (7.20). It can be inferred from Eqs. (4.1) and (4.2) that

$$\frac{\mathrm{d}}{\mathrm{d}z}\left(\ln\frac{\sinh\alpha}{\alpha}\right)=\frac{\mathrm{d}}{\mathrm{d}\alpha}\left(\ln\frac{\sinh\alpha}{\alpha}\right)\cdot\frac{\mathrm{d}\alpha}{\mathrm{d}H_z}\cdot\frac{\mathrm{d}H_z}{\mathrm{d}z}=L(\alpha)\cdot\frac{\mu_0 m_\mathrm{p}}{k_\mathrm{B}T}\cdot\frac{\mathrm{d}H_z}{\mathrm{d}z}$$

Applying this equation and Eq. (7.20), the third term on the right-hand side of Eq. (7.15) can be expressed as

$$\mu_0 M_z\frac{\mathrm{d}H_z}{\mathrm{d}z}=\mu_0 N m_\mathrm{p} L(\alpha)\frac{\mathrm{d}H_z}{\mathrm{d}z}=k_\mathrm{B}TN\frac{\mathrm{d}}{\mathrm{d}z}\left(\ln\frac{\sinh\alpha}{\alpha}\right)\tag{7.21}$$

Regarding the coefficient of the final term on the right-hand side of Eq. (7.15), by applying Eqs. (7.20), (4.1), and (4.2), one obtains

$$\frac{\mu_0}{4r}\frac{\tau M_z H_z}{1+\frac{\mu_0\tau}{4\zeta}M_z H_z}=\frac{1}{4r}\frac{\mu_0\tau N m_\mathrm{p} L(\alpha)H_z}{1+\frac{\mu_0\tau}{4\zeta}M_z H_z}$$

Substituting Eqs. (7.16) and (7.18) into this expression, and applying Eqs. (7.17) and (7.19) gives

$$\frac{\mu_0}{4r}\frac{\tau M_z H_z}{1+\frac{\mu_0\tau}{4\zeta}M_z H_z}=\frac{1}{4r}\frac{\mu_0\tau N m_\mathrm{p} L(\alpha_\mathrm{max})H_\mathrm{max}L^* H_z^*}{1+\frac{\mu_0\tau}{4\zeta}M_{0\mathrm{max}}H_\mathrm{max}M_z^* H_z^*}=\frac{1}{r}\frac{\zeta\hbar L^* H_z^*}{1+\hbar M_z^* H_z^*}\tag{7.22}$$

Substitution of the results from Eqs. (7.21) and (7.22) into Eq. (7.15) yields

$$\frac{\mathrm{d}p}{\mathrm{d}z}=\eta\frac{1}{r}\frac{\mathrm{d}v_z}{\mathrm{d}r}+\eta\frac{\mathrm{d}^2 v_z}{\mathrm{d}r^2}+k_\mathrm{B}TN\frac{\mathrm{d}}{\mathrm{d}z}\left(\ln\frac{\sinh\alpha}{\alpha}\right)+\frac{1}{r}\frac{\zeta\hbar L^* H_z^*}{1+\hbar M_z^* H_z^*}\frac{\mathrm{d}v_z}{\mathrm{d}r}\tag{7.23}$$

Using the definitions from Eq. (7.16) in this equation and simplifying, one obtains the dimensionless motion equation,

$$\frac{\mathrm{d}p^*}{\mathrm{d}z^*}=\frac{1}{r^*}\frac{\mathrm{d}v_z^*}{\mathrm{d}r^*}+\frac{\mathrm{d}^2 v_z^*}{\mathrm{d}r^{*2}}+\frac{R_0 k_\mathrm{B}TN}{\eta v_m}\frac{\mathrm{d}}{\mathrm{d}z^*}\left(\ln\frac{\sinh\alpha}{\alpha}\right)+\frac{1}{r^*\eta}\frac{\zeta\hbar L^* H_z^*}{1+\hbar M_z^* H_z^*}\frac{\mathrm{d}v_z^*}{\mathrm{d}r^*}$$

Letting

$$p^{*\prime}=p^*-\frac{R_0 k_\mathrm{B}TN}{\eta v_m}\ln\frac{\sinh\alpha}{\alpha}\tag{7.24}$$

$$\Delta\eta=\frac{\zeta\hbar L^* H_z^*}{1+\hbar M_z^* H_z^*}\tag{7.25}$$

then Eq. (7.23) can be transformed into a general partial differential equation (Cunha and Sobral, 2005)

$$\frac{d^2 v_z^*}{dr^{*2}} + \left(1 + \frac{\Delta\eta}{\eta}\right)\frac{1}{r^*}\frac{dv_z^*}{dr^*} = \frac{dp^{*\prime}}{dz^*} \tag{7.26}$$

To solve Eq. (7.26), letting

$$f\left(r^*\right) = \frac{dv_z^*}{dr^*} \tag{7.27}$$

then Eq. (7.26) becomes

$$f'\left(r^*\right) + \left(1 + \frac{\Delta\eta}{\eta}\right)\frac{1}{r^*} f\left(r^*\right) = \frac{dp^{*\prime}}{dz^*} \tag{7.28}$$

The general solution of the corresponding homogeneous equation is

$$f_h\left(r^*\right) = C_1 e^{-\int\left(1+\frac{\Delta\eta}{\eta}\right)\frac{1}{r^*}dr^*} = C_1 e^{-\left(1+\frac{\Delta\eta}{\eta}\right)\ln r^*} = C_1 r^{*-\left(1+\frac{\Delta\eta}{\eta}\right)} \tag{7.29}$$

The special solution of Eq. (7.28) is

$$\begin{aligned} f_p\left(r^*\right) &= e^{-\int\left(1+\frac{\Delta\eta}{\eta}\right)\frac{1}{r^*}dr^*} \int \frac{dp^{*\prime}}{dz^*} e^{\int\left(1+\frac{\Delta\eta}{\eta}\right)\frac{1}{r^*}dr^*} dr^* \\ &= r^{*-\left(1+\frac{\Delta\eta}{\eta}\right)} \frac{dp^{*\prime}}{dz^*} \int r^{*\left(1+\frac{\Delta\eta}{\eta}\right)} dr^* \\ &= \frac{1}{2+\frac{\Delta\eta}{\eta}} \frac{dp^{*\prime}}{dz^*} r^* \end{aligned} \tag{7.30}$$

Based on Eqs. (7.29) and (7.30), the solution to Eq. (7.28) can be obtained

$$f\left(r^*\right) = f_h\left(r^*\right) + f_p\left(r^*\right) = C_1 r^{*-\left(1+\frac{\Delta\eta}{\eta}\right)} + \frac{1}{2+\frac{\Delta\eta}{\eta}} \frac{dp^{*\prime}}{dz^*} r^* \tag{7.31}$$

Utilizing the axial symmetry of both the geometric model and the magnetic field, the boundary condition can be derived

$$\frac{dv_z^*}{dr^*} = 0, \text{when } r^* = 0 \tag{7.32}$$

Substituting this condition into Eq. (7.31), one obtains $C_1 = 0$, which leads to

$$f\left(r^*\right) = \frac{1}{2+\frac{\Delta\eta}{\eta}} \frac{dp^{*\prime}}{dz^*} r^* \tag{7.33}$$

Substituting Eq. (7.33) into Eq. (7.27) yields the equation

$$\frac{dv_z^*}{dr^*} = \frac{1}{2+\frac{\Delta\eta}{\eta}} \frac{dp^{*\prime}}{dz^*} r^* \tag{7.34}$$

The general solution of this equation is given by

$$v_z^* = \frac{1}{2} \frac{1}{2+\frac{\Delta\eta}{\eta}} \frac{dp^{*\prime}}{dz^*} r^{*2} + C_2$$

Applying the boundary condition

$$v_z^* = 0, \text{when } r^* = 1 \tag{7.35}$$

the value of C_2 can be obtained. Then, the velocity distribution of the ferrofluid flow in a uniform axial magnetic field when $\tau\Omega \ll 1$ can be determined

$$v_z^* = -\frac{1}{4\left(1+\frac{\Delta\eta}{2\eta}\right)} \frac{dp^{*\prime}}{dz^*}\left(1-r^{*2}\right) \tag{7.36}$$

When compared to the flow patterns of ordinary fluids in identical geometric models, the formal addition of $(1 + \Delta\eta/2\eta)$ in Eq. (7.36) indicates an increase in the apparent viscosity of ferrofluids by $\Delta\eta/2$ under the influence of an axially uniform magnetic field. This means that ferrofluids have non-Newtonian properties, which is characterized by the presence of a rotational viscosity $\eta_r = \Delta\eta/2$.

Equation (7.36) can be expressed as

$$\frac{dp^{*\prime}}{dz^*} = \frac{4}{r^{*2}-1}\left(1+\frac{\Delta\eta}{2\eta}\right) v_z^* \tag{7.37}$$

Integrating this equation with respect to z^*, and then applying Eq. (7.24), one obtains

$$p_2^* - p_1^* = \frac{R_0 k_B T N}{\eta v_m}\left(\ln\frac{\sinh\alpha}{\alpha}\right)_{z_1^*}^{z_2^*} + \frac{4v_z^*}{r^{*2}-1}\left(z_2^* - z_1^*\right) + \frac{2v_z^*}{r^{*2}-1}\int_{z_1^*}^{z_2^*} \frac{\Delta\eta}{\eta} dz^* \tag{7.38}$$

where subscripts 1 and 2 denote any two points within the ferrofluid, respectively. In Eq. (7.38), the first term on the right-hand side represents the pressure due to magnetic body forces, the second term accounts for the pressure resulting from internal fluid friction in the absence of a magnetic field, and the third term represents the pressure loss due to changes in apparent viscosity induced by the magnetic field.

7.1.1.2 Flow Velocity and Pressure in a Uniform Axial Magnetic Field When $\tau\Omega \geq 1$

When $\tau\Omega \geq 1$, the final term on the right-hand side of Eq. (7.14) becomes significant and cannot be disregarded. Under uniform magnetic field conditions, substituting the quantities from Eq. (7.16) and Eq. (7.5) into Eq. (7.14) yields the corresponding dimensionless equation

$$\frac{Pe^2 M_z^*}{4\left(1+\hbar M_z^* H_z^*\right)^2}\left(\frac{dv_z^*}{dr^*}\right)^2 + M_z^* - M_0^* = 0 \tag{7.39}$$

where Pe is the Péclet number, denoted as

$$Pe = \frac{\tau v_m}{R_0} \tag{7.40}$$

Under the same assumptions, the dimensionless equation corresponding to the motion equation (7.15) is

$$\frac{d^2 v_z^*}{dr^{*2}} + \left(1+\frac{\Delta\eta'}{\eta}\right)\frac{1}{r^*}\frac{dv_z^*}{dr^*} - \frac{dp^*}{dz^*} = 0 \tag{7.41}$$

where

$$\Delta\eta' = \frac{\zeta\hbar H_z^* M_z^*}{1+\hbar M_z^* H_z^*}$$

Solving the system of equations composed of Eqs. (7.39) and (7.41) reveals the flow characteristics within a ferrofluid pipe flow under a uniform axial magnetic field. Notably, these equations exhibit mutual coupling. Assuming that M_z^* remains constant with respect to r^* and both M_z^* and r^* possess axial symmetry, Eq. (7.41) has a solution resembling that of Eq. (7.36)

$$v_z^* = -\frac{1}{4\left(1+\dfrac{\Delta\eta'}{2\eta}\right)}\frac{dp^*}{dz^*}\left(1-r^{*2}\right) \tag{7.42}$$

In this formula, the term $\Delta\eta'$ incorporates the unknown variable M_z^*. From Eq. (7.42), one has

$$\frac{dv_z^*}{dr^*} = \frac{r^*}{2\left(1+\frac{\Delta\eta'}{2\eta}\right)}\frac{dp^*}{dz^*} \tag{7.43}$$

Substitution of it into Eq. (7.39) gives

$$\frac{\left(\eta Per^*\right)^2}{4}\left(\frac{dp^*}{dz^*}\right)^2 \frac{M_z^*}{\left[2\eta+\left(2\eta+\zeta\right)\hbar H_z^* M_z^*\right]^2} + M_z^* - M_0^* = 0 \tag{7.44}$$

Solving the nonlinear Eq. (7.44) numerically yields M_z^*, which can then be substituted into Eq. (7.42) to obtain the distribution of velocity v_z^*.

Figures 7.2 and 7.3 depict the velocity distributions in ferrofluid pipe flows for different magnetization relaxation times and different axial magnetic field intensities, respectively, where the parameters of the ferrofluid are identical to those listed in Table 6.1 (except for the magnetization relaxation time). Evidently, an increase in either magnetization relaxation time or magnetic field intensity leads to a decrease in the maximum velocity within the pipe flow. When compared to the flow of a regular fluid without a magnetic field, the application of an axial magnetic field exhibits a hindrance effect on the ferrofluid pipe flow. This hindrance becomes more pronounced with longer magnetization relaxation times or stronger magnetic fields. For instance, at magnetic field intensities of 10^5 A/m and 10^6 A/m, the maximum velocities in the ferrofluid pipe flow decrease by 5.2% and 7.7%, respectively, compared to the velocities observed in the absence of a magnetic field. The underlying cause of this hindrance effect is the interaction between fluid flow and the magnetic particles within the ferrofluid. Specifically, fluid motion induces the particles to rotate along the

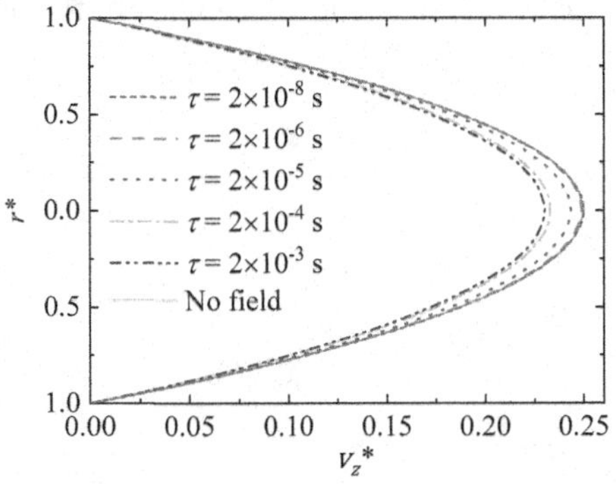

FIGURE 7.2 Velocity distribution of ferrofluid pipe flows under the influence of a uniform constant axial magnetic field varies with the magnetization relaxation time ($H_z = 10^4$ A/m, $Pe = 1$).

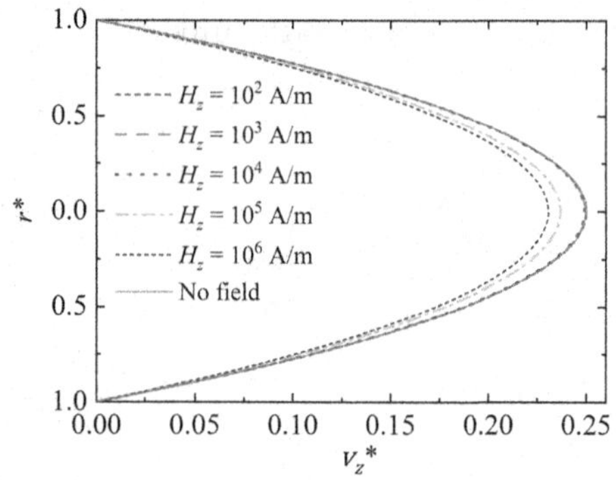

FIGURE 7.3 Velocity distribution of ferrofluid pipe flows varies with the axial magnetic field intensity ($\tau = 2 \times 10^{-6}$ s, $Pe = 1$).

direction of flow vorticity. However, a uniform magnetic field opposes this rotation, resulting in energy dissipation and a subsequent decrease in fluid kinetic energy and velocity.

The energy consumption resulting from the magnetic field's hindrance of particle rotation can be characterized by the rotational viscosity η_r, which can be derived by analogy with the relationship for the velocity distribution in the pipe flow of ordinary fluids using the following equation:

$$v_z^* = -\frac{1}{4\left(1+\dfrac{\eta_r}{\eta}\right)}\frac{dp^*}{dz^*}\left(1-r^{*2}\right) \tag{7.45}$$

The coefficients present in Eq. (7.45) can be determined by fitting the velocity distribution curves of ferrofluid pipe flows, and the η_r/η values can be calculated based on these coefficients. Figures 7.4 and 7.5 illustrate the relationships between rotational viscosity and both axial magnetic field intensity and magnetization relaxation time. It is evident from these figures that the rotational viscosity increases with both magnetic field intensity and

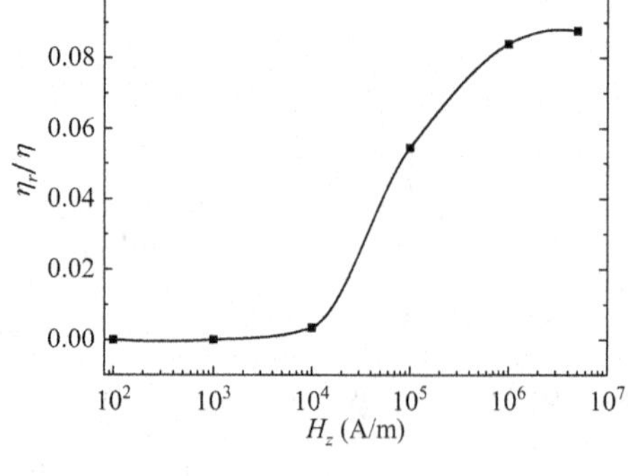

FIGURE 7.4 Variation of rotational viscosity in the ferrofluid pipe flows as a function of axial magnetic field intensity ($\tau = 2 \times 10^{-6}$ s, $Pe = 1$).

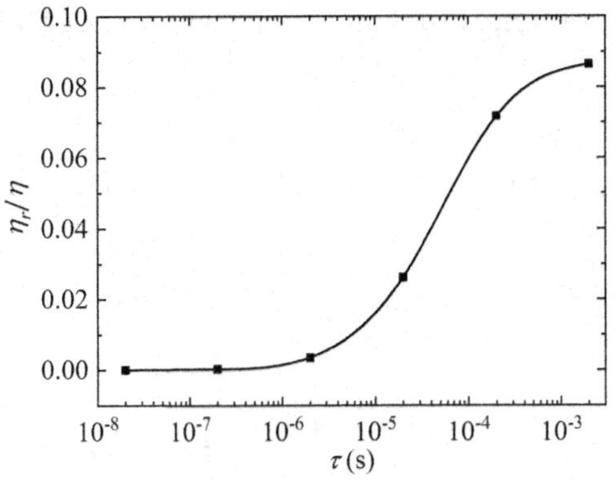

FIGURE 7.5 Variations in rotational viscosity with magnetization relaxation time in the ferrofluid pipe flows under an application of a uniform constant axial magnetic field ($H_z = 10^4$ A/m, $Pe = 1$).

magnetization relaxation time. This observation is consistent with the following relationship derived from Eq. (7.42)

$$\eta_r = \frac{\Delta\eta'}{2} = \frac{\zeta}{2\left[1/\left(\hbar H_z^* M_z^*\right)+1\right]}$$

The maximum rotational viscosity η_r within the range depicted in both figures can reach up to 8.8% of the dynamic viscosity η of the ferrofluids.

7.1.2 Flow Characteristics under the Influence of Radial Magnetic Fields

In the pipe flow depicted in Figure 7.1, when an externally radial uniform magnetic field is applied, the magnetic field intensity can be represented as

$$\boldsymbol{H} = \left(H_r, 0, 0\right) \tag{7.46}$$

Under the same assumptions as outlined in Section 7.1.1, the motion equation (7.3) can be simplified to

$$\frac{\mathrm{d}p}{\mathrm{d}z} = \eta\frac{1}{r}\frac{\mathrm{d}v_z}{\mathrm{d}r} + \eta\frac{\mathrm{d}^2 v_z}{\mathrm{d}r^2} + \frac{\mu_0}{2}\frac{1}{r}\frac{\partial}{\partial r}\left(rM_z H_r\right) \tag{7.47}$$

Then Eqs. (7.7) to (7.9) can be simplified as follows:

$$0 = \omega_{\mathrm{p},\theta} M_z - \frac{1}{\tau}M_r + \frac{1}{\tau}M_0 \tag{7.48}$$

$$0 = -\omega_{\mathrm{p},\theta} M_r - \frac{1}{\tau}M_z \tag{7.49}$$

$$\omega_{\mathrm{p},\theta} = \Omega + \frac{\mu_0}{4\zeta}M_z H_r \tag{7.50}$$

Substituting Eq. (7.50) into Eq. (7.49) and simplifying, one obtains

$$M_z = -\frac{\tau\Omega M_r}{1+\frac{\mu_0\tau}{4\zeta}M_rH_r} = \frac{\tau M_r}{2\left(1+\frac{\mu_0\tau}{4\zeta}M_rH_r\right)}\frac{dv_z}{dr} \tag{7.51}$$

Introducing Eqs. (7.50) and (7.51) into Eq. (7.48) yields

$$\left(1-\frac{\mu_0\tau}{4\zeta}\frac{M_rH_r}{1+\frac{\mu_0\tau}{4\zeta}M_rH_r}\right)\frac{(\tau\Omega)^2 M_r}{1+\frac{\mu_0\tau}{4\zeta}M_rH_r}+M_r-M_0=0 \tag{7.52}$$

Substitution of Eq. (7.51) into Eq. (7.47) gives

$$\frac{d^2v_z}{dr^2}+\frac{1}{r}\frac{dv_z}{dr}-\frac{1}{\eta+\frac{\mu_0\tau M_rH_r}{4\left(1+\frac{\mu_0\tau}{4\zeta}M_rH_r\right)}}\frac{dp}{dz}=0 \tag{7.53}$$

Substituting the respective physical quantities from Eq. (7.16) into Eqs. (7.52) and (7.53), one obtains the dimensionless equations

$$\frac{Pe^2M_r^*}{4\left(1+\hbar M_r^*H_r^*\right)^2}\left(\frac{dv_z^*}{dr^*}\right)^2+M_r^*-M_0^*=0 \tag{7.54}$$

$$\frac{d^2v_z^*}{dr^{*2}}+\frac{1}{r^*}\frac{dv_z^*}{dr^*}-\frac{1}{1+\frac{\Delta\eta''}{\eta}}\frac{dp^*}{dz^*}=0 \tag{7.55}$$

where

$$\Delta\eta''=\frac{\zeta\hbar M_r^*H_r^*}{1+\hbar M_r^*H_r^*} \tag{7.56}$$

Subject to the assumption that M_r^* remains constant with respect to r^*, the solution to Eq. (7.55) can be derived by applying the boundary conditions represented by Eqs. (7.32) and (7.35)

$$v_z^*=-\frac{1}{4\left(1+\frac{\Delta\eta''}{\eta}\right)}\frac{dp^*}{dz^*}\left(1-r^{*2}\right) \tag{7.57}$$

When $\tau\Omega \ll 1$, it follows from Eq. (7.52) that $M_r \approx M_{0,r}$. Under these conditions, $\Delta\eta''$ solely depends on the applied magnetic field, and specifically, $\Delta\eta'' = \Delta\eta$. By comparing Eqs. (7.57) and (7.36), it is observed that the rotational viscosity influenced by a radial magnetic field is twice as high as that influenced by an axial magnetic field.

When $\tau\Omega \geq 1$, it is necessary to solve Eqs. (7.54) and (7.57) simultaneously to obtain the velocity distribution. After taking the derivative of Eq. (7.57) with respect to r^*, the resultant expression is then substituted into Eq. (7.54) along with the expression for $\Delta\eta''$ given in Eq. (7.56) to obtain

$$\frac{\left(\eta Per^*\right)^2}{16}\left(\frac{dp^*}{dz^*}\right)^2 \frac{M_r^*}{\left[\eta+\left(\eta+\zeta\right)\hbar M_r^* H_r^*\right]^2} + M_r^* - M_0^* = 0 \tag{7.58}$$

By numerically solving the nonlinear equation, one can obtain M_r^*. Substituting this result into Eq. (7.57) yields the distribution of the velocity v_z^*. Notably, the rotational viscosity in this case is twice as high as that observed under an axial magnetic field with the same conditions. Furthermore, compared to the situation with an axial magnetic field, the flow velocity in the ferrofluid pipe flow decreases to some extent. Figure 7.6 illustrates the comparison of flow velocity distributions under radial and axial uniform magnetic fields at three different magnetic field intensities. At magnetic field intensities of 10^5 A/m and 10^6 A/m, the maximum flow velocities decrease by 4.9% and 7.2%, respectively. When compared to the flow velocity without a magnetic field, the radial magnetic fields of both intensities cause the maximum flow velocities to decrease by 9.8% and 14.4%, respectively.

7.1.3 Flow Rate and Drag Coefficient

Under the influence of uniform magnetic fields acting in both directions, the flow rate of ferrofluid through a tube is also smaller than that of an ordinary fluid. Figure 7.7 illustrates the relationship between flow rate and magnetic field intensity, as well as magnetization

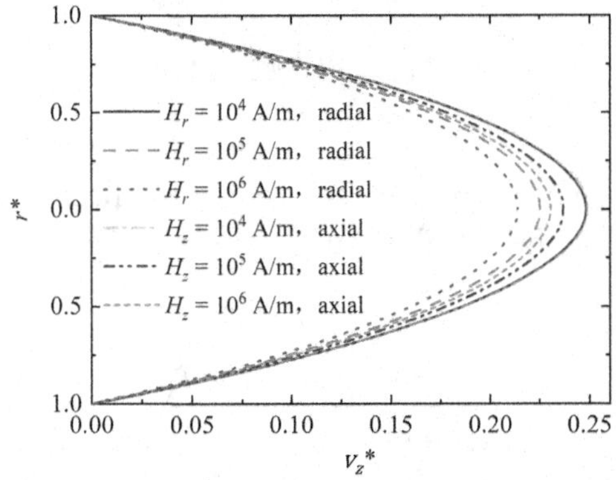

FIGURE 7.6 Comparison of velocity distributions in ferrofluid pipe flows under the influence of radial and axial magnetic fields ($\tau = 2 \times 10^{-6}$ s, $Pe = 1$).

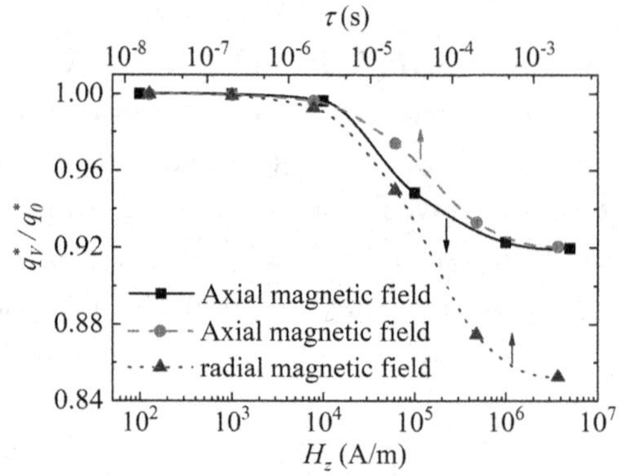

FIGURE 7.7 Flow rate of ferrofluid pipe flows under the action of radial and axial uniform magnetic field ($\tau = 2 \times 10^{-6}$ s, $Pe = 1$).

relaxation time. In this figure, q_0^* and q_V^* represent the dimensionless flow rates of regular fluid and ferrofluid, respectively, defined as

$$q_V^* = \frac{Q}{\pi R_0^2 v_m} = \frac{2}{\pi R_0^2 v_m} \int_0^1 v_z^* r^* \mathrm{d}r^* \tag{7.59}$$

The flow rate depicted in the figure decreases with both magnetic field intensity and magnetization relaxation time, with the radial magnetic field causing a greater reduction in flow rate. Specifically, when $\tau = 10^{-3}$ s, the flow rate decreases by 14.7%.

The resistance coefficient λ_r characterizes the frictional losses in pipe flows, and the resistance coefficient in laminar pipe flow of an ordinary Newtonian fluid is

$$\lambda_r = \frac{64}{Re}$$

For pipe flow, the Reynolds number is

$$Re = \frac{2\rho v_m R_0}{\eta}$$

Under the influence of a uniform magnetic field, it has been experimentally confirmed that the drag coefficient can be calculated using the modified Reynolds number, denoted as Re^*

$$\lambda = \frac{64}{Re^*}, \; Re^* = \frac{2\rho v_m R_0}{\eta + \eta_r} \tag{7.60}$$

Based on the conclusions drawn in Section 7.1.2, it is evident that in a uniformly constant magnetic field, the relative viscosity η_r of ferrofluids increases with both the magnetic field intensity and the magnetization relaxation time. Consequently, as indicated by Eq. (7.60), when the flow of ferrofluid within a pipe is laminar, the drag coefficient also increases with both the magnetic field intensity and the magnetization relaxation time.

Additionally, the distribution of particle mass fraction within the ferrofluid has the potential to influence the flow characteristics within the pipe, as demonstrated by the studies conducted by Shimada and Kamiyama (2000) and Shuchia et al. (2002).

7.2 STEADY FLOW IN CONSTANT NON-UNIFORM MAGNETIC FIELDS

7.2.1 Flow Characteristics at Weak Non-Equilibrium States

This section delves into the flow characteristics of a ferrofluid in a pipe under the influence of a constant non-uniform axial magnetic field, particularly when only weak non-equilibrium properties are present. By positioning permanent magnets at the inlet or outlet of a circular pipe, with their magnetization direction aligned along the axial direction of the pipe, it is feasible to achieve such a magnetic field effect (Papadopoulos et al., 2012). Under the assumption of weak non-equilibrium states, the ferrofluid's magnetic stress tensor is symmetric, allowing for the approximation of $\boldsymbol{M} \times \boldsymbol{H} = \mathbf{0}$. Some literature also refers to fluids with such stress tensors as symmetric fluids. Consequently, the last term on the right-hand side of both the motion equation (2.219) and the magnetization equation (4.18) can be neglected. In a cylindrical coordinate system, the component equations of these two equations become

$$\eta \frac{\mathrm{d}^2 v_z}{\mathrm{d}r^2} + \eta \frac{1}{r}\frac{\mathrm{d}v_z}{\mathrm{d}r} + \mu_0 M_z \frac{\mathrm{d}H_z}{\mathrm{d}z} = \frac{\partial p}{\partial z} \tag{7.61}$$

$$\frac{1}{2} M_z \frac{\mathrm{d}v_z}{\mathrm{d}r} + \frac{1}{\tau}\left(M_r - M_{0,r}\right) = 0 \tag{7.62}$$

$$\frac{1}{2} M_r \frac{\mathrm{d}v_z}{\mathrm{d}r} - \frac{1}{\tau}\left(M_z - M_{0,z}\right) = 0 \tag{7.63}$$

where assumptions are made that $M_r \ll M_z$ and $H_r \ll H_z$.

To facilitate the solution of Eqs. (7.61) to (7.63), this subsection employs dimensionless quantities distinct from those used in Eq. (7.16),

$$r^* = \frac{r}{R_0}, z^* = \frac{z}{R_0}, v_z^* = \frac{v_z}{v_m}, H_z^* = \frac{H_z}{H_0}, M_z^* = \frac{M_z}{H_0}, M_r^* = \frac{M_r}{H_0}, p^* = \frac{p}{\rho v_m^2} \tag{7.64}$$

The following dimensionless constants are defined

$$Re = \frac{\rho v_m R_0}{\eta}, Re_m = \frac{\mu_0 H_0^2}{\rho v_m^2}, Pe = \frac{\tau}{R_0 / v_m} \tag{7.65}$$

where Re_m, also known as the magnetic Reynolds number, represents the relative magnitude of the inertial and magnetic forces (Rosa et al., 2016).

By substituting the respective physical quantities from Eq. (7.64) into the equations ranging from Eqs. (7.61)–(7.63), and applying the definition from Eq. (7.65), a dimensionless system of equations is obtained

$$\frac{d^2 v_z^*}{dr^{*2}} + \frac{1}{r^*}\frac{dv_z^*}{dr^*} + Re\,Re_m\,M_z^*\frac{dH_z^*}{dz^*} = Re\frac{\partial p^*}{\partial z^*} \tag{7.66}$$

$$\frac{1}{2}M_z^*\frac{dv_z^*}{dr^*} + \frac{1}{Pe}M_r^* = 0 \tag{7.67}$$

$$\frac{1}{2}M_r^*\frac{dv_z^*}{dr^*} - \frac{1}{Pe}\left(M_z^* - M_{0,z}^*\right) = 0 \tag{7.68}$$

where dH_z^* / dz^* is the dimensionless magnetic field gradient.

7.2.1.1 *Flow Velocity and Rotational Viscosity When Pe ≪ 1*

When $Pe \ll 1$, from Eq. (7.68), one obtains $M_z^* = M_{0,z}^*$, indicating that the magnetization remains constant regardless of position. This allows one to solve Eq. (7.66) through direct integration. By applying the boundary conditions (7.32) and (7.35), the solution of Eq. (7.66) can be obtained

$$v_z^* = -\frac{Re}{4}\left(\frac{\partial p^*}{\partial z^*} - Re_m\,M_{0,z}^*\frac{dH_z^*}{dz^*}\right)\left(1 - r^{*2}\right) \tag{7.69}$$

Comparing Eq. (7.69) with the velocity distribution observed in a conventional pipe flow without a magnetic field, an expression for the relative rotational viscosity can be derived

$$\frac{\eta_r}{\eta} = \frac{1}{1 - \left(Re_m\,M_{0,z}^*\frac{dH_z^*}{dz^*}\right) / \frac{\partial p^*}{\partial z^*}} - 1 \tag{7.70}$$

7.2.1.2 *Flow Velocity and Rotational Viscosity When Pe ≥ 1*

When $Pe \geq 1$, the vorticity term $\left(\frac{1}{2}\frac{dv_z^*}{dr^*}\right)$ in Eqs. (7.67) and (7.68) cannot be neglected, and the flow of ferrofluid will affect the magnetization. From Eq. (7.67) one gets

$$M_r^* = -\frac{1}{2}Pe\,M_z^*\frac{dv_z^*}{dr^*} \tag{7.71}$$

Substituting it into Eq. (7.68) and simplifying gives

$$M_z^* = M_{0,z}^*\left[\frac{1}{4}Pe^2\left(\frac{dv_z^*}{dr^*}\right)^2 + 1\right]^{-1} \tag{7.72}$$

Utilizing the binomial theorem to expand Eq. (7.72) and approximating M_z^* by considering soley the first two terms of the expansion, one obtains

$$M_z^* \approx M_{0,z}^*\left[1-\frac{1}{4}Pe^2\left(\frac{dv_z^*}{dr^*}\right)^2\right] \tag{7.73}$$

Substituting Eq. (7.73) into Eq. (7.66) yields

$$\frac{d^2v_z^*}{dr^{*2}}+\frac{1}{r^*}\frac{dv_z^*}{dr^*}-\frac{Pe^2}{4}Re\,Re_m\,M_{0,z}^*\frac{dH_z^*}{dz^*}\left(\frac{dv_z^*}{dr^*}\right)^2=Re\left(\frac{\partial p^*}{\partial z^*}-Re_m\,M_{0,z}^*\frac{dH_z^*}{dz^*}\right) \tag{7.74}$$

This equation describes the coupling between vorticity and magnetization, presenting a nonlinear ordinary differential equation as it contains squared velocity derivative terms.

To solve Eq. (7.74), letting

$$\epsilon=\frac{1}{4}Pe^2\,Re\,Re_m\,M_{0,z}^*\frac{dH_z^*}{dz^*} \tag{7.75}$$

and if ϵ is a small parameter, that is $\epsilon \ll 1$, the method of regular perturbation can be applied to solve Eq. (7.74). Let the second-order approximation of v_z^* be

$$v_z^*=v_{z,0}^*+\epsilon v_{z,1}^*+\epsilon^2 v_{z,2}^* \tag{7.76}$$

Substituting it into Eq. (7.74), ignoring terms of second order and higher in ϵ and letting

$$\gamma=Re\left(\frac{\partial p^*}{\partial z^*}-Re_m\,M_{0,z}^*\frac{dH_z^*}{dz^*}\right) \tag{7.77}$$

one gets

$$\left(\frac{d^2v_{z,0}^*}{dr^{*2}}+\frac{1}{r^*}\frac{dv_{z,0}^*}{dr^*}-\gamma\right)+\epsilon\left[\frac{d^2v_{z,1}^*}{dr^{*2}}+\frac{1}{r^*}\frac{dv_{z,1}^*}{dr^*}-\left(\frac{dv_{z,0}^*}{dr^*}\right)^2\right]$$
$$+\epsilon^2\left(\frac{d^2v_{z,2}^*}{dr^{*2}}+\frac{1}{r^*}\frac{dv_{z,2}^*}{dr^*}-2\frac{dv_{z,0}^*}{dr^*}\frac{dv_{z,1}^*}{dr^*}\right)=0$$

Then the solution of Eq. (7.74) is converted to solving the following system of linear partial differential equations:

$$\begin{cases} \dfrac{d^2 v_{z,0}^*}{dr^{*2}} + \dfrac{1}{r^*}\dfrac{dv_{z,0}^*}{dr^*} - \gamma = 0 \\ \dfrac{d^2 v_{z,1}^*}{dr^{*2}} + \dfrac{1}{r^*}\dfrac{dv_{z,1}^*}{dr^*} - \left(\dfrac{dv_{z,0}^*}{dr^*}\right)^2 = 0 \\ \dfrac{d^2 v_{z,2}^*}{dr^{*2}} + \dfrac{1}{r^*}\dfrac{dv_{z,2}^*}{dr^*} - 2\dfrac{dv_{z,0}^*}{dr^*}\dfrac{dv_{z,1}^*}{dr^*} = 0 \end{cases}$$

Applying the direct integration method individually to each of these equations and incorporating the boundary conditions (7.32) and (7.35) yields the solution

$$\begin{cases} v_{z,0}^* = \dfrac{1}{4}\gamma r^{*2} - \dfrac{1}{4}\gamma \\ v_{z,1}^* = \dfrac{1}{64}\gamma^2 r^{*4} - \dfrac{1}{64}\gamma^2 \\ v_{z,2}^* = \dfrac{1}{576}\gamma^3 r^{*6} - \dfrac{1}{576}\gamma^3 \end{cases}$$

Substituting these results into Eq. (7.76) yields the velocity distribution

$$v_z^* = \frac{1}{4}\gamma\left(r^{*2} - 1\right) + \frac{1}{64}\gamma^2\epsilon\left(r^{*4} - 1\right) + \frac{1}{576}\gamma^3\epsilon^2\left(r^{*6} - 1\right) \tag{7.78}$$

where γ is given by Eq. (7.77). According to Eq. (7.78), when there is no external magnetic field ($\epsilon = 0$), the velocity distribution represented by Eq. (7.78) simplifies to a parabolic distribution analogous to that of an ordinary fluid.

To obtain the rotational viscosity, Eq. (7.78) is further rewritten as

$$\begin{aligned} v_z^* &= -\frac{1}{4}\gamma\left(1 - r^{*2}\right)\left[1 + \frac{1}{16}\gamma\epsilon\left(1 + r^{*2}\right) + \frac{1}{144}\gamma^2\epsilon^2\left(1 + r^{*2} + r^{*4}\right)\right] \\ &= -\frac{1}{4}Re\left(\frac{\partial p^*}{\partial z^*} - Re_m\, M_{0,z}^*\frac{dH_z^*}{dz^*}\right)\left(1 - r^{*2}\right)\left[1 + \frac{1}{16}\gamma\epsilon\left(1 + r^{*2}\right) + \frac{1}{144}\gamma^2\epsilon^2\left(1 + r^{*2} + r^{*4}\right)\right] \end{aligned} \tag{7.79}$$

Substituting Eq. (7.65) for *Re* into Eq. (7.79) and comparing the resulting expression with the distribution of flow velocity in an ordinary pipe flow in the absence of a magnetic field, an expression for the rotational viscosity is obtained

$$\frac{\eta_r}{\eta} = \frac{1}{\left(1 - Re_m\, M_{0,z}^*\dfrac{dH_z^*}{dz^*} \Big/ \dfrac{\partial p^*}{\partial z^*}\right)\Theta} - 1 \tag{7.80}$$

where

$$\Theta = 1 + \frac{1}{16}\gamma\epsilon\left(1 + r^{*2}\right) + \frac{1}{144}\gamma^2\epsilon^2\left(1 + r^{*2} + r^{*4}\right) \tag{7.81}$$

In the absence of a magnetic field, it is known from Eq. (7.80) that $\eta_r = 0$. However, upon exposing the ferrofluid to an external magnetic field, a rotational viscosity is generated, resulting in a change in the apparent viscosity.

If the parameter ϵ represented by Eq. (7.75) is relatively large, the regular perturbation method becomes unsuitable for solving Eq. (7.74). Instead, numerical methods must be employed. In this context, the fourth-order Runge–Kutta method is utilized to solve the nonlinear ordinary differential Eq. (7.74).

To assess the validity range of analytical solutions (7.69) and (7.78), three distinct approaches are employed to calculate the rotational viscosity for various values of *Pe*. Figure 7.8 illustrates the corresponding results. Notably, for $Pe \leq 2$, the direct integration method exhibits an error of less than 0.26% in rotational viscosity, while the regular perturbation method demonstrates an error of less than 0.11%. However, as *Pe* exceeds 2, the error in rotational viscosity computed using the direct integration method increases significantly. Furthermore, for *Pe* values exceeding 6, the ϵ value also surpasses 1, leading to a noticeable increase in error in the rotational viscosity calculated by regular perturbation method. Given that *Pe* values in practical applications of ferrofluids are typically less than 1, the errors in flow velocity and rotational viscosity obtained from all three methods are relatively minor. Therefore, subsequent calculations in this section will not differentiate between these methods.

As shown in Figures 7.9 and 7.10, the velocity distribution in a ferrofluid pipe flow under the action of a constant non-uniform magnetic field varies with both the magnetic Reynolds number Re_m and the magnetic field gradient. Since Re_m represents the relative magnitude of the reference magnetic field intensity, an increase in Re_m corresponds to a stronger magnetic field and a subsequent stronger promoting effect on fluid flow at a given gradient. In the example depicted in Figure 7.9, as Re_m increases to 100, the peak velocity increases to

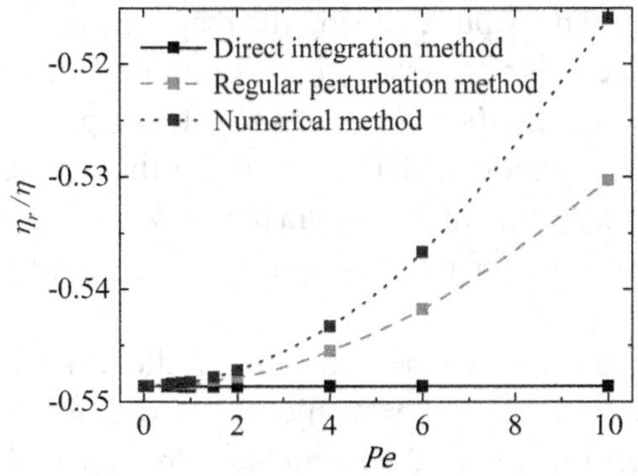

FIGURE 7.8 Variation of relative rotational viscosity with *Pe* calculated by different methods ($Re_m = 100$, $Re = 0.1$, $\partial p^*/\partial z^* = -1$, $\mathrm{d}H_z^*/\mathrm{d}z^* = 0.01$).

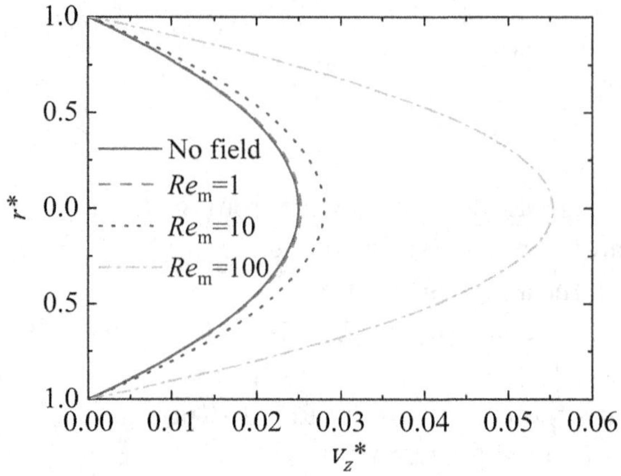

FIGURE 7.9 Flow velocity distributions at different Re_m ($Pe = 0.1$, $Re = 0.1$, $\partial p^*/\partial z^* = -1$, and $dH_z^* / dz^* = 0.01$).

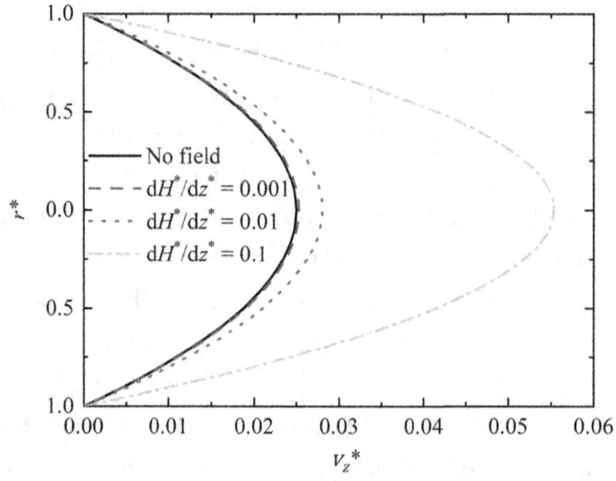

FIGURE 7.10 Flow velocity distributions at different magnetic field gradients ($Pe = 0.1$, $Re = 0.1$, $\partial p^*/\partial z^* = -1$, and $Re_m = 10$).

2.2 times its value in the absence of a magnetic field. Similarly, maintaining a constant reference magnetic field intensity but elevating the magnetic field gradient also leads to an augmentation in flow velocity. As evidenced in the example shown in Figure 7.10, when the dimensionless magnetic field gradient increases to 0.1, the peak velocity increases to 2.3 times its value without a magnetic field. Overall, when the magnetic field gradient aligns with the flow direction, it enhances the ferrofluid pipe flow, and the strength of this enhancing effect is related to both the reference magnetic field intensity and the magnetic field gradient.

The underlying cause of the enhancing effect of gradient magnetic fields on the ferrofluid pipe flow lies in the significant alteration of the ferrofluid's apparent viscosity by the magnetic field. When the gradient of the magnetic field aligns with the flow velocity, the apparent viscosity diminishes below the ferrofluid's intrinsic viscosity, effectively inducing a negative rotational viscosity in the ferrofluid. Figure 7.11 illustrates the relationship

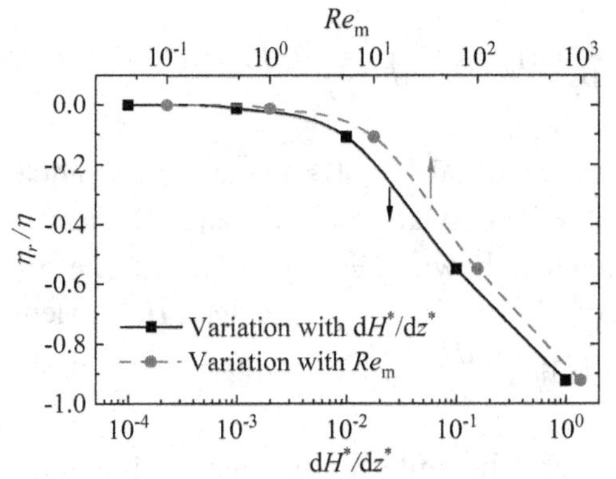

FIGURE 7.11 Variation of rotational viscosity with Re_m ($dH_z^*/dz^* = 0.01$) and magnetic field gradient ($Re_m = 10$) ($Pe = 0.1$, $Re = 0.1$, and $\partial p^*/\partial z^* = -1$).

between rotational viscosity and both Re_m and the magnetic field gradient. Both increasing the magnetic field gradient and Re_m lead to an increase in the absolute value of rotational viscosity. In the example depicted, when Re_m increases to 1000 or the magnetic field gradient rises to 1, the rotational viscosity can reach 92% of the ferrofluid's intrinsic viscosity.

7.2.2 Flow Characteristics at Non-Equilibrium States

Under general non-equilibrium conditions, the magnetic torque term $\mu_0 \mathbf{M} \times \mathbf{H}$ is non-zero in a flowing ferrofluid. Neglecting the radial components of magnetic field intensity and magnetization resulting from magnetization relaxation, the component equations of the motion equation (2.219) and magnetization equation (4.18) respectively become

$$-\frac{\partial p}{\partial z} + \eta\left(\frac{1}{r}\frac{\partial v_z}{\partial r} + \frac{\partial^2 v_z}{\partial r^2}\right) + \mu_0 M_z \frac{\partial H_z}{\partial z} - \frac{\mu_0}{2r}\frac{\partial}{\partial r}\left[r\left(M_r H_z - M_z H_r\right)\right] = 0 \tag{7.82}$$

$$-\frac{1}{2}M_z\frac{dv_z}{dr} + \frac{1}{\tau}\left(M_{0,r} - M_r\right) + \frac{\mu_0}{4\zeta}M_z\left(M_z H_r - M_r H_z\right) = 0 \tag{7.83}$$

$$\frac{1}{2}M_r\frac{dv_z}{dr} + \frac{1}{\tau}\left(M_{0,z} - M_z\right) - \frac{\mu_0}{4\zeta}M_r\left(M_z H_r - M_r H_z\right) = 0 \tag{7.84}$$

Using the variables defined in Eq. (7.64), the corresponding dimensionless equations for Eqs. (7.82)–(7.84) are obtained as follows:

$$\frac{d^2 v_z^*}{dr^{*2}} + \frac{1}{r^*}\frac{dv_z^*}{dr^*} + \frac{Re}{Re_m}M_z^*\frac{dH_z^*}{dz^*} - \frac{1}{2}Re\,Re_m\frac{1}{r^*}\frac{\partial}{\partial r^*}\left[r^*\left(M_r^* H_z^* - M_z^* H_r^*\right)\right] = Re\frac{\partial p^*}{\partial z^*} \tag{7.85}$$

$$-\frac{1}{2}M_z^*\frac{dv_z^*}{dr^*} + \frac{1}{Pe}\left(M_{0,r}^* - M_r^*\right) + \frac{1}{6\phi}Re\,Re_m\,M_z^*\left(M_z^* H_r^* - M_r^* H_z^*\right) = 0 \tag{7.86}$$

$$\frac{1}{2}M_r^*\frac{dv_z^*}{dr^*}+\frac{1}{Pe}\left(M_{0,z}^*-M_z^*\right)-\frac{1}{6\phi}Re\,Re_m\,M_r^*\left(M_z^*H_r^*-M_r^*H_z^*\right)=0 \tag{7.87}$$

Numerical methods are required to solve the system of equations comprised of Eqs. (7.85) to (7.87). The effect of the axial gradient magnetic field on flow characteristics is investigated using the scenario shown in Figure 7.12. Here, the circular tube has a radius of R_0 = 1 mm and a length of 15 mm. The magnetic field, $\boldsymbol{H}$, is oriented along the z-direction and exhibits a constant gradient $\frac{dH_z^*}{dz^*}$ in the z-direction. The field's action extends over the range from z_0 to $z_0 + L$.

To solve the above problem, the finite volume method is applied. To prevent oscillations caused by abrupt changes in the magnetic field during the computational process, the Heaviside function is incorporated to facilitate a smooth transition in magnetic field intensity variations. In this case, the magnetic field intensity is represented as a segmentation function

$$H_z(z)=\begin{cases} 0, z>z_0+L+2\hat{d} \text{ or } z<z_0-2\hat{d} \\ H_0\left[\dfrac{z-z_0+2\hat{d}}{2\hat{d}}+\dfrac{1}{2\pi}\sin\left(\dfrac{\pi\left(z-z_0+\hat{d}\right)}{\hat{d}}\right)\right], z_0-2\hat{d}\le z<z_0 \\ H_0+\dfrac{\partial H_z}{\partial z}(z-z_0), z_0\le z<z_0+L \\ \left(H_0+\dfrac{\partial H_z}{\partial z}L\right)\left[1-\dfrac{z-z_0-L}{2\hat{d}}-\dfrac{1}{2\pi}\sin\left(\dfrac{\pi\left(z-z_0-L-\hat{d}\right)}{\hat{d}}\right)\right], z_0+L\le z<z_0+L+2\hat{d} \end{cases} \tag{7.88}$$

where $\hat{d}$ is the width of the magnetic field transition region, let $\hat{d}=R_0$.

Figure 7.13 illustrates the axial pressure distribution within a ferrofluid pipe flow at Re = 1 and $\partial p^*/\partial z^*$ = − 1, varying with different magnetic field gradients G_H and Reynolds numbers Re_m. Here, G_H represents the gradient of the magnetic field strength along the axial direction, defined as $G_H = dH_z^*/dz^*$. Figure 7.13(a) depicts the case of a uniform magnetic field with a constant field strength of H_m. When the magnitude of the magnetic field intensity exceeds a certain threshold, a significant pressure increase occurs within the ferrofluid pipe flow, specifically in the region influenced by the magnetic field. In a uniform axial magnetic field, the increased pressure remains uniform as shown in Figure 7.13(a).

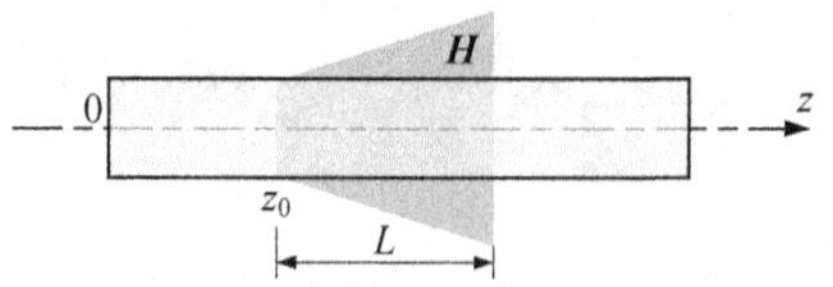

FIGURE 7.12 Ferrofluid pipe flow in an axial gradient magnetic field.

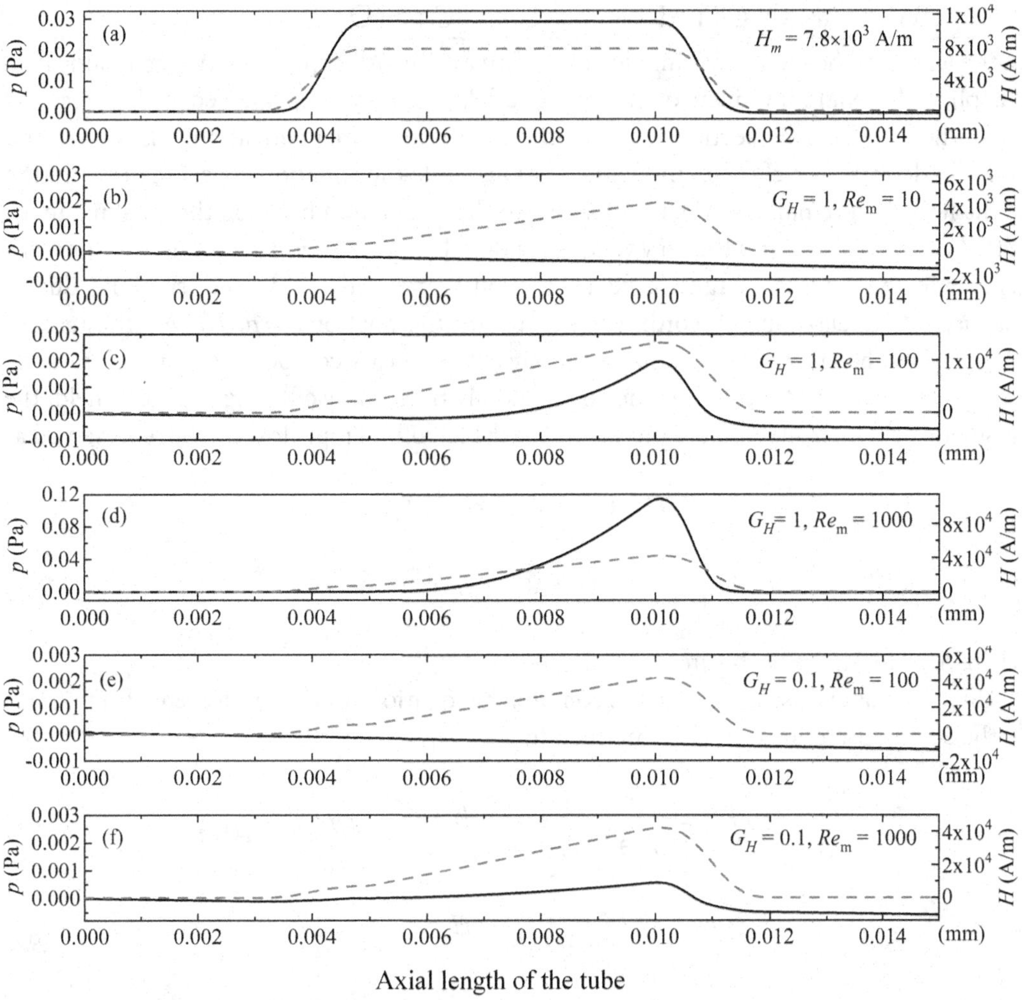

FIGURE 7.13 Pressure distribution along the axial direction in ferrofluid pipe flows under the action of axial gradient magnetic fields (solid line: pressure distribution; dashed line: magnetic field intensity distribution). (a) Uniform magnetic field; (b) $G_H = 1$, $Re_m = 10$; (c) $G_H = 1$, $Re_m = 100$; (d) $G_H = 1$, $Re_m = 1000$; (e) $G_H = 0.1$, $Re_m = 100$; (f) $G_H = 0.1$, $Re_m = 1000$.

However, in a gradient magnetic field, the pressure monotonically increases and peaks at the location of the maximum magnetic field strength as shown in Figure 7.13(c), (d), and (f). Comparing Figure 7.13(b)–(d), it can be observed that at a constant magnetic field gradient, an increase in Re_m leads to a stronger magnetic field at the same location, resulting in a more pronounced pressure increase and a higher maximum pressure. Additionally, by comparing Figure 7.13(c) with (e) and 7.13(d) with (f), it can be seen that, at the same Re_m value, the greater magnetic field gradient results in a more significant pressure variation.

The velocity distribution within a ferrofluid pipe flow, under the influence of an axially gradient magnetic field, varies with G_H and Re_m, exhibiting a similar trend to the results in Section 7.2.1.

7.3 FLOW IN AN OSCILLATING MAGNETIC FIELD

In the ferrofluid flow depicted in Figure 7.1, if an axially uniform oscillating magnetic field is applied, the axial component of the magnetic field intensity is represented as $H_z = H_0 \cos wt$ (Felderhof, 2001). This section employs the microscopic magnetization Eqs. (4.47) and (4.45) to describe the changes in magnetization, while the equation governing the motion of the ferrofluid remains as Eq. (2.219). Due to the interaction between the axial magnetization component M_z and the flow vorticity, a radial magnetization component M_r perpendicular to the external magnetic field direction is induced within the ferrofluid due to magnetization relaxation. According to the constitutive relation $\boldsymbol{B} = \mu_0(\boldsymbol{H} + \boldsymbol{M})$, the absence of B_r implies the presence of a radial magnetic field intensity component H_r, which satisfies $H_r = -M_r$. Thus, under the influence of an axially uniform oscillating magnetic field, the magnetic field intensity and magnetization in the ferrofluid pipe flow can be represented as

$$\boldsymbol{H} = \left(-M_r, 0, H_0 \cos(wt)\right) \tag{7.89}$$

$$\boldsymbol{M} = \left(M_r, 0, M_z\right) \tag{7.90}$$

where M_r and M_z are both functions of r and t.

Under the above assumptions, the component equations of the motion equation (2.219) in the z and r directions can be simplified to

$$\rho \frac{\partial v_z}{\partial t} = -\frac{\partial p}{\partial z} + \eta \left(\frac{1}{r}\frac{\partial v_z}{\partial r} + \frac{\partial^2 v_z}{\partial r^2} \right) - \frac{\mu_0}{2r}\frac{\partial}{\partial r}\left[r\left(M_r H_z - M_z H_r\right)\right] \tag{7.91}$$

$$\frac{\partial p}{\partial r} = \mu_0 M_r \frac{\partial H_r}{\partial r} \tag{7.92}$$

The pressure gradient can be obtained from Eq. (7.92)

$$\nabla p = \left(\mu_0 M_r \frac{\partial H_r}{\partial r}, 0, -\frac{\Delta p}{L_0} \right) \tag{7.93}$$

where

$$\frac{\partial p}{\partial z} = -\frac{\Delta p}{L_0}$$

Δp represents the pressure drop across a circular pipe with a length of L_0. Using Eqs. (4.45), (7.89) and (7.90), the final term on the right-hand side of Eq. (7.91) can be rewritten as

$$\frac{\partial}{\partial r}\left[r\left(M_r H_z - M_z H_r\right)\right] = \frac{\partial}{\partial r}\left[r\left(\frac{M_S L(\varsigma)\varsigma_r}{\varsigma} H_0 \cos(wt) + \frac{M_S^2 L^2 \varsigma_z \varsigma_r}{\varsigma^2} \right)\right]$$

Then Eq. (7.91) becomes

$$\rho\frac{\partial v_z}{\partial t}=-\frac{\partial p}{\partial z}+\eta\left(\frac{1}{r}\frac{\partial v_z}{\partial r}+\frac{\partial^2 v_z}{\partial r^2}\right)-\frac{\mu_0}{2r}\frac{\partial}{\partial r}\left[r\left(\frac{M_S L(\varsigma)\varsigma_r}{\varsigma}H_0\cos(wt)+\frac{M_S^2 L^2\varsigma_z\varsigma_r}{\varsigma^2}\right)\right] \quad (7.94)$$

Similarly, the component equations for the magnetization in the z and r directions are given as follows:

$$\begin{aligned}\frac{\partial M_z}{\partial t}&=\frac{1}{2}M_r\frac{\partial v_z}{\partial r}-\frac{1}{\tau_B}\left(M_z-M_S\frac{L(\varsigma)}{\varsigma}\alpha_z\right)\\&-\frac{\mu_0}{L(\varsigma)}\left(\frac{1}{L(\varsigma)}-\frac{3}{\varsigma}\right)\frac{M_r\left(M_zH_r-M_rH_z\right)}{6\eta\phi}\end{aligned} \quad (7.95)$$

$$\frac{\partial M_r}{\partial t}=-\frac{1}{2}M_z\frac{\partial v_z}{\partial r}-\frac{1}{\tau_B}\left(M_r-M_S\frac{L(\varsigma)}{\varsigma}\alpha_r\right)+\frac{\mu_0}{L(\varsigma)}\left(\frac{1}{L(\varsigma)}-\frac{3}{\varsigma}\right)\frac{M_z\left(M_zH_r-M_rH_z\right)}{6\eta\phi} \quad (7.96)$$

After expressing the magnetization components using Eq. (4.45), the two equations become

$$\begin{aligned}\frac{\partial}{\partial t}\left(\frac{L\varsigma_z}{\varsigma}\right)&=\frac{1}{2}\frac{L\varsigma_r}{\varsigma}\frac{\partial v_z}{\partial r}-\frac{1}{\tau_B}\frac{L}{\varsigma}(\varsigma_z-\alpha_z)\\&+\mu_0 M_S\left(\frac{1}{L}-\frac{3}{\varsigma}\right)\frac{L\varsigma_r^2}{\varsigma^3}\frac{\left[M_S L\varsigma_z+\varsigma H_0\cos(wt)\right]}{6\eta\phi}\end{aligned} \quad (7.97)$$

$$\begin{aligned}\frac{\partial}{\partial t}\left(\frac{L\varsigma_r}{\varsigma}\right)&=-\frac{1}{2}\frac{L\varsigma_z}{\varsigma}\frac{\partial v_z}{\partial r}-\frac{1}{\tau_B}\frac{L}{\varsigma}(\varsigma_r-\alpha_r)\\&-\mu_0 M_S\left(\frac{1}{L}-\frac{3}{\varsigma}\right)\frac{L\varsigma_z\varsigma_r}{\varsigma^3}\frac{\left[M_S L\varsigma_z+\varsigma H_0\cos(wt)\right]}{6\eta\phi}\end{aligned} \quad (7.98)$$

The following variables are utilized to non-dimensionalize Eqs. (7.94), (7.97), and (7.98)

$$r^*=\frac{r}{R_0},\ z^*=\frac{z}{R_0},\ t^*=\frac{t}{\tau_B},\ v_z^*=\frac{v_z}{\frac{\Delta p}{l}\frac{R_0^2}{4\eta}},\ L^*=\frac{L}{\varsigma} \quad (7.99)$$

Then the corresponding dimensionless equations are obtained:

$$\frac{\rho R_0^2}{\eta\tau_B}\frac{\partial v_z^*}{\partial t^*}=4+\frac{1}{r^*}\frac{\partial}{\partial r^*}\left(r^*\frac{\partial v_z^*}{\partial r^*}\right)-\frac{2L_0\mu_0 M_S^2}{R_0\Delta p}\frac{1}{r^{*2}}\frac{\partial}{\partial r^*}\left[r^*L^*\varsigma_r\left(\frac{H_0}{M_S}\cos\left(w\tau_B t^*\right)+L^*\varsigma_z\right)\right] \quad (7.100)$$

$$\frac{\partial}{\partial t^*}\left(L^*\varsigma_z\right)=\frac{1}{2}\frac{\Delta p}{L_0}\frac{R_0\tau_B}{4\eta}L^*\varsigma_r\frac{\partial v_z^*}{\partial r^*}-L^*\left[\varsigma_z-\alpha_0\cos\left(w\tau_B t^*\right)\right]$$
$$+\frac{\mu_0 M_S^2\tau_B}{6\eta\phi}\left(1-3L^*\right)\frac{\varsigma_r^2}{\varsigma^2}\left[L^*\varsigma_z+\frac{H_0}{M_S}\cos\left(w\tau_B t^*\right)\right] \tag{7.101}$$

$$\frac{\partial}{\partial t^*}\left(L^*\varsigma_r\right)=-\frac{1}{2}\frac{\Delta p}{L_0}\frac{R_0\tau_B}{4\eta}L^*\varsigma_z\frac{\partial v_z^*}{\partial r^*}-L^*\varsigma_r\left(1+\alpha_0\frac{M_S}{H_0}L^*\right)$$
$$-\frac{\mu_0 M_S^2\tau_B}{6\eta\phi}\left(1-3L^*\right)\frac{\varsigma_z\varsigma_r}{\varsigma^2}\left[L^*\varsigma_z+\frac{H_0}{M_S}\cos\left(w\tau_B t^*\right)\right] \tag{7.102}$$

According to Eqs. (4.1) and (4.8), α_0 can be expressed as

$$\alpha_0=\frac{\mu_0 m_p^2 N}{3k_B T}\frac{3H_0}{m_p N}=\chi_0\frac{3H_0}{M_S} \tag{7.103}$$

Also, using Eqs. (4.2), (4.8), and (4.13), a simplified expression can be obtained

$$\frac{\mu_0 M_S^2\tau_B}{18\eta\phi\chi_0}=\frac{M_S^2\tau_B}{6\eta\phi}\frac{k_B T}{m_p^2 N}=\frac{\tau_B N k_B T}{6\eta\phi}=\frac{1}{2} \tag{7.104}$$

Substituting the results of Eqs. (7.103) and (7.104) into the equations ranging from Eqs. (7.100)–(7.102) obtains the corresponding equations

$$\frac{1}{\hat{\alpha}}\frac{\partial v_z^*}{\partial t^*}=4+\frac{1}{r^*}\frac{\partial}{\partial r^*}\left(r^*\frac{\partial v_z^*}{\partial r^*}\right)-\hat{\beta}\frac{1}{r^{*2}}\frac{\partial}{\partial r^*}\left[r^*L^*\varsigma_r\left(\alpha_0\cos\left(\hat{\gamma}t^*\right)+3\chi_0 L^*\varsigma_z\right)\right] \tag{7.105}$$

$$\frac{\partial}{\partial t^*}\left(L^*\varsigma_z\right)=\frac{1}{2}\Omega_0\tau_B L^*\varsigma_r\frac{\partial v_z^*}{\partial r^*}-L^*\left[\varsigma_z-\alpha_0\cos\left(\hat{\gamma}t^*\right)\right]+\frac{1}{2}\left(1-3L^*\right)\frac{\varsigma_r^2}{\varsigma^2}\left[3\chi_0 L^*\varsigma_z+\alpha_0\cos\left(\hat{\gamma}t^*\right)\right] \tag{7.106}$$

$$\frac{\partial}{\partial t^*}\left(L^*\varsigma_r\right)=-\frac{1}{2}\Omega_0\tau_B L^*\varsigma_z\frac{\partial v_z^*}{\partial r^*}-L^*\varsigma_r\left(1+3\chi_0 L^*\right)$$
$$-\frac{1}{2}\left(1-3L^*\right)\frac{\varsigma_z\varsigma_r}{\varsigma^2}\left[3\chi_0 L^*\varsigma_z+\alpha_0\cos\left(\hat{\gamma}t^*\right)\right] \tag{7.107}$$

where

$$\hat{\alpha}=\frac{\eta\tau_B}{\rho R_0^2},\ \hat{\beta}=\frac{2\mu_0 L_0 M_S^2}{3\chi_0 R_0\Delta p},\ \hat{\gamma}=w\tau_B,\Omega_0=\frac{\Delta p}{L_0}\frac{R_0}{4\eta} \tag{7.108}$$

The system of equations consisting of Eqs. (7.105)–(7.107) is solved numerically at the given parameters: A tube radius of $R_0 = 0.5$ mm, a ferrofluid density of $\rho = 2200$ kg/m^3, a viscosity of $\eta = 0.077$ Pa·s, a magnetic particle volume fraction of $\phi = 0.2$, a magnetization relaxation time of $\tau_B = 1.6$ ms, and an initial susceptibility of $\chi_0 = 0.22$.

As shown in Figure 7.14, the relative change in flow rate, $\Delta Q/Q_0$, is depicted within the plane defined by $(w\tau_B, \alpha_0)$. Here, Q_0 represents the flow rate of the ferrofluid in the absence of a magnetic field

$$Q_0 = \frac{\pi R_0^4}{8\eta}\frac{\Delta p}{L_0} \tag{7.109}$$

The positive ΔQ depicted in the figure, indicating an increase in flow rate, arises from a decrease in the apparent viscosity of the ferrofluid. Conversely, the negative ΔQ corresponds to an increase in apparent viscosity. Figure 7.15 illustrates the nonlinear

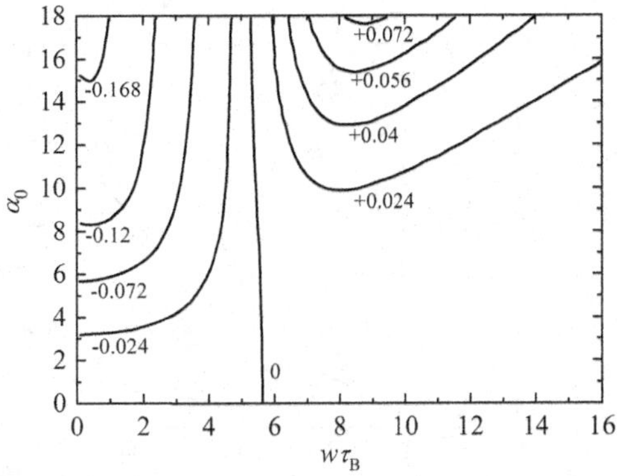

FIGURE 7.14 Contour lines of relative flow rate (with the numbers below the curves indicating the values of relative flow rate $\Delta Q/Q_0$, $\Omega_0\tau_B = 5$).(Reprinted from Krekhov et al. (2005) with permission from AIP.)

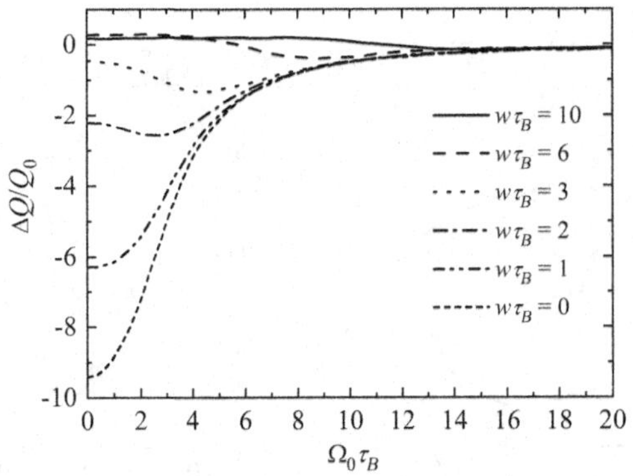

FIGURE 7.15 Relationship between the relative variation in flow rate and $\Omega_0\tau_B$ for different values of $w\tau_B$ ($\alpha_0 = 3$). (Reprinted from Krekhov et al. (2005) with permission from AIP.)

relationship between the relative change in flow rate, $\Delta Q/Q_0$, and $\Omega_0\tau_B$. At lower values of the flow vorticity $\Omega_0\tau_B$, the high-frequency magnetic field induces a negative rotational viscosity, leading to an increase in flow rate, as observed in Figure 7.15 for $w\tau_B = 6$ and $w\tau_B = 10$. Conversely, low-frequency magnetic fields produce a positive rotational viscosity, resulting in a decrease in flow rate.

The distribution of rotational viscosity along the radial direction of a circular pipe is depicted in Figure 7.16. When $\Omega_0\tau_B$ is sufficiently small, the rotational viscosity η_r remains negative across the entire cross-section of the pipe, as exemplified by the case of $\Omega_0\tau_B = 1$ in the figure. However, with an increase in the flow vorticity, negative rotational viscosity only occurs near the axis of the pipe. This spatial variation in apparent viscosity causes deviations in the velocity distribution from the typical flow profile within a pipe. The deviation in velocity between ferrofluid flow and traditional pipe flow is denoted as v_1, and it varies with the frequency of the magnetic field, as shown in Figure 7.17. The figure further illustrates the relationship between the spatial variation of the time-averaged value $\langle v_1 \rangle$ and

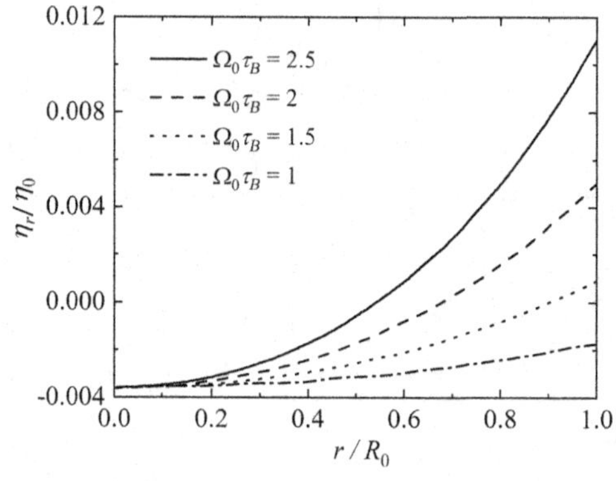

FIGURE 7.16 Variation of relative rotational viscosity along the radial direction of the circular pipe ($w\tau_B = 4$). (Reprinted from Krekhov et al. (2005) with permission from AIP.)

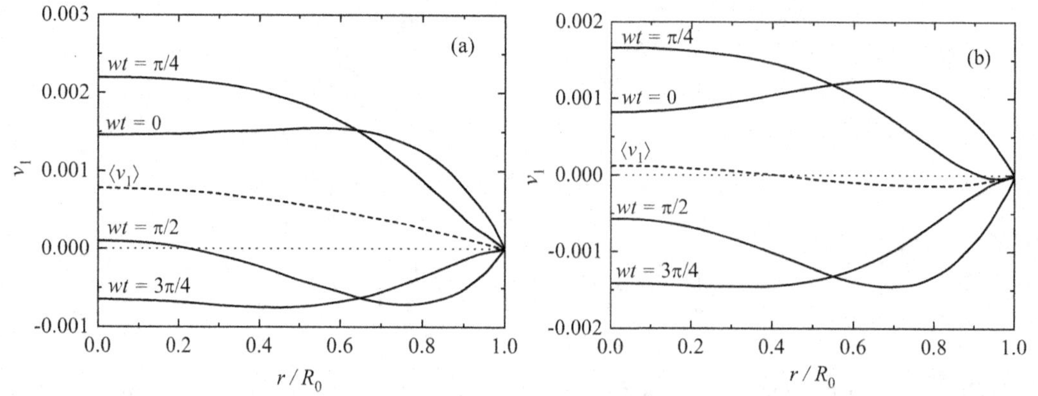

FIGURE 7.17 Distributions of additional velocity v_1 at different time instances ($\alpha_0 = 3$, $w\tau_B = 4$). (a) $\Omega_0\tau_B = 0.5$; (b) $\Omega_0\tau_B = 1.8$;

(Continued)

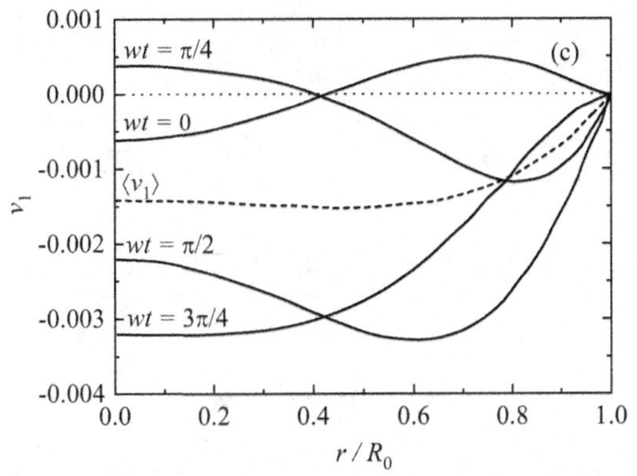

FIGURE 7.17 (Continued) and (c) $\Omega_0\tau_B = 3$. (Reprinted from Krekhov et al. (2005) with permission from AIP.)

its sign, which is closely related to the magnitude of $\Omega_0\tau_B$. Specifically, when $\Omega_0\tau_B$ is small, $\langle v_1 \rangle$ is positive, indicating a positive relative change in flow rate due to the magnetic field, as depicted in Figure 7.17(a). Conversely, at higher values of $\Omega_0\tau_B$, $\langle v_1 \rangle$ turns negative, resulting in a negative relative change in flow rate due to the magnetic field.

7.4 SPONTANEOUS ROTATION IN FERROFLUID PIPE FLOWS

In previous sections of this chapter, it was assumed that the flow velocity of ferrofluids only has an axial component. However, Krekhov and Shliomis (2017) removed this limitation and discovered the existence of a circumferential flow velocity component in ferrofluid pipe flows under the influence of a constant axial magnetic field. This velocity component leads to a helical motion or rotational instability. This section will elaborate on this characteristic of ferrofluid pipe flow.

The emergence of rotational phenomena in ferrofluid pipe flow is attributed to the magneto viscous effect, which is related to the fluid's shear rate. Their relationship originates from the interplay between magnetic torque and viscous torque acting on the magnetic particles. In the absence of a magnetic field, each particle within the ferrofluid pipe flow rotates at an angular velocity Ω_θ along a cylindrical surface coaxial with the pipe. When an axial magnetic field is applied, the magnetic moment of the particles tends to align with the direction of the external magnetic field, hindering their free rotation. As a result of the combined action of magnetic torque and viscous torque, the magnetization deviates from the direction of the external magnetic field, resulting in a radial component of magnetization and a corresponding magnetic field component in the opposite direction. This magnetic field component is perpendicular to the applied magnetic field and provides positive feedback for the instability of the ferrofluid pipe flow.

7.4.1 Governing Equations and Their Solutions

This section focuses on the magnetization variations in weak and moderate external magnetic fields as described by Eq. (4.52). The governing equations are comprised of Eqs. (2.219), (4.52), (2.207), and (2.82). To non-dimensionalize these equations, the following variables are employed

$$r^* = \frac{r}{R_0}, z^* = \frac{z}{R_0}, t^* = \frac{t}{\tau_B}, \boldsymbol{v}^* = \frac{\boldsymbol{v}}{R_0 / \tau_B}, p^* = \frac{p}{\eta / \tau_B}, \boldsymbol{\Omega}^* = \boldsymbol{\Omega}\tau_B$$
$$\mathcal{L} = \frac{3L}{\varsigma}, \boldsymbol{M}^* = \frac{\boldsymbol{M}}{M_S / 3}, \boldsymbol{H}^* = \frac{\boldsymbol{H}}{M_S / 3\chi_0} \quad (7.110)$$

Based on these definitions, by applying Eqs. (4.50) and (4.45), respectively, it can be derived that

$$\boldsymbol{M}^* = \mathcal{L}\boldsymbol{\varsigma}, \boldsymbol{H}^* = \boldsymbol{\alpha} \quad (7.111)$$

Substituting Eqs. (7.110) and (7.111) into Eqs. (2.219), (4.52), (2.207), and (2.82) respectively, one obtains the dimensionless forms of the governing equations

$$\frac{1}{\hat{\alpha}}\left[\frac{\partial \boldsymbol{v}^*}{\partial t^*} + \left(\boldsymbol{v}^* \cdot \nabla\right)\boldsymbol{v}^*\right] = -\nabla p^* + \nabla^2 \boldsymbol{v}^* + \hat{\mathrm{E}}\left[2\mathcal{L}\left(\boldsymbol{\varsigma}\cdot\nabla\right)\boldsymbol{\alpha} + \nabla\times\left(\mathcal{L}\boldsymbol{\varsigma}\times\boldsymbol{\alpha}\right)\right] \quad (7.112)$$

$$\frac{\mathrm{D}\left(\mathcal{L}\boldsymbol{\varsigma}\right)}{\mathrm{D}t} = \mathcal{L}\boldsymbol{\Omega}^*\times\boldsymbol{\varsigma} - \mathcal{L}\left(\boldsymbol{\varsigma} - \boldsymbol{\alpha}\right) - \frac{3}{2\varsigma^2}\left(1 - \mathcal{L}\right)\boldsymbol{\varsigma}\times\left(\boldsymbol{\varsigma}\times\boldsymbol{\alpha}\right) \quad (7.113)$$

$$\nabla\times\boldsymbol{\alpha} = \boldsymbol{0}, \nabla\cdot\left(\boldsymbol{\alpha} + \mathcal{L}\chi_0\boldsymbol{\varsigma}\right) = 0 \quad (7.114)$$

where

$$\hat{\alpha} = \frac{\eta\tau_B}{\rho R_0^2}, \hat{\mathrm{E}} = \frac{\mu_0\tau_B M_S^2}{18\eta\chi_0}$$

Assuming the existence of an axial velocity component v_θ in a cylindrical coordinate system (r, θ, z), the respective variables become

$$\boldsymbol{v}^* = \left(0, v_\theta, v_z\right), \boldsymbol{\Omega}^* = \left(0, \Omega_\theta, \Omega_z\right), \boldsymbol{\varsigma} = \left(\varsigma_r, \varsigma_\theta, \varsigma_z\right), \boldsymbol{\alpha} = \left(\alpha_r, 0, \alpha_0\right)$$

where the components of each variable are dimensionless, with the superscript asterisk omitted for brevity. In the steady-state flow, the components of the various variables depend solely on r. The dimensionless magnetic field intensity, denoted as α_0, is represented by Eq. (7.103). The relationship between the components of vorticity and the velocity components is

$$\Omega_\theta = -\frac{1}{2}\frac{\mathrm{d}v_z}{\mathrm{d}r}, \Omega_z = \frac{1}{2r}\frac{\mathrm{d}\left(rv_\theta\right)}{\mathrm{d}r} \quad (7.115)$$

The static magnetic field equation (7.114) can be simplified as

$$\frac{\partial}{\partial r}\left[r\left(\alpha_r + \mathcal{L}\chi_0\varsigma_r\right)\right] = 0$$

The solution of this equation is

$$\alpha_r = -\chi_0 \mathcal{L} \varsigma_r = -(\mu_r - 1)\mathcal{L}\varsigma_r \tag{7.116}$$

where μ_r is the initial relative permeability of the ferrofluid.

The function $\mathcal{L}(\varsigma)$ can be expanded as

$$\mathcal{L}(\varsigma) = 1 - \frac{\varsigma^2}{15} + O(\varsigma^4)$$

For weak to moderate intensities of the applied magnetic field, $\mathcal{L}(\varsigma) \approx 1$. The magnetization equation (7.113) will be simplified under this condition. For steady-state flows, the term on the left-hand side of the equation becomes

$$\begin{aligned}\frac{\mathrm{D}\varsigma}{\mathrm{D}t} &= (\boldsymbol{v}^* \cdot \nabla)\varsigma = (v_\theta \boldsymbol{e}_\theta + v_z \boldsymbol{e}_z)\cdot\left(\boldsymbol{e}_r \frac{\partial}{\partial r} + \boldsymbol{e}_\theta \frac{1}{r}\frac{\partial}{\partial \theta} + \boldsymbol{e}_z \frac{\partial}{\partial z}\right)(\varsigma_r \boldsymbol{e}_r + \varsigma_\theta \boldsymbol{e}_\theta + \varsigma_z \boldsymbol{e}_z) \\ &= \left(\frac{v_\theta}{r}\frac{\partial}{\partial \theta} + v_z \frac{\partial}{\partial z}\right)(\varsigma_r \boldsymbol{e}_r + \varsigma_\theta \boldsymbol{e}_\theta + \varsigma_z \boldsymbol{e}_z) = \frac{1}{r} v_\theta \varsigma_r \boldsymbol{e}_\theta - \frac{1}{r} v_\theta \varsigma_\theta \boldsymbol{e}_r\end{aligned} \tag{7.117}$$

where $\boldsymbol{e}_r$, $\boldsymbol{e}_\theta$, and $\boldsymbol{e}_z$ are unit vectors along the radial, azimuthal, and axial coordinates in the cylindrical coordinate system, respectively. This derivation utilizes the identities for the cylindrical coordinate system

$$\frac{\partial \boldsymbol{e}_r}{\partial \theta} = \boldsymbol{e}_\theta, \frac{\partial \boldsymbol{e}_\theta}{\partial \theta} = -\boldsymbol{e}_r \tag{7.118}$$

Substituting the result from Eq. (7.117) into Eq. (7.113) yields the component equations

$$\begin{cases} -\frac{1}{r} v_\theta \varsigma_\theta = \Omega_\theta \varsigma_z - \Omega_z \varsigma_\theta - \varsigma_r + \alpha_r \\ \frac{1}{r} v_\theta \varsigma_r = \Omega_z \varsigma_r - \varsigma_\theta \\ -\Omega_\theta \varsigma_r - \varsigma_z + \alpha_0 = 0 \end{cases}$$

After substituting Eq. (7.116) into this system of equations, the solutions are obtained

$$\varsigma_r = \frac{\Omega_\theta \alpha_0}{\Lambda^2 + \Omega_\theta^2 + \mu_r}, \varsigma_\theta = \Lambda \varsigma_r, \varsigma_z = \frac{\Lambda^2 + \mu_r}{\Omega_\theta} \varsigma_r \tag{7.119}$$

where

$$\Lambda = \Omega_z - \frac{1}{r} v_\theta = \frac{r}{2}\frac{\mathrm{d}}{\mathrm{d}r}\left(\frac{v_\theta}{r}\right) \tag{7.120}$$

According to Eq. (7.120), Λ equals zero when the ferrofluid exhibits no rotational motion ($v_\theta = 0$) or when the entire ferrofluid within the circular tube rotates with a constant angular velocity (v_θ/r is constant). Only when the annular layer of the ferrofluid within the tube rotates with varying angular velocities will Λ be non-zero.

To obtain the component equations of the motion equation (7.112), now simplify the various terms within this equation

$$\begin{aligned}\left(\boldsymbol{v}^*\cdot\nabla\right)\boldsymbol{v}^* &= \left(v_\theta\boldsymbol{e}_\theta + v_z\boldsymbol{e}_z\right)\cdot\left(\boldsymbol{e}_r\frac{\partial}{\partial r} + \boldsymbol{e}_\theta\frac{1}{r}\frac{\partial}{\partial\theta} + \boldsymbol{e}_z\frac{\partial}{\partial z}\right)\left(v_\theta\boldsymbol{e}_\theta + v_z\boldsymbol{e}_z\right)\\ &= \left(\frac{v_\theta}{r}\frac{\partial}{\partial\theta} + v_z\frac{\partial}{\partial z}\right)\left(v_\theta\boldsymbol{e}_\theta + v_z\boldsymbol{e}_z\right)\\ &= -\frac{1}{r}v_\theta^2\boldsymbol{e}_r\end{aligned} \tag{7.121}$$

$$\begin{aligned}\nabla^2\boldsymbol{v}^* &= \left(\frac{\partial^2}{\partial r^2} + \frac{1}{r}\frac{\partial}{\partial r} + \frac{1}{r^2}\frac{\partial^2}{\partial\theta^2} + \frac{\partial^2}{\partial z^2}\right)\left(v_\theta\boldsymbol{e}_\theta + v_z\boldsymbol{e}_z\right)\\ &= \left(\frac{\partial^2 v_\theta}{\partial r^2} + \frac{1}{r}\frac{\partial v_\theta}{\partial r} - \frac{v_\theta}{r^2}\right)\boldsymbol{e}_\theta + \left(\frac{\partial^2 v_z}{\partial r^2} + \frac{1}{r}\frac{\partial v_z}{\partial r}\right)\boldsymbol{e}_z\end{aligned} \tag{7.122}$$

$$\begin{aligned}\left(\boldsymbol{\varsigma}\cdot\nabla\right)\boldsymbol{\alpha} &= \left(\varsigma_r\boldsymbol{e}_r + \varsigma_\theta\boldsymbol{e}_\theta + \varsigma_z\boldsymbol{e}_z\right)\cdot\left(\boldsymbol{e}_r\frac{\partial}{\partial r} + \boldsymbol{e}_\theta\frac{1}{r}\frac{\partial}{\partial\theta} + \boldsymbol{e}_z\frac{\partial}{\partial z}\right)\left(\alpha_r\boldsymbol{e}_r + \alpha_0\boldsymbol{e}_z\right)\\ &= \left(\varsigma_r\frac{\partial}{\partial r} + \frac{1}{r}\varsigma_\theta\frac{\partial}{\partial\theta} + \varsigma_z\frac{\partial}{\partial z}\right)\left(\alpha_r\boldsymbol{e}_r + \alpha_0\boldsymbol{e}_z\right)\\ &= \varsigma_r\frac{\partial\alpha_r}{\partial r}\boldsymbol{e}_r + \frac{1}{r}\varsigma_\theta\alpha_r\boldsymbol{e}_\theta\end{aligned} \tag{7.123}$$

$$\begin{aligned}\nabla\times\left(\boldsymbol{\varsigma}\times\boldsymbol{\alpha}\right) &= \frac{1}{r}\begin{vmatrix}\boldsymbol{e}_r & r\boldsymbol{e}_\theta & \boldsymbol{e}_z\\ \dfrac{\partial}{\partial r} & \dfrac{\partial}{\partial\theta} & \dfrac{\partial}{\partial z}\\ \alpha_0\varsigma_\theta & r\left(\varsigma_z\alpha_r - \alpha_0\varsigma_r\right) & -\varsigma_\theta\alpha_r\end{vmatrix}\\ &= \boldsymbol{e}_\theta\frac{\partial}{\partial r}\left(\varsigma_\theta\alpha_r\right) + \boldsymbol{e}_z\frac{\partial}{\partial r}\left[r\left(\varsigma_z\alpha_r - \alpha_0\varsigma_r\right)\right]\end{aligned} \tag{7.124}$$

Substituting Eqs. (7.121)–(7.124) into Eq. (7.112) yields the corresponding steady-state component equations in the z and θ directions, respectively,

$$-\frac{\partial p^*}{\partial z^*} + \frac{1}{r}\frac{\mathrm{d}}{\mathrm{d}r}\left(r\frac{\mathrm{d}v_z}{\mathrm{d}r}\right) + \frac{\hat{E}}{r}\frac{\mathrm{d}}{\mathrm{d}r}\left[r\left(\varsigma_z\alpha_r - \alpha_0\varsigma_r\right)\right] = 0 \tag{7.125}$$

$$\frac{1}{r}\frac{\mathrm{d}}{\mathrm{d}r}\left(r\frac{\mathrm{d}v_\theta}{\mathrm{d}r}\right) - \frac{v_\theta}{r^2} + \hat{E}\left[\frac{2}{r}\varsigma_\theta\alpha_r + \frac{\mathrm{d}}{\mathrm{d}r}\left(\varsigma_\theta\alpha_r\right)\right] \tag{7.126}$$

Letting

$$\frac{\partial p}{\partial z} = -\frac{\Delta p}{L_0}$$

one gets

$$\frac{\partial p^*}{\partial z^*} = \frac{\tau_B R_0}{\eta}\frac{\partial p}{\partial z} = -\frac{\tau_B R_0}{\eta}\frac{\Delta p}{L_0} = -4P'$$

where

$$P' = \frac{\tau_B R_0}{4\eta}\frac{\Delta p}{L_0} \tag{7.127}$$

Using Eq. (7.115), one can rewrite Eq. (7.125) as

$$\frac{\mathrm{d}}{\mathrm{d}r}\left\{r\left[2\Omega_\theta - \hat{E}\left(\varsigma_z\alpha_r - \alpha_0\varsigma_r\right)\right]\right\} = 4rP' \tag{7.128}$$

For the terms in Eq. (7.126), one has

$$\frac{1}{r}\frac{\mathrm{d}}{\mathrm{d}r}\left(r\frac{\mathrm{d}v_\theta}{\mathrm{d}r}\right) - \frac{v_\theta}{r^2} = \frac{2}{r^2}\frac{\mathrm{d}}{\mathrm{d}r}\left(r^2\Lambda\right)$$

$$\frac{2}{r}\varsigma_\theta\alpha_r + \frac{\mathrm{d}}{\mathrm{d}r}\left(\varsigma_\theta\alpha_r\right) = \frac{1}{r^2}\frac{\mathrm{d}}{\mathrm{d}r}\left(r^2\varsigma_\theta\alpha_r\right)$$

Utilizing these two formulas, Eq. (7.126) is written as

$$\frac{\mathrm{d}}{\mathrm{d}r}\left[r^2\left(2\Lambda + \hat{E}\varsigma_\theta\alpha_r\right)\right] = 0 \tag{7.129}$$

Integrating both Eqs. (7.128) and (7.129) directly with respect to r, and utilizing the boundary conditions

$$v_\theta = v_z = 0, \text{when } r = 1$$

and Eq. (7.116), one obtains

$$\Lambda\left[1 - \frac{\Omega_\theta^2\hat{h}^2}{\left(\Lambda^2 + \Omega_\theta^2 + \mu_r\right)^2}\right] = 0 \tag{7.130}$$

$$\Omega_\theta\left[1+\frac{\hat{h}^2\left(\mu_r\Lambda^2+\Omega_\theta^2+\mu_r^2\right)}{\left(\mu_r-1\right)\left(\Lambda^2+\Omega_\theta^2+\mu_r\right)^2}\right]=rP' \tag{7.131}$$

where

$$\hat{h}=\frac{\alpha_0 M_S}{6}\sqrt{\frac{\mu_0\tau_B}{\eta}}=\frac{\chi_0 H_0}{2}\sqrt{\frac{\mu_0\tau_B}{\eta}}$$

7.4.2 Conditions for Generating Rotational Motion

As can be seen from Eq. (7.130), when there is a rotational motion of ferrofluid ($\Lambda \neq 0$), it satisfies

$$\Lambda^2+\Omega_\theta^2+\mu_r=\Omega_\theta\hat{h} \tag{7.132}$$

This formula determines the function Λ related to Ω_θ, which is expressed as

$$\Lambda^2=\left(\Omega_\theta-\Omega_{\theta,1}\right)\left(\Omega_{\theta,2}-\Omega_\theta\right) \tag{7.133}$$

where

$$\Omega_{\theta,1}=\frac{1}{2}\left(\hat{h}-\sqrt{\hat{h}^2-4\mu_r}\right),\ \Omega_{\theta,2}=\frac{1}{2}\left(\hat{h}+\sqrt{\hat{h}^2-4\mu_r}\right)$$

Substituting Eq. (7.132) into Eq. (7.131) and simplifying, one obtains

$$\frac{\mu_r\hat{h}}{\mu_r-1}=rP' \tag{7.134}$$

As the condition for the validity of Eq. (7.132) is $\Lambda \neq 0$, it follows from Eq. (7.134) that rotational motion only exists on the surface of a cylindrical ferrofluid with a radius of

$$\tilde{r}=\frac{\mu_r\hat{h}}{\left(\mu_r-1\right)P'}$$

Furthermore, on this surface, the value of Λ ranges from 0 to $\sqrt{\frac{\hat{h}^2}{4}-\mu_r}$, reaching its minimum and maximum values at $\Omega_\theta=\Omega_{\theta,1}$, $\Omega_\theta=\Omega_{\theta,2}$, and $\frac{1}{2}\left(\Omega_{\theta,1}+\Omega_{\theta,2}\right)$ respectively.

The vorticity expression for ordinary fluid flow in pipes is given by

$$\Omega=\frac{1}{4\eta}\frac{\Delta p}{L_0}r$$

where the variables Ω and r are dimensional quantities that need to be converted to dimensionless forms according to the definition given in Eq. (7.110),

$$\Omega^* = \frac{\tau_B R_0}{4\eta}\frac{\Delta p}{L_0}r^* = P'r^*$$

Comparing this equation with Eq. (7.131), it can be observed that the apparent viscosity, η_{eff}, in the ferrofluid pipe flow is

$$\eta_{\text{eff}} = \eta\left[1 + \frac{\hat{h}^2\left(\mu_r\Lambda^2 + \Omega_\theta^2 + \mu_r^2\right)}{\left(\mu_r - 1\right)\left(\Lambda^2 + \Omega_\theta^2 + \mu_r\right)^2}\right] \tag{7.135}$$

Substituting Eq. (7.131) into it and utilizing Eq. (7.134), one obtains

$$\eta_{\text{eff}} = \eta\frac{rP'}{\Omega_\theta} = \eta\frac{\mu_r\hat{h}}{\Omega_\theta\left(\mu_r - 1\right)} \tag{7.136}$$

When the ferrofluid element within the circular tube is slightly displaced from the surface with a radius of $\tilde{r}$, Λ becomes zero, corresponding to Ω_θ values of $\Omega_{\theta,1}$ and $\Omega_{\theta,2}$. Consequently, on both sides of the surface with a radius of $\tilde{r}$, the ferrofluid exhibits distinct apparent viscosities, given by $\eta\frac{\mu_r\hat{h}}{\Omega_{\theta,1}\left(\mu_r - 1\right)}$ and $\eta\frac{\mu_r\hat{h}}{\Omega_{\theta,2}\left(\mu_r - 1\right)}$, respectively. In other words, there is a viscosity jump across this surface

$$\left[\eta_{\text{eff}}\right]_{\tilde{r}+0}^{\tilde{r}-0} = \eta\frac{\mu_r\hat{h}}{\mu_r - 1}\left(\frac{1}{\Omega_{\theta,1}} - \frac{1}{\Omega_{\theta,2}}\right) = \eta\frac{2\hat{h}}{\mu_r - 1}\sqrt{\frac{\hat{h}^2}{4} - \mu_r} \tag{7.137}$$

To guarantee the existence of real solutions for Eq. (7.133), it is necessary to satisfy the condition that $\hat{h}^2 > 4\mu_r$, while simultaneously ensuring that $\tilde{r} < 1$. This yields the conditions for generating rotational motion

$$2\sqrt{\mu_r} < \hat{h} < \frac{\left(\mu_r - 1\right)P'}{\mu_r}$$

Thus, the entire ferrofluid pipe flow is divided into a cylindrical liquid section with a radius of $\tilde{r}$ and an annular liquid layer with a thickness of $\left(1 - \tilde{r}\right)$, both coaxial with the circular pipe. The cylindrical liquid section rotates with a constant angular velocity of v_θ/r alongside its axial motion v_z, resulting in helical trajectories for the liquid elements within this section. In contrast, the fluid elements within the annular liquid layer follow straight trajectories as illustrated in Figure 7.18.

The emergence of rotational motion in ferrofluid pipe flow can be observed from the flow rate variation curve as illustrated in Figure 7.19, which depicts the relationship between

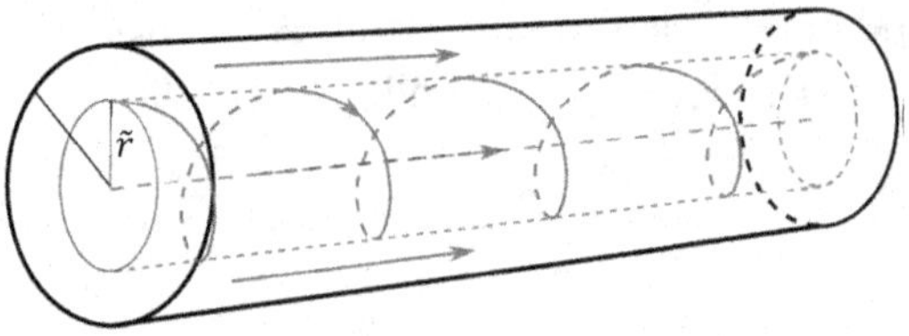

FIGURE 7.18 Rotational flow in a ferrofluid pipe flow. (Reprinted from Krekhov and Shliomis (2017) with permission from APS.)

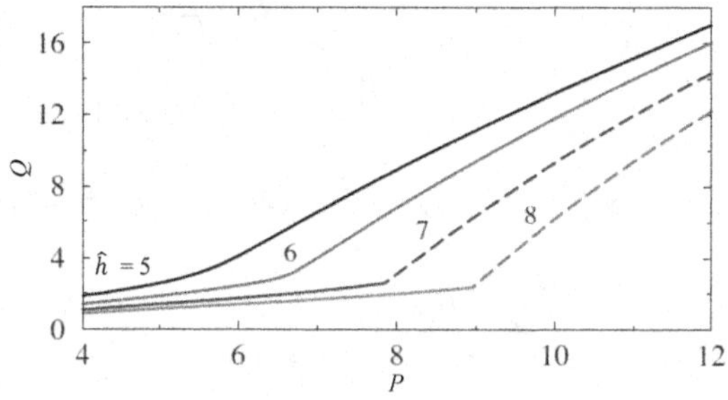

FIGURE 7.19 Variation of volumetric flow rate with pressure gradient P' in ferrofluid pipe flows ($\mu_r = 9$, the solid line represents pure axial flow and the dashed line represents rotational motion). (Reprinted from Krekhov and Shliomis (2017) with permission from APS.)

volumetric flow rate Q and pressure gradient P' in a ferrofluid pipe flow. Here, Q is defined as $Q = 2\pi \int_0^1 v_z r dr$. The figure clearly demonstrates the occurrence of rotational motion. When $\hat{h} \le \hat{h}_c = 2\sqrt{\mu_r}$, the ferrofluid exhibits pure axial flow, as shown in cases when $\hat{h} = 5$ and $\hat{h} = 6$. However, when $\hat{h} > \hat{h}_c$ and the pressure gradient P' exceeds a certain threshold, rotational motion emerges as exemplified by cases when $\hat{h} = 7$ and $\hat{h} = 8$. This transition from pure axial flow to rotational motion is clearly marked by an inflection point on the flow rate curve.

The emergence of rotational motion in ferrofluid pipe flow also leads to a significant increase in axial velocity compared to the case without rotation as illustrated in Figure 7.20. The inset in this figure shows the distribution of $v_\theta(r)$ and $\Lambda(r)$ functions near $\tilde{r} = 0.8931$ at $\hat{h} = 8.8$. The $\Lambda(r)$ width measures 0.0138, peaking at 2.940. As can be seen from Figure 7.20, when $r < \tilde{r}$, a circumferential velocity component appears in the pipe flow and an inflection point appears at $r = \tilde{r}$ on the $v_z(r)$ curve. When disregarding the circumferential velocity component, the $v_z(r)$ curve remains smooth.

The rotational motion in the ferrofluid pipe flow also modifies the distribution of magnetization as demonstrated in Figure 7.21, where $\tilde{\boldsymbol{M}} = \mathcal{L}(\varsigma)\varsigma$. When neglecting the circumferential component of velocity, both $\tilde{M}_z$ and $\tilde{M}_\theta$ exhibit smooth curves as indicated by the dashed lines in the figure. However, in the presence of rotational motion, a discontinuity occurs in $\tilde{M}_z$ and $\tilde{M}_\theta$ at $r = \tilde{r}$.

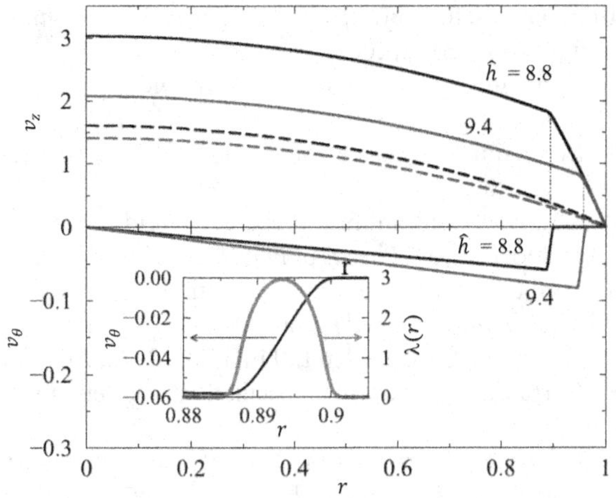

FIGURE 7.20 Axial and circumferential velocity components in a ferrofluid pipe flow (μ_r = 9, P' = 11, the dashed line represents the calculation results when the circumferential speed is set to zero). (Reprinted from Krekhov and Shliomis (2017) with permission from APS.)

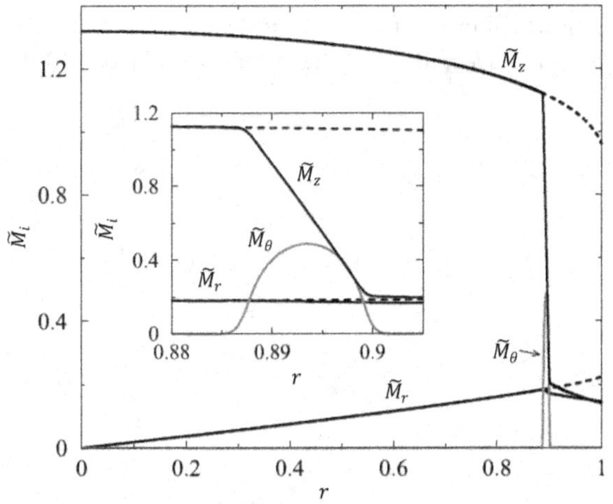

FIGURE 7.21 Components of magnetization of ferrofluids (μ_r = 9, P' = 11, and $\hat{h}$ = 8.8, the dashed line represents the calculation result when the circumferential speed is set to zero). (Reprinted from Krekhov and Shliomis (2017) with permission from APS.)

REFERENCES

Bianco V., Trubatch A. D., Wei H., Yecko P., Quantifying volume loss of a magnetically localized ferrofluid bolus in pulsatile pipe flow, *Journal of Magnetism and Magnetic Materials*, 524: 167595, 2021.

Cunha F. R., Sobral Y. D., Asymptotic solution for pressure-driven flows of magnetic fluids in pipes, *Journal of Magnetism and Magnetic Materials*, 289: 314–317, 2005.

Dalvi S., van der Meer T. H., Shahi M., Numerical evaluation of the ferrofluid behaviour under the influence of three-dimensional non-uniform magnetic field, *International Journal of Heat and Fluid Flow*, 94: 108901, 2022.

Felderhof B. U., Flow of a ferrofluid down a tube in an oscillating magnetic field, *Physical Review E*, 64: 021508, 2001.

Kamiyama S., Ferrohydrodynamics, in *Encyclopedia of materials: Science and technology (second edition)*, Amsterdam: Elsevier, 3116–3127, 2001.

Krekhov A., Shliomis M., Spontaneous core rotation in ferrofluid pipe flow, *Physical Review Letters*, 118: 114503, 2017.

Krekhov A. P., Shliomis M., Kamiyama S., Ferrofluid pipe flow in an oscillating magnetic field, *Physics of Fluids*, 17: 033105, 2005.

Papadopoulos P. K., Vafeas P., Hatzikonstantinou P. M., Ferrofluid pipe flow under the influence of the magnetic field of a cylindrical coil, *Physics of Fluids*, 24: 122002, 2012.

Rosa A. P., Gontijo R. G., Cunha F. R., Laminar pipe flow with drag reduction induced by a magnetic field gradient, *Applied Mathematical Modelling*, 40: 3907–3918, 2016.

Schumacher K. R., Sellien I., Knoke G. S., Cader T., Finlayson B. A., Experiment and simulation of laminar and turbulent ferrofluid pipe flow in an oscillating magnetic field, *Physical Review E*, 67: 026308, 2003.

Shimada K., Kamiyama S., Numerical analysis of effect of distribution of mass concentration on steady magnetic fluid flow in a straight tube, *IEEE Transactions on Magnetics*, 36(5): 3706–3708, 2000.

Shuchia S., Shimada K., Kamiyama S., Yamaguchi H., Hydrodynamic characteristics of steady magnetic fluid flow in a straight tube by taking into account the non-uniform distribution of mass concentration, *Journal of Magnetism and Magnetic Materials*, 252: 166–168, 2002.

Wang H., Monroe J. G., Kumari S., Leontsev S. O., Vasquez E. S., Thompson S. M., Berg M. J., Walters D. K., Walters K. B., Analytical model for electromagnetic induction in pulsating ferrofluid pipe flows, *International Journal of Heat and Mass Transfer*, 175: 121325, 2021.

Laminar Circular Flow of Ferrofluids

The circular flow of ferrofluids refers to the motion of the fluid within the micro gap between two coaxial cylinders. When the cylinders rotate, they initiate a flow within the ferrofluid, causing fluid elements to trace concentric circles centred around the cylinder axis (Yang et al., 2022). A uniform static magnetic field, H_0, is applied perpendicular to the cylinder axis, as illustrated in Figure 8.1. The outer and inner coaxial cylinders have outer and inner diameters of R_1 and R_2, respectively, and angular velocities of Ω_1 and Ω_2, with the counterclockwise direction designated as positive.

8.1 VELOCITY AND VORTICITY OF FERROFLUIDS

Assuming a narrow gap between the two cylinders, the effects of a uniform magnetic field and gravity on the flow of ferrofluid are neglected. Within this gap, the axial velocity component of the ferrofluid is zero, maintaining pure circular motion. In a cylindrical coordinate system (r, θ, z), fluid elements trace concentric circles. Then one has

$$v_r = 0, v_z = 0$$

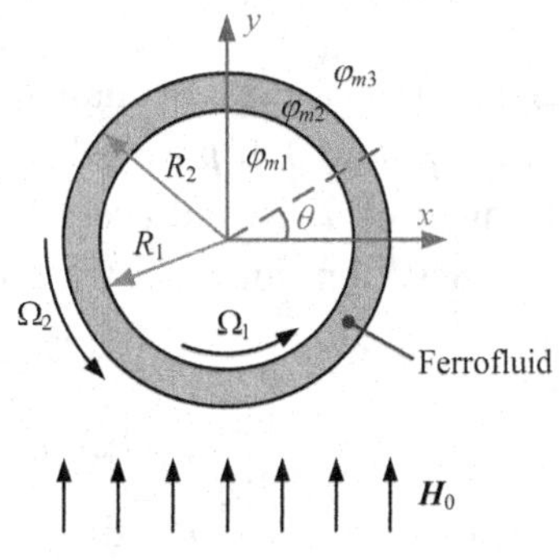

FIGURE 8.1 Schematic diagram of ferrofluid concentric flows.

DOI: 10.1201/9781003540342-8

Based on the assumption of incompressible ferrofluids, the continuity equation is simplified as

$$\partial v_\theta / \partial \theta = 0$$

Maintaining constant angular velocities for both the inner and outer cylinders, the flow field eventually attains a steady-state. In this steady flow condition, the component equations of the motion equation (2.219) can be simplified to

$$\begin{cases} \dfrac{v_\theta^2}{r} = \dfrac{1}{\rho}\dfrac{\partial p}{\partial r} \\ 0 = -\dfrac{1}{r\rho}\dfrac{\partial p}{\partial \theta} + \mu\left(\dfrac{\partial^2 v_\theta}{\partial r^2} + \dfrac{1}{r}\dfrac{\partial v_\theta}{\partial r} + \dfrac{\partial^2 v_\theta}{\partial z^2} - \dfrac{v_\theta}{r^2}\right) \\ 0 = \dfrac{\partial p}{\partial z} \end{cases} \tag{8.1}$$

The first equation demonstrates that the centripetal force maintaining the circular motion of fluid elements is provided by the radial variation of pressure. By taking the derivative of the first equation with respect to z and incorporating the third equation, it can be derived that

$$\partial v_\theta / \partial z = 0$$

It demonstrated that the concentric circular flow of incompressible ferrofluids can be described as planar parallel motion, with the velocity component v_θ solely dependent on the radial distance r. This allows for the replacement of partial derivatives with ordinary differential equations. By taking the derivative of the first equation in Eq. (8.1) with respect to θ and applying the continuity equation, one obtains

$$\frac{\partial}{\partial \theta}\left(\frac{\partial p}{\partial r}\right) = \frac{\partial}{\partial r}\left(\frac{\partial p}{\partial \theta}\right) = 0$$

It shows that $\partial p/\partial \theta$ remains constant regardless of changes in r. Based on the uniqueness of p at $r = 0$, it can be deduced that $\partial p/\partial \theta = 0$. If p varied with θ, meaning different values of p corresponding to different angles of θ, it would result in multiple pressure values at $r = 0$, contradicting the laws of physics. Consequently, the second equation in Eq. (8.1) can be simplified to

$$\frac{\mathrm{d}^2 v_\theta}{\mathrm{d}r^2} + \frac{1}{r}\frac{\mathrm{d}v_\theta}{\mathrm{d}r} - \frac{v_\theta}{r^2} = \frac{\mathrm{d}}{\mathrm{d}r}\left[\frac{1}{r}\frac{\mathrm{d}(rv_\theta)}{\mathrm{d}r}\right] = 0 \tag{8.2}$$

The corresponding boundary conditions are

$$v_\theta = \Omega_1 R_1,\ \text{when}\ r = R_1$$
$$v_\theta = \Omega_2 R_2,\ \text{when}\ r = R_2$$

The general solution of Eq. (8.2) is given by

$$v_\theta = C_1 r + \frac{C_2}{r}$$

After determining the integration constants using the boundary conditions, the velocity distribution can be obtained

$$v_\theta = \Omega\left(r + \tilde{s}\frac{R_2^2}{r}\right) \tag{8.3}$$

where the vorticity is

$$\boldsymbol{\Omega} = \Omega \boldsymbol{e}_z = \frac{1}{2}\nabla \times \boldsymbol{v} = \left(\frac{1}{r}\frac{\partial(r v_\theta)}{\partial r} - \frac{1}{r}\frac{\partial v_r}{\partial r}\right)\boldsymbol{e}_z = \boldsymbol{e}_z\frac{1}{r}\frac{\partial(r v_\theta)}{\partial r}$$

and

$$\Omega = \frac{\Omega_2 R_2^2 - \Omega_1 R_1^2}{R_2^2 - R_1^2} \tag{8.4}$$

$$\tilde{s} = \frac{R_1^2}{R_2^2 - R_1^2}\left(\frac{\Omega_1 - \Omega_2}{\Omega}\right) \tag{8.5}$$

Moreover, the frictional torque on a unit-length cylindrical surface is (Landau and Lifshitz, 2013)

$$L = -4\pi\mu\frac{(\Omega_1 - \Omega_2)R_1^2 R_2^2}{R_2^2 - R_1^2} \tag{8.6}$$

Clearly, although the entire flow field is established through viscous transport and vorticity diffusion, velocity distribution (Eq. 8.3) in the steady-state condition is independent of viscosity. This is due to the fact that, in the steady-state, both vorticity and frictional torque remain constant, independent of position. Consequently, the viscous term $\mu\nabla^2\boldsymbol{v}$ in the fluid motion equation vanishes, making the fluid behave as if it is an ideal fluid. Additionally, the vorticity of the ferrofluid within the gap is uniformly distributed.

The dimensionless parameter $\tilde{s}$ represents the relative magnitude of shear stress or elongational flow (Odenbach, 2009).

In the cylindrical coordinate, the deformation velocity tensor of ferrofluids is represented as

$$\mathbf{S}=\begin{pmatrix} \frac{\partial v_r}{\partial r} & \frac{1}{2}\left(\frac{1}{r}\frac{\partial v_r}{\partial \theta}+\frac{\partial v_\theta}{\partial r}-\frac{v_\theta}{r}\right) & \frac{1}{2}\left(\frac{\partial v_z}{\partial r}+\frac{\partial v_r}{\partial z}\right) \\ \frac{1}{2}\left(\frac{1}{r}\frac{\partial v_r}{\partial \theta}+\frac{\partial v_\theta}{\partial r}-\frac{v_\theta}{r}\right) & \frac{1}{r}\frac{\partial v_\theta}{\partial \theta}+\frac{v_r}{r} & \frac{1}{2}\left(\frac{\partial v_\theta}{\partial z}+\frac{1}{r}\frac{\partial v_z}{\partial \theta}\right) \\ \frac{1}{2}\left(\frac{\partial v_z}{\partial r}+\frac{\partial v_r}{\partial z}\right) & \frac{1}{2}\left(\frac{\partial v_\theta}{\partial z}+\frac{1}{r}\frac{\partial v_z}{\partial \theta}\right) & \frac{\partial v_z}{\partial z} \end{pmatrix} \tag{8.7}$$

After substituting the velocity component (8.3) into Eq. (8.7), one obtains

$$\mathbf{S}=\frac{1}{2}\left(\frac{\partial v_\theta}{\partial r}-\frac{v_\theta}{r}\right)\begin{pmatrix} 0 & 1 & 0 \\ 1 & 0 & 0 \\ 0 & 0 & 0 \end{pmatrix}=-\Omega\tilde{s}\frac{R_2^2}{r^2}\begin{pmatrix} 0 & 1 & 0 \\ 1 & 0 & 0 \\ 0 & 0 & 0 \end{pmatrix} \tag{8.8}$$

The symmetric tensor **S** governs the elongational flow in ferrofluids and significantly impacts their magnetization. As evident from Eq. (8.8), a larger value of $\tilde{s}$ at a given position r corresponds to increased values of the deformation velocity tensor components.

8.2 MAGNETIC FIELD DISTRIBUTION AND EQUILIBRIUM MAGNETIZATION IN FERROFLUIDS WHEN THE CYLINDERS ARE STATIONARY

Assuming that the axial lengths of the two cylinders far exceed their diameters and disregarding the influence of demagnetization fields, the ferrofluid within the gap is in a state of equilibrium magnetization when both cylinders are stationary. Under the assumption of a weak magnetic field, the equilibrium magnetization and the local magnetic field intensity within the ferrofluid exhibit a nearly linear relationship, characterized by a constant susceptibility χ. This relationship is expressed as $\boldsymbol{M}_0=\chi\boldsymbol{H}$. In this case, the scalar magnetic potential within the inner cylinder, the ferrofluid layer, and the outer region outside the cylinder all satisfy the Laplace equation

$$\nabla^2\varphi_m=0 \tag{8.9}$$

For ease of representation, the three areas are designated as 1, 2, and 3, respectively, as shown in Figure 8.1.

In the cylindrical coordinate, the general solution of Eq. (8.9) is given by

$$\varphi_m=\sum_{n=1}^{\infty}\left\{r^n\left[A_n\sin(n\theta)+B_n\cos(n\theta)\right]+r^{-n}\left[C_n\sin(n\theta)+D_n\cos(n\theta)\right]\right\} \tag{8.10}$$

For φ_{m3}, as $r \to \infty$, the magnetic field intensity equals the intensity of the applied magnetic field, and the corresponding scalar magnetic potential can be represented as

$$\varphi_{m3}\left(r \to \infty\right) = -H_0 r \sin\theta \tag{8.11}$$

Additionally, as satisfies $r \to \infty$, φ_{m3} satisfies

$$\sum_{n=1}^{\infty}\left\{r^n\left[A_n \sin(n\theta) + B_n \cos(n\theta)\right] + r^{-n}\left[C_n \sin(n\theta) + D_n \cos(n\theta)\right]\right\} = -H_0 r \sin\theta$$

It is evident that in order for the equation to hold true, the condition $n = 1$ must be satisfied, along with the requirement that all coefficients of the cosine terms must be zero. This leads to

$$\varphi_{m3} = A_{1,3} r \sin\theta + \frac{C_{1,3}}{r}\sin\theta$$

Based on this equation and in conjunction with Eq. (8.11), one gets $A_{1,3} = -H_0$. Therefore, the scalar magnetic potential in the region outside the outer cylinder is

$$\varphi_{m3} = -H_0 r \sin\theta + \frac{C_3}{r}\sin\theta \tag{8.12}$$

Here, for the sake of convenience, C_3 is used as a substitute for $C_{1,3}$.

The scalar magnetic potential φ_{m1} inside the inner cylinder must have a finite value at $r = 0$, thus eliminating any negative power terms from Eq. (8.10). This condition, which is a natural boundary condition, leads to

$$\varphi_{m1} = \sum_{n=1}^{\infty}\left\{r^n\left[A_n \sin(n\theta) + B_n \cos(n\theta)\right]\right\} \tag{8.13}$$

Next, the scalar magnetic potential in each region will be solved further utilizing the boundary conditions. Specifically, at the interface between the ferrofluid and the inner wall of the outer cylinder, a boundary condition exists

$$\varphi_{m2} = \varphi_{m3}, \text{when } r = R_2 \tag{8.14}$$

This leads to

$$\sum_{n=1}^{\infty}\left\{r^n\left[A_n \sin(n\theta) + B_n \cos(n\theta)\right] + r^{-n}\left[C_n \sin(n\theta) + D_n \cos(n\theta)\right]\right\}_{r=R_2}$$
$$= -H_0 R_2 \sin\theta + \frac{C_3}{R_2}\sin\theta$$

To guarantee the validity of this equation, it is necessary to satisfy the condition $n = 1$ within the ferrofluid region and ensure that all cosine terms are zero, namely,

$$\varphi_{m2} = A_2 r \sin\theta + \frac{C_2}{r}\sin\theta \tag{8.15}$$

Applying the boundary condition (8.14) yields

$$A_2 R_2 + \frac{C_2}{R_2} = -H_0 R_2 + \frac{C_3}{R_2} \tag{8.16}$$

Using the boundary condition

$$(1+\chi)\frac{\partial \varphi_{m2}}{\partial r} = \frac{\partial \varphi_{m3}}{\partial r}, \text{when } r = R_2 \tag{8.17}$$

and Eqs. (8.12) and (8.15) yields

$$(1+\chi)\left(A_2 - \frac{C_2}{R_2^2}\right) = -H_0 - \frac{C_3}{R_2^2} \tag{8.18}$$

The equation satisfied by A_2 and C_2 can be derived from Eqs. (8.16) and (8.18),

$$(2+\chi)A_2 - \chi\frac{C_2}{R_2^2} = -2H_0 \tag{8.19}$$

At the interface between the ferrofluid and the wall of the inner cylinder, a boundary condition exists

$$\varphi_{m1} = \varphi_{m2}, \text{when } r = R_1 \tag{8.20}$$

This leads to

$$\sum_{n=1}^{\infty}\left\{r^n\left[A_n \sin(n\theta) + B_n \cos(n\theta)\right]\right\} = A_2 r \sin\theta + \frac{C_2}{r}\sin\theta$$

In this formula, only the coefficient of the sine term when $n = 1$ is non-zero, namely,

$$\varphi_{m1} = rA_1 \sin\theta \tag{8.21}$$

$$A_1 = A_2 + \frac{C_2}{R_1^2} \tag{8.22}$$

Then from the boundary condition

$$(1+\chi)\frac{\partial \varphi_{m2}}{\partial r} = \frac{\partial \varphi_{m1}}{\partial r}, r = R_1 \tag{8.23}$$

and Eqs. (8.15) and (8.21), one obtains

$$(1+\chi)\left(A_2 - \frac{C_2}{R_1^2}\right) = A_1 \tag{8.24}$$

Combining Eqs. (8.22) and (8.24) yields

$$\chi A_2 - (2+\chi)\frac{C_2}{R_1^2} = 0 \tag{8.25}$$

Upon combining Eqs. (8.19) and (8.25), it is found that

$$A_2 = \frac{-2(2+\chi)H_0}{4(1+\chi)+\left(1-R_1^2/R_2^2\right)\chi^2} \tag{8.26}$$

$$C_2 = \frac{-2\chi R_1^2 H_0}{4(1+\chi)+\left(1-R_1^2/R_2^2\right)\chi^2} \tag{8.27}$$

Letting

$$\tilde{a} = \frac{-2(2+\chi)}{4(1+\chi)+\left(1-R_1^2/R_2^2\right)\chi^2} \tag{8.28}$$

$$\tilde{b} = a\frac{\chi R_1^2}{2+\chi} \tag{8.29}$$

one gets

$$A_2 = \tilde{a}H_0, C_2 = \tilde{b}H_0 \tag{8.30}$$

$$\varphi_{m2} = H_0\left(\tilde{a}r + \frac{\tilde{b}}{r}\right)\sin\theta \tag{8.31}$$

and the distribution of magnetic field in the ferrofluid,

$$\begin{aligned} \boldsymbol{H}_2 &= -\nabla\varphi_{m2} = -\frac{\partial \varphi_{m2}}{\partial r}\boldsymbol{e}_r - \frac{1}{r}\frac{\partial \varphi_{m2}}{\partial \theta}\boldsymbol{e}_\theta \\ &= -H_0\left(\tilde{a} - \frac{\tilde{b}}{r^2}\right)\sin\theta \boldsymbol{e}_r - H_0\left(\tilde{a} + \frac{\tilde{b}}{r^2}\right)\cos\theta \boldsymbol{e}_\theta \end{aligned} \tag{8.32}$$

Based on Eq. (8.32), the equilibrium magnetization of the ferrofluid at rest can be determined

$$\boldsymbol{M}_0 = \chi \boldsymbol{H}_2 = -\chi H_0 \left(\left(\tilde{a} - \frac{\tilde{b}}{r^2} \right) \sin\theta \ \left(\tilde{a} + \frac{\tilde{b}}{r^2} \right) \cos\theta \right) \begin{pmatrix} \boldsymbol{e}_r \\ \boldsymbol{e}_\theta \end{pmatrix} \tag{8.33}$$

In addition, from Eq. (8.22), one gets

$$A_1 = \left(\tilde{a} + \frac{\tilde{b}}{R_1^2} \right) H_0 \tag{8.34}$$

Then from Eq. (8.21), one can obtain the scalar magnetic potential in the internal region of the cylinder

$$\varphi_{m1} = r \left(\tilde{a} + \frac{\tilde{b}}{R_1^2} \right) H_0 \sin\theta \tag{8.35}$$

and the magnetic field distribution within the cylindrical region

$$\begin{aligned} \boldsymbol{H}_1 &= -\nabla \varphi_{m1} \\ &= -\left(\tilde{a} + \frac{\tilde{b}}{R_1^2} \right) H_0 \sin\theta \boldsymbol{e}_r - \left(\tilde{a} + \frac{\tilde{b}}{R_1^2} \right) H_0 \cos\theta \boldsymbol{e}_\theta \\ &= -\left(\tilde{a} + \frac{\tilde{b}}{R_1^2} \right) H_0 \boldsymbol{e}_y = -\tilde{a} H_0 \frac{2(1+\chi)}{2+\chi} \boldsymbol{e}_y \end{aligned} \tag{8.36}$$

Clearly, when the cylinders are at rest, the magnetic field intensity within the inner region of the cylinder remains aligned with the direction of the external magnetic field. This is due to the fact that, at this point, the magnetization of the ferrofluid reaches a state of equilibrium, with the magnetization aligned in the same direction as the external magnetic field.

8.3 MAGNETIZATION IN FERROFLUIDS WHILE THE CYLINDERS ARE ROTATING

When the two cylinders rotate, the flow of ferrofluid induces the magnetization relaxation phenomenon within it. Assuming a relatively weak external magnetic field, the change in magnetization is described by the phenomenological magnetization equation I (Eq. 4.18). Neglecting higher-order terms pertaining to H_0 and retaining solely the steady-state term, the magnetization equation becomes

$$(\boldsymbol{v} \cdot \nabla) \boldsymbol{M} - \boldsymbol{\Omega} \times \boldsymbol{M} = -\frac{1}{\tau} (\boldsymbol{M} - \boldsymbol{M}_0) \tag{8.37}$$

Assuming that the gradient of the ferrofluid flow along the radial direction is much smaller than the reciprocal of the magnetization relaxation time, that is $\tau|\nabla v_\theta| \ll 1$, the deviation of the magnetization from its equilibrium state is relatively minor. Additionally, the influence of the magnetic field on the flow is assumed to be negligible. To solve Eq. (8.37), equilibrium magnetization $\boldsymbol{M}_0$ is substituted for $\boldsymbol{M}$ in the left-hand side terms of the equation, while the velocity in these terms is still represented by Eq. (8.3).

In the cylindrical coordinate system, magnetization $\boldsymbol{M}$ is represented as

$$\boldsymbol{M} = M_r \boldsymbol{e}_r + M_\theta \boldsymbol{e}_\theta \tag{8.38}$$

For the various terms in Eq. (8.37), one has

$$(\boldsymbol{v}\cdot\nabla)\boldsymbol{M}_0 = \frac{v_\theta}{r}\frac{\partial M_{0r}}{\partial \theta}\boldsymbol{e}_r + \frac{v_\theta}{r}\frac{\partial M_{0\theta}}{\partial \theta}\boldsymbol{e}_\theta$$

$$\boldsymbol{\Omega}\times\boldsymbol{M}_0 = \frac{1}{r}\begin{vmatrix} \boldsymbol{e}_r & r\boldsymbol{e}_\theta & \boldsymbol{e}_z \\ 0 & 0 & \Omega \\ M_{0r} & rM_{0\theta} & 0 \end{vmatrix} = -\Omega M_{0\theta}\boldsymbol{e}_r + \Omega M_{0r}\boldsymbol{e}_\theta$$

Substituting these two equations into Eq. (8.37) yields a system of equations in component form

$$\begin{cases} \dfrac{v_\theta}{r}\dfrac{\partial M_{0r}}{\partial \theta} + \Omega M_{0\theta} = -\dfrac{1}{\tau}\left(M_r - M_{0r}\right) \\ \dfrac{v_\theta}{r}\dfrac{\partial M_{0\theta}}{\partial \theta} - \Omega M_{0r} = -\dfrac{1}{\tau}\left(M_\theta - M_{0\theta}\right) \end{cases} \tag{8.39}$$

Solving this system of equations obtains the components of magnetization

$$\begin{cases} M_r = M_{0r} - \tau\left(\dfrac{v_\theta}{r}\dfrac{\partial M_{0r}}{\partial \theta} + \Omega M_{0\theta}\right) \\ M_\theta = M_{0\theta} - \tau\left(\dfrac{v_\theta}{r}\dfrac{\partial M_{0\theta}}{\partial \theta} - \Omega M_{0r}\right) \end{cases} \tag{8.40}$$

Substituting Eqs. (8.3) and (8.33) into Eq. (8.40), one obtains the magnetization of the ferrofluid when the cylinders are rotating

$$\begin{cases} M_r = -\chi H_0\left(\tilde{a} - \dfrac{\tilde{b}}{r^2}\right)\sin\theta + \tau\Omega\chi H_0\left[\left(1 + s\dfrac{R_2^2}{r^2}\right)\left(\tilde{a} - \dfrac{\tilde{b}}{r^2}\right) + \left(\tilde{a} + \dfrac{\tilde{b}}{r^2}\right)\right]\cos\theta \\ M_\theta = -\chi H_0\left(\tilde{a} + \dfrac{\tilde{b}}{r^2}\right)\cos\theta - \tau\Omega\chi H_0\left[\left(1 + s\dfrac{R_2^2}{r^2}\right)\left(\tilde{a} + \dfrac{\tilde{b}}{r^2}\right) + \left(\tilde{a} - \dfrac{\tilde{b}}{r^2}\right)\right]\sin\theta \end{cases} \tag{8.41}$$

8.4 MAGNETIC FIELD INTENSITY PERPENDICULAR TO THE EXTERNAL MAGNETIC FIELD INSIDE THE CYLINDER WHEN THE CYLINDERS ROTATE

As the cylinders rotate, the scalar magnetic potential within the inner region of the cylinder still satisfies the Laplace equation (8.9), whereas the scalar magnetic potential within the ferrofluid must be described by the Poisson equation (6.9) instead of the Laplace equation (Yang et al., 2022). Given the difficulty in solving the Poisson equation, the magnetization results within the ferrofluid during cylinder rotation (8.41) and the magnetic field distribution there when the cylinders are stationary (8.32) are utilized to determine the magnetic field distribution within the inner region of the cylinders. Once again, the influence of ferrofluid flow on magnetic field distortion is neglected.

The general solution for the scalar magnetic potential within the inner cylinder region is assumed to retain the form given by Eq. (8.13), and accordingly, the corresponding magnetic field distribution is

$$\boldsymbol{H}_1 = -\boldsymbol{e}_r \sum_{n=1}^{\infty} \left\{ n r^{n-1} \left[A_n' \sin(n\theta) + B_n' \cos(n\theta) \right] \right\} - \boldsymbol{e}_\theta \sum_{n=1}^{\infty} \left\{ n r^{n-1} \left[A_n' \cos(n\theta) - B_n' \sin(n\theta) \right] \right\} \tag{8.42}$$

At the interface between the ferrofluid and the outer wall of the inner cylinder, a boundary condition exists

$$B_{r1} = B_{r2}, \text{ when } r = R_1 \tag{8.43}$$

Based on the constitutive equation, $\boldsymbol{B} = \mu_0(\boldsymbol{H} + \boldsymbol{M})$, one can deduce that

$$H_{r1} = H_{r2} + M_r \tag{8.44}$$

Substituting Eqs. (8.41), (8.32), and (8.42) into Eq. (8.44), respectively, one obtains

$$\begin{aligned} &-\sum_{n=1}^{\infty} \left\{ n r^{n-1} \left[A_n' \sin(n\theta) + B_n' \cos(n\theta) \right] \right\}_{r=R_1} \\ &= -H_0 \left(\tilde{a} - \frac{\tilde{b}}{r^2} \right) \sin\theta - \chi H_0 \left(\tilde{a} - \frac{\tilde{b}}{r^2} \right) \sin\theta \\ &+ \tau\Omega\chi H_0 \left[\left(1 + \tilde{s} \frac{R_2^2}{r^2} \right) \left(\tilde{a} - \frac{\tilde{b}}{r^2} \right) + \left(\tilde{a} + \frac{\tilde{b}}{r^2} \right) \right] \cos\theta \end{aligned} \tag{8.45}$$

The validity of this formula necessitates $n = 1$, and the corresponding coefficients are

$$A_1' = H_0 \left(\tilde{a} - \frac{\tilde{b}}{R_1^2} \right) (1 + \chi) \tag{8.46}$$

$$B_1' = -\tau\Omega\chi H_0\left[\left(1+\tilde{s}\frac{R_2^2}{R_1^2}\right)\left(\tilde{a}-\frac{\tilde{b}}{R_1^2}\right)+\left(\tilde{a}+\frac{\tilde{b}}{R_1^2}\right)\right] \tag{8.47}$$

In the coordinate system depicted in Figure 8.1, the scalar magnetic potential within the cylindrical region can be represented as

$$\varphi_{m1} = rA_1'\sin\theta + rB_1'\cos\theta = A_1'y + B_1'x \tag{8.48}$$

The magnetic field intensity within the corresponding cylindrical region, along the direction of the external magnetic field, is

$$H_{y1} = -\frac{\partial\varphi_{m1}}{\partial y} = -A_1 = -H_0\left(\tilde{a}-\frac{\tilde{b}}{R_1^2}\right)(1+\chi)$$

Substituting Eq. (6.29) into it yields

$$H_{y1} = -H_0\tilde{a}\frac{2(1+\chi)}{2+\chi} \tag{8.49}$$

The results of Eqs. (8.49) and (8.36) are identical, indicating that in relatively weak magnetic fields, when the cylinders rotate and the influence of ferrofluid flow on magnetic field distortion is neglected, the magnetic field intensity along the direction of the external magnetic field remains unchanged within the cylinder's inner region, regardless of whether the cylinders are rotating or stationary.

The magnetic field intensity perpendicular to the direction of the external magnetic field inside the cylinder, induced by magnetization, is

$$H_{x1} = -\frac{\partial\varphi_{m1}}{\partial x} = -B_1 \tag{8.50}$$

Substituting Eqs. (8.47), (8.28), and (8.29) into Eq. (8.50) and simplifying, one obtains

$$\begin{aligned}H_{x1} &= \tau\Omega\chi H_0\left[\left(1+\tilde{s}\frac{R_2^2}{R_1^2}\right)\left(\tilde{a}-\frac{\tilde{b}}{R_1^2}\right)+\left(\tilde{a}+\frac{\tilde{b}}{R_1^2}\right)\right]\\ &= -\tau\Omega H_0\frac{2\chi(2+\chi)R_2^2}{4(1+\chi)R_2^2+\left(R_2^2-R_1^2\right)\chi^2}\left[\left(1+\tilde{s}\frac{R_2^2}{R_1^2}\right)\left(1-\frac{\chi}{2+\chi}\right)+\left(1+\frac{\chi}{2+\chi}\right)\right]\\ &= -\tau\Omega H_0\frac{4\chi(2+\chi)R_2^2}{4(1+\chi)R_2^2+\left(R_2^2-R_1^2\right)\chi^2}\left(1+\frac{\tilde{s}}{2+\chi}\frac{R_2^2}{R_1^2}\right)\end{aligned} \tag{8.51}$$

Equation (8.51) demonstrates that when the cylinders rotate counterclockwise, the magnetization relaxation of the ferrofluid generates a magnetic field intensity that is perpendicular to the external magnetic field within the inner cylinder region, along the negative x-axis. Once the susceptibility of the ferrofluid and the dimensions of the cylinders are given, the magnetic field intensity along the x-direction becomes a function of the parameters Ω and s, assuming the external magnetic field remains constant.

Utilizing the expression given by Eq. (8.51) for magnetization perpendicular to the external magnetic field within the cylindrical region, the magnetization relaxation time τ can be determined through experimental measurements of the magnetization as the cylinders rotate (Yang et al., 2022). For instance, to facilitate measurements, the ferrofluid can be rotated rigidly, ensuring identical rotational speeds for both the inner and outer cylinders. Under these conditions, where $\tilde{s} = 0$ in Eq. (8.51), the expression for the magnetic field intensity along the x-direction becomes

$$H_{x1} = -\tau\Omega H_0 \frac{4\chi(2+\chi)R_2^2}{4(1+\chi)R_2^2 + \left(R_2^2 - R_1^2\right)\chi^2} \tag{8.52}$$

Once the magnetic permeability χ of the ferrofluid is determined, magnetization time τ can be calculated using Eq. (8.52) by measuring the magnetic field intensity component H_{x1}.

This chapter has focused solely on the annular Couette flow when the rotational speed difference between the two cylinders is relatively small and has not yet delved into the vortex flow, known as Taylor–Couette flow, that emerges at larger rotational speed differences. For further research on Taylor–Couette flow, please refer to the studies conducted by Altmeyer et al. (2010), Altmeyer et al. (2017), and Altmeyer (2019).

REFERENCES

Altmeyer S., Agglomeration effects in rotating ferrofluids, *Journal of Magnetism and Magnetic Materials*, 482: 239–250, 2019.

Altmeyer S., Do Y., Ryu S., Transient behavior between multi-cell flow states in ferrofluidic Taylor-Couette flow, *Chaos*, 27: 113112, 2017.

Altmeyer S., Hoffmann C., Leschhorn A., Lücke M., Influence of homogeneous magnetic fields on the flow of a ferrofluid in the Taylor-Couette system, *Physical Review E*, 82: 016321, 2010.

Landau L. D., Lifshitz E. M., *Fluid dynamics*. Beijing: Higher Education Press. 2013.

Odenbach S., Stationary off-equilibrium magnetization in ferrofluids under rotational and elongational flow, *Physical Review Letters*, 89(3): 037202, 2009.

Yang W. M., Jin Y., Fang B., A method and apparatus for measuring the magnetization relaxation times of ferrofluids, *IEEE Transactions on Instrumentation and Measurement*, 71: 6002707, 2022.

CHAPTER 9

Spin-Up Flow of Ferrofluids

THE SPIN-UP FLOW IS a unique flow phenomenon observed in ferrofluids. When a vertically stationary cylindrical container filled with ferrofluid is exposed to a horizontally applied uniform magnetic field rotating around its axis, the fluid within the container also exhibits a rotational motion (Moskowitz and Rosensweig, 1967; Zaitsev and Shliomis, 1969; Felderhof, 2011). The rotational velocity can reach several millimetres per second (Chaves et al., 2006), and the direction of rotation can be either concurrent or counter to the direction of the rotating magnetic field. The rotational velocity at the fluid surface is directly proportional to the radial distance from the centre to the container wall. Additionally, the free surface of the ferrofluid displays a concave or convex meniscus shape. When the meniscus appears concave, the rotation direction is the same as the magnetic field's rotation, while for a convex meniscus, the rotation direction is opposite (Rosensweig, 2023). This flow phenomenon in ferrofluids is referred to as spin-up flow (Finlayson, 2013).

The physical mechanism underlying the spin-up flow of ferrofluids remains inconclusive. The spin-diffusion theory, which has been used for decades based on the assumption of relatively large spin viscosity values (Finlayson, 2013), cannot be applied due to the very low spin viscosity values exhibited by ferrofluids (10^{-20} kg·m/s, Shliomis, 2021). This chapter solely presents the experimentally validated spin-up flow theory.

9.1 ROTATIONAL VELOCITY AT THE FERROFLUID SURFACE

In the spin-up flow of ferrofluids, the mechanisms causing the rotational motion differ between the upper surface layer and the bulk of the fluid contained within a cylindrical container. The motion of the surface layer is attributed to surface stress, whereas the bulk flow is driven by body stress (Shliomis, 2021). Therefore, it is necessary to consider these two components separately when discussing the underlying flow mechanisms. This section focuses on the flow characteristics of the surface layer of ferrofluid.

Using the spin-up flow with a concave surface shape depicted in Figure 9.1 as an illustrative example, when subjected to a horizontally radial rotating magnetic field, the rotating surface ascends along the container wall. The ascending distance, denoted as h_s, is typically

DOI: 10.1201/9781003540342-9

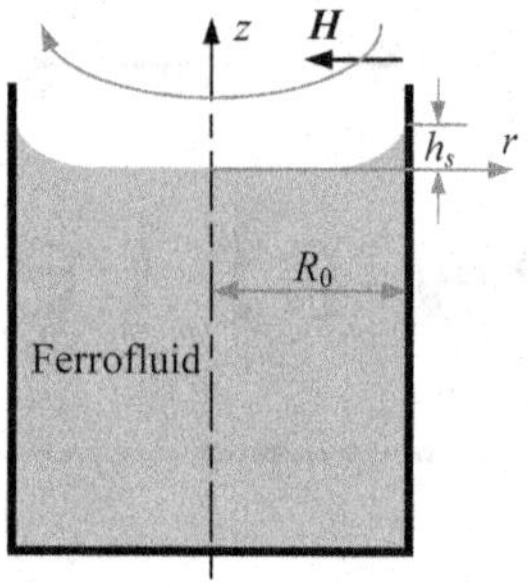

FIGURE 9.1 Ferrofluid spin-up flow (when the surface is concave).

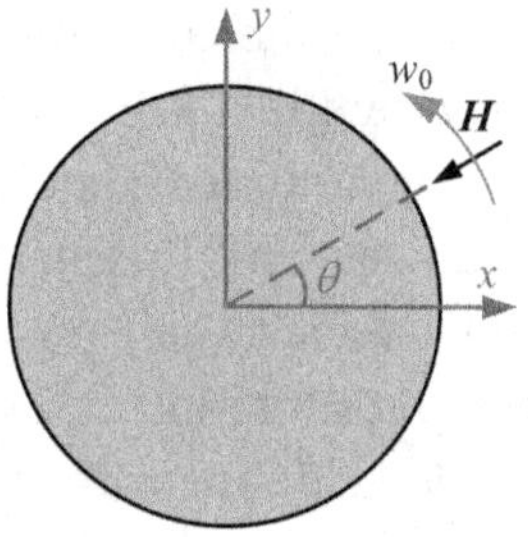

FIGURE 9.2 Top-down view of the ferrofluid spin-up flows.

less than 2 to 3 millimetres, with the thickness of the surface layer being comparable in magnitude. Beneath the free surface, the rotation speed rapidly decreases in the $(-z)$ direction.

9.1.1 Magnetic Torque Acting on the Ferrofluid

Assuming that the rotational velocity of the magnetic field is w_0 in magnitude and directed along the z-axis, as shown in Figure 9.2, the intensity of the applied magnetic field can be represented as

$$\boldsymbol{H} = \left(H_0 \cos w_0 t, H_0 \sin w_0 t, 0\right) \tag{9.1}$$

where H_0 is the magnetic field intensity. Under the influence of this magnetic field, the magnetization in the ferrofluid rotates at the same angular frequency w_0, but with a phase lag of ϑ.

Ignoring the spin viscosity of ferrofluids, phenomenological magnetization equation II (4.60) is used to describe the change in magnetization. According to Eq. (4.45), the effective field $\boldsymbol{H}_e$ rotates in the same direction as magnetization $\boldsymbol{M}$. Therefore, the effective field $\boldsymbol{H}_e$ can be expressed as

$$\boldsymbol{H}_e = \left(H_e \cos\left(w_0 t - \vartheta\right), H_e \sin\left(w_0 t - \vartheta\right), 0\right) \tag{9.2}$$

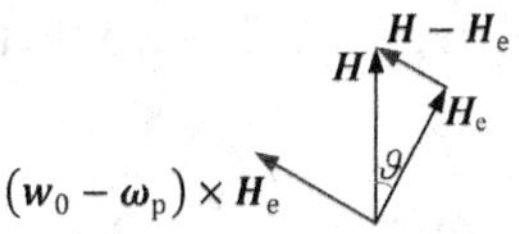

FIGURE 9.3 Geometric relationships among variables.

Thereby, the terms on the left-hand side of Eq. (4.60) can be expressed as

$$\frac{\mathrm{D}\boldsymbol{H}_{\mathrm{e}}}{\mathrm{D}t} = \boldsymbol{w}_0 \times \boldsymbol{H}_{\mathrm{e}} \tag{9.3}$$

Substituting Eq. (9.3) into Eq. (4.60) yields

$$\boldsymbol{H} - \boldsymbol{H}_{\mathrm{e}} = \tau_B \left(\boldsymbol{w}_0 - \boldsymbol{\omega}_{\mathrm{p}}\right) \times \boldsymbol{H}_{\mathrm{e}} \tag{9.4}$$

The geometric relationship among the variables in Eq. (9.4) is illustrated in Figure 9.3. Since $(\boldsymbol{H} - \boldsymbol{H}_e)$ is parallel to $(\boldsymbol{w}_0 - \boldsymbol{\omega}_{\mathrm{p}}) \times \boldsymbol{H}_e$, it follows that $(\boldsymbol{H} - \boldsymbol{H}_e)$ is perpendicular to $\boldsymbol{H}_e$. Consequently, it can be deduced that

$$H_{\mathrm{e}} = H \cos \vartheta \tag{9.5}$$

Since $(\boldsymbol{w}_0 - \boldsymbol{\omega}_{\mathrm{p}})$ is perpendicular to $\boldsymbol{H}_e$, and $\boldsymbol{w}_0$ and $\boldsymbol{\omega}_{\mathrm{p}}$ are coaxial, then one has

$$\tan \vartheta = \frac{\left|\boldsymbol{H} - \boldsymbol{H}_{\mathrm{e}}\right|}{H_{\mathrm{e}}} = \frac{\left|\tau_B \left(\boldsymbol{w}_0 - \boldsymbol{\omega}_{\mathrm{p}}\right) \times \boldsymbol{H}_{\mathrm{e}}\right|}{H_{\mathrm{e}}} = \tau_B \left(w_0 - \omega_{\mathrm{p}}\right) \tag{9.6}$$

Based on Eqs. (9.5) and (9.6), the magnetic torque density acting on the ferrofluid can be derived as

$$\mu_0 \boldsymbol{M} \times \boldsymbol{H} = \mu_0 M H \sin \vartheta \boldsymbol{e}_z = \mu_0 M H_{\mathrm{e}} \tan \vartheta \boldsymbol{e}_z = \mu_0 \tau_B M H_{\mathrm{e}} \left(\boldsymbol{w}_0 - \boldsymbol{\omega}_{\mathrm{p}}\right) \tag{9.7}$$

Substituting Eqs. (4.13) and (4.45) into Eq. (9.7), one obtains

$$\mu_0 \boldsymbol{M} \times \boldsymbol{H} = \frac{3V_{\mathrm{h}}\eta_0}{k_{\mathrm{B}}T} \frac{\varsigma k_{\mathrm{B}}T}{m_{\mathrm{p}}} M_{\mathrm{S}} L(\varsigma) \left(\boldsymbol{w}_0 - \boldsymbol{\omega}_{\mathrm{p}}\right) \tag{9.8}$$

Then by substituting Eqs. (4.2) and (4.4) into Eq. (9.8), and applying Eq. (2.213), one gets

$$\mu_0 \boldsymbol{M} \times \boldsymbol{H} = 2\zeta\varsigma L(\varsigma) \left(\boldsymbol{w}_0 - \boldsymbol{\omega}_{\mathrm{p}}\right) \tag{9.9}$$

On the other hand, when neglecting the spin viscosity of the ferrofluid and assuming it is at rest ($\boldsymbol{v} = \boldsymbol{\Omega} = \boldsymbol{0}$), the magnetic torque density experienced by the ferrofluid can be obtained from Eq. (2.217)

$$\mu_0 \boldsymbol{M} \times \boldsymbol{H} = 4\zeta \boldsymbol{\omega}_{\mathrm{p}} \tag{9.10}$$

Comparing Eqs. (9.10) and (9.9), it can be derived that

$$\boldsymbol{\omega}_{\mathrm{p}} = \frac{\varsigma L(\varsigma)}{2+\varsigma L(\varsigma)}\boldsymbol{w}_0 \tag{9.11}$$

Letting

$$\Gamma(\varsigma) = \frac{\varsigma L(\varsigma)}{2+\varsigma L(\varsigma)} \tag{9.12}$$

and substituting Eqs. (9.11) back into Eq. (9.10), the magnetic torque density is obtained

$$\mu_0 \boldsymbol{M}\times\boldsymbol{H} = 4\zeta\Gamma(\varsigma)\boldsymbol{w}_0 = 6\eta\phi\boldsymbol{\omega}_{\mathrm{p}} \tag{9.13}$$

Moreover, one has

$$\tan\vartheta = \tau_B w_0\left[1-\Gamma(\varsigma)\right] \tag{9.14}$$

9.1.2 Rotational Velocity at the Ferrofluid Surface

Assuming that the concave meniscus shape of the ferrofluid as depicted in Figure 9.1 can be represented by (Shliomis, 2021)

$$z(r) = h_s\frac{\cosh(r/h)}{\cosh(R/h)} \tag{9.15}$$

This representation ensures that the height of the ferrofluid decreases exponentially from a value of h_s at the container wall to approximately zero at the axis. Similarly, for a convex-shaped meniscus, it can be represented as (Shliomis, 2021)

$$z(r) = h_s\left[1-\frac{\cosh(r/h)}{\cosh(R/h)}\right]$$

Based on Eqs. (9.15) and (9.13), the average magnetic torque density acting on the ferrofluid layer with a thickness of h_s is obtained

$$\mu_0\overline{\boldsymbol{M}\times\boldsymbol{H}} = 6\eta\phi\Gamma(\varsigma)\boldsymbol{w}_0\frac{\cosh(r/h_s)}{\cosh(R_0/h_s)} \tag{9.16}$$

To determine the circumferential rotational velocity of the ferrofluid, it is necessary to write the θ-directional component equation of Eq. (2.219) in the cylindrical coordinate system depicted in Figures 9.1 and 9.2. Assuming steady-state conditions and neglecting inertial forces, this component equation is

$$\frac{\mathrm{d}^2 v_\theta}{\mathrm{d}r^2}+\frac{1}{r}\frac{\mathrm{d}v_\theta}{\mathrm{d}r}-\frac{v_\theta}{r^2}=3\phi w_0\Gamma(\varsigma)\frac{\mathrm{d}}{\mathrm{d}r}\left[\frac{\cosh(r/h_s)}{\cosh(R_0/h_s)}\right] \tag{9.17}$$

As the magnetic field intensity remains unchanged with respect to the θ position, the magnetic body force is zero. Furthermore, the average magnetic torque density represented by Eq. (9.16) is substituted for the magnetic torque density, disregarding the impact of pressure p.

Assuming that $R_0 \gg h_s$, the influence of the container wall curvature is neglected. To simplify the solution process, only the highest-order derivative term related to v_θ is retained in Eq. (9.17), which is thereby simplified to

$$\frac{\mathrm{d}^2 v_\theta}{\mathrm{d}r^2}=3\phi w_0\Gamma(\varsigma)\frac{\mathrm{d}}{\mathrm{d}r}\left[\frac{\cosh(r/h_s)}{\cosh(R_0/h_s)}\right] \tag{9.18}$$

Solving this equation and applying the boundary condition

$$v_\theta = 0, \text{ when } r = 0 \text{ or } r = R_0 \tag{9.19}$$

the rotational velocity of the ferrofluid surface layer is obtained,

$$v_\theta(r)=-3\phi h_s w_0\Gamma(\varsigma)\tanh\left(\frac{R_0}{h_s}\right)\left[\frac{r}{R_0}-\frac{\sinh(r/h_s)}{\sinh(R_0/h_s)}\right] \tag{9.20}$$

Since $R_0 \gg h_s$, then $\tanh(R_0/h_s) \approx 1$. Therefore, Eq. (9.20) can be expressed approximately as

$$v_\theta(r)=-3\phi h_s w_0\Gamma(\varsigma)\left[\frac{r}{R_0}-\frac{\sinh(r/h_s)}{\sinh(R_0/h_s)}\right] \tag{9.21}$$

Equation (9.20) demonstrates that the magnitude of the rotational velocity at the surface of the ferrofluid is independent of its viscosity. Letting

$$v_c = -3\phi h_s w_0\Gamma(\varsigma)$$

it follows that

$$\frac{v_\theta(r)}{v_c}=\frac{r}{R_0}-\frac{\sinh(r/h_s)}{\sinh(R_0/h_s)} \tag{9.22}$$

As shown in Figure 9.4, a comparison is presented between the rotational velocity of the ferrofluid surface in a spin-up flow, which is calculated using Eq. (9.22), and the experimental results. It is observed that there is a good agreement between them.

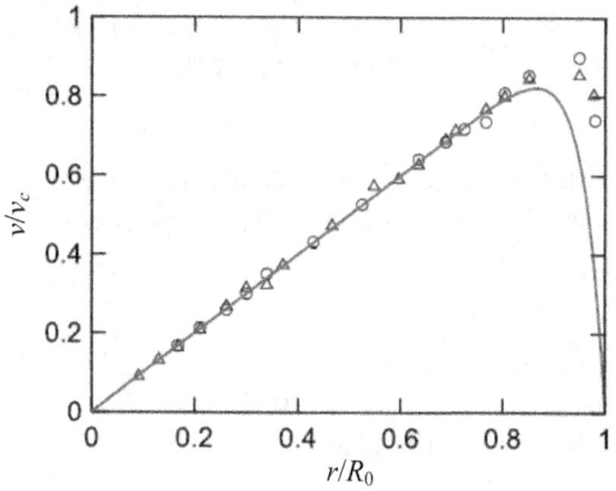

FIGURE 9.4 Rotational velocity of the ferrofluid surface in a spin-up flow. (R_0 = 23.3 mm, $w_0/2\pi$= 60 Hz). The solid lines represent the calculated results from Eq. (9.22), while the symbols O and Δ represent experimental results corresponding to $H_0 = -7.96 \times 10^3$ A/m and $H_0 = -1.59 \times 10^4$ A/m, respectively. (Reprinted from Shliomis, 2021 with permission from APS.)

9.2 TORQUE ACTING ON THE FERROFLUID

Although there are concerns regarding the overestimation of spin viscosity assumed in the spin-diffusion theory when explaining the spin-up phenomenon of ferrofluids, the torque acting on the cylindrical container wall obtained based on this theory is in good agreement with experimental measurements (Pshenichnikov and Lebedev, 2000; Torres-Diaz et al., 2014; Chaves et al., 2008). Therefore, this section introduces a torque calculation method based on the spin-diffusion theory.

9.2.1 Rotational Velocity of Ferrofluid and Particle Spin Velocity Based on Spin-Diffusion Theory

To determine the rotational velocity of the ferrofluid and the particle spin velocity within the spin-up flow, a method similar to that described in Section 6.4.3 is employed. The governing equations for this problem encompass the mass conservation equation (2.205), the motion equation (2.218), the angular momentum equation (2.210) with the inclusion of the spin viscosity term, the magnetization equation (4.17), and the static magnetic field equations (2.82), (2.108), and (2.207).

9.2.1.1 Dimensionless Governing Equations

To solve the system of governing equations, the following dimensionless quantities are applied to non-dimensionalize each equation:

$$w^* = w\tau, t^* = wt, r^* = \frac{r}{R_0}, \boldsymbol{v}^* = \frac{\boldsymbol{v}}{v_0}, \boldsymbol{\omega}_p^* = \frac{\boldsymbol{\omega}_p}{\omega_0}$$
$$\boldsymbol{\Omega}^* = \frac{\boldsymbol{\Omega}}{\omega_0}, \boldsymbol{H}^* = \frac{\boldsymbol{H}}{H_0}, \boldsymbol{M}^* = \frac{\boldsymbol{M}}{\chi_0 H_0}, p^* = \frac{p}{\eta\omega_0}, \nabla^* = R_0\nabla \tag{9.23}$$

where v_0 is the characteristic velocity and ω_0 is the characteristic spin velocity. Letting

$$\omega_0 = \frac{\mu_0 \chi_0 H_0^2 w^*}{\zeta} \tag{9.24}$$

and one has $v_0 = R_0\omega_0$, where w and H_0 are the angular frequency and the intensity magnitude of the applied magnetic field, respectively. The intensity of the rotating magnetic field can be represented using complex notation as

$$\boldsymbol{H} = \text{Re}\left[H_0 e^{\text{i}(wt+\phi_0)}\right]\boldsymbol{e}_z \tag{9.25}$$

where ϕ_0 is the initial phase angle. Based on the representation provided in Eq. (9.24), it follows that

$$\boldsymbol{v}^* = \frac{\zeta \boldsymbol{v}}{\mu_0 \chi_0 R_0 H_0^2 w^*}, \boldsymbol{\omega}_\text{p}^* = \frac{\zeta \boldsymbol{\omega}_\text{p}}{\mu_0 \chi_0 H_0^2 w^*}, p^* = \frac{\zeta p}{\mu_0 \eta \chi_0 H_0^2 w^*} \tag{9.26}$$

Substituting Eqs. (9.23) and (9.26) into the respective governing equations and simplifying, one obtains

$$\nabla^* \times \boldsymbol{H}^* = \boldsymbol{0} \tag{9.27}$$

$$\nabla^* \cdot \left(\chi_0 \boldsymbol{M}^* + \boldsymbol{H}^*\right) = 0 \tag{9.28}$$

$$\frac{\zeta+\eta}{\eta}\nabla^{*2}\boldsymbol{v}^* + \frac{\zeta}{\eta w^*}\boldsymbol{M}^* \cdot \nabla^* \boldsymbol{H}^* + 2\frac{\zeta}{\eta}\nabla^* \times \boldsymbol{\omega}_\text{p}^* - \nabla^* p^* = \boldsymbol{0} \tag{9.29}$$

$$\frac{\eta'}{\zeta R_0^2}\nabla^{*2}\boldsymbol{\omega}_\text{p}^* + 2\nabla^* \times \boldsymbol{v}^* - 4\boldsymbol{\omega}_\text{p}^* + \frac{1}{w^*}\boldsymbol{M}^* \times \boldsymbol{H}^* = \boldsymbol{0} \tag{9.30}$$

$$w^* \frac{\partial \boldsymbol{M}^*}{\partial t^*} + \omega_0 \tau \left(\boldsymbol{v}^* \cdot \nabla^*\right)\boldsymbol{M}^* = \omega_0 \tau \boldsymbol{\omega}_\text{p}^* \times \boldsymbol{M}^* + \left(\boldsymbol{M}_0^* - \boldsymbol{M}^*\right) \tag{9.31}$$

Rewriting Eq. (9.30) by letting

$$\kappa = \left[\frac{4\eta\zeta R_0^2}{(\zeta+\eta)\eta'}\right]^{\frac{1}{2}} \tag{9.32}$$

leads to the following expression

$$\frac{4\eta}{(\zeta+\eta)\kappa^2}\nabla^{*2}\boldsymbol{\omega}_\text{p}^* + 2\nabla^* \times \boldsymbol{v}^* - 4\boldsymbol{\omega}_\text{p}^* + \frac{1}{w^*}\boldsymbol{M}^* \times \boldsymbol{H}^* = \boldsymbol{0} \tag{9.33}$$

In Eq. (9.31), the coefficient $\omega_0\tau$ appears twice. Letting $\omega_0\tau = \epsilon w^*$, where ϵ is given by

$$\epsilon = \frac{\mu_0 \chi_0 H_0^2 \tau}{\zeta} \tag{9.34}$$

according to Eq. (9.24). This representation is different from that in Section 6.4.3. The Brownian relaxation time τ_B represented by Eq. (4.13) is applied in Eq. (9.34) to replace τ, and χ_0, obtained under a weak magnetic field, is represented by Eq. (4.8). Through simplification, one obtains

$$\epsilon = \frac{2}{3}\alpha_0^2 \tag{9.35}$$

where α_0 is the Langevin parameter corresponding to the magnetic field intensity H_0,

$$\alpha_0 = \frac{\mu_0 m_p H_0}{k_B T}$$

From Eq. (4.1), it can be inferred that $\alpha_0/H_0 = \alpha/H$, and given that $H = H_0 H^*$, the parameter ϵ can also be expressed as

$$\epsilon = \frac{2}{3}\frac{\alpha^2}{H_0^2} \tag{9.36}$$

Due to the representation given in Eq. (9.34), Eq. (9.31) can be further expressed as

$$w^* \frac{\partial \boldsymbol{M}^*}{\partial t^*} + \epsilon w^* \left(\boldsymbol{v}^* . \nabla^*\right) \boldsymbol{M}^* = \epsilon w^* \boldsymbol{\omega}_p^* \times \boldsymbol{M}^* + \left(\boldsymbol{M}_0^* - \boldsymbol{M}^*\right) \tag{9.37}$$

where $\boldsymbol{M}_0^* = \boldsymbol{M}_0 / \chi_0 H_0$. Based on Eq. (4.1) and the power series expansion of the Langevin function under weak magnetic fields

$$L(\alpha) = \frac{\alpha}{3} - \frac{\alpha^3}{45} + \frac{4\alpha^5}{1890} + O\left(\alpha^7\right)$$

and by applying Eqs. (4.8) and (9.36), $\boldsymbol{M}_0^*$ can be expressed as

$$\begin{aligned}
\boldsymbol{M}_0^* &= \frac{\boldsymbol{M}_0}{\chi_0 H_0} \approx \frac{\phi M_d}{\chi_0 H_0} \frac{\boldsymbol{H}}{H}\left(\frac{\alpha}{3} - \frac{\alpha^3}{45} + \frac{4\alpha^5}{1890} + O\left(\alpha^7\right)\right) \\
&= \frac{\phi M_d}{\chi_0} \frac{\mu_0 m_p}{3 k_B T} \boldsymbol{H}^* \left(1 - \frac{\alpha^2}{15} + \frac{4\alpha^4}{630} + O\left(\alpha^6\right)\right) \\
&= \boldsymbol{H}^* \left(1 - \frac{\epsilon}{10} H_0^2 + \frac{\epsilon^2}{70} H_0^4 + O\left(\epsilon^3\right)\right)
\end{aligned} \tag{9.38}$$

Thus, Eq. (9.37) becomes

$$w^*\frac{\partial \boldsymbol{M}^*}{\partial t^*}+\epsilon w^*\left(\boldsymbol{v}^*\cdot\nabla^*\right)\boldsymbol{M}^*=\epsilon w^*\boldsymbol{\omega}_{\mathrm{p}}^*\times\boldsymbol{M}^*-\boldsymbol{M}^*+\boldsymbol{H}^*\left[1-\frac{\epsilon}{10}H_0^2+\frac{\epsilon^2}{70}H_0^4+O\left(\epsilon^3\right)\right] \quad (9.39)$$

9.2.1.2 *Solving the Governing Equations Using Regular Perturbation Method*

To solve the coupled system of equations consisting of Eqs. (9.27)–(9.29), (9.33), and (9.39), the regular perturbation method is applied. It is assumed that all field variables can be represented as power series in the small parameter ϵ, that is

$$\boldsymbol{M}^*=\sum_{n=0}^{\infty}\epsilon^n\boldsymbol{M}_n^* \quad (9.40)$$

$$\boldsymbol{H}^*=\sum_{n=0}^{\infty}\epsilon^n\boldsymbol{H}_n^* \quad (9.41)$$

$$\boldsymbol{v}^*=\sum_{n=0}^{\infty}\epsilon^n\boldsymbol{v}_n^* \quad (9.42)$$

$$\boldsymbol{\omega}_{\mathrm{p}}^*=\sum_{n=0}^{\infty}\epsilon^n\boldsymbol{\omega}_{\mathrm{p},n}^* \quad (9.43)$$

$$p^*=\sum_{n=0}^{\infty}\epsilon^n p_n^* \quad (9.44)$$

Each nth-order term in the expansions exhibits the same time-dependent relationship as the field variable itself.

After substituting Eqs. (9.40)–(9.44) into the respective governing equations, organizing the terms with the same power of ϵ into a single term, and then setting the coefficients of ϵ^0 and ϵ^1 to zero in each equation, one obtains the zeroth-order and first-order equation sets, respectively. Specifically, the zeroth-order equation set is as follows:

$$\nabla^*\times\boldsymbol{H}_0^*=\boldsymbol{0} \quad (9.45)$$

$$\nabla^*\cdot\left(\chi_0\boldsymbol{M}_0^*+\boldsymbol{H}_0^*\right)=0 \quad (9.46)$$

$$\frac{\zeta+\eta}{\eta}\nabla^{*2}\boldsymbol{v}_0^*+\frac{\zeta}{\eta w^*}\boldsymbol{M}_0^*\cdot\nabla^*\boldsymbol{H}_0^*+2\frac{\zeta}{\eta}\nabla^*\times\boldsymbol{\omega}_{\mathrm{p},0}^*-\nabla^*p_0^*=\boldsymbol{0} \quad (9.47)$$

$$\frac{4\eta}{(\zeta+\eta)\kappa^2}\nabla^{*2}\boldsymbol{\omega}_{\mathrm{p},0}^* + 2\nabla^*\times\boldsymbol{v}_0^* - 4\boldsymbol{\omega}_{\mathrm{p},0}^* + \frac{1}{w^*}\boldsymbol{M}_0^*\times\boldsymbol{H}_0^* = \boldsymbol{0} \tag{9.48}$$

$$w^*\frac{\partial \boldsymbol{M}_0^*}{\partial t^*} = -\boldsymbol{M}_0^* + \boldsymbol{H}_0^* \tag{9.49}$$

The system of first-order equations is given by

$$\nabla^*\times\boldsymbol{H}_1^* = \boldsymbol{0} \tag{9.50}$$

$$\nabla^*\cdot\left(\chi_0\boldsymbol{M}_1^* + \boldsymbol{H}_1^*\right) = 0 \tag{9.51}$$

$$\frac{\zeta+\eta}{\eta}\nabla^{*2}\boldsymbol{v}_1^* + \frac{\zeta}{\eta w^*}\left(\boldsymbol{M}_0^*\cdot\nabla^*\boldsymbol{H}_1^* + \boldsymbol{M}_1^*\cdot\nabla^*\boldsymbol{H}_0^*\right) + 2\frac{\zeta}{\eta}\nabla^*\times\boldsymbol{\omega}_{\mathrm{p},1}^* - \nabla^* p_1^* = \boldsymbol{0} \tag{9.52}$$

$$\frac{4\eta}{(\zeta+\eta)\kappa^2}\nabla^{*2}\boldsymbol{\omega}_{\mathrm{p},1}^* + 2\nabla^*\times\boldsymbol{v}_1^* - 4\boldsymbol{\omega}_{\mathrm{p},1}^* + \frac{1}{w^*}\left(\boldsymbol{M}_0^*\times\boldsymbol{H}_1^* + \boldsymbol{M}_1^*\times\boldsymbol{H}_0^*\right) = \boldsymbol{0} \tag{9.53}$$

$$w^*\frac{\partial \boldsymbol{M}_1^*}{\partial t^*} + w^*\boldsymbol{v}_0^*\cdot\nabla^*\boldsymbol{M}_0^* = w^*\boldsymbol{\omega}_{\mathrm{p},0}^*\times\boldsymbol{M}_0^* - \boldsymbol{M}_1^* + \boldsymbol{H}_1^* - \frac{1}{10}\left(H_0^*\right)^2\boldsymbol{H}_1^* \tag{9.54}$$

The next step is to solve the zeroth-order system of equations ranging from Eq. (9.45) to Eq. (9.49), in which the magnetic field intensity and magnetization can be solved independently. According to the representation given in Eq. (9.25), the zeroth-order magnetic field intensity and magnetization can be expressed separately since their identical time dependence with the field variables themselves

$$\boldsymbol{H}_0^* = \mathrm{Re}\left\{\left[\hat{H}_{r,0}\left(r^*\right)\boldsymbol{e}_r + \hat{H}_{\theta,0}\left(r^*\right)\boldsymbol{e}_\theta\right]e^{\mathrm{i}(wt+\phi_0)}\right\} \tag{9.55}$$

$$\boldsymbol{M}_0^* = \mathrm{Re}\left\{\left[\hat{M}_{r,0}\left(r^*\right)\boldsymbol{e}_r + \hat{M}_{\theta,0}\left(r^*\right)\boldsymbol{e}_\theta\right]e^{\mathrm{i}(wt+\phi_0)}\right\} \tag{9.56}$$

Here, the quantities $\hat{H}_{r,0}\left(r^*\right)$, $\hat{H}_{\theta,0}\left(r^*\right)$, $\hat{M}_{r,0}\left(r^*\right)$, and $\hat{M}_{\theta,0}\left(r^*\right)$ are all complex numbers.

Equation (9.49) can be expressed in its component form

$$\begin{cases} w^*\dfrac{\partial M_{r,0}^*}{\partial t^*} = -M_{r,0}^* + H_{r,0}^* \\ w^*\dfrac{\partial M_{\theta,0}^*}{\partial t^*} = -M_{\theta,0}^* + H_{\theta,0}^* \end{cases} \tag{9.57}$$

As can be seen from Eqs. (9.55) and (9.56) that

$$H_{r,0}^* = \text{Re}\left[\hat{H}_{r,0}\left(r^*\right)e^{\mathrm{i}(wt+\phi_0)}\right], M_{r,0}^* = \text{Re}\left[\hat{M}_{r,0}\left(r^*\right)e^{\mathrm{i}(wt+\phi_0)}\right]$$

Substituting these two equations into the first equation of the system (9.57) yields

$$\text{Re}\left[\mathrm{i}w^*\hat{M}_{r,0}\left(r^*\right)e^{\mathrm{i}(wt+\phi_0)}\right] = -\text{Re}\left[\hat{M}_{r,0}\left(r^*\right)e^{\mathrm{i}(wt+\phi_0)}\right] + \text{Re}\left[\hat{H}_{r,0}\left(r^*\right)e^{\mathrm{i}(wt+\phi_0)}\right]$$

Thus, one obtains

$$\hat{M}_{r,0}\left(r^*\right) = \frac{\hat{H}_{r,0}}{1+\mathrm{i}w^*} \tag{9.58}$$

The same reasoning leads to

$$\hat{M}_{\theta,0} = \frac{\hat{H}_{\theta,0}}{1+\mathrm{i}w^*} \tag{9.59}$$

Expanding Eq. (9.46) into its components

$$\frac{1}{r^*}\frac{\partial}{\partial r^*}\left(r^*\chi_0 M_{r,0}^* + r^* H_{r,0}^*\right) + \frac{1}{r^*}\frac{\partial}{\partial \theta}\left(\chi_0 M_{\theta,0}^* + H_{\theta,0}^*\right) = 0 \tag{9.60}$$

Substituting Eqs. (9.58) and (9.59) into Eq. (9.60) gives

$$\text{Re}\left[\left(\frac{\chi_0}{1+\mathrm{i}w^*}+1\right)\frac{\partial\left(r^*\hat{H}_{r,0}\right)}{\partial r^*}e^{\mathrm{i}(wt+\phi_0)}\right] = \text{Re}\left[\mathrm{i}\left(\frac{\chi_0}{1+\mathrm{i}w^*}+1\right)\hat{H}_{\theta,0}e^{\mathrm{i}(wt+\phi_0)}\right]$$

It can be concluded from this equation that

$$\frac{\partial\left(r^*\hat{H}_{r,0}\right)}{\partial r^*} = \mathrm{i}\hat{H}_{\theta,0} \tag{9.61}$$

Equation (9.45) can be expanded into

$$\left[\frac{1}{r^*}\frac{\partial}{\partial r^*}\left(r^* H_{\theta,0}^*\right) - \frac{1}{r^*}\frac{\partial H_{r,0}^*}{\partial \theta}\right]\mathbf{e}_z = \mathbf{0} \tag{9.62}$$

Substituting Eqs. (9.58) and (9.59) into Eq. (9.62) yields

$$\text{Re}\left[\frac{\partial\left(r^*\hat{H}_{\theta,0}\right)}{\partial r^*}e^{\mathrm{i}(wt+\phi_0)}\right] + \text{Re}\left[\mathrm{i}\hat{H}_{r,0}e^{\mathrm{i}(wt+\phi_0)}\right] = 0$$

From this equation, one gives

$$\frac{\partial\left(r^*\hat{H}_{\theta,0}\right)}{\partial r^*} = -\mathrm{i}\hat{H}_{r,0} \tag{9.63}$$

The system of equations consisting of Eqs. (9.61) and (9.63) admits a general solution

$$\begin{cases} \hat{H}_{r,0} = C_{01} + \dfrac{C_{02}}{r^{*2}} \\ \hat{H}_{\theta,0} = -\mathrm{i}\left(C_{01} - \dfrac{C_{02}}{r^{*2}}\right) \end{cases} \tag{9.64}$$

As the solution for magnetic field intensity should be finite when $r^* \to 0$, it follows that $C_{02} = 0$. Furthermore, regarding the tangential component continuity boundary condition for magnetic field intensity, one has

$$\hat{H}_{\theta,0} = -1, \text{ when } r^* = 1$$

From this, one can deduce that $C_{01} = -\,\mathrm{i}$. Substituting this value back into the system of Eq. (9.64) yields

$$\hat{H}_{r,0} = -\mathrm{i}, \hat{H}_{\theta,0} = -1 \tag{9.65}$$

Substituting Eq. (9.65) into both Eq. (9.58) and Eq. (9.59) yields

$$\hat{M}_{r,0}\left(r^*\right) = \frac{-\mathrm{i}}{1+\mathrm{i}w^*} = \frac{-w^* - \mathrm{i}}{1+w^{*2}}, \hat{M}_{\theta,0}\left(r^*\right) = \frac{-1}{1+\mathrm{i}w^*} = \frac{-1+\mathrm{i}w^*}{1+w^{*2}} \tag{9.66}$$

Utilizing Eqs. (9.65) and (9.66), the zeroth-order magnetic body force and magnetic moment can be computed accordingly.

$$\boldsymbol{M}_0^* \cdot \nabla^* \boldsymbol{H}_0^* = \boldsymbol{0} \tag{9.67}$$

$$\begin{aligned} \boldsymbol{M}_0^* \times \boldsymbol{H}_0^* &= \left(M_{r,0}^* H_{\theta,0}^* - M_{\theta,0}^* H_{r,0}^*\right)\boldsymbol{e}_z \\ &= \frac{1}{4}\left[\hat{M}_{r,0}e^{\mathrm{i}(wt+\phi)} + \overline{\hat{M}_{r,0}e^{-\mathrm{i}(wt+\phi)}}\right]\left[\hat{H}_{\theta,0}e^{\mathrm{i}(wt+\phi)} + \overline{\hat{H}_{\theta,0}e^{-\mathrm{i}(wt+\phi)}}\right]\boldsymbol{e}_z \\ &\quad - \frac{1}{4}\left[\hat{M}_{\theta,0}e^{\mathrm{i}(wt+\phi)} + \overline{\hat{M}_{\theta,0}e^{-\mathrm{i}(wt+\phi)}}\right]\left[\hat{H}_{r,0}e^{\mathrm{i}(wt+\phi)} + \overline{\hat{H}_{r,0}e^{-\mathrm{i}(wt+\phi)}}\right]\boldsymbol{e}_z \\ &= \frac{1}{4}\left[\hat{M}_{r,0}\hat{H}_{\theta,0}e^{2\mathrm{i}(wt+\phi)} + \overline{\hat{M}_{r,0}\hat{H}_{\theta,0}e^{-2\mathrm{i}(wt+\phi)}} + \hat{M}_{r,0}\overline{\hat{H}_{\theta,0}} + \overline{\hat{M}_{r,0}}\hat{H}_{\theta,0}\right]\boldsymbol{e}_z \\ &\quad - \frac{1}{4}\left[\hat{M}_{\theta,0}\hat{H}_{r,0}e^{2\mathrm{i}(wt+\phi)} + \overline{\hat{M}_{\theta,0}\hat{H}_{r,0}e^{-2\mathrm{i}(wt+\phi)}} + \hat{M}_{\theta,0}\overline{\hat{H}_{r,0}} + \overline{\hat{M}_{\theta,0}}\hat{H}_{r,0}\right]\boldsymbol{e}_z \\ &= \frac{w^*}{1+w^{*2}}\boldsymbol{e}_z \end{aligned} \tag{9.68}$$

Based on Eqs. (9.67) and (9.68), it can be inferred that the zeroth-order magnetic body force equals zero, while the zeroth-order magnetic moment remains constant.

To determine the zero-order velocity and spin velocity, it needs to derive the component equations in the θ and z directions for Eqs. (9.47) and (9.48), respectively,

$$\frac{\zeta+\eta}{\eta}\frac{\mathrm{d}}{\mathrm{d}r^*}\left[\frac{1}{r^*}\frac{\mathrm{d}\left(r^* v^*_{\theta,0}\right)}{\mathrm{d}r^*}\right]-2\frac{\zeta}{\eta}\frac{\mathrm{d}\omega^*_{\mathrm{p},z,0}}{\mathrm{d}r^*}=0 \tag{9.69}$$

$$\frac{4\eta}{(\zeta+\eta)\kappa^2}\frac{1}{r^*}\frac{\mathrm{d}}{\mathrm{d}r^*}\left(r^*\frac{\mathrm{d}\omega^*_{\mathrm{p},z,0}}{\mathrm{d}r^*}\right)+\frac{2}{r^*}\frac{\mathrm{d}\left(r^* v^*_{\theta,0}\right)}{\mathrm{d}r^*}-4\omega^*_{\mathrm{p},z,0}+\frac{1}{1+w^{*2}}=0 \tag{9.70}$$

Since

$$\Omega^*_{z,0}=\frac{1}{2}\frac{1}{r^*}\frac{\mathrm{d}\left(r^* v^*_{\theta,0}\right)}{\mathrm{d}r^*}\boldsymbol{e}_z \tag{9.71}$$

then Eq. (9.69) can be written as

$$\frac{\zeta+\eta}{\eta}\frac{\mathrm{d}\Omega^*_{z,0}}{\mathrm{d}r^*}-\frac{\zeta}{\eta}\frac{\mathrm{d}\omega^*_{\mathrm{p},z,0}}{\mathrm{d}r^*}=0 \tag{9.72}$$

Integrating Eq. (9.72) once leads to zero-order vorticity

$$\Omega^*_{z,0}=\frac{\zeta}{\zeta+\eta}\omega^*_{\mathrm{p},z,0}+\frac{\eta}{\zeta+\eta}C_{11} \tag{9.73}$$

Substituting Eqs. (9.71) and (9.73) into Eq. (9.70) yields an alternative form of the equation

$$\frac{1}{r^*}\frac{\mathrm{d}}{\mathrm{d}r^*}\left(r^*\frac{\mathrm{d}\omega^*_{\mathrm{p},z,0}}{\mathrm{d}r^*}\right)-\kappa^2\omega^*_{\mathrm{p},z,0}+\kappa^2 C_{11}+\frac{(\zeta+\eta)\kappa^2}{4\eta\left(1+w^{*2}\right)}=0 \tag{9.74}$$

Equation (9.74) represents a transformed version of the Bessel equation, and its general solution is given by

$$\omega^*_{\mathrm{p},z,0}\left(r^*\right)=C_{11}+C_{12}J_0\left(\kappa r^*\right)+\frac{\zeta+\eta}{4\eta\left(1+w^{*2}\right)} \tag{9.75}$$

where $J_0(\kappa r^*)$ is the zeroth-order first-kind Bessel function.

Substitute Eq. (9.75) into Eq. (9.73), and subsequently integrating the resulting expression, one obtains an expression for the rotational velocity. Then, with the aid of the boundary condition,

$$v_{\theta,0}^{*}=0, \omega_{p,z,0}^{*}=\frac{\gamma_0}{2}\frac{1}{r^{*}}\frac{d\left(r^{*}v_{\theta,0}^{*}\right)}{dr^{*}}, \text{when } r^{*}=1$$

where $\gamma_0 \in [0, 1]$ is the adjustable parameter, the zero-order velocity and spin velocity can be obtained (Chaves et al., 2008).

$$v_{\theta,0}^{*}\left(r^{*}\right)=\frac{\zeta}{2\kappa\tilde{\eta}\left(1+w^{*2}\right)}\frac{J_1(\kappa)}{J_0(\kappa)}\left[r^{*}-\frac{J_0\left(\kappa r^{*}\right)}{J_1(\kappa)}\right] \tag{9.76}$$

$$\omega_{p,z,0}^{*}\left(r^{*}\right)=\frac{1}{4\tilde{\eta}\left(1+w^{*2}\right)}\left\{(\zeta+\eta)\left[1-\frac{J_0\left(\kappa r^{*}\right)}{J_0(\kappa)}\right]-\gamma_0\zeta\frac{J_2(\kappa)}{J_0(\kappa)}\right\} \tag{9.77}$$

where the functions $J_1(\kappa)$ and $J_2(\kappa)$ represent the first- and second-order first-kind Bessel functions, respectively, and

$$\tilde{\eta}=\eta-\zeta\left(\gamma_0-1\right)\frac{J_2(\kappa)}{J_0(\kappa)}$$

From Eqs. (9.76) and (9.77), it is evident that both the rotational velocity of the ferrofluid and the particle spin velocity are strongly correlated with κ, which is intimately linked to the value of the spin viscosity η'.

To solve the first-order problem consisting of Eqs. (9.50)–(9.54), one can employ a similar approach as that used for the zeroth-order problem. The magnetization Eq. (9.54) can be solved first. Given that the zeroth-order magnetization is a uniform field, the second term on the left-hand side of the equation vanishes. With $H_0^{*}=1$, one can obtain the first-order magnetization (Chaves et al., 2008).

$$\hat{M}_{r,1}=\frac{1}{1+iw^{*}}\left(\hat{H}_{r,1}-w^{*}\hat{\omega}_{p,z,0}\hat{M}_{\theta,0}-\frac{\hat{H}_{r,0}}{10}\right) \tag{9.78}$$

$$\hat{M}_{\theta,1}=\frac{1}{1+iw^{*}}\left(\hat{H}_{\theta,1}+w^{*}\hat{\omega}_{p,z,0}\hat{M}_{r,0}-\frac{\hat{H}_{\theta,0}}{10}\right) \tag{9.79}$$

By substituting Eqs. (9.78) and (9.79) into the expanded forms of Eqs. (9.50) and (9.51), a system of equations similar to Eqs. (9.61) and (9.63) is obtained. However, the difference is that this system of equations can be simplified into a differential equation solely concerning the circumferential component of the magnetic field intensity

$$\frac{\mathrm{d}}{\mathrm{d}r^*}\left[r^*\frac{\mathrm{d}\left(r^*\hat{H}_{\theta,1}\right)}{\mathrm{d}r^*}\right]-\mathrm{i}\hat{H}_{\theta,1}=\tilde{A}r^*J_1\left(\kappa r^*\right) \tag{9.80}$$

where

$$\tilde{A}=\frac{\left(\zeta+\eta\right)\chi_0\kappa w^*}{4\tilde{\eta}\left(1+\chi_0+\mathrm{i}w^*\right)\left(w^*-\mathrm{i}\right)^2\left(w^*+\mathrm{i}\right)J_0\left(\kappa\right)}$$

Solving Eq. (9.80) leads to (Chaves et al., 2008)

$$\hat{H}_{\theta,1}\left(r^*\right)=\frac{\tilde{A}}{\kappa^2}\left[\frac{J_1\left(\kappa r^*\right)}{r^*}-J_1\left(\kappa\right)\right] \tag{9.81}$$

$$\hat{H}_{r,1}\left(r^*\right)=\mathrm{i}\frac{\tilde{A}}{2\kappa}\left[J_0\left(\kappa r^*\right)+J_2\left(\kappa r^*\right)-\frac{2J_1\left(\kappa\right)}{\kappa}\right] \tag{9.82}$$

Substituting Eqs. (9.81) and (9.82) into Eqs. (9.78) and (9.79), respectively, yields the components of the first-order magnetization.

Similarly, the first-order magnetic moment is (Chaves et al., 2008)

$$\boldsymbol{M}_0^*\times\boldsymbol{H}_1^*+\boldsymbol{M}_1^*\times\boldsymbol{H}_0^*=\tilde{B}+\tilde{C}J_0\left(\kappa r^*\right) \tag{9.83}$$

where

$$\tilde{B}=\left[\left(\gamma_0-1\right)\zeta-\eta\right]\frac{\left(7-w^{*2}+2w^{*4}\right)w^*}{20\tilde{\eta}\left(1+w^{*2}\right)^3}$$
$$-2w^*J_1\left(\kappa\right)\frac{5\left(\zeta+\eta\right)\chi_0\left(2+\chi_0\right)w^{*2}+\zeta\left[\left(1+\chi_0\right)^2+w^{*2}\right]\left[7\gamma_0-2-\left(4+\gamma_0\right)w^{*2}+2\left(\gamma_0-1\right)w^{*4}\right]}{20\tilde{\eta}\kappa\left(1+w^{*2}\right)^3\left[\left(1+\chi_0\right)^2+w^{*2}\right]J_0\left(\kappa\right)} \tag{9.84}$$

$$\tilde{C} = \frac{(\zeta+\eta)w^*\left[(1+\chi_0)^2 - w^{*4}\right]}{4\tilde{\eta}\left(1+w^{*2}\right)^3\left[(1+\chi_0)^2 + w^{*2}\right]J_0(\kappa)} \tag{9.85}$$

Analogous to the approach employed in solving for the zero-order velocity and zero-order spin velocity, the modified Bessel equation regarding the spin velocity can be derived from Eqs. (9.52) and (9.53).

$$\frac{1}{r^*}\frac{\mathrm{d}}{\mathrm{d}r^*}\left(r^*\frac{\mathrm{d}\omega^*_{\mathrm{p},z,1}}{\mathrm{d}r^*}\right) - \kappa^2\omega^*_{\mathrm{p},z,1} + \kappa^2 C_{21} + \frac{(\zeta+\eta)\kappa^2}{4\eta w^*}\left[\tilde{B} + \tilde{C}J_0(\kappa r^*)\right] = 0 \tag{9.86}$$

The particular solution of this equation is given by (Chaves et al., 2008)

$$\omega^*_{\mathrm{p},z,1}(r^*) = C_{21} + C_{22}J_0(\kappa r^*) + \frac{\zeta+\eta}{4\eta w^*}\left[\tilde{B} - \frac{\kappa r^*\tilde{C}}{2\kappa}J_1(\kappa r^*)\right] \tag{9.87}$$

After applying boundary conditions to obtain C_{21} and C_{22}, the first-order spin velocity and velocity are respectively derived as

$$\omega^*_{\mathrm{p},z,1}(r^*) = \frac{1}{8\eta\tilde{\eta}w^*J_0(\kappa)}\left\{\begin{array}{c} 2\zeta\left[(\gamma_0-1)\zeta-\eta\right]\left[J_1(\kappa)^2\tilde{C} + J_2(\kappa)\tilde{B}\right] \\ +(\zeta+\eta)J_0(\kappa r^*)\left[-2\eta\tilde{B} + \left[(1-\gamma_0)\zeta+\eta\right]\kappa J_1(\kappa)\tilde{C} + 2\zeta(\gamma_0-1)J_2(\kappa)\tilde{C}\right] \\ +J_0(\kappa)\left[(\zeta+\eta)\tilde{\eta}\left(2\tilde{B} - r^*J_1(\kappa r^*)\kappa\tilde{C}\right) + 2\zeta\left[(1-\gamma_0)\zeta+\eta\right]J_2(\kappa)\tilde{C}\right] \end{array}\right\} \tag{9.88}$$

$$v^*_{\theta,1}(r^*) = \frac{\zeta}{4\eta w^*}\left\{\begin{array}{c} \dfrac{J_1(\kappa r^*)\left[-2\eta\tilde{B} + \left[(1-\gamma_0)\zeta+\eta\right]\kappa J_1(\kappa)\tilde{C} + 2\zeta(\gamma_0-1)J_2(\kappa)\tilde{C}\right] + \tilde{B}}{\tilde{\eta}\kappa J_0(\kappa)} - \dfrac{\tilde{C}}{\tilde{\eta}\kappa} \\ + \dfrac{r^*\left[(\gamma_0-1)\zeta-\eta\right]\left\{J_1(\kappa)^2\tilde{C} + J_2(\kappa)\left[\tilde{B} - J_0(\kappa)\tilde{C}\right]\right\}}{\tilde{\eta}J_0(\kappa)} + r^*\left[\tilde{B} - J_2(\kappa r^*)\tilde{C}\right] \end{array}\right\} \tag{9.89}$$

Upon acquiring the zeroth-order and first-order variables, the rotational velocity of the ferrofluid and the particle spin velocity in the spin-up flow can be approximately represented as

$$\boldsymbol{v}^* = \left[v^*_{\theta,0}(r^*) + \epsilon v^*_{\theta,1}(r^*)\right]\boldsymbol{e}_\theta \tag{9.90}$$

$$\boldsymbol{\omega}^*_{\mathrm{p}} = \left[\omega^*_{\mathrm{p},z,0}(r^*) + \epsilon\omega^*_{\mathrm{p},z,1}(r^*)\right]\boldsymbol{e}_z \tag{9.91}$$

where the respective components are given by Eqs. (9.76), (9.77), (9.88), and (9.89).

9.2.2 Torque Acting on the Container Wall

In the spin-up flow of ferrofluid, the torque acting on the surface of a cylindrical container can be expressed as

$$L = \oint_S \mathrm{d}\boldsymbol{S} \cdot (-\mathbf{T} \times \boldsymbol{r} + \mathbf{C}) \tag{9.92}$$

where stress tensor $\mathbf{T}$ and couple stress tensor $\mathbf{C}$ of the ferrofluid are given by Eqs. (2.221) and (2.183), respectively.

The magnetic stress tensor component $\mathbf{T}_m$ within the stress tensor $\mathbf{T}$ is

$$\mathbf{T}_m = -\frac{\mu_0 H^2}{2}\mathbf{I} + \boldsymbol{BH}$$

Due to the boundary conditions of the continuous tangential component of magnetic field intensity and the continuous normal component of magnetic flux density, the circumferential component of magnetic force caused by the magnetic stress tensor is zero. Therefore, the magnetic stress tensor does not contribute to the torque exerted on the container wall.

For the viscous stress tensor within the stress tensor $\mathbf{T}$, it is expressed as

$$\mathbf{T}_v = \eta\left[\nabla \boldsymbol{v} + (\nabla \boldsymbol{v})^{\mathrm{T}}\right] + 2\zeta \epsilon \cdot (\boldsymbol{\Omega} - \boldsymbol{\omega}_{\mathrm{p}}) \tag{9.93}$$

For the flow depicted in Figure 9.1, this tensor can be expanded into

$$\mathbf{T}_v = \eta\left[\left(\frac{\mathrm{d}v_\theta}{\mathrm{d}r} - \frac{v_\theta}{r}\right)\boldsymbol{e}_r\boldsymbol{e}_\theta + \left(\frac{\mathrm{d}v_\theta}{\mathrm{d}r} - \frac{v_\theta}{r}\right)\boldsymbol{e}_\theta\boldsymbol{e}_r\right] + 2\zeta\left(\frac{1}{2r}\frac{\mathrm{d}(rv_\theta)}{\mathrm{d}r} - \omega_{\mathrm{p},z}\right)\boldsymbol{e}_r\boldsymbol{e}_\theta \tag{9.94}$$

From Eq. (9.94), it can be derived that

$$\mathbf{T} \times \boldsymbol{r} = \mathbf{T}_v \times \boldsymbol{e}_r r = \left[-\eta r^2 \frac{\mathrm{d}}{\mathrm{d}r}\left(\frac{v_\theta}{r}\right) - 2\zeta\left(\frac{1}{2}\frac{\mathrm{d}(rv_\theta)}{\mathrm{d}r} - r\omega_{\mathrm{p},z}\right)\right]\boldsymbol{e}_r\boldsymbol{e}_z \tag{9.95}$$

Additionally, regarding the flow illustrated in Figure 9.1, Eq. (2.183) can be expanded as

$$\mathbf{C} = \eta'\left(\frac{\mathrm{d}\omega_{\mathrm{p},z}}{\mathrm{d}r}\boldsymbol{e}_r\boldsymbol{e}_z + \frac{\mathrm{d}\omega_{\mathrm{p},z}}{\mathrm{d}r}\boldsymbol{e}_z\boldsymbol{e}_r\right) \tag{9.96}$$

Substituting Eqs. (9.95) and (9.96) into Eq. (9.92), and applying

$$\mathrm{d}\boldsymbol{S} = r\mathrm{d}\theta\mathrm{d}z\boldsymbol{e}_r$$

one can obtain the torque expression

$$L = e_z \int_0^l \int_0^{2\pi} \left[\eta r^2 \frac{\mathrm{d}}{\mathrm{d}r}\left(\frac{v_\theta}{r}\right) + \zeta \frac{\mathrm{d}(r v_\theta)}{\mathrm{d}r} - 2\zeta r \omega_{\mathrm{p},z} + \eta' \frac{\mathrm{d}\omega_{\mathrm{p},z}}{\mathrm{d}r} \right] r\mathrm{d}\theta \mathrm{d}z \tag{9.97}$$

where l is the length of the cylindrical ferrofluid. By substituting the variables from Eq. (9.23) into Eq. (9.97) and calculating the integral of the resulting equation, one can obtain (Chaves et al., 2008)

$$L^* = e_z 2\chi_0 w^* \left[\frac{\eta}{\zeta} r^{*2} \frac{\mathrm{d}}{\mathrm{d}r^*}\left(\frac{v_\theta^*}{r^*}\right) + \frac{\mathrm{d}(r^* v_\theta^*)}{\mathrm{d}r^*} - 2r^* \omega_{\mathrm{p},z}^* + \frac{4\eta}{(\zeta+\eta)\kappa^2} \frac{\mathrm{d}\omega_{\mathrm{p},z}^*}{\mathrm{d}r^*} \right]_{r^*=1} \tag{9.98}$$

where

$$L^* = \frac{L}{2\pi\mu_0 H_0^2 R_0^2 l}$$

Substituting Eqs. (9.90) and (9.91) into Eq. (9.98) yields a first-order approximate expression for the dimensionless torque

$$L^* \approx -e_z \chi_0 \left\{ \frac{w^*}{1+w^{*2}} + \left[2\tilde{B} + \frac{\tilde{C} J_1(\kappa)}{\kappa} \right] \epsilon \right\} \tag{9.99}$$

where $\tilde{B}$ and $\tilde{C}$ are given by Eqs. (9.84) and (9.85), respectively.

For the results given by Eq. (9.99), if the spin viscosity $\eta' \to 0$, then $\kappa \to \infty$, and the corresponding first-order approximate dimensionless torque becomes

$$L^*(\eta' \to 0) \approx -e_z \frac{\chi_0 w^*}{1+w^{*2}} \left[1 + \frac{7 - w^{*2} + 2w^{*4}}{10\tilde{\eta}\left(1+w^{*2}\right)^2} \epsilon \right] \tag{9.100}$$

In an experiment conducted by Chaves et al. (2008), the torque exerted on a cylindrical container within a spin-up ferrofluid flow was measured. Figure 9.5 depicts the ratio of the experimental measurements to the zeroth-order torque values predicted by Eq. (9.99). In this figure, V_f represents the volume of the ferrofluid, while L_{exp} and L_{th} correspond to the experimentally measured torque and the calculated first-order torque, respectively. The various volume fractions of the same ferrofluid were obtained by diluting the original ferrofluid. It is evident from the data that Eq. (9.99) can accurately predict the torque experienced by the cylindrical container within the ferrofluid spin-up flow.

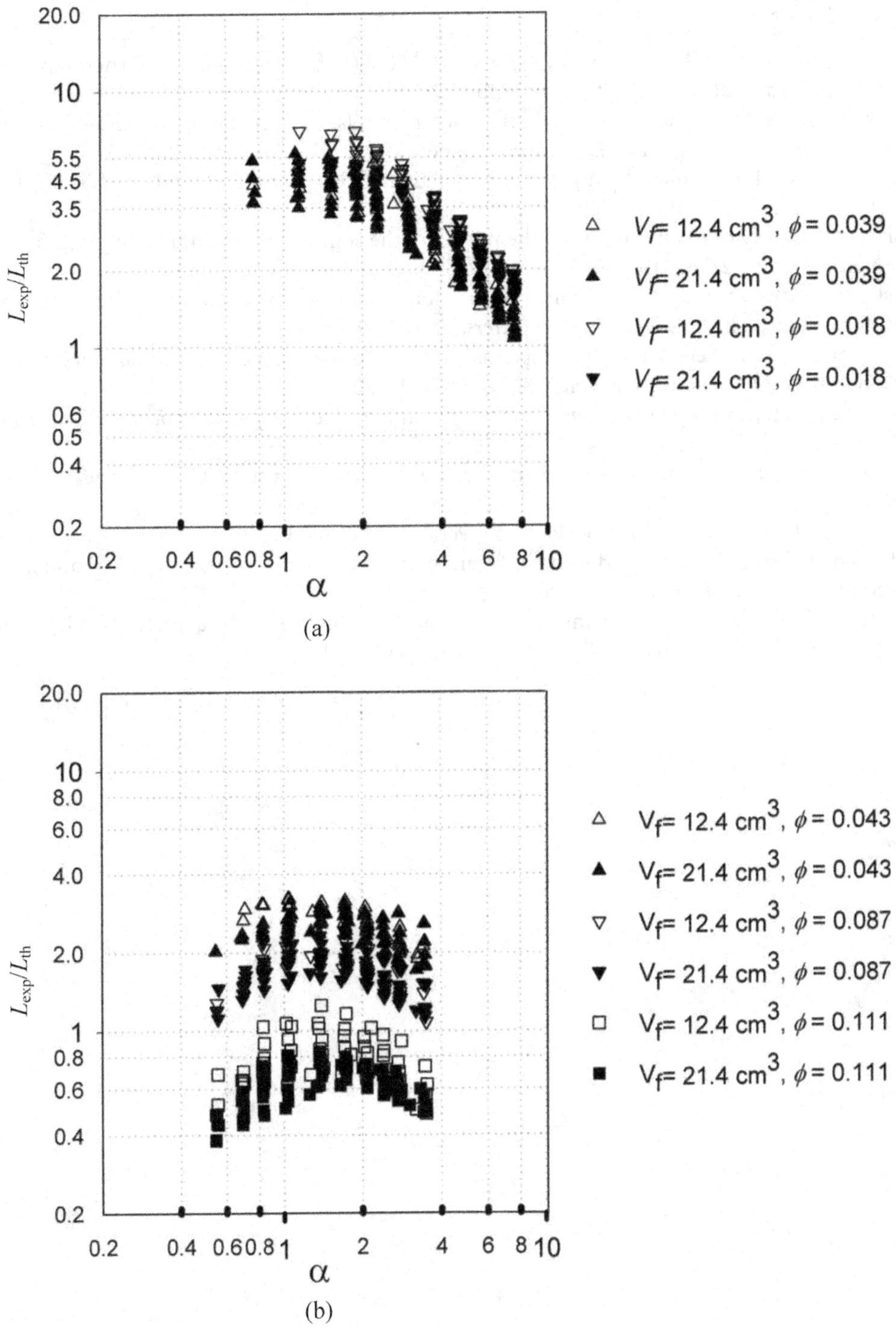

FIGURE 9.5 Torque experienced by the cylindrical container in a spin-up flow of ferrofluids. (a) Ferrofluid EMG705 and (b) ferrofluid EMG900. (Reprinted from Chaves (2008) with permission from AIP.)

REFERENCES

Chaves A., Rinaldi C., Elborai S., He X., Zahn M., Bulk flow in ferrofluids in a uniform rotating magnetic field, *Physical Review Letters*, 96: 194501, 2006.

Chaves A., Zahn M., Rinaldi C., Spin-up flow of ferrofluids: Asymptotic theory and experimental measurements, *Physics of Fluids*, 20: 053102, 2008.

Felderhof B. U., Entrainment by a rotating magnetic field of a ferrofluid contained in a cylinder, *Physical Review E*, 84: 026312, 2011.

Finlayson B. A., Spin-up of ferrofluids: The impact of the spin viscosity and the Langevin function, *Physics of Fluids*, 25: 073101, 2013.

Moskowitz R., Rosensweig R. E., Nonmechanical torque-driven flow of a ferromagnetic fluid by an electromagnetic field, *Applied Physics Letters*, 11(10): 301–303, 1967.

Pshenichnikov A. F., Lebedev A. V., Tangential stresses on the magnetic fluid boundary and rotational effect, *Magnetohydrodynamics*, 36(4): 317–326, 2000.

Rosensweig R. E., Spin-up flow in ferrofluids: A toy model, *The European Physical Journal E*, 46: 83, 2023.

Shliomis M. I., How a rotating magnetic field causes ferrofluid to rotate, *Physical Review Fluids*, 6: 043701, 2021.

Torres-Diaz I., Cortes A., Cedeño-Mattei Y., Perales-Perez O., Rinaldi C., Flows and torques in Brownian ferrofluids subjected to rotating uniform magnetic fields in a cylindrical and annular geometry, *Physics of Fluids*, 26: 012004, 2014.

Zaitsev V. M., Shliomis M. I., Entrainment of ferromagnetic suspension by a rotating field, *Journal of Applied Mechanics and Technical Physics*, 10: 696–700, 1969.

Index

Pages in *italics* refer to figures, pages in **bold** refer to tables, and pages followed by "n" refer to notes.

A

Alternating magnetic field 116, 125–126, 128–139, 184–185
Angular momentum equation 41–46, 190, 208, 223
Apparent electric field intensity 21
Apparent magnetic field intensity 21
Apparent viscosity 2, 117, 255
Average particle size 100, **153**
Axial magnetic field 221
Axial magnetic levitation force 82–96, *89*, *96*

B

Bernoulli equation for ferrofluids 51–52
Body couple density 16, 19
Brownian relaxation time 44, 101–103

C

Cauchy's equations of motion 15–16
Cauchy's principle of stress 14–15
Circular flow of ferrofluids 259, *259*
Clausius-Duhem inequality 34–35
Complex representation of time-varying magnetic fields 184–185
Conservation equation of angular momentum 16–21
Conservation equation of energy 21–29
Conservation equation of mass 14
Constitutive equations 35, 268
Constitutive relationships for electric fields 27
Constitutive relationships for magnetic fields 27
Continuity equation 14
Couette flow 116–117, *117*, 142
Couette–Poiseuille flow 142, *142–143*, *163*, *169–170*
Couple stress tensor 18, 38, 287
Current image method 89–94, *93*
Cylindrical composite magnets *74*, 74–75

D

Deformation velocity tensor of ferrofluids 262
Dimensionless pressure gradient 143
Drag coefficient 234–235

E

Effective field 103, 109–110, 121–122, 135, 272–273
Effective viscosity 118
Einstein formula 114
Electromagnetic field momentum density 23
Electromagnetic momentum 14
Entropy growth equation 33
Equation of entropy growth rate 30–33
Equation of motion 39–41, 43–46
Equilibrium magnetization equation 98–101
Equivalent magnetic stress tensor 51, 65
Eulerian method 11–12
Extensive property 12
Extrinsic magnetization 2
Extrinsic superparamagnetism 103

F

Ferrofluid damper 4–5, *5*
Ferrofluid dynamic seals 3–4, *4*
Ferrofluids 1–3
Flow rate 144, 152, 158, 160, 162, 176–177, 233–234, *234*, 247–249, *247*, 255–256, *256*
Flow vorticity 33, 115–117, 128, 144, *155*, 159, *159*, *168*, *171*, 183, 222, 250, 254, 261, 283
Fluid rotation rate 33

Force exerted on the upper plate 152–153
Frictional torque 261

G

Gibbs equation 29–30
Gradient magnetic field 163–166, 242–243

H

Heaviside function 242

I

Initial magnetic levitation force *73*, 73, 81–82, *82*
Initial susceptibility of ferrofluids 100–101
Intensity quantity 12
Intrinsic magnetization 2
Intrinsic superparamagnetism 103
Intrinsic viscosity of ferrofluids 114–116

K

Kelvin force 50, 52–53

L

Lagrangian method 11–12
Lamé coefficients 57
Langevin function 50, 52, 98, 100, 278
Langevin parameter 98–99, 278
Laplace equation 57, 262

M

Magnetic charge image method 82–87, *86*
Magnetic charge surface density 82–83
Magnetic levitation forces in ferrofluids 53–96, *80*
Magnetic moment of a particle 50, 99, 101, 103, 107–108
Magnetic relaxation 101–103
Magnetic Reynolds number 235, 239
Magnetic stress tensor 31, 40, 51, 65, 287
Magnetic torque 101, 107, 130, 138, 189, 191, 204–208, 271, 273–274
Magnetization current *90*, 90–92
Magnetization current surface density 89–90
Magnetization of ferrofluid 98–101, *99*, *155*, *166*, 185, *257*
Magnetization perpendicular to the magnetic field 118, 153–154, *154*, *156*, *158*, *161–162*, 173–174, *173–174*, 268–270
Magnetization relaxation effect 154–157
Magnetization relaxation time *102*, 102–103, 111, **153**, 161–162, 229–230, 270
Material derivative 11–12
Material volume 12
Maxwell's equations 23–24
Microscopic magnetization equation 107–111, 134
Modified Bessel equation 286
Moment of inertia 20
Momentum conservation equation 15–16

N

Néel relaxation time 101–103
"Negative" viscosity 132–133
Non-dimensional magnetic field gradient 165
Non-equilibrium ferrohydrodynamics 6–7
Non-polar fluids 16
Normalized ferrohydrodynamic equations 144–148

O

Orbital/external angular momentum 16
Oscillating magnetic field 244

P

Péclet number 145, 228
Phenomenological magnetization equation I 103–107
Phenomenological magnetization equation II 111–112
Plane bipolar coordinate system *54*, 54–55
Planar Couette–Poiseuille flow 142–144
Poiseuille flow 142

Q

Quasi-equilibrium ferrohydrodynamic equation 49–51

R

Radial magnetic field 231
Radial magnetic levitation force 54–82, *72*
Regular perturbation method 150–151, 201–203, 239, 279
Resistance coefficient 234–235
Reynolds number 145, 234
Reynolds' transport theory 11–14
Root mean square vorticity 153, *160*, *172*
Rotating magnetic field 126, 179, 184–185, 189, 197–199, 210–219

Rotational velocity at ferrofluid surface 274–276, *276*
Rotational viscosity 115–140, *120*, *123–124*, *132*, 213–215, *214–216*, 230–231, *230*, *231*, 236–241, *239*, *241*, *248*

S

Saturation magnetization 2, 45, 50, 98–99, **153**
Scalar magnetic potential 56–57, 63–65, 75, 84–85, 262–266, 269
Sinusoidal time-varying magnetic field 184–185
Small parameter 134–135, 201, 203, 237, 278–279
Spin angular momentum density 20
Spin velocity *156*, 156–157, 159–162, *161*, 175–176, *175–176*, 184, 195, 198–199, *199*, 211–214, *211–212*, *214*, 276–286
Spin velocity/vorticity matching boundary condition 213
Spin viscosity 38, 44–45, 209, 213
Spin-diffusion theory 271, 276
Spin-up flow 271, *272*
Spontaneous rotation 249
Strain rate tensor 33
Stress tensor of ferrofluids 31, 39, 46–47, 77–78, 182, 287
Surface couple density 16, 18
Surface couple stress tensor 38
Surface stress vector 14–15

T

Tangential stress on walls 47, 147, 158, *158*, 177–178, *178*, 183, 215–220, *217*, *219*
Time-averaged magnetic force 187–189
Time-averaged magnetic moment 187–194, *189*, 204–208, *205*, *207*
Time-varying magnetic field 179
Torque acting on container wall 287–289, *289*
Total stress tensor 15

V

Viscous stress tensor 31, 38–39, 287
Volume force per unit mass 14
Vortex viscosity 38, 44, 209

W

Weak field condition 112, 119, 148–149